The other world
Spiritualism and psychical research i ~~England~~
1850–1914

The other world

*Spiritualism and psychical research
in England, 1850–1914*

Janet Oppenheim

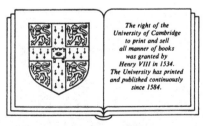

The right of the
University of Cambridge
to print and sell
all manner of books
was granted by
Henry VIII in 1534.
The University has printed
and published continuously
since 1584.

Cambridge University Press

Cambridge

London   New York   New Rochelle

Sydney   Melbourne

Published by the Press Syndicate of the University of Cambridge
The Pitt Building, Trumpington Street, Cambridge CB2 1RP
32 East 57th Street, New York, NY 10022, USA
10 Stamford Road, Oakleigh, Melbourne 3166, Australia

First published 1985

Library of Congress Cataloging in Publication Data
Oppenheim, Janet, 1948–
The other world.
Includes index.
1. Spiritualism – England – History – 19th century.
2. Spiritualism – England – History – 20th century.
3. Psychical research – England – History – 19th century.
4. Psychical research – England – History – 20th century.
I. Title.
BF1242.G7066   1985   133.9′0942   84-5887
ISBN 0 521 26505 3
ISBN 0 521 34767 X

Transferred to digital printing 2002

To

John and Ashley

CONTENTS

ILLUSTRATIONS

ACKNOWLEDGMENTS

Among the many people and organizations who helped me in writing this book, I want first to thank the Society for Psychical Research, London, not only for permission to quote from the papers of Sir William Barrett and Sir Oliver Lodge, but also for the aid that Eleanor O'Keeffe offered during two summers of research in the Society's library and archives. I am also grateful to the College of Psychic Studies and the Spiritualist Association of Great Britain, both in London, for the opportunity to use their libraries and records. A. H. Wesencraft, Honorary Curator of the Harry Price Library, University of London, was extremely helpful in making that library's collections available to me. I drew on the expertise of the staffs of the British Library and the Library of Congress, and am particularly indebted to Leonard N. Beck, Subject Collections Specialist, Rare Book and Special Collections Division, Library of Congress. The Houdini Collection, which I consulted extensively, fell under Mr. Beck's supervision, and he made using it a pleasure. I gratefully acknowledge permission from the Yale University Library to quote from letters in the Thomas Davidson Papers, and from the Master and Fellows of Trinity College, Cambridge, to quote from letters in the Papers of Henry Sidgwick and of F. W. H. Myers.

Research grants from the National Endowment for the Humanities and the American Council of Learned Societies are similarly acknowledged with gratitude.

I have benefited from conversation or correspondence with Joseph Baylen, Elizabeth Eisenstein, Alan Gauld, Peter Marsh, Seymour Mauskopf, Terry Parssinen, Harold Perkin, Joan Richards, John D. Root, Frank M. Turner, James Winter, Anthony Wohl, and Stephen Koss, who also read and commented on much of the manuscript. Scott Parker was generous in lending me books for use in Chapter 6. Indeed, my colleagues at The American University, Washington, D.C., have always been generous with their time and have provided me with stimulating ideas whenever I confronted them with psychics and spirits.

Michael Dauman Gnat's enthusiastic assistance at all stages of the book's production has been much appreciated.

Finally, I need especially to thank my daughter Ashley and my husband, John H. M. Salmon. In their different ways, they have both been enormously helpful.

Introduction

Loss of Christian faith, despite all the volumes that have been devoted to the subject, must always remain fundamentally inaccessible to the historian's probe. The renunciation of particular creeds and rituals may be documented, but the process of ceasing to believe can never be fully chronicled. That the Victorian age was a period of religious uncertainty is beyond question, but the reasons for that doubt are not capable of precise explanation. The onslaughts of reason and science played a part, of course, in what has been called "the secularization of the European mind in the nineteenth century," but they were not alone responsible.[1] Nor was rationalism uniform in its impact and consequences. It prompted some to rethink particular aspects of the biblical narratives without jeopardizing an underlying adherence to Christianity. For others, it demanded not only a repudiation of the Christian religion, but the denial of God's very existence. Still others paid scant attention to the arguments raised by scientific, historical, and biblical scholarship. Faith could fade gradually, less the consequence of logical deduction than a matter of indifference, of changing habits, of new social patterns and economic demands. Yet, whatever the complex relationships of cause and effect, Victorians themselves were fully aware that the place of religion in the cultural fabric of their times was scarcely secure. In an effort to counter that insecurity, to calm their fears, and to seek answers where contemporary churches were ambiguous, thousands of British men and women in the Victorian and Edwardian eras turned to spiritualism and psychical research.

I have purposely imposed geographical and chronological boundaries on this study. Although many similar developments vividly color American and European history of the nineteenth century, there is such a wealth of available material that it seemed sufficient challenge to try to make good sense out of the British evidence alone. It is also true that the strength of the Nonconformist tradition, the intensity of the evangelical experience in Britain, and the particular influence of the Anglican church provided a context for spiritualism and psychical research that was distinct from continental or American varieties. I hope, nonetheless, that I have remembered to point out from time to time that even England is never truly an island where the exchange of ideas and cultural trends is concerned. I have also excluded from this work people whose interest in spiritualism or psychical research flowered after 1914, not because they were few in number – on the contrary, the Great War substantially enhanced the appeal of psychics and sé-ances – but because the context had changed. The new converts were responding, not to the intellectual and emotional crises of the mid- and late-Victorian decades, but to the unprecedented horrors of World War.

In recognizing the perplexity that many of their contemporaries felt about Christian beliefs, the British spiritualists and psychical research-ers before World War I perceived in science the main source of tension. They may have erred in emphasizing so heavily the threat of science, but their extensive writings leave no doubt that, for them, scientific modes of thought posed the outstanding challenge to the foundations of Christianity. It was science in all its manifestations, they believed, that was broadcasting a materialistic philosophy. They opposed that tendency of modern thought with a bold affirmation of spiritualism, the assertion that spirit exists and functions in the universe as surely as matter. Indeed no group of people more zealously threw their energies into the effort to discredit materialism than the men and women who endorsed spiritualism in the second half of the nineteenth century. Their response to the Victorian crisis of faith was vigorous and heart-felt. It was, furthermore, far more representative of contemporary re-ligious attitudes than the agnosticism embraced by the comparatively few intellectuals who have dominated the historical record of that cri-sis.

It goes without saying that not every person who ever attended a séance in nineteenth-century Britain was seeking to combat materi-alism. Inevitably, there were participants who played with spiritualism, as they played with other fashionable and passing fads. The impetus behind modern spiritualism came, nevertheless, from the thousands who looked to spiritualism for far more urgent reasons than mere tit-illation. It came from the men and women who searched for some incontrovertible reassurance of fundamental cosmic order and purpose, especially reassurance that life on earth was not the totality of human existence. While Victorians of a scholarly bent found relief disputing theologically among themselves in the periodical press, spiritualists found their comfort at the séance table. There, in the spirit voices, the spirit hands, faces, and bodies, the messages rapped out on walls, floors, and furniture, or scribbled on slate, spiritualists received proof that the human spirit survives bodily death. With that proof, they lib-

erated themselves from the religious anxiety and emotional bewilderment that had afflicted them and continued to torment countless numbers of their contemporaries.

In the chapters that follow, the terms "spiritualist" and "psychical
researcher" recur repeatedly. They are not intended to be synonymous, for they designate distinct approaches to psychic phenomena.
The spiritualists, on the one hand, were likely to attend séances in an
accepting frame of mind. Believing, as they firmly did, in human survival after death and in the possible activity of disembodied human
spirits, they did not hesitate to assert the reality of communication with
the dead and to accept as genuine most of the phenomena that they
witnessed at séances. Psychical researchers, on the other hand, trod
with greater circumspection and even, in some cases, skepticism.
Eager to investigate the allegedly spiritualistic phenomena as exhaus-
tively as possible, they did not consider a critical mind inappropriate
in the séance room. They were attracted to the subject, not only because it apparently offered a chance to prove immortality, but also
because it presented the opportunity to explore the mysteries of the
human mind. In all their investigations, psychical researchers claimed
to be gathering information objectively, collecting the facts needed for
strict scientific evaluation, and harboring no preconceived explanatory
theories. Ironically, some psychical researchers were as eager as the
spiritualists to force the methods of science into the service of an unseen, immaterial world. Their work, instead of building on scientific
discoveries, misunderstood, misapplied, and distorted them.

No doubt every age in human history has felt the lure of the occult.
Ever since science began establishing its claim to epistemological supremacy, there have been people drawn to phenomena that apparently
defied rational, scientific explanation. Recent years have been no exception, with psychokinesis, biorhythms, horoscopes, psychic surgery,
Kirlian auras, out-of-body experiences, and that supreme example of
occult silliness, pyramid power. There is, nonetheless, a significant
distinction between spiritualism and psychical research in the late nineteenth century and parapsychology in the late twentieth. Now, after
decades of disappointments, few influential or renowned scholars endorse the claims of parapsychology. Then, a century ago, spiritualism
and psychical research loomed as very serious business to some very
serious and eminent people, such as the Fellows of the Royal Society,
university professors, and Nobel prize-winning scientists who supported the Society for Psychical Research. Together with the industrious middle-class professionals and self-educated artisans who joined
spiritualist clubs both in London and the provinces, these intellectuals
turned to psychic phenomena as courageous pioneers hoping to dis-

cover the most profound secrets of the human condition and of man's
place in the universe. With psychology in its infancy, it still seemed
in the late nineteenth century that psychical research, if not spiritualism, might play a legitimate and important role in the growth of a
new science.

The Victorians and Edwardians who appear in this volume varied
greatly in their degrees of credulity. Some greeted the occult crazes
of their day with derision, whereas for others nothing evidently taxed
too greatly their capacities for belief. None, however, could blandly
accept God's absence from the universe. They had not had time to
adjust to an amoral world that neither cared about humanity nor made
manifest an ultimate meaning in life. If they turned to spiritualism and
psychical research as refuge from bleak mechanism, emptiness, and
despair, they did so as part of a widespread effort in this period to
believe in *something*. Their concerns and aspirations placed them –
far from the lunatic fringe of their society – squarely admidst the cultural, intellectual, and emotional moods of the era.

PART I
The setting

1

Mediums

No study of the spiritualist boom in the second half of the nineteenth century would be complete without the mediums – the men, women, and children who claimed to function as channels of communication between the living and the dead. They attracted thousands to their séances and ensured substantial publicity for spiritualist phenomena. Indeed, there could have been no spiritualist movement whatsoever without these conduits of spirit power, and it is only appropriate to begin an examination of spiritualism and psychical research by first allowing the mediums some time under the spotlight at center stage.

Perhaps the theatrical metaphor is unfair. Certainly a number of mediums were actors, consciously playing roles, purposely deceiving their audiences, and giving public performances worthy of any trained thespian. In fact, Emma Hardinge Britten, one of the best known trance lecturers both in Britain and the United States, had first tried unsuccessfully to launch a career as an actress. Not a few mediums were caught, at one time or another in their careers, practicing trickery, and professional conjurors had no trouble reproducing many of their allegedly spiritualist manifestations. But purposeful fraud – whether for financial gain, the need for excitement and public attention, or other psychological motives – did not discredit the activities of every Victorian and Edwardian medium. There were many conducting public séances, and far greater numbers working in private circles, who believed in their powers and thought that they were serving humanity through their mediumship. The degree to which the conscious and the subconscious, the voluntary and the involuntary, combined to produce results in the cases of most mediums cannot be determined today with accuracy or justice. "As Huxley said of the crayfish, to know how a medium feels and thinks one must become a medium."[1]

Under the label *medium*, furthermore, many subspecies and variations gathered in the nineteenth century, as today. Some took payment for their services, others refused. Some could produce phenomena in front of hundreds of strangers, others exclusively in the intimacy of their own homes. Some spoke English with a foreign accent, others like a native, and quite a few like Americans. Some specialized in particular effects, whereas others offered a broad repertoire of manifestations. That repertoire might include the materialization of entire spirit bodies – "full-form materializations" – in addition to the more commonplace rapping, table tilting, and emergence of spirit hands. Reports of séances also told of furniture cavorting around the room, objects floating in the air, mediums levitating, musical instruments playing tunes by themselves, bells ringing, tambourines jangling, strange breezes blowing, weird lights glowing, alluring fragrances and ethereal music wafting through the air. From the bodies of some mediums a strange foamy, frothy, or filmy substance, dubbed ectoplasm, might be seen to condense.

Mediums could relay the words of the departed through the laborious process of alphabet rappings or through the more efficient trance utterance. They might find their hands writing automatically, preside over the baffling appearance of messages written on slates, and ask the planchette or Ouija board for answers to questions posed by eager sitters. The communicating spirit might be that of Benjamin Franklin, Plato, the archangel Gabriel, or the sitter's Aunt Nellie. The possibilities were limitless.

As a medium's reputation spread, he or she would acquire a circle of devoted sitters, a fan club composed of men and women who might seek out the medium's powers as often as several times a week and whose faith in the medium's gifts was unshakable. Any professional medium worthy of the name in the second half of the nineteenth century could point to one or more prominent persons deeply impressed by the phenomena produced at his or her séances, and perhaps even converted to spiritualism as a result. Nonetheless, the significant spread of spiritualism in Victorian Britain probably owed far less to the exertions of professionals than to the development of hundreds of private, or amateur, mediums who discovered their spiritualist powers in their own drawing rooms. As television today, by its location in the home, has an impact that the theater cannot begin to rival, so the work of private mediums, in small domestic circles, brought spiritualism more intimately into the lives of countless believers than could the public sittings of professional mediums. It is, after all, far more compelling to see one's own dining table in motion than to read about the antics of someone else's furniture.

Although the séances of well-known professional mediums have been described and documented in some detail, the activities of most private mediums are, by their very nature, less accessible to the historian's curious eye. One finds references in memoirs, diaries, and letters to the shock experienced at the discovery of unsuspected spiritualist affinities in spouse, parent, offspring, sibling, friends, and neighbors. One reads of domestic circles developing around the mediumistic talents of a household servant, and the spiritualist press overflowed with stories of men, women, and children who, to their own immense surprise, spoke, wrote, or drew pictures automatically, beheld visions, or caused chairs and tables to gyrate. Spiritualistic prowess often spread among the members of a single family, for mediumship was catching, it would seem, and certainly the power of suggestion and example must have played a substantial part in the rapid multiplication of private mediums during the second half of the nineteenth century. Many of them, however, somewhat in awe of their abilities, shunned publicity, and it is impossible to compute with any precision the numbers of people, of all ages and social strata, who became convinced spiritualists without ever venturing beyond their domestic séances.

Particularly striking is the number of middle-class housewives who discovered powers of trance communication, clairvoyance, and furniture relocation during the 1850s, 1860s, and 1870s. Many of their husbands shared these skills, but it was the womenfolk who predominated in the ranks of the amateur mediums. Indeed mediumship could be, in its fashion, as domesticated and feminine an art as embroidery. Thus while Cromwell Varley, an electrical engineer of some renown, was busy with the Atlantic cable, his wife developed her gifts of trance utterance, clairvoyance, and automatic writing. (Interestingly enough, so did her maid.) Mrs. Augustus De Morgan, wife of the eminent mathematician, saw visions. The wife of a Mr. Fusedale regularly beheld spirits, as did numerous other women, married and single, who communicated their experiences in the columns of the spiritualist press.[2] The testimony is very difficult to evaluate, coming as it does from the medium herself, an enthusiastic relative, or admiring friend. What safeguards against deception were possible? How reliable were the witnesses? Such obvious questions must, of course, be asked and answered before any adequate conclusions can be reached about private mediumship in nineteenth-century Britain, and it is, unfortunately, most unlikely that any replies will be forthcoming to satisfy the skeptical inquirer a century later.

The comment of a contemporary skeptical inquirer at the turn of the century, that called attention to the unfathomed levels of self-deception "in persons of unquestioned good faith," is suggestive, but evasive.[3] It does not explain why so many housebound women embraced mediumistic pursuits in this period. The religious motivations will be explored in subsequent chapters, but surely another part of the explanation lurks in the very word "housebound." Recent work in women's history has underscored the frustrations experienced by uncounted mid-Victorian women, barred from gainful and stimulating employment by social conventions, with horizons limited by the predictable routine that domestic responsibilities imposed. If fewer men than women developed into private mediums, the reason may well lie primarily in the male duty to earn a living and support a family. There was nothing financially remunerative about home séances, but they offered other rewards that must have appealed mightily to the bored housewife or spinster. (And how much greater must have been the appeal to the household servant, drudging away at endless domestic chores, longing for personal significance and status in a walk of life that provided neither.) If professional mediumship enticed hardier women with career opportunities and glimmering possibilities of excitement and fame, private or amateur mediumship must have served a similar purpose, on a more modest scale, in homes across the country.[4] Domestic séances may have offered something of the escapism so abundantly supplied by soap operas today, but with one important difference: In the spiritualist home circle, the medium was not simply an outside observer; she was the crucial participant in the unfolding drama.

Sir Edward Bulwer-Lytton, the successful novelist and politician who investigated spiritualism for twenty years before his death in 1873, distinguished "mediums of probity and honor supported by people of like character," from "paid professionals" in the business of providing séances.[5] His distinction would appear to be between amateur mediums who used their powers without thought of material reward, and those who sought to support themselves by their talents. Yet such a distinction is too simple. There were mediums who accepted no payment for their séances, but who must surely rank among the most professional of their times, and whose probity and honor were, from time to time, suspect. Likewise, there were mediums who regularly charged a fee for sittings, but whose character was not thereby noticeably blemished. The only distinction that seems generally valid is between those mediums, on the one hand, whose séances were open only to family, friends, and a few strangers introduced by friends, and those, on the other, who sat for a much larger public. Some vignettes of the latter at work may help to suggest their special skills and the milieu in which they operated.

DANIEL DUNGLAS HOME

Of all nineteenth-century mediums, none provoked more commentary – not only in the spiritualist press, but also in the leading periodicals of Victorian England – than Daniel Dunglas Home. He was an enigma to contemporaries and has remained a puzzle to students of spiritualism ever since. His talents were enormous, and no medium had more zealous disciples. His séances were free, but he accepted lavish gifts from wealthy patrons and was charged in 1868 with trying to defraud a rich old widow of thousands of pounds. *Punch* might relish snide observations at Home's expense, informing readers that "Spirit-hands, at his bidding, will come, touch, and go/ (But you mustn't peep under the table, you know)." The editors of the *Mask* might observe: "That he has sharp eyes, a cunning wit, and quick, long fingers, there is no denying – so has a fox."[6] Nevertheless, no one ever proved that Home was a charlatan, and perhaps the best way to describe this unusual man is the simple label of "human oddity."[7] John Truesdell, in *The Bottom Facts Concerning the Science of Spiritualism*, expressed the opinion that "nearly every fairly-intelligent person is known to possess some latent mediumistic qualities, . . . though remarkable mediums, like true poets or great musicians, are by no means common."[8] By all accounts, Home was the most uncommon of all remarkable mediums.

He was not, however, the first of his profession to arouse public curiosity in England, nor the first medium from America. When he arrived in London in the spring of 1855, two American women, Mrs. Hayden and Mrs. Roberts, had already preceded him across the Atlantic in 1852 and 1853. In their wake, domestic circles all across Britain had blossomed, and table tilting became, for a brief time, a national hobby. Home's appearance in England in 1855, and his return in the autumn of 1859, helped rekindle and sustain the public's enthusiasm, and by the 1860s the spiritualist movement had gained a momentum quite independent of the exertions of any one medium.

It was no coincidence that the heralds of the spiritualist movement in Britain came from the United States. In 1848, Hydesville, New York, was the site of the first series of disturbances that inaugurated what is known as modern spiritualism. News of the "Rochester Rappings" spread quickly across the eastern United States, and the young Fox sisters, Margaret and Kate, in whose presence the rappings occurred, became celebrities. In an atmosphere prepared by widespread interest in mesmerism and phrenology, religious unorthodoxy, mysticism, and social utopianism, spiritualism found a ready audience in numerous American communities. As spiritualism steadily moved westward across the United States, expansion to the east, across the ocean, was only a matter of time. There was a virgin audience in Britain, primed by news of the American phenomena, and ready to be impressed.

Home was not, strictly speaking, American. He was born in Scotland in 1833, had emigrated to the United States with members of his family in the 1840s, probably toward the end of the decade, and had lived for a few years in Connecticut. His mediumistic powers manifested themselves in 1850, and during the next five years he traveled throughout New York and New England, giving séances and collecting patrons. It was a group of the latter who paid for his passage to England in 1855. He had been away from Britain long enough to be received, by some as a representative of the latest American madness, by others as harbinger of America's newly revealed spiritual truths.

Between 1855 and the early 1870s, Home collected circles of adoring followers, not only in London, but throughout Europe as well. He was a thoroughly cosmopolitan medium, equally at ease in English drawing rooms, Florentine villas, and royal palaces. He presided over séances at the Tuileries with Napoleon III and Empress Eugenie, with Queen Sophia of Holland at The Hague, and with Tsar Alexander II at Peterhof. Moving gracefully and mysteriously through European society, this slim, pale, and tubercular young man cultivated the rich, the famous, and the powerful wherever he went.

The séances that Home gave in England, particularly after his return from the continent in 1859 and throughout the 1860s, attracted prominent and titled sitters as honey draws bees. Famous hostesses, such as Lady Waldegrave and Lady Hastings, opened their salons to him. Lord Stafford, Lord Dufferin, Lady Dunsany, the Marchioness of Salisbury, Lord Lyndhurst, the Duchess of Somerset, the Marchioness of Ely, the Duchess of Sutherland, the Dowager Duchess of St. Alban's, the Marchioness of Londonderry, the Countess of Medina Pomar (subsequently the Countess of Caithness), and Lady Fairfax figured among the aristocrats who entertained D. D. Home. His most assiduous hostess was Mrs. Thomas Milner Gibson, wife of the Liberal President of the Board of Trade. Although Milner Gibson himself did not participate in the frequent séances at his Hyde Park home, his fellow free trade crusader, John Bright, could not resist an opportunity to meet the renowned medium in 1864. Robert Owen, the venerable utopian visionary, was one of Home's earliest visitors in 1855 – so was the former Lord Chancellor and Whig reformer, Lord Brougham, then approaching his eightieth birthday.[9] Home gave séances for luminaries of the literary world, including Bulwer-Lytton, Robert and Elizabeth Barrett Browning, Mrs. Frances Trollope, Thackeray, and Ruskin. He cooperated when scientists sought to investigate his powers. Certainly no other medium could boast of so illustrious an audience, but Home, to his credit, was also accessible, without charge, to men and women who lacked any ostensible claim to fame whatsoever.

By no means all his visitors, eminent or humble, were impressed by the phenomena that they witnessed in Home's presence. Robert Browning, for one, despite his wife's enthusiasm, had a low opinion

of the mediumistic profession in general and of Home in particular –
an opinion that he registered in "Mr. Sludge, 'The Medium.'" Ruskin,
too, after an initial period of hopefulness, began to have second
thoughts. "In all the manifestations of this new power," he wrote to
Mrs. Cowper (later Lady Mount-Temple) on 6 August 1866, concerning
an evening with Home,

> I have great sense of a wrongness and falseness somewhere – It
> seems, in the *best* people, to mean some slight degree of nervous
> disease: while in most of the instances I have heard of – or seen – it
> has not been manifested at all to the best people, – or the wisest.[10]

Doubtless, too, many of Home's sitters attended his séances merely
for a stylish lark, with no deep convictions to be confirmed or challenged, and only a desire for amusement and novelty to motivate them.
Although some remained to become spiritualist true believers, others
in time drifted away to sample fresh sources of entertainment and di-
version.

With a few striking exceptions, Home's séances were not markedly
different from those of numerous other mediums communicating spirit
messages in the mid-Victorian decades. He frequently fell into a state
of trance prior to making contact with the spirits. "At regular séances
preparations and certain formulae were usually observed, sitting round
a table, touching hands and so on," reminisced the Earl of Dunraven,
who as the young Lord Adare was Home's close friend in the late
1860s.[11] In Home's presence, furniture trembled, swayed, and rose
from the floor (often without disturbing the objects on its surface);
diverse articles soared through the air; the séance room itself might
appear to shake with quivering vibrations; raps announced the arrival
of the communicating spirits; spirit arms and hands emerged, occasionally to write messages or distribute favors to the sitters; musical
instruments, particularly Home's celebrated accordion, produced their
own music; spirit voices uttered their pronouncements; spirit lights
twinkled, and cool breezes chilled the sitters. If Home announced his
own levitation, as he did from time to time, the sitters might feel their
hair ruffled by the soles of his feet. Their clothing might be pulled by
playful spirits; they might even be poked and pinched. But this much
could have been observed and experienced under the auspices of a
number of mediums, and Home's fame rested on more than the typical
mediumistic repertoire. Certain aspects of his séances set him apart
from the commonplace and underscored his distinction from the throng
of contemporary mediums.

Home's greatest distinction was, apparently, his attitude toward
light. Although there is some contradictory evidence about the degree

of light permitted at his séances, many of his sitters concurred that Home was willing to proceed in conditions other than total blackness. After straining their eyes to perceive, with difficulty, the manifestations produced in the presence of other mediums, the participants at Home's séances were startled to observe phenomena occurring in fair light. This is not to say that all Home's manifestations appeared in brightly lit rooms. That was far from the case, and, indeed, the light in the séance room frequently had to be dim to enable spirit lights and luminous arms to emerge. But by allowing some light, whether natural or artificial, to penetrate his séance room, Home emphasized the difference between himself and the majority of mediums who found virtually complete darkness most conducive to spiritualist phenomena. With characteristic flair for the dramatic, he explained his attitude in bold declarative terms: "'Light!' was the dying cry of Goethe. 'Light' should be the demand of every Spiritualist . . . By no other [tests] are scientific inquirers to be convinced. Where there is darkness there is the possibility of imposture, and the certainty of suspicion."[12]

Two particular phenomena, developed rather late in his career, further placed Home in a class beyond ordinary mediumship. One unusual accomplishment was known as elongation, whereby, witnesses claimed, Home's body grew in length, anywhere from several inches to nearly a foot. The other featured Home nonchalantly handling red-hot coals drawn from the fire, with no apparent discomfort or after-effects. This version of the age-old fire test, or fire ordeal, produced a great effect when performed in nineteenth-century English drawing rooms.[13] It was hardly a run-of-the-mill parlor game, particularly when Home shared his talent with sitters, as he did in the case of this eye-witness account from the winter of 1869.

> I saw Mr. Home take out of our drawing-room fire a red-hot coal, a little less in size than a cricket-ball, and carry it up and down the room. He said to Lord Adare . . . "Will you take it from me? It will not hurt you." Lord Adare took it from him and held it in his hand for about half a minute, and before he threw it back in the fire I put my hand close to it and felt the heat like that of a live coal.[14]

In all the annals of modern spiritualism, few performances have occasioned more analysis, controversy, and bewilderment than Home's Ashley House levitation, alleged to have occurred in London, in December 1868. The specific details of this extraordinary event are subject to dispute, as the accounts left by the witnesses are inconsistent and somewhat garbled. What they thought they had observed, however, is reasonably certain. According to all three of them – the Master of Lindsay (subsequently the Earl of Crawford), Lord Adare, and his cousin Captain Charles Wynne – an entranced Home floated out the window of one room overlooking Victoria Street, and entered through the window of the adjacent room where the startled witnesses were assembled. Although the fact that none of the three gentlemen actually saw Home's exit from the first window may seem significant today, news of his spectacular achievement must have vastly increased Home's already considerable contemporary reputation as a medium. The problems facing the historian, or psychical researcher for that matter, who wants to explain the Ashley House levitation are merely one aspect of the many problems posed by Home's mediumship in general. How does one elucidate in rational terms his tremendous power to capture the emotions and imagination of audiences over a period of nearly twenty years? If one is not a convinced spiritualist, how can the modern investigator make sense out of D. D. Home?[15]

A wealth of possible replies to that question has been forthcoming since Home's own day. In some cases, they draw on broad assumptions about the human potential for egregious self-deception, on the possibility of group hypnosis, on the sheer dexterity of the skilled medium, or on the venality of servants ready to be bribed for their help in preparing hidden mechanical devices around the séance room. Other explanations have focused on specific phenomena associated with Home. William Henry Harrison, a London journalist and spiritualist during the 1870s, suggested, for example, the following hypothesis to account for the medium's ability to elongate himself: "The facts indicate the existence of some general inter-atomic repulsive force, which under certain conditions can push asunder the multifarious atoms of various descriptions, which build up the different parts of the human body." Harrison's theory was similar to a number of pseudophysiological hypotheses embraced in the mid-nineteenth century, sometimes as supplement to the straight spiritualist line, sometimes as alternative.[16] Vague in terminology and confused in concepts of causation, the arguments put forward in these hypotheses were as elusive as the proverbial greased pig.

Fraser's Magazine did not have to go to such lengths to find a reasonable explanation for Home's achievements. Comparing him to the Mormon leader Joseph Smith, a contributor to *Fraser's* observed in January 1865:

> Mr. Home beats Joe Smith hollow; for he persuades people that they hear what they do not hear; that they see what they do not see; that an accordion, which makes an irregular noise, is playing a popular tune; and that he is floating near the ceiling when he is simply standing on his chair with one foot touching a disciple's shoulder.[17]

The psychical researcher Frank Podmore, pursuing the same thoughts a few decades later, commented that Home apparently saved his special effects for a limited circle of select admirers – men and women who could be psychologically primed to anticipate miracles, and who would subsequently swear that they had indeed witnessed them. Not a few observers of the spiritualist scene attributed most of the phenomena to an unstable, neurotic element in the personality of mediums, an observation that included Home as prime example. Perhaps the most curious theory, proposed to account for Home's authoritative influence over his sitters, was that of the early anthropologist E. B. Tylor who tentatively suggested that Home might be a werewolf with "the power of acting on the minds of sensitive spectators."[18]

That Home's presence had the power of acting on the minds of sensitive spectators there can be no doubt whatsoever. Whether that power stemmed from a form of what we would today call mind control, whether from the religious messages and inspirational announcements that habitually punctuated Home's séances, from his disarmingly disingenuous manner, or, as his admirers argued, from the indubitable authenticity of the manifestations associated with him, is clearly a question that defies solution. Disciples and critics would, naturally, offer very different answers. But if it is true that nothing succeeds like success, Home's enduring success can at least in part be explained just that simply. Even so consistent a skeptic as Podmore recorded, with grudging admiration, that "Home was never publicly exposed as an impostor; there is no evidence of any weight that he was even privately detected in trickery."[19] At a time when fraudulent mediums were being caught out right and left, Home's record fully entitled him to the pinnacle of eminence that he occupied among the ranks of his profession. When he retired in the early 1870s, spiritualism in Europe and the United States lost one of its most intriguing figures.

FULL-FORM MATERIALIZATIONS

British scientists were inevitably provoked by Home's alleged spiritualist manifestations. Some ridiculed his pretensions and regretted his popularity, but others sought to examine, not only Home's phenomena, but the accomplishments of other mediums as well. No scientist was more engaged in this investigation during the 1870s than William Crookes, physicist, chemist, and future President of the Royal Society. His inquiries involved several lovely young women who claimed to produce the full-form materializations that intrigued him. His relationship with one in particular – Florence Cook, a teenaged medium from Hackney – has caused considerable controversy among spiritualists, psychical researchers, and their critics. When Trevor Hall argued in *The Spiritualists* (1962) that Cook was Crookes's mistress, and he her confidant in a blatantly fraudulent series of séances, Hall earned the wrath of all those who preferred to hold Crookes's name in high esteem. Unfortunately for Crookes, Florence Cook's record as a medium was by no means as untarnished as Home's.

Cook was merely an adolescent when her "full-form materializations became the talk of every Spiritualistic circle in London and even abroad."[20] The phenomenon of full-form materialization did not figure prominently in Home's repertoire, but it was Cook's specialty, and her embodied control, or spirit guide, "Katie King," quickly became a familiar name in Victorian spiritualist households. Although Home almost always remained among, or above, his sitters (the Ashley House maneuver was a rare exception), Cook's phenomena demanded that she withdraw for a while from the séance circle, into a cabinet, cupboard, or portion of the room separated from the sitters by a curtain. There, seated on a chair and tied up with a string whose knots were sealed, Cook would remain for a varying length of time, while the sitters composed themselves, often with spiritualist hymns, for the upcoming encounter. Eventually, if conditions were favorable, "Katie's" face would emerge at an opening in the curtains or at the top of the cabinet. One sitter, who recorded his impressions of "Katie" in 1872, noted that

> Her face is of waxen whiteness, and bears a certain resemblance to that of the Medium . . . it is a sweetly-pretty face, and talks away quite freely in the broadest gas-light. She maintains a lively conversation with the circle, and is greatly delighted if her beauty and cleverness are praised. She is intensely human in that respect. Indeed it is very hard to believe that the bright smiling face one sees is not flesh and blood, but only a materialized something about to be resolved into immaterial nothing.[21]

Frequently at Florence Cook's séances, several different faces would appear in turn at the aperture in curtain or cupboard, and swarthy visages alternated with "Katie's" pallor. Sometimes the accompanying spirit hands would accept paper and pencil from the sitters and scribble messages for their enlightenment.

Much more astonishing and impressive, however, were the occasions when "Katie" would sally forth from behind her protective cover and mingle among the sitters with easy familiarity. Crookes investigated these full-form materializations over an extended period of time and was privileged to walk arm in arm with the attractive spirit. She had, it seems, an irresistibly solid quality. In any case, Florence Cook's séances, whether for faces or full forms, always reached their appointed end, some time after the last spirit had vanished, when the sitters opened cabinet doors or curtains to find the medium entranced, often huddled in the corner, and exhausted from the spirit materializations. Since there was widespread agreement in spiritualist circles that the materialized spirit forms emanated in some way from the medium – drew their temporary substance from the medium's own vital forces – it was natural at the close of such séances to find the medium utterly drained of energy. What was less readily explained was the fabric in which the spirits emerged draped from head to toe. Described variously as coarse calico, muslin, cotton net, or rough cotton, it was clearly not as other-worldly as the spiritualists would have liked. *Punch* sneered that "a piece of cloth cut off by a female spirit from her materialised skirt was found to have been dressed with lime in the Manchester fashion."[22]

Even more damaging to the public reputation of the spirit visitors was the occasional telltale appearance of what looked strikingly like a terrestrial corset or stays beneath the spirit gown. It was a glimpse of such paraphernalia that prompted Sir George Sitwell to grab the alleged full-form materialization during a séance with Florence Cook (then Mrs. Corner) at the rooms of the British National Association of Spiritualists in January 1880. In seizing the materialized spirit, Sir George transgressed the bounds of good behavior expected of séance participants, but his action certainly proved to all but the most blindly faithful that "the vivacious and apparently youthful ghost" was "a common cheat." Supporting evidence was secured when the curtains forming the medium's retreat were pulled aside to disclose "the medium's chair empty, and with the knot of the rope slipped, while . . . stockings, boots and other discarded garments of the medium lay about in ungraceful confusion."[23]

So glaring were the opportunities for deception in the production of full-form materializations that some mediums resented the ill repute that the shady phenomena brought to the entire profession. It was D. D. Home, in fact, who wrote the most scathing contemporary denunciation of what he called the "simulation" of spirit forms. His *Lights and Shadows of Spiritualism* (1877) contains a chapter on "Trickery and its Exposure" that could serve as a "how-to" guide for fraudulent mediumship. In the course of the chapter, Home quoted from a recent letter written to him by Edward W. Cox, a levelheaded observer of the spiritualist scene, who was highly critical of the conditions under which materialization séances were held. Cox had seen a letter from one female medium to another, explaining how the full-form materialization trick was turned, and he passed the information along to Home in a communication dated 8 March 1876. The critical aspect of the stunt was that the fabric, used to represent spirit robes, had to be thin enough to fold into a very compact little bundle. This was particularly necessary since the medium carried the cloth in her drawers, a hiding place where no Victorian gentleman would publicly dare to search. After retiring to the cabinet, or behind the curtains, the medium undressed, removing all her clothes except two shifts (the double layer being another necessary feature of the medium's séance attire). Then, having draped the clothes over a chair in the area to which she had withdrawn, in such a way as to suggest that she was resting on it, the medium emerged, wrapped in a thin veil and ready to perform.[24]

In his exposé, Home was not aiming his slings and arrows at Florence Cook. On the contrary, the older retired medium paid tribute to the younger star when he went out of his way to cite "the carefully-conducted experiences of Mr. Crookes with Miss Cook" as proof of the possibility of genuine spirit materializations.[25] One may suspect that Crookes's well-publicized experiments with Home, prior to those with Cook, persuaded Home to endorse the latter series, but, be that as it may, Home did not need to cast aspersions on the young woman in order to make his point about charlatans promenading as materialized spirits. Other examples came readily to hand. Frank Herne, for instance, a young professional medium whose séances featured spirit faces and bodies, was caught red-handed at a sitting in Liverpool in 1875. He was seized in the midst of impersonating a spirit, and when "the gas [was] turned up, he was found clothed in about two yards of stiffened muslin, wound round his head and hanging down as far as his thigh."[26] By most accounts, Herne was an unabashed rascal, and it clearly undermines Home's endorsement of Florence Cook to know that Herne was her friend, that he helped to develop her mediumship, and even held joint séances with her.[27]

Through Florence Cook, one glimpses the seamier side of spiritualism. One also confronts the provocative question of why the movement conspicuously attracted nubile young women in this period. For Cook was by no means the only fetching female medium of the 1870s. As the Reverend C. M. Davies observed in 1875:

> Of the many fluctuations to which spiritualistic society has been exposed of late is a very prominent irruption of young lady mediums. The time seems to have gone by for portly matrons . . . or elderly spinsters . . . and we anxious investigators can scarcely complain of the change which brings us face to face with fair young maidens in their teens.[28]

Davies came face to face with several such spiritualist practitioners whose physical appearance no doubt gave considerable extra charm to their séances.

Mary Rosina Showers, for one, emerged as a promising medium in the early 1870s. In her late teens, she was the daughter of a high-ranking officer in the Indian Army and, unlike Cook, enjoyed both social standing and useful connections. As her powers developed, full-form materializations appeared at her private séances – "Lenore Fitzwarren," for example, or "Florence Maple," both of whom evidently enjoyed nibbling the sweets offered to them by solicitous sitters. Another spirit, "Peter," used Mary Showers as a channel for his direct voice communications, which included songs performed in a pleasant baritone. Like Cook, Showers found her way to Crookes's laboratory for investigation by that zealous scientist. The two young women soon became comrades in mediumship, and on at least one occasion, in March 1874, Cook served as sole witness to the genuineness of a Showers materialization.[29]

Cook was not always on hand, however, to act as her friend's troubleshooter, and, shortly afterward in April, Showers suffered the humiliation of exposure during a séance at the home of Edward Cox. Like Sitwell a few years later, one sitter failed to observe the rules of the game and prematurely pulled aside the curtains through whose opening the spirit-face of "Florence Maple" had appeared. As "Florence" struggled to close the curtains, "the head-gear fell off, and betrayed the somewhat voluminous chignon of Miss S. herself." Furthermore, as Podmore commented, "the chair, where the medium should have been sitting, was seen to be empty." In Showers's defense, the best that her admirers could do for her reputation was to claim that, in a state of "somnambulism," she was not responsible for her actions, and certainly not guilty of willful deception.[30]

Annie Eva Fay was yet another comely young woman who made a name for herself in British spiritualist circles during the mid-1870s. She descended on London from her native United States in June 1874, and advertisements declared her to be nothing short of an "Indescribable Phenomenon." Her act initially featured the calico or cotton bandage test, during which the seated medium was secured by thin bandages to a pole or stanchion placed behind her. Bandages at her wrists and around her neck were passed through rings attached to the pole, and her ankles were tied with a long cord whose end was held by a member of the audience. With the medium thus, in theory, held immobile behind curtains, or in a cabinet, musical instruments placed on her lap nonetheless began to play, and other objects, such as hoops and hats, landed outside the medium's enclosed sanctuary.[31]

This was a vaudeville performance, no doubt of considerable skill and polish, but not the least meriting attention for its spiritual significance. Yet such were the attitudes in spiritualist circles in this period that almost any baffling performance was eagerly hailed as proof of the spirits at work. Fay, who had emphasized only the enigmatic nature of her talents, soon found herself in the curious position of being touted for mediumistic powers that she had not hitherto claimed in her publicity. Once it became clear, however, that Cambridge intellectuals and eminent men of science, among them Crookes, were anxious to investigate her mediumship, Fay was quick to bill herself "the Celebrated Spiritual Medium." Full-form materializations naturally became part of her second series of London performances in the fall and winter of 1874–5.[32]

Cook, Showers, Fay – not in appropriately did Davies remark that together they "would have made a very pretty model for a statuette of the Three Graces."[33] Their success as mediums owed much to their graceful, charming ways and airs of innocence. Yet it was more than sweet girlishness that produced such extraordinary attention to so many young female mediums who flourished in these years, some scarcely out of childhood, and some whose lights never shone beyond the confines of provincial circles. All the motives discussed earlier to suggest why women turned to mediumship of course apply here in considering the specific case of adolescent girls and young women in early adulthood. For those who sought a wider public than relatives and neighbors, the desire for celebrity status and personal recognition may have combined with individual dislocations and developmental crises of adolescence to prompt young women to discover their mediumistic powers. There may also have been a potent element of sensual enjoyment, possibly subconscious, that enhanced the séances. Without exaggerating the extent of sexual repression in Victorian society, one can surmise that the holding of hands and the caressing of spirit forms might have been stimulating not only to the sitters, but also to the young women whose emerging sexuality was denied natural means of expression. That the sexual potential beneath the séance routine was apparently perceived by contemporaries is implied not only in the rumors circulating about Crookes and Cook, but also in allegations that séances could serve as a cover for assignations.[34]

SPIRIT-WRITERS AND MAGICIANS

Although full-form materializations were undoubtedly the most gripping of spiritualist phenomena, and the manifestations most fraught with danger for the presiding medium, they were by no means necessary to elevate a medium to the top of the profession. There were several avenues to success in spiritualist circles, and numerous mediums earned reputations for diverse skills. Some, for example, excelled in trance speaking, whereas others could obtain written messages from spirits under seemingly impossible circumstances. Of the former, Emma Hardinge Britten became famous, not only as a chronicler of contemporary spiritualism on both sides of the Atlantic, but as a tireless propagandist whose speaking tours after the mid-1860s took her all over England and introduced large audiences to inspirational lectures delivered with great fervor.

Equally, if not even more, famous on the trance lecture circuit was Cora Lavinia Victoria Scott Hatch Daniels Tappan Richmond, a woman of substantial talents and several husbands. Achieving fame as a spiritualist speaker while still in her teens in the United States, she visited England, with splendid success, in the 1870s. Her pensive, serene face graced the cover of the spiritualist weekly, *Medium and Daybreak*, on 2 January 1874, and the greater part of that issue was devoted to a discussion of her mediumship. Her public addresses, it seems, followed the same pattern as Mrs. Hardinge Britten's. Upon entering the lecture hall, Mrs. Tappan Richmond would be informed of a topic selected by the audience, and, after a brief pause to settle into the trance condition, she proceeded to speak on the subject at great length. She had no opportunity to prepare the address in advance, and her orations appear to have been genuinely improvised on the spot. There is little doubt that trance utterances bear the unmistakable markings of the medium's mind, through which they have filtered, and that trance pronouncements are of very dubious merit as evidence of spirit power. Nor is it likely, as Podmore commented, that after uttering three thousand orations in a fifteen-year period, this lecturer would have been surprised by any subject chosen for her. Her long speeches were certainly monotonous, crammed with excess verbiage and devoid of literary elegance. Yet even Podmore was impressed with Mrs. Tappan Richmond and admitted that her trance utterances were characterized by an unusual degree of coherence.[35] One need not accept her claim to speak under the influence of spirit guides in order to find her mediumship noteworthy. Highly suggestive of the prowess of the subconscious mind, it deserved the inquiries of psychical researchers far more than the visually impressive manifestations produced by materialization mediums who relied heavily on external props.

Of the many mediums in this period who specialized in spirit writing, "Dr." Henry Slade ranks among the most notorious. Still another American exploiter of the spiritualist movement, he arrived in London in the summer of 1876 and immediately established himself as an expert at obtaining spirit messages written on slate. He had perfected this skill to such a remarkable degree that some hardheaded psychical investigators were baffled by Slade's phenomena. A few particularly obser-

vant sitters, however, sufficed to blow the whistle on him. John Truesdell, for one, was not deceived when he sat with "Dr." Slade in New
York and reported that he "could plainly see the movements of the
cords in the Doctor's wrist," working busily at the very same time that
the alleged spirit message was being scratched on the surface of a slate
held by Slade under the séance table. Furthermore, Truesdell suggested
that a subsequent knock on the door during his séance with Slade was
opportunely timed to allow the medium to leave the table and switch
slates when the sitter was not paying close attention.[36] Needless to
say, the message found on the slate after the interruption was both
longer and written in a much clearer hand than the previous brief message, which Slade had had to scrawl himself with a bit of slate pencil
lodged in a fingernail.

Slade's exposure in England came at the hands of E. Ray Lankester,
FRS, professor of zoology at University College, London, and Dr.
Horatio B. Donkin, a physician at Westminster Hospital. In September
1876, both men wrote letters to the *Times*, describing their séances
with Slade. They had ferreted out his tricks, noted the telltale movements of the wrist tendons, and decided to press charges. He was
accused under the Vagrancy Act of illegally employing "certain subtle
craft and devices to deceive and impose on certain of her Majesty's
subjects," and the case was heard at Bow Street early in October.[37]
The famous naturalist Alfred Russel Wallace gave testimony on Slade's
behalf; so did Edward Cox, a Serjeant at Law himself. Their efforts
met with no success. The spiritualist press might thunder that "the
Slade case has had no parallel since the days of Galileo,"[38] but the
presiding magistrate took a less exalted view of the affair. Slade was
sentenced to three months in jail with hard labor, but he escaped that
fate when his conviction was overturned because of a technical problem
with the wording of the statute under which he was found guilty.[39] The
medium thereafter quickly departed England's inhospitable shores, but
his admirers, who insisted that Slade had not received a fair trial, continued to profess their belief in the validity of his mediumship.

These sketches of several prominent mediums only suggest the variety and range of mediumistic talents that blossomed as spiritualism
emerged in Great Britain during the second half of the nineteenth century. Uncounted numbers of other men and women successfully exercised their powers, both as professionals and amateurs, in these decades. If none ever quite attained Home's international heights, several
became widely acclaimed on both sides of the Atlantic and the English
Channel. As the leading mediums of the 1860s and 1870s retired from
public view, others appeared who tended the flame of the spiritualist
faith and continued to puzzle even cautious psychical researchers.

Yet a medium's lot was not always a happy one. Not only were they subject to physical strains – whether fraudulent or not, their performances could be fatiguing – but the law, as Slade discovered, took a dim view of their profession. In the United States, as early as 1860, mediums joined together in a Mutual Aid Association in Boston, but a benefit society for mediums was not successfully launched in Great Britain until 1900 with the establishment of the Lancashire Mediums' Union, subsequently known as the British Mediums' Union.[40] Even a sympathetic union, however, could not protect mediums from the gibes of professional conjurors, the magicians who could expose the tricks of dishonest mediums, and who took a positive delight in doing so.

The distinction between conjuring and mediumship was, indeed, of fairly recent origin. Throughout the Middle Ages and down through the seventeenth century, conjurors were assumed to possess supernatural powers that might include, as the very name implied, the conjuring up of spirits. The other names used in sixteenth- and seventeenth-century England to designate village magicians – charmers, blessers, sorcerers, witches, cunning men, wise women – all connote something more than the ability to perform clever tricks. They imply a deeper knowledge of nature, a certain affiliation with forces, both natural and supernatural, far beyond any prowess ever attributed to a nineteenth-century conjuror. For centuries, the Catholic church resolutely maintained that supernatural effects were either divine or diabolical in origin; they could not be "theologically neutral." Protestantism inherited the struggle against conjuring, and it took a long time to strip magic of its other-wordly aura.[41]

By the nineteenth century, however, magicians had become purely secular figures, entertainers whose function was to amuse and baffle an audience, but certainly not to heal the sick, predict the future, locate lost articles, or apprehend a criminal. Nineteenth-century mediums did not, in most cases, claim the full range of powers formerly attributed to the village wizard, but they were in a sense the modern successors to those aspects of the wizard's craft that had fallen by the wayside of rationalist, technological progress. It is no wonder that, even in the late nineteenth century, the confusion of terminology persisted and that conjurors and mediums were occasionally mistaken for each other. Not only was there the ancient historical relationship, but, more to the point for modern audiences, mediums who were supposed to be in touch with spirit agencies might, if caught cheating, prove themselves no different than magicians, whose performances depended only on the right equipment. Equally, the exaggerated credulity of many spiritualists often assigned to spirit power any conjuring stunt that they could not fathom. Not a few professional magicians refused to recognize the possibility of genuine mediumship at all and insisted that mediums were merely stage performers who offered a highly diverting routine, largely by sleight of hand, assorted conjuring tricks, and a magician's paraphernalia. The manifestations produced at séances, the Victorian conjurors scoffed, were no more spiritual in origin than their own mundane achievements.

Magic acts were an extremely popular form of entertainment and amusement in the Victorian era, and the more commercially minded mediums were not above capitalizing on a ready-made audience. There is no denying that séances could be jolly good shows, in the best conjuring tradition, and that a number of mediums employed such conjuror's devices as the *temps*, a movement or activity designed to distract the audience's attention from the critically important actions going on at the same time. Discussing these aids to chicanery, the *Academy* warned its readers in 1878 that they "should be carefully studied, since they form the stock-in-trade of spiritualist mediums, who seldom, if ever, possess much power of sleight-of-hand, but who invariably cultivate the science of making *temps*. Slade was a very successful hand at this manoeuvre."[42]

Indeed, it is not unlikely that the success of magic shows in Britain before midcentury helped prepare the ground for spiritualism after 1850. There were, of course, other preparatory influences, but it is worthwhile to consider for a moment the magician's role in stimulating segments of the public to accept as genuine the so-called spiritualist phenomena. In the 1840s, British audiences could choose from a rich assortment of conjuring acts, presented by native and foreign prestidigitators alike. There were, among many others, J. H. Anderson from Scotland who styled himself "The Wizard of the North," Ludwig Döbler from Germany, the Frenchman Phillippe, and a performer dubbed "The Mysterious Lady" whose back was always turned to the audience. At the end of the decade, another French conjuror, the celebrated Jean Robert-Houdin, perfected an especially impressive routine with his son Emile. The boy, blindfolded and facing away from the audience, was nonetheless able to describe the contents of closed containers, including purses, boxes, and wrapped packages handed to his father from the audience. The father's role, obviously, was crucial, not only for the sharp thumbnail that he cultivated expressly to make tiny slits in wrapping paper, but for the verbal cues that he slipped his son during lively conversation with the audience.

The Mysterious Lady's somewhat similar act probably also depended on a confederate, for one gentleman regularly introduced her to audiences at the Egyptian Hall, Piccadilly. Whether or not anyone in the audience guessed the man's function, it is certainly fair to surmise that numbers of baffled men and women were indeed "persuaded to believe that it was all done, as the playbill put it, 'by the exertion of a faculty hitherto unknown.' "[43] There is, unfortunately, no way of ascertaining whether those same people were subsequently attracted to the spiritualist explanation of mediumistic powers. The assumption seems safe, however, that conjurors in Britain during the first half of the nineteenth century convinced some observers that forces and faculties were at work with which humanity had never yet come to terms. In the second half of the century, mediums could reap where magicians had sown.

Not a few magicians complained about this arrangement and resented the acclaim accorded to mediums for performing old, stale conjuring tricks. Harry Houdini, with a professional name derived from Robert-Houdin's, won fame in the early twentieth century not only as a master escape artist, but also as a debunker of spiritualist pretensions. Even earlier, however, British conjurors had been challenging the claims of mediums and had tried to alert the public to the humbug that masqueraded as spiritualism. In the 1870s, the Egyptian Hall served as a kind of headquarters for this antispiritualist activity. On the first floor, John Nevil Maskelyne and his conjuring partner, George Cooke, regularly entertained their audiences by exposing mediumistic tricks in the course of their act. Elsewhere in the building, Dr. Lynn[44] was for once in substantial agreement with his rivals, Maskelyne and Cooke, as he also found that scoring off mediums enlivened his own performance.

The spiritualist response to these assaults was curious. Some outraged believers sought to return tit for tat by exposing conjurors. George Sexton, for example, the spiritualist editor, lecturer, and writer, took great pleasure in showing audiences how Lynn, Maskelyne, and Cooke produced their special effects. His endeavor seemed particularly pointless; all it accomplished was to prove that conjurors were exactly what they claimed to be – conjurors.[45] More significant was the response of spiritualists who excused exposures of mediumistic trickery with the reassuring explanation that the magicians were themselves mediums. Maskelyne was an obvious target for such wishful thinking. He had launched his public career in 1865 by reproducing the manifestations of the American professional mediums, Ira and William Davenport, and had missed no opportunities thereafter to ridicule spiritualism. It was far less shattering for a devout spiritualist to believe Maskelyne a cryptomedium than to admit that the Davenports were awfully good at wriggling out of the knots that bound them hand and foot in a large cabinet. Benjamin Coleman, a prominent figure in London spiritualist circles, came to just that conclusion. After seeing Maskelyne and Cooke perform, Coleman became convinced of their spirit power and described them in 1875 as "very powerful mediums, who find it much more profitable to pander to the prejudices of the multitude by pretending to expose Spiritualism than by honestly taking their proper place in our ranks."[46] Even Reverend Davies, a very different sort of spiritualist, and one who regularly applied his critical faculties to the phenomena that he witnessed, could not accept the exposure of mediums by conjurors as definitive proof of anything. "The magicians produced many of the same phenomena as Moses;" he observed, "but, even so, if we are orthodox we must believe the source of such manifestations to have been utterly different."[47] Critical faculties could not hope to compete with such belief.

There is, needless to say, another tradition that provided an even more important foundation than magic shows for modern spiritualism in the nineteenth and twentieth centuries. That tradition is man's age-old belief in witchcraft and spirit possession, a belief enhanced by countless legends concerning strange powers of second sight and the visits of ghosts, apparitions, and poltergeists. This is not the place to trace the lingering strength of superstition in the modern consciousness, but it should not be necessary to point out that stories of this nature have by no means vanished from Great Britain, or from anywhere else for that matter. England, in fact, seems to revel in ghostliness, with its *Ghost Who's Who* and assorted tourist activities organized around the haunts of famous spooks.

British spiritualism, clearly, did not emerge out of nowhere and nothing in the 1850s. In part it drifted over from the United States, but it also drew much strength from indigenous cultural and anthropological traditions, in which religious faith and magic rituals were closely intertwined. In the second half of the nineteenth century, nourished on rich deposits of spiritual lore from previous ages, spiritualism in Britain flourished in the specific conditions created by the troubled relations of science and religion. Within this context, however, the mediums were the catalysts. Whether they were sincere or calculating, well-intentioned or corrupt, skillful or inept, it was upon their shoulders that the fondest hopes and deepest longings of thousands of spiritualists rested. Societies and associations established to propagate the spiritualist faith, or committees of inquiry designed to investigate the spiritualist claims – all were ultimately dependent on the quality and integrity of the mediums to whom they turned.

2

Membership

Looking back from the grim vantage point of World War I, George Bernard Shaw described English society during the decades before 1914 in the preface to *Heartbreak House*. It was, he wrote,

> addicted to table-rapping, materialization séances, clairvoyance, palmistry, crystal-gazing and the like to such an extent that it may be doubted whether ever before in the history of the world did soothsayers, astrologers, and unregistered therapeutic specialists of all sorts flourish as they did during this half century of the drift to the abyss.

Although spiritualists would have taken issue with the bleakness of Shaw's interpretation – his vision of a helpless society turning to quacks for relief and release – they would certainly have concurred that fascination with spiritualism and psychic phenomena reached a high point in Great Britain during the late nineteenth century. A rich diversity of people during that period shared the fascination, formed organizations to pursue the subject systematically, and patronized a spiritualist press that served to publicize the activities of spiritualist circles around the country.

Spiritualism and psychical research were never monopolized by any one class of British society. While lords and ladies were lionizing D. D. Home, factory hands in the north of England flocked to hear trance speakers as they made their way on circuit tours around the country. There were no social prerequisites for participating at séances, forming domestic circles, seeing apparitions, or receiving spirit messages. A textile worker might have as profound a spiritualist experience as a duchess, though not, of course, at the same séance.

Upper-class interest in spiritualism did not end with Home's retirement in the 1870s. The possibilities of communing with the dead and divining the future intrigued the idle rich right through the Edwardian period. Although some approached the matter with a touch of religious uneasiness, others continued to treat spiritualism purely as a fashionable pastime, and one chronicler of Edwardian England has observed that among the ways in which society ladies whiled away their hours were "palmistry, necromancy, crystal-gazing, and betting on horses. These . . . had taken the place of charity and good works."[1] Unsubstantiated rumors persisted that Queen Victoria herself had sought to contact Albert beyond the grave, but there was no doubt whatsoever that numerous aristocratic ladies regularly frequented séances. At the less elevated, but nonetheless affluent, level of the upper-middle class, zealous supporters of spiritualism included Charles Blackburn, the gentleman from Manchester whose financial subsidies helped to launch Florence Cook's career as a medium. J. C. Luxmoore, scion of an old Devon county family, admired and promoted both Mary Showers and Florence Cook; he pleased the second young lady mightily when he invited her, suitably escorted by her mother, "to go with him for a run in his yacht."[2] It was neither among the titled aristocracy nor the untitled ranks of leisured ladies and gentlemen, however, that the most active spiritualists and psychical researchers were found in the Victorian and Edwardian decades. These were drawn, instead, from both the professional middle class and the better educated strata of the working class.

Members of the professional middle class, with all the variety of background, education, and income which that amorphous category includes, abounded in the spiritualist and psychical research organizations established after midcentury. Much of the time-consuming and often tedious work of these groups was accomplished by people who combined such chores with the responsibilities of lawyers, doctors, clergymen, politicians, professors, journalists, and literati of all sorts. As the authors of *The Year-Book of Spiritualism for 1871* recorded: "Among investigators we may number divines, logicians, and teachers in our schools of learning; physicians and lawyers; men of note in the arts, sciences, and literature; statesmen."[3] Clearly, there is no room in this study to portray all these energetic men and women in any detail, but a few miniature portraits are irresistible. They are needed, in any case, to give bodily dimensions to the abstractions of spiritualist belief and psychical inquiry, to put features on the largely faceless crowd of people who attended séances and subscribed to the spiritualist press. As some divines, physicians, academicians, and statesmen will be featured in subsequent chapters, examples here are limited to the legal and literary professions.

Edward William Cox, for one, brought to his spiritualist investigations the welcome combination of professional standing and personal wealth. Although his career as MP for his native Taunton was cut short in 1869, when he was unseated on petition, Cox did not find time hanging heavily on his hands. Serjeant at Law, Recorder of Portsmouth, Chairman of the second court of Middlesex Sessions from 1870 until his death in 1879, he launched and for many years edited the *Law Times*. He was also the founder of *Crockford's Clerical Directory*, while owning several other papers that furnished him with a comfortable income. Apparently he found no difficulty in mixing his different interests, and the list of his publications presents a curious blend indeed. Among much else, thirteen volumes of *Reports of Cases in Criminal Law*, four volumes of *Reports of All the Cases . . . Relating to the Law of Joint-Stock Companies*, and six editions of *The Law and Practice of Joint-Stock Companies* rub elbows with the more philosophically inclined *Spiritualism Answered by Science, The Mechanism of Man: an Answer to the Question, What am I?*, and *A Monograph of Sleep and Dreams.*[4]

Cox's flirtation with spiritualism endured for years despite a strong tendency towards skepticism. Indeed, after much diligent inquiry, he at length concluded that he could not rightly call himself a spiritualist at all. Although certain that many séance manifestations were genuine, he preferred to ascribe them to the workings of some psychic force rather than the activities of spirits.[5] Caution was perhaps necessary to safeguard his judicial reputation; nonetheless, Cox tirelessly continued to attend séances, sometimes hosting them himself at his London home in Russell Square. After one such séance in October 1872, when the Reverend William Stainton Moses served as medium, Moses wrote in his notebooks: "Cox displayed the greatest possible interest in my experiences, . . . and especially he was glad to have found an educated Psychic who could observe and record his sensations. He is a shrewd observer, a clever little man, and most ardently interested in this subject."[6]

During the last years of his life, Cox channeled that ardent interest into the Psychological Society of Great Britain, which he founded in 1875 and whose president he remained until his death. Although the title Psychological Society would today carry a very different meaning, in the 1870s the name was widely used to signify groups devoted to spiritualist inquiry. Cox's society, however, emerged as something of a departure from the typical "psychological" organization of this time, for it represented "a reaction against the slovenly acquiescence of the great body of English Spiritualists in the belief in spirits of the dead as the sufficient and exclusive agencies in the production of the phenomena."[7] Unfortunately, this admirable principle did not translate itself into concrete performance, and by the time of its demise, shortly after Cox's, the Psychological Society of Great Britain had accomplished little of note.

Charles Carleton Massey, a barrister like Cox, shared Cox's strong doubts about the validity of the physical phenomena produced at séances. Yet, although Massey considered the physical manifestations "certainly not edifying, either morally or intellectually," other attractions drew him powerfully to spiritualism. Born in 1838, he had begun to question the bases of the Christian religion while still a schoolboy, and he spent most of his adult life, until his death in 1905, searching for a religious and philosophical truth that could transcend his intellectual scruples and satisfy his emotional needs. Unlike Cox, he found that the demands of a legal career left inadequate time for metaphysical inquiries, and he abandoned his successful practice at an early age. Family and friends were disappointed. The grandnephew of Lord Bolton, the cousin of Lady Dorchester, and the son of the Rt. Hon. William N. Massey, a prominent Liberal MP for more than twenty years and Minister of Finance for India during the 1860s, Charles Carleton by any estimate could have readily attained public office. Private pursuits, however, were more to his liking, and he was lucky to have an understanding, supportive father.

In virtual retirement, Massey read widely, not only in the literature of theology, Eastern mysticism, and philosophy, but also in the emerging study of psychology. He had already read Swedenborg as a young man; now he took up astrology and pored over astrological charts until the early hours of the morning. For a time, he embraced Theosophy. All the while, he was gradually working toward a position of Christian mysticism, in which love of Christ, "the Divine and unifying Principle of the Race," would allow each believer to shed the particularity separating him from his fellow human beings and to partake of universality, or the universal personality of mankind.[8]

Given Massey's interests, it was natural that his reading quickly led him to spiritualism. The possibility of human communication in defiance of time, space, and death appealed strongly to his mystic vision of oneness, and he joined several organizations dedicated to exploring that possibility. He was a member of Cox's Psychological Society, served on the first council of the Society for Psychical Research (SPR) when it was launched in 1882, and a few years earlier had been active in the affairs of the British National Association of Spiritualists (BNAS), holding office as one of its vice-presidents and serving on the Experimental Research and General Purposes Committees. In June 1878, Massey was appointed to a BNAS subcommittee established, on his own motion, to "consider the general question of the Lunacy Laws."[9] His legal expertise was not altogether irrelevant to his studies of the psyche.

It was, in fact, particularly relevant when Henry Slade was brought to trial in 1876. In this the only time that Massey broke his self-imposed retirement from the Bar, he appeared with F. K. Munton (another member of the Psychological Society) before the Bow Street magistrate as counsel for the defence. Massey was convinced that Slade was innocent of imposture, and his vigorous efforts to clear Slade's name prompted Alfred Russel Wallace, himself an enthusiastic spiritualist, to call Massey "one of the most intelligent and able of the Spiritualists."[10] Massey continued to worry about Slade's reputation, even after the medium had left England. When Johann Zöllner, Professor of Physical Astronomy at Leipzig, subsequently investigated Slade during 1877–8 and, unable to detect his deceptive technique, published a glowing account of Slade's supernormal powers, it was Massey who translated Zöllner's *Transcendental Physics* into English in 1880. He took advantage of the translator's preface to reiterate his faith in Slade's honesty, while discussing the importance of evidence and the evaluation of testimony as if he were still arguing the case in court. Seven years later, Massey's determination to vindicate Slade had not abated when he undertook to refute – again with legal precision lightened by refreshing moments of wry humor – the charges that Zöllner had been disturbed, deranged, or on the verge of insanity during his experiments with the American medium.[11]

Massey's other contributions to the literature of spiritualism, psychical research, and his own religiophilosophical concerns were limited. He translated into English Eduard von Hartmann's *Spiritism* and Carl Du Prel's *Philosophy of Mysticism*; he published numerous articles in the spiritualist weekly, *Light*; and, by his death, he had filled twenty notebooks with unfinished metaphysical essays. But if his friends were sorry never to receive from his pen the great original speculative work that they had anticipated, they did not appreciate any the less the studious bachelor whose regular habits included an afternoon rubber of whist at the Athenaeum. As the widow of Frederic Myers, a leading member of the Society for Psychical Research, wrote to another of the Society's founding fathers, shortly before Massey's death: "I am so sorry that dear Mr. Massey has been so ill – Fred used to say to me of him 'he will be very high in the Kingdom of Heaven' – he is such a true and *faithful* friend."[12]

Cox and Massey were joined by a number of professional colleagues who likewise sought to probe the mysteries of mind and spirit. There was Henry D. Jencken, barrister, who earned a permanent niche among British spiritualists in 1872 when he married the famous American medium Kate Fox. He had already won his spiritualist spurs, however, by helping D. D. Home out of financial difficulties in the mid-1860s and by serving, with Cox, as a member of the committee, formed by the London Dialectical Society in 1869, to investigate "Phenomena alleged to be Spiritual Manifestations."[13] There was Francis William Percival, barrister and examiner in the Education Department, who joined Massey on the first council of the SPR, and John S. Rymer, the Ealing solicitor whose home was the setting for many of D. D. Home's earliest séances in England in 1855. William M. Wilkinson was not only Home's solicitor; he also doubled as ghostwriter for the medium's autobiography, *Incidents in My Life*. Among an early group of convinced spiritualists in England during the 1850s, Wilkinson published *Spirit Drawings* in 1858, as a record of the literary and artistic talent that blossomed, under spirit influence, among his own family and circle of friends. Shortly thereafter, he became proprietor and coeditor of the *Spiritual Magazine*, the first successful journal of British spiritualism. Never a man to mince words, Wilkinson informed the Committee of the London Dialectical Society, when asked for his views on spiritualism: "The public will think you are a set of asses if you report in favour of it, and I shall think you are not very wise if you go the other way."[14]

The names of writers recur as frequently as those of lawyers in the records of British spiritualism and psychical research during the sixty years before World War I. "Men of note in . . . literature" was the phrase used in *The Year-Book of Spiritualism for 1871*, and its authors ought to have added "women of note" as well. Poets, journalists, novelists, and literary hangers-on of both sexes contributed energetically to spiritualism and psychical research in these decades, often capitalizing on public familiarity with their names in order to advertise their spiritualist interests and commitment. A few even jeopardized professional standing in pursuit of a truth that they valued far more dearly.

W. T. Stead is an excellent case in point. As editor of the *Pall Mall Gazette* during the 1880s, Stead was one of the most influential journalists of his day. Through the columns of the *Gazette*, he had a hand in sending General Gordon to Khartoum and in prompting Gladstone's government to bolster its plans for strengthening the navy. In 1885, his exposure of child prostitution in England was a cause célèbre. Throughout the decade, he was in the vanguard of provocative, crusading journalism, but in the 1890s his stature began to decline, and by the time he drowned on the *Titanic* in 1912, his reputation was foundering too.

This decline, in part, merely reflected the fact that by 1890 Stead, having left the fracas of daily journalism, had lost his daily soap box; in part, it reflected public disapproval of his opposition to the Boer War. It was, however, also accelerated by Stead's endorsement of spiritualist beliefs that represented an acute departure from his previous career as a hardheaded newspaperman. Having first attended a séance in 1881, he became profoundly attracted to spiritualism during the 1890s, and from 1893 to 1897, while he was busy editing the successful monthly *Review of Reviews*, he also edited *Borderland*, a quarterly journal of psychical inquiry. In 1892, Stead discovered his ability to write automatically, a skill that he soon put to use in transmitting spirit messages from the late Julia Ames, an American journalist and temperance reformer whom he had met briefly in 1890, shortly before her death. In 1897, he published a collection of these communications, *Letters from Julia*, and in 1909, apparently at her request, he founded "Julia's Bureau," an office whose sole function was *"to enable those who had lost their dead, who were sorrowing over friends and relatives, to get into touch with them again."*[15] Julia's was the guiding intelligence behind the bureau's work, and if difficulties arose over the best means of handling any applicant's request for help, "appeal was made to Julia in Council, when her decision, received by a clairvoyant, was final." Stead had no doubts whatsoever that Julia Ames was getting in touch, and almost weekly messages from a beloved son who died in 1907 further confirmed Stead's certainty that death was only a transformation, not a termination, of the human personality.[16]

The investigative zeal that characterized Stead's earlier work as a journalist did not color his later attitudes as a spiritualist. Evidently he wanted no part of critical skepticism where psychical research was concerned and was downright hostile toward Frank Podmore, a man whose work was steeped in that quality of doubt. In a letter to Oliver Lodge, physicist and psychical researcher, in which he described possible "evidence in favour of the genuineness of spirit photography," Stead commented: "If you care for particulars I shall be very glad to send them to you. But I absolutely decline to have anything to do with Mr. Podmore in any shape or form. He and the spirit which he represents have been and are the bane of the SPR."[17] It is no wonder that Lodge subsequently observed "W. T. Stead was a man whose energy, honesty, and enthusiasm were abundantly conspicuous, but whose judgment was sometimes in default. He was occasionally imposed upon by unscrupulous knaves."[18]

Enthusiasm and credulity were also salient traits of Samuel Carter Hall and his wife Anna Maria. This team of Irish Protestant writers became D. D. Home's staunchest champions in the 1860s; in their eyes, Home was "the greatest of the mediums God has given to humanity in later times."[19] The Halls' London residence was frequently the scene of Home's séances, and at one such gathering,

> Home flung himself back in his chair, looking wild and white; and
> then rising slowly and solemnly, went to the still bright fire, into which

he thrust his unprotected hands, and taking out a double handful of live coals, placed them – as a fire offering – upon Mr. Hall's snow-white head, combing the hair over them with his fingers, all which our host appeared to receive more than patiently – religiously.[20]

In 1864, Ruskin attended "an evening with Home at the Halls," and although no spirits manifested themselves on that occasion, S. C. Hall was customarily a great favorite with visitors from the other side. His success with spirits prompted Ruskin, at a later date, to complain of their poor aesthetic taste. "I like Mr. S. C. Hall," Ruskin told Mrs. Cowper in August 1866,

> but he has assuredly all his life been doing mischief in his own editorial business, – he knows nothing about art, yet talks and works at it – in a wholly harmful and mistaken manner. And the spirits come to *him*. I am bold to say that *I* do know my business . . . and the spirits *don't* come to me![21]

Ruskin was referring to Hall's position as editor of the *Art Journal*, which for most of Victoria's reign was the leading publication devoted to art. Hall ran the *Art Journal* for forty years, from 1839 until 1880, and also collaborated with his wife on numerous literary projects, such as *Ireland, Its Scenery and Character*, in three volumes. Anna Maria Fielding Hall, like her husband, was born and raised in Ireland, and she turned to her native land for the subject matter of many books, including *Lights and Shadows of Irish Life* and *Stories of the Irish Peasantry*. Mrs. Hall was a novelist, playwright, author of children's stories, and, again like her spouse, journalist. She contributed to numerous periodicals and edited a few as well.[22]

Anna Maria Hall not only shared her husband's professional interests; she was as keen in her spiritualist convictions as he was in his, and their joint commitment to the spiritualist vision of life after death, and to the reality of spirit communication, sustained Samuel during the years between her death in 1881 and his in 1889. He was certain that he received scores of messages from her, both through automatic writing and through several mediums, including Mrs. Kate Fox Jencken and the retired Home. Hall went so far as to claim that he occasionally found little slips of paper by his bedside, bearing words of love and encouragement from his wife, which her spirit had written while he slept. Whatever the channel of communication, Anna Maria's posthumous messages were constant sources of hope and reassurance to her bereaved husband. Her happiness in heaven was complete, she informed him, particularly since she expected him to join her soon. She basked in golden sunshine, surrounded by friends, relatives, and favorite pets, with flowers blooming in abundance. Hers (or her husband's) was not a very original vision of heaven, but originality was not necessarily the most potent weapon against the cold question mark deposited by science in the firmament. The spiritualist alternative unabashedly tugged at the heartstrings and addressed itself to the universal human capacity for grief.

If the sorrow of a husband mourning his deceased wife were not enough, Hall gave his wife's alleged spirit letters an even broader applicability. He described her joy at meeting in heaven their one child, an infant daughter who died when ten days old, but who had blossomed into a lovely young woman in the spiritualist afterlife. He hastened to explain that he was publishing fragments of his wife's messages specifically "in the hope that I may thus console parents whose children have been taken from them." The vision of a lost child grown into fair spirit adulthood was something of a convention in spiritualist literature, but there is no reason to doubt the sincerity with which such stories were repeated.[23] The very fact of the convention suggests much about the needs which spiritualism was designed to meet, and there is no doubt that those needs were amply filled for S. C. Hall.

It was in the Highgate home of William and Mary Howitt, another husband and wife team of prolific writers, that the Halls "became assured there was more than [they] had hitherto 'dreamt' of in the mysteries of Spiritualism."[24] The Howitts, like their friends the Halls, were versatile literary professionals, who could turn their hands readily to novels, poetry, history, children's stories, travel sketches, translations, or journalism. Their religious experiences were almost as variegated, for they migrated from the Quaker faith of their youth at the start of the nineteenth century, through a rapturous espousal of spiritualism, to end finally – in Mary's case – with conversion to Roman Catholicism in 1882, six years before her death. By the 1880s, Mary had repudiated spiritualism, telling her niece that she considered it "one of the greatest misfortunes that ever visited us; it was false, all false and full of lies."[25] Throughout the 1850s and 1860s, however, she was closely identified with spiritualism, as was her husband, who died in 1879 without, apparently, ever denying his fundamental spiritualist beliefs. William, a dissenting Protestant to the last, evolved his own highly individualistic creed that included belief in the Triune God and conviction that spiritualism was "the seal and servant of Christianity." While disgusted with what he deemed the degrading vogue for physical materializations that characterized British spiritualism in the 1870s, he could still write in 1875 to Thomas Shorter, a fellow spiritualist, that they ought "not to be dismayed at the attempts of low spirits to damage [the] clearness and fairness" of spiritualism. "Flies and wasps, too, are sure to collect about a honey-pot, but that is precisely because it *is* a honey-pot.''[26]

The Howitts began seriously investigating spiritualism in the mid-1850s, as part of a group of earnest middle-class inquirers that included, among others, Professor and Mrs. De Morgan, William Wilkinson and his brother, Dr. J. J. Garth Wilkinson, Benjamin Coleman, Professor Nenner (Hebrew Professor at the Dissenters' College, St. John's Wood), and Mrs. Nenner. Shortly after the Howitts' introduction to séances within their private circle, mediumistic powers began to develop among their children, and finally William and Mary, too, found themselves writing and drawing automatically. Their daughter Anna Mary, author of the memoir of her father's life, was in fact quite troubled by her psychic powers, which prompted her to see apparitions and to suffer from other disturbances. In 1859, she married Alaric Alfred Watts, another convinced spiritualist.[27]

William and Mary Howitt, with lives that spanned the greater part of the nineteenth century and pens that had to keep busy to support a large family, became entangled in numerous causes and campaigns of the day. They spoke out against alcohol, slavery, vivisection, corn laws, game laws, and poor laws. William lambasted the corruptions of the Anglican church in *A Popular History of Priestcraft in all Ages and Nations*. Mary, like so many Victorian authors, concerned herself with the morals of the young and wrote improving tales and poems by the dozen. William went prospecting for gold down under in 1852 and later published *The History of Discovery in Australia, Tasmania, and New Zealand*. Mary introduced English readers to Hans Christian Andersen and received a silver medal from the Literary Academy of Stockholm for her translations of Scandinavian poets.[28] Clearly, spiritualism was not the only activity that kept these prodigiously industrious people occupied, and yet from about 1856 until the Howitts left England for milder continental climates in 1870, spiritualism was absolutely central to their interests and pursuits. They went to séances with Home and a number of private mediums; they admired the trance eloquence of Mrs. Hardinge Britten; and they willingly put their literary skills to work on behalf of their beliefs.

Even before they had wholeheartedly embraced spiritualism, they had collaborated on a work that suggested their future concerns. In the early 1850s, William translated from the German Joseph Ennemoser's *History of Magic*, and when it was published in Bohn's Scientific Library in 1854, it appeared with an appendix selected by Mary which, as the subtitle boasted, contained "the most remarkable and best authenticated stories of apparitions, dreams, second sight, somnambulism, predictions, divination, witchcraft, vampires, fairies, table-turning, and spirit-rapping.'' Mary, together with Mrs. Hall, also wrote a tribute to D. D. Home's first wife, Sacha, for the end of *Incidents in My Life*, the first volume of memoirs that the medium published in 1863.

It was William Howitt, however, who became the apologist par excellence for the spiritualist point of view and a principal contributor to the spiritualist press that emerged in the late 1850s and 1860s. The *British Spiritual Telegraph*, whose brief life ran from 1857 through 1859, received several pieces from Howitt, while the *Spiritual Magazine* printed over a hundred of his articles between 1860 and 1873.[29] These ranged from specific commentaries on personalities and issues in contemporary spiritualism to more general essays that underscored the historical and philosophical underpinnings of the movement. The same combination of perspectives marked Howitt's major work on the subject of spiritualism, his two volume *History of the Supernatural* (1863). It was a masterful job that very nearly achieved its claim to survey the topic in all ages, countries, and religions.

In commenting on Howitt's *History of the Supernatural, Fraser's* lamented that its substantial virtues were "defaced by occasional inaccuracies and misstatements, for which nothing but inordinate credulity can account."[30] If Howitt's comparatively restrained work stands convicted of inordinate credulity, the spiritualist writings of Florence Marryat are positively drowning in eager belief. The daughter of Frederick Marryat, a dashing hero of the Royal Navy and a novelist himself, Florence produced dozens of forgettable novels, scores of short stories, articles, essays, and a few plays before her death in 1899. She edited *London Society* in the 1870s, toured in *Patience* with the D'Oyly Carte Company in 1882, and in the early 1890s published three highly enthusiastic endorsements of spiritualism: *There is No Death, The Risen Dead*, and *The Spirit World*.[31] A Catholic, she found that her religion failed to reassure her on the fundamental question of life after death, as she wrote at the beginning of *The Spirit World*.

> From my earliest and most unthinking days I have always felt that the one, great, unfulfilled want of this world is the undeniable proof that, when we leave it, we shall live again, or, rather, that we shall *never cease* to live. There must be a big screw loose somewhere in the various religions presented to us, which profess to give us everything but this – vague hopes – threatening fears – promises of reward and dread of punishment – but not *an atom of proof.*[32]

That proof Florence Marryat found abundantly in her twenty-five years of séances and spiritualist experiences. Introduced to the subject in 1873, she became a familiar figure in spiritualist circles, under her married names of Mrs. Ross Church and, after 1879, Mrs. Lean. She attended sittings assiduously, with Florence Cook and Mary Showers among many other mediums, and rarely failed to be overwhelmed by the manifestations that she witnessed. She believed that she herself exerted some kind of supportive influence that helped mediums produce their best results in her company. "I may mention," she commented in *There is No Death*, "that Miss Showers and I were so much en rapport that her manifestations were always much stronger in my presence." In the materialized figures that greeted her regularly at séances, Marryat claimed to recognize departed friends and family, including her own deformed baby, transformed into a fair-haired, blue-eyed maiden of seventeen (but still bearing the identifying deformity). All these she hastened to proclaim as indisputable evidence of survival after death. Whether or not her gullibility, as manifest as S. C. Hall's, tempted mediums to stray from the straight and narrow bears some consideration. Certainly the vivid detail that enlivens her séance accounts owes something to a novelist's imagination.[33]

One could go on multiplying examples of professional writers, both obscure and eminent, who ventured into the uncharted realms of spiritualism and psychical research during the second half of the nineteenth century. The names of the prominent literary figures who sat with Home have already been mentioned; there are other literary names galore in the records of the movement before World War I. Frederick Tennyson, a poet very much overshadowed by his younger, laureate brother; Andrew Lang, poet, classicist, essayist, and anthropologist of religion; Laurence Oliphant, travel writer, novelist, journalist, and mystic; John Addington Symonds, poet and scholar of the Italian Renaissance; and, of course, W. B. Yeats – the lives of these authors, and many more, were profoundly touched, even in some instances altered, by spiritualism and psychical research.

THE INDUSTRIOUS WORKING CLASS

As with their counterparts in the white-collar professions, members of the working classes were not slow to turn to spiritualism after 1850. Emma Hardinge Britten's record of her tours throughout England, particularly in the north, is peppered with references to working-class spiritualists. "In Leeds, Halifax, Keighley, Bradford, and nearly all the principal towns and villages of Yorkshire," for example, she found numerous men and women who functioned as "resident Mediums." "All these are working people," she reported, "toiling during the week in their several vocations, but giving cheerfully, without stint, and often at the cost of labour and fatigue to themselves, their best service every

Sunday to platform utterances, and that most commonly with little or
no remuneration." When Mrs. Hardinge Britten visited the collieries
of the Newcastle region, she "was deeply moved by the sight of the
earnest-looking sons of toil massed in serried groups around her." Indeed, as Dr. Gauld has aptly observed:

> The typical venue of a Spiritualist meeting was not a genteel parlour,
> but a Mechanics' Institute or a Temperance Hall. By the mid-eigh-
> teen-sixties Spiritualism was gaining a considerable foothold amongst
> artisans and the more educated working men, and these pretty soon
> came to constitute the bulk of its supporters.[34]

One of the principal stimuli behind the spread of working-class spiritualism was the conversion of Robert Owen to a belief in spirit communication. The octogenarian social reformer endorsed the new movement in 1853, five years before his death, and hailed the spirit
revelations as harbingers of "a great moral revolution . . . about to be
effected for the human race."[35] Ever on the lookout for ways to im-
prove man's earthly situation, Owen seized upon spiritualism as merely
the last in a long series of visionary schemes. Through his influence
on the British labor movement, a number of working-class Owenites
followed him into the spiritualist fold, where they enthusiastically continued their ongoing search for the "new moral world."[36]

The career of Thomas Shorter clearly illustrates this connection. The
coeditor of the *Spiritual Magazine* in the 1860s, Shorter reached spiritualism after years of active involvement in working-class causes.
Born in Clerkenwell in 1823, he was apprenticed as an adolescent to
a watchcase finisher and became one by profession. He espoused Char-
tism and served as secretary to a group of Finsbury Owenites in the
mid-1840s. Subsequently he became secretary of the Society for Promoting Working Men's Associations and, thanks to the influence of F.
D. Maurice, embraced Christian Socialism. When Christian Socialist
endeavors launched the London Working Men's College in 1854,
Shorter was named its first secretary. Like many of his fellow Owenites
– whether blue- or white-collar – he nurtured dreams of the millennium,
but was nonetheless eager to ameliorate the human condition in the
more immediate future as well. He had been raised in a strict evangelical home, and a deep religious strain underlay his secular human-
itarianism, so that in the 1850s, Owen's example combined with Shorter's own emotional needs to draw him to spiritualism.[37] In addition to
his extensive journalistic contributions to the movement, he wrote long
speculative essays, such as *The Two Worlds, The Natural and the
Spiritual: Their Intimate Connexion and Relation Illustrated by Examples and Testimonies, Ancient and Modern,* published in 1864 under

the transparent pseudonym, Thomas Brevior. His ideas were scarcely compelling, largely because he rarely subjected them to critical analysis. Having satisfied himself almost at once "as to the genuineness and spiritual origin" of the phenomena he witnessed in the early days of his spiritualist investigations, he informed the Committee of the London Dialectical Society that he was inclined to suspend "rigid scrutiny" at the séances that he continued to attend.[38] Blindness in later life made scrutiny altogether impossible, but did not hinder him from writing poetry and concerning himself with the broad historical, philosophical, and religious implications of spiritualism.

A similar route characterized Gerald Massey's passage to spiritualism. Massey – no relation to Charles Carleton – was very much a man of the working classes, the son of a poor canal boatman, barely educated, but determined to spend his scant leisure time in study and self-improvement. Like Shorter, Gerald Massey, who was born in 1828, went through a rationalist phase in early adulthood, became a dedicated Chartist, and, in 1850, turned to Christian Socialism. By the mid-1850s, Massey had established himself as a leading poet of the people and enjoyed considerable popularity for a period. His unshakable faith in spiritualism dated from the early 1860s, and it led him along some peculiar paths. A little volume, *Concerning Spiritualism* (1871) was innocuous enough, but for much of the last thirty years of his life, he wasted his literary talents producing book after arcane book that purported to find in ancient Egypt the key to the psychic mysteries puzzling modern man.[39] He was extremely proud of these works, which he considered his greatest, but few contemporaries were similarly impressed.

That elusive place on the social scale where the upper echelons of the working class adopt the attitudes, and often the income, of the lower middle class proved a fertile ground for the care and feeding of future spiritualists. The trance medium Edward Walter Wallis, for example, came from a family of Twickenham grocers who experimented with table rapping in 1866, until business suffered: "Folks would 'not go to Wallis's,' lest they should get 'spirits in their tea.'" Wallis's own mediumship began to develop in the early 1870s, when he was still an adolescent, and in 1876 he left his job at a W. H. Smith bookstall in Vauxhall Station to concentrate his energies on spiritualism. Shortly thereafter he married Miss Eagar, another medium, and together they decided to place themselves "at the disposal of the Movement," traveling around the country as missionary mediums, and helping to launch spiritualist circles from Plymouth to Glasgow.[40]

Even more important for the development and consolidation of British spiritualism was the work of Edmund Dawson Rogers. The son of poor Norfolk parents and fatherless from a very young age, Rogers was in many ways the archetypal upwardly mobile working-class lad. During his apprenticeship to a local pharmacist, he busily applied himself to mastering Pitman's shorthand. He also became an avid naturalist and earned the thanks of the Royal Botanical Society "for the discovery of the habitat of a very rare species of fern." After his marriage, Rogers took a job as a surgeon's dispenser at Wolverhampton, but, finding it hard to support a wife, child, and scholarly habits on 25 shillings a week, was glad to change careers by becoming a reporter for the *Staffordshire Mercury*. In 1848, he joined the staff of the weekly *Norfolk News*, where he remained until 1870 when he helped to found the *Eastern Daily Press*. Two years later, he left Norwich for London where early in 1873 he established the National Press Agency to distribute news to provincial papers in the Liberal interest. He continued to manage the press agency for twenty years.

Rogers arrived in London just when the organizing bee had lodged in the spiritualist bonnet, and he was to play major roles in the principal London spiritualist associations for the next thirty-seven years, until his death in 1910. Although Rogers had been practicing mesmerism since the mid-1840s, he did not seriously turn his attention to spiritualism until the late 1860s. Thanks to an acquaintance with fellow journalist S. C. Hall, Rogers was invited to attend a séance with Home in March 1869, and the following year Rogers began a long friendship with the Pentonville tailor, Thomas Everitt, and his wife, a well-known private medium. It was through his connection with Everitt that Rogers became a founding member of the British National Association of Spiritualists in 1873, and in 1881, when that body was on its last legs, Rogers took the initiative in founding the weekly spiritualist paper, *Light*. Early in 1882, he played a central part in organizing the Society for Psychical Research – indeed the original idea for such a group appears to have been his own – and he remained on the Society's governing council until 1885. In 1884, he joined the London Spiritualist Alliance, founded that year by Stainton Moses; on Moses's death in 1892, Rogers succeeded him as president of the Alliance and remained so until his own death.[41] Rogers, clearly, was an organization man.

So, in a different fashion, was James Burns. While Rogers's "good services to the cause of Spiritualism" were performed with "a total absence of personal display," Burns's were executed with all the flair for self-promotion of a born entrepreneur. He was actually born into the family of a Scots farmer craftsman, but migrated to London as an adolescent. In time, he built up his own business as a leading publisher and vendor of books, tracts, pamphlets, and newspapers, not only about spiritualism, but about a range of other subjects that he deemed progressive in one way or another. His gifts of showmanship were combined with an "indomitable energy" and enthusiasm that made his establishment, the Progressive Library and Spiritual Institution, "a meeting place and centre for Spiritualists in London." Founded in 1863, and located first in Camberwell and then in Holborn, this augustly named enterprise was, even in its heyday, not much more than a "dark little shop at 15, Southampton Row," but it served nevertheless as the site both of Burns's publishing business and of assorted séances, lectures, and demonstrations. A number of young mediums made their spiritualist debuts under Burns's auspices. James J. Morse, for example, one of the earliest native-born trance mediums to gain wide recognition in Great Britain, left his job in a city pub and began his public career giving weekly séances at the Spiritual Institution in the late 1860s. E. W. Wallis met his future wife while she was similarly employed.[42]

Burns's headquarters in Holborn was not designed for the elite spiritualist practitioners who could congregate in Mayfair drawing rooms. It catered instead to popular spiritualism, and although fashionable ladies and gentlemen were by no means unwelcome, one was more likely to find working-class and lower-middle-class enthusiasts at 15 Southampton Row, where admission might be as comparatively inexpensive as a shilling a head. Audiences there were exposed to a wide array of Burns's interests, including phrenology, which he investigated as zealously as communication with the dead. He was also a vigorous crusader on behalf of temperance and dietary reform, and although temperance may be firmly associated today with Victorian respectability, Burns made no effort to serve as champion of middle-class morality. On the contrary, "the columns of Burns' papers were always open, and his personal help always ready, for the spokesmen of minorities, the smaller the better, from the advocates of divided skirts to the exponents of the newest theologies."[43] Nor did Burns endeavor to mold his spiritualist beliefs into an acceptably Christian form; indeed "progressive" spiritualism, of Burns's variety, became synonymous with non-Christian spiritualism.

The exuberant reformist aspect of Burns's spiritualism – even if combined with sharp tactics against his rivals in the spiritualist publishing business – was far more characteristic of the movement on a popular than a fashionable level. The Owenite connection gave the working-class wing of British spiritualism an inclination toward socialism; although that inclination was diffuse and unfocused, it did help to shed a progressive light on many spiritualist activities in Great Britain during the late nineteenth century. Popular provincial spiritualism, in particular, was likely to affiliate itself with the campaigns for temperance and vegetarianism and to support protests against vivisection and capital punishment. These concerns strongly underscored the conviction, shared by spiritualists throughout the Victorian and Edwardian years, that their beliefs implied far more than communication with the dead alone.[44] Whether they deplored the effects of alcohol and the habit of eating meat, or excoriated the practice of experimenting on live animals and the taking of human lives, numerous spiritualists were articulating fundamental principles of their movement. They were unequivocally asserting the sanctity of life, the worth and dignity of the physical frame enclosing an immortal soul.

THE SPIRITUALIST PRESS

As with virtually every other aspect of Victorian life, newspapers and magazines helped to shape the world of nineteenth-century British spiritualism. The entire gamut of interests, obsessions, and convictions held by spiritualists, from whatever social background, found expression in the spiritualist press. Before World War I, groups or individuals devoted to diverse aspects of spiritualism had launched well over a dozen newspapers and magazines, of which about half enjoyed considerable success and longevity. Some purveyed news for a national readership, while others served purely local needs. Their columns provided information about lecture tours, public meetings, the whereabouts of mediums, and important developments abroad. Readers wrote to share their significant spiritualist experiences with each other, frequently claiming to offer indubitable proof of spirit identity. Editors regularly thundered against all nonspiritualists who refused to investigate the subject with an open mind. The tone of voice ranged from the magisterial – when, for example, cases of spiritualism in the Bible were under examination – to the chatty, informal announcements that so-and-so, a prominent member of such-and-such a spiritualist circle, had left town to spend the month of August by the sea.

Of the influential organs of the British spiritualist press, five dominated the field: the *Spiritual Magazine,* the *Medium and Daybreak,* the *Spiritualist Newspaper, Light,* and the *Two Worlds.* The first of these has already been mentioned as the joint effort of William M. Wilkinson and Thomas Shorter, with William Howitt as chief contributor. It was, on the whole, a sober-minded paper, bearing the impress of Howitt's scholarship, his European perspective, and his occasional misinformation, and its spiritualism was firmly grounded in the Christian faith. After Shorter's death, one obituary notice announced, in no uncertain terms, that it was ''mainly due to the labours of Mr. Howitt and Mr. Shorter, who were intimate friends, that we are indebted for the sound, sensible Spiritualism that prevails in this country, free from the vagaries that abound elsewhere."[45] This reassuring verdict was perhaps too complacent for accuracy, but it nevertheless reflected the goal of the editors and leading writer of the *Spiritual Magazine* – to present spiritualism as a thoroughly respectable and plausible phenomenon. Established as a monthly in 1860, the *Spiritual Magazine* was virtually without rival throughout the decade, but by the early 1870s control of the periodical had changed hands, Wilkinson and Shorter were no longer editors, Howitt had left the country, and the pioneering magazine was spluttering toward its demise in 1877.[46]

For a few years, from 1869 to 1874, the *Spiritual Magazine* was published as one of James Burns's many enterprises, but it was never the jewel in his crown of spiritualist publications. Pride of place belonged to the *Medium and Daybreak*, a weekly paper that flourished from 1870 until Burns's death in 1895. In founding the newspaper, Burns absorbed a provincial paper, *Daybreak*, that had appeared a year or two earlier under the editorship of the Reverend John Page Hopps. So successful was Burns's venture that by 1871 *The Year-Book of Spiritualism* described it as "a live periodical, aglow with startling phenomena . . . It is broad and tolerant, and rapidly increasing in circulation." In fact, Podmore reported that *Medium and Daybreak* "for years had the largest circulation, chiefly in the provinces, of any English Spiritualist paper."[47]

The *Spiritualist,* or *Spiritualist Newspaper* as it was eventually called, was founded almost at the same time as *Medium and Daybreak* and ran from late 1869 until 1882. It began as a fortnightly publication, but soon became weekly, selling for twopence. For its entire duration, the *Spiritualist Newspaper* was under the editorship of William Henry Harrison, who wrote "chiefly in connection with scientific subjects, [for] five or six influential London newspapers," including the *Daily Telegraph*. Having encountered in the press a "vast amount of prejudice against spiritualism" and finding it impossible either to read or write about the subject as fully as he wanted in existing periodicals, he resolved "to start a scientific journal" of his own "to meet the requirements of minds of a scientific order."[48] Harrison insisted on the investigative, analytic nature of his paper, which was accordingly subtitled *A Record of the Progress of the Science and Ethics of Spiritualism*, and Podmore gave it credit for maintaining "a more critical standpoint than any of its predecessors or contemporaries." Harrison's critical faculty was not always vigilant – he was wildly enthusiastic about Florence Cook in the early 1870s – but his record for exposing chicanery placed him well outside the ranks of blind believers. Unlike the editors of the *Spiritual Magazine* and *Medium and Daybreak*, for example, Harrison did not hesitate in 1872 to reveal the trickery of the popular spirit photographer, F. A. Hudson, whose use of double exposure and fancy dress ghost costumes had convinced some clients that he could produce snapshots of deceased friends and relatives.[49]

The *Spiritualist Newspaper* benefited for a few years from its affiliation with the British National Association of Spiritualists. Between 1874 and 1879, the paper functioned as the official organ of the Association, carrying its advertisements and all reports of its proceedings. Many of the leading members of the Association were regular contributors to Harrison's paper, and until 1879 he published it out of Association headquarters at 38 Great Russell Street. As a member of the council of the BNAS, however, Harrison inevitably became entangled in Association politics and earned the disfavor of Stainton Moses, E. D. Rogers, and others in the small core of spiritualists who controlled the organization. In 1879, he was unceremoniously ejected from Great Russell Street, and the BNAS briefly transferred its advertising and announcements to a new publication, *Spiritual Notes*.[50] The *Spiritualist Newspaper* declined rapidly thereafter.

One of the reasons for the failure of Harrison's paper in the early 1880s was competition from the new spiritualist weekly, *Light*. Rogers established the paper in January 1881 with the official blessing of the BNAS, and after the Association had disbanded, *Light* soon became affiliated with its successor organization, the London Spiritualist Alliance. From the start, Rogers supervised business arrangements for *Light*, but the editorship changed hands a few times before he assumed the entire management of the newspaper. Its first editor was John Stephen Farmer, spiritualist, journalist, and apparently author of a famous dictionary of slang. Stainton Moses took over from Farmer in 1886, and not long after Moses's death, Rogers followed him in combining the editorship of *Light* with the presidency of the London Spiritualist Alliance.[51] Unlike any of its predecessors in the British spiritualist press, *Light* found the formula for enduring, not just a decade or two, but for a century. Although Podmore complained in 1902 that "its pages furnish rather dull reading,"[52] it is still being published today, as a quarterly review.

In 1887, with the establishment of the penny weekly *Two Worlds*, the British spiritualist press received a shot in the arm. Not only did the success of a paper published in Manchester signal the end of London's monopoly of major spiritualist publications, but it was conducted with a vigor and initiative lacking in the *Spiritual Magazine, Spiritualist Newspaper*, or *Light*. While these served a largely middle-class, educated metropolitan audience, *Two Worlds* appealed to provincial readers with fewer social and intellectual pretensions. With the termination of Burns's publishing activities in 1895, *Two Worlds* became the leading organ of reform-minded and "progressive," or non-Christian, spiritualism. Proudly *Two Worlds* proclaimed itself "A Journal Devoted to Spiritualism, Occult Science, Ethics, Religion and Reform," and this heady blend has proved as long-lasting as the staid journalism of *Light*.

Under the editorship of Emma Hardinge Britten until 1892, followed by E. W. Wallis, and J. J. Morse from 1906, *Two Worlds* fairly pulsated with ideas for spiritualist organizations and reform crusades. It was indicative of Mrs. Hardinge Britten's interest in spiritualism at the local level that in 1888 *Two Worlds* undertook a census of spiritualist associations throughout the country. After 1890, the newspaper was also active in the effort to form a truly national and lasting organization of British spiritualists to connect and coordinate the work of the small local groups. Continuing the alliance of progressive spiritualism and temperance reform, *Two Worlds* pushed for the foundation of a Spiritualists' League of Total Abstinence around the turn of the century. Although that effort met with little success, it was typical of a publication whose shareholders were known to convene, for their annual general meeting, in a vegetarian restaurant.[53]

These five periodicals were only the proverbial tip of the iceberg, the more highly visible and widely circulated of the spiritualist publications in Victorian and Edwardian Britain. There were others that attained a briefer prominence, such as Burns's monthly *Human Nature*, which appeared from 1867 to 1877. Linked briefly with the British Association of Progressive Spiritualists, in which Burns was active, *Human Nature* exhibited the full eclecticism of its publisher's tastes and included within the scope of its subtitle "Physiology, Phrenology, Psychology, Spiritualism, Philosophy, the Laws of Health, and Sociology," as well as "Popular Anthropology." *Borderland*, although short-lived, was noteworthy in the 1890s as the handiwork of W. T. Stead. His assistant editor was Ada Goodrich Freer, a well-known folklorist, clairvoyant medium, and for many years a member of the Society for Psychical Research.[54] In their hands, *Borderland* featured a potpourri of occultism, mysticism, and spiritualism, with occasional forays into the harder sciences. The issue of July 1894, for example, blithely sandwiched an explanation of Alfred Binet's work on human thought processes between articles on palmistry and Theosophy.

Finally, there were a number of strictly ephemeral additions to the British spiritualist press that lasted a few years at most – often only a few months – and attracted little attention. Some of these now have historical interest, such as the *Yorkshire Spiritual Telegraph*, published at Keighley between 1855 and 1857, which claimed to be the first spiritualist periodical in England and clearly reflected Owen's role in molding provincial spiritualism. The *Christian Spiritualist* of the early 1870s followed the *Spiritual Magazine* in underscoring the vital connections between Christianity and spiritualism. There was the *Herald of Progress*, published at Newcastle-on-Tyne, the *Spiritual Record* in Glasgow, Lancashire's *Spiritual Reporter*, Burnley's *Medium*. London fleetingly had a spiritualist *Herald, Messenger,* and *Times*.[55] The presses rolled on, as spiritualists hastened to inform their colleagues around the country of their unflagging zeal and untiring labors on behalf of the cause. Minimizing setbacks and pitying disbelievers, the spiritualist newspapers and magazines of nineteenth-century Britain were a constant source of encouragement and hope to thousands of readers.

It would, of course, be a mistake to assume that only spiritualist publications addressed themselves to the subject that aroused such curiosity during the second half of the nineteenth century. As early as 1862, a Newcastle spiritualist reported that articles and letters concerning spiritualism had already appeared in *Blackwood's, Bentley's, Punch, Cornhill, Saturday Review, Lancet, Nonconformist, Morning Advertiser,* and *Daily Chronicle*, not to mention the *Times*, among others.[56] The trouble was that most of these notices, like the many that appeared after 1862, were critical of the movement, deploring public fascination with table rapping, enjoying a good chuckle at the spiritualists' expense, and sometimes even charging spiritualists with criminally deceptive activities. Once in a while, a favorable article might come along, such as "Stranger than Fiction," published in the *Cornhill Magazine* in August 1860, and written by Robert Bell, an Irish journalist. Bell attended séances with D. D. Home and recorded for the *Cornhill's* readers the reasons why he accepted as genuine the phenomena he witnessed, including the medium's levitation. But the *Cornhill* was edited at the time by Bell's friend, Thackeray, who was himself interested in exploring these odd phenomena, and such fortunate journalistic circumstances did not frequently occur. Much more common was the scorn poured on spiritualism in 1865, for example, in the columns of *Fraser's*, where it was treated as one of the "popular delusions" of the period, or, a decade earlier, the amusement with which Dickens's *Household Words* reported the "Latest Intelligence from Spirits."[57]

By and large, the nonspiritualist press reported the world of séances and spirits in a tone of condescension, repeatedly questioning the judgment and critical faculties – not to mention the honesty – of spiritualists in general. With the shady and ridiculous side of spiritualism thus publicly emphasized, to the extent that many spiritualists feared to lose professional and social status by openly pursuing their convictions, subscribers looked to the spiritualist press to correct the balance by proclaiming the integrity of their fellow spiritualists and the feasibility of spiritualist phenomena. Unfortunately for their hopes, the spiritualist press preached largely to the converted.

ORGANIZATIONS AND SOCIETIES

As there was a newspaper or magazine to cater to every shade of spiritualist opinion in Victorian and Edwardian Britain, so was there an organization. More than two hundred such groups developed in this period, both provincial societies and London-based associations, that provided a meeting place, social club, debating ground, and séance rooms for spiritualists and psychical researchers. Some of these, as we have seen, were substantial enough to have their own journals and organs of publicity. Others included only a handful of members and were scarcely larger than a private circle of close friends and neighbors. Outside London, the greatest concentration of societies was found in Lancashire, Yorkshire, the West Midlands, and in the northeast around Newcastle, but there was a scattering of spiritualist activity throughout the country. As the trance medium J. J. Morse remarked about his travels to spread the spiritualist message: "I receive constant calls from every part of England, Wales, and Scotland, but my most active sphere of operation is the Midlands and the North of England."[58]

Membership figures for these organizations do not, however, begin to give satisfying statistical tallies of the numbers of British spiritualists in this period. Working-class spiritualists could ill afford the subscription fee, often as high as a guinea, required to join many of the clubs and associations; when one compares the hundreds of men and women who respectfully gathered to hear Mrs. Hardinge Britten, J. J. Morse, and other trance speakers on the public lecture circuit with the mere dozens that most local societies could boast, it is easy to conclude that the lion's share of British spiritualists did not formally join any society, but met informally, in local Mechanics' Institutes or even more frequently in their own homes. Nor was the price of admission the only reason why many spiritualists declined to join the local organization. Members of the middle classes, for whom a guinea represented no insuperable obstacle, might prefer the privacy of their parlors lest they jeopardize jobs, or position in the community, through acknowledged affiliation with spiritualism. These are only some of the difficulties involved in trying to arrive at reasonable estimates of the British spiritualist population in the late nineteenth and early twentieth century. It was, furthermore, not only an elusive population, but also a shifting one. For numbers of participants, spiritualism was just a temporary enthusiasm, a way station en route to another theological or metaphysical cast of mind. Nor were theological and metaphysical labels then any more precise than they are today. Were Swedenborgians spiritualists? Were Theosophists? And how should one categorize the mildly curious investigator who attended a séance every now and again without throwing his full energies into the movement? With all these problems left unresolved, it is simply important to recognize that the societies, alliances, associations, and committees only attracted a portion of British spiritualists in the late nineteenth and early twentieth century. Their total number can at best be approximated, and that very roughly, somewhere between ten thousand and one hundred thousand.[59]

The pages of the spiritualist press are crammed with information about the organizations formed by like-minded enthusiasts across Britain. One reads, for example, about "Psychological Societies" in Brixton, Liverpool, Newcastle, Halifax, Derby, and Churwell; Associations of Spiritualists in Nottingham, Darlington, Glasgow, Hull, and the East Riding of Yorkshire; societies in Keighley, Cardiff, Blackburn, Bradford, Colne, Exeter, Huddersfield, Rochdale, Leeds, Leicester, Manchester, Yarmouth, Sowerby Bridge, and Gawthorpe; Ossett had its Spiritual Institution, and Birmingham its Midland Spiritual Institute. Other communities had developing circles where the powers of neophyte mediums were encouraged. London, naturally, had a little bit of everything, from the developing circle, supervised by a Mr. Cogman, in the East End to the organizations that claimed, with varying degrees of justification, to include a national membership. After 1883, there was even a hotel run exclusively for spiritualists by J. J. Morse's family, and conveniently located near Regent's Park.[60]

Lest spiritualist organizations in London should be thought to have eclipsed their provincial counterparts, something must first be said about developments outside the metropolis, for they were not merely a pale reflection of activities in the nation's capital. The millenarian, visionary aspect of the movement, scarcely evident in London, cropped up in the provinces, particularly in spiritualist circles formed in the 1850s while Owen's influence was still strong. The spiritualist society in Keighley, for one, "combined Owenite socialist ideas with preparations for the millennium,"[61] while in Nottingham, a circle formed around J. G. H. Brown who combined astrological revelations with a critique of England's social system. Brown, the author of numerous spiritualist pronouncements, advertised himself as a spiritual healer and informed his readers that he was in touch with the archangels, the Duke of Wellington, and Oliver Cromwell, among many others. In 1854, he formed the Universal Church of Christ in Nottingham, and a few years later, he and his followers proclaimed the foundation of the Great Universal Organization whose function was to rectify political and economic injustice prior to the Second Coming. Brown's group was distinct from Nottingham's Association of Spiritualists, which operated in a somewhat tamer environment.[62]

For many years, Newcastle-upon-Tyne had one of the most active of the provincial spiritual associations, one that provided the initiative for much spiritualist activity in northeastern England. In 1876, it called the conference that established the North of England Spiritualists Central Committee in an attempt to organize spiritualist societies in Northumberland and County Durham. The attempt failed, and in the following decade Newcastle was the center of an effort to launch the North-Eastern Federation of Spiritualists. Confusingly, in the 1890s, yet another such organization was formed with the title of North-Eastern Spiritual Federation.[63]

Indeed a plethora of district organizations formed, collapsed, and formed again under new names in Lancashire, Yorkshire, and the Midlands from the 1870s until World War I. Early in the twentieth century, the drive toward organization even affected spiritualists in Devonshire and Wales. The various district councils and unions that emerged "were all mainly concerned with coordinating the activities of local societies, and with the provision of speakers and mediums for affiliated societies." They foundered wherever the member societies were not vigorous enough to support the work of a district committee, and even where district unions survived, the local societies generally enjoyed complete autonomy. Many spiritualists, in fact, opposed organization beyond the local level for fear of losing the spontaneity that was so important to the practice of spiritualism. Bureaucracy, its opponents argued, was absolutely incompatible with the essence of their spiritualist faith.[64]

Victorian England was, however, an organized and organizing society, and the spiritualists were no exception. Not only did the pattern of district organization gradually take hold from the 1890s, but plans for organization on a national scale had been afoot since the 1860s. In that decade the British Association of Progressive Spiritualists struggled to sustain national ties for three years before disintegrating in 1868. Its advocacy of progressive spiritualism earned it the enthusiastic support of James Burns, but Britain in the 1860s was not quite ready for a national association of non-Christian spiritualists; nor was there as yet adequate local organization to buttress any genuinely national society.[65] During the next three decades, as that local work was slowly going forward, the idea of a national network, based outside metropolitan London, remained in abeyance while London spiritualists tried their collective hand at national coordination. It was only in the 1890s that provincial spiritualists once again turned to central organization.

In 1891 the Spiritualists National Federation (SNF) was accordingly constituted, with the endorsement of Emma Hardinge Britten and *Two Worlds*, and with forty-two affiliated local spiritualist societies. By 1913 that number had expanded to 141, and the SNF, after much debate, had been transformed into the legally incorporated Spiritualists' National Union (SNU).[66] The change in name and legal status did not, however, change the fundamental character of the organization. From the start, it had drawn most of its strength from the north of England and the Midlands, and later, when a number of London spiritualist groups joined the ranks of the affiliated, it continued to resist the lure of the metropolis. Its headquarters was for many years in Manchester, and although some years ago it moved to more stately quarters in Essex, it remains basically a lower-middle- and working-class organization. It also remains non-Christian and, in old age, still as subject to passionate controversies and heated antagonisms as it was in infancy and youth.

The history of spiritualism in London is likewise fraught with fierce conflicts and nasty personal animosities, intensified by the sheer diversity of opportunities available to the metropolitan spiritualist. From January 1857, the Charing Cross Spirit-Power Circle publicized its activities, and the internal squabbles and divisions that it suffered during the following year or so were thoroughly prophetic of things to come for the movement in general. In the 1860s and 1870s, London saw the emergence of a number of spiritualist societies: the Christian Spiritual Enquirers, East London Association of Spiritualists, St. John's Association, Serjeant Cox's Psychological Society of Great Britain, the Dalston Association of Inquirers into Spiritualism, Marylebone Spiritualist Association, and British National Association of Spiritualists.[67] Some, as their names implied, were purely neighborhood gatherings, which occasionally combined spiritualist inquiry with the functions of a friendly, or mutual aid, society; a few sought national status, without any great success. There was a high degree of shared membership among the several London spiritualist groups and a marked similarity among the services and programs that each offered to its members. Apparently London's active spiritualists never tired of attending social evenings, hearing papers on such subjects as "Our Duties as Spiritualists to Opponents, Inquirers, and Ourselves," participating at séances under club auspices, collecting reports of phenomena, and helping to stock spiritualist libraries.

Of all the London societies, the Marylebone Spiritualist Association went furthest on the least promising foundations. Although it was established in 1872, its early years were precarious, and it was not until the 1890s that it seemed assured of continued existence. From 1872, when a dozen or so Marylebone friends and neighbors decided to organize a spiritualist society in their community, until 1889, the list of members had only increased to about thirty. The following year, however, Mrs. Everitt, the medium, joined the Association, as did her husband Thomas, who became its president for the next fifteen years. The Everitts were a great catch for Marylebone, and their participation in the Association's activities soon attracted more members and badly needed publicity. When the Association found a new hall for its meetings in 1891, Florence Marryat gave the opening address. Three years later, Emma Hardinge Britten performed the same service when the group moved to different rooms. Gradually, the Marylebone Association became a fixed feature of London spiritualism, and so tenacious of life did the once struggling society prove to be that today, as the Spiritualist Association of Great Britain, it occupies a handsome building in Belgrave Square.[68]

As throughout the rest of the country, London spiritualist groups attempted now and then to collaborate in forming a more broadly based metropolitan body; names such as the London Spiritualist Confederation, the London District Council, and the Union of London Spiritualists adorn the records of British spiritualism around the turn of the century.[69] An earlier attempt, the British National Association of Spiritualists, had not only sought city-wide affiliation, but also more ambitiously courted national and even international connections. The tale of its rise, decline, and fall is illustrative of the problems besetting spiritualist organizations, in the Victorian era as today.

The British National Association of Spiritualists emerged from a meeting in Liverpool, in August 1873, sponsored by the local Psychological Society. Attendance was not confined to spiritualists from the immediate area, and among the participants were W. H. Harrison and Thomas Everitt from London. The meeting heard several papers advocating the benefits of national organization for the expansion and consolidation of British spiritualism, and these arguments carried the day. The conference resolved to form a national association, and initiative then passed to London, where the following year the BNAS commenced its activities. From 1875, it was comfortably housed at 38 Great Russell Street, the scene of its numerous séances, both public and private, committee meetings, lectures, and social gatherings.[70]

The inauguration of the Association failed to arouse the enthusiasm of all British spiritualists. For every reason that Harrison marshaled in favor of central organization, James Burns, for one, could find good grounds for opposition. To some extent, the persistent spiritualist fear of heavy-handed dogmatic authority operated against the BNAS in 1873–4, but in Burns's case there was also intense resentment against any metropolitan organization that might encroach on the domain of his Spiritual Institution. For rather different reasons, William Howitt likewise lamented the establishment of the BNAS. He regretted in its early policy statements the attempt to avoid any specifically Christian identification and to adopt a purely nonsectarian, even secular, position. To Howitt, such an attempt was unthinkable and such a position wholly untenable; for him, spiritualism was an entirely religious phenomenon, meaningful only in a Christian context. The new association was forced by the outcries of such staunch crusaders as Howitt and J. Enmore Jones, another well-known Christian spiritualist, to modify its stand. It withdrew from its constitution the Declaration of Principles and Purposes, which had included the offending announcement:

> The Association, whilst cordially sympathizing with the teachings of Jesus Christ, will hold itself entirely aloof from all dogmatism or finalities, whether religious or philosophical, and will content itself with the establishment and elucidation of well-attested facts, as the only basis on which any true religion or philosophy can be built up.[71]

Nor were religious issues the only source of disagreement in the BNAS. After a few years, problems of internal organization came to plague the Association and, in fact, ultimately hounded it to death. These difficulties arose because, despite the early opposition it provoked, the BNAS soon became the dominant spiritualist society in Britain. It attracted an impressive list of members, drawn largely from middle-class Londoners – although a few resided outside the metropolis – and including some of the most energetic and prominent British spiritualists. They were, for the most part, a well-educated and eminently respectable group. With such a pool of talent to choose from, the selection of a mere handful of officers seemed invidious, and the result was a herd of vice-presidents and a council of unwieldy proportions. Among the men and women who served the BNAS in these capacities were Charles Blackburn, Benjamin Coleman, the Countess of Caithness, Thomas Everitt, Florence Marryat, H. D. Jencken, W. H. Harrison, C. C. Massey, E. D. Rogers, and the Reverend William Stainton Moses. The president, for much of the Association's ten-year history, was Alexander Calder, a city merchant.[72]

Among the general membership, the BNAS could also boast some noteworthy figures. Frank Podmore, a young man in his early twenties, became active in the Association at the end of the 1870s, after a successful academic performance at Oxford. A civil servant at the General

Post Office, he shortly lost his spiritualist convictions and in 1902 published the classic cool appraisal of the movement, *Modern Spiritualism*.[73] Not coincidentally, he was also the biographer of Robert Owen. Captain Richard F. Burton, scholar of Arabic literature and lore, renowned traveler, and amateur anthropologist, was an honorary member of the Association. Burton, who according to the Earl of Dunraven "prided himself on looking like Satan," was not in the least convinced that spirit messages could bridge the gulf between life and death, but the subject intrigued him for its obvious links with certain Eastern occult practices.[74]

Over the years the work of the BNAS was divided among a number of committees, including Experimental Research, Finance, Soirees, Library, Correspondence, General Purposes, and House and Offices. The first of these dealt with matters of weighty significance to spiritualists around the world; the last had to cope with urgent practical considerations for the BNAS alone. In March 1879, for instance, the House and Offices Committee met to authorize the purchase of "three small teapots, two milk jugs, three candlesticks, scrubbing brush, and a toilette service."[75] For a time, this organization functioned smoothly, and the Association prospered. In 1875, membership surpassed four hundred, a sizable total for a spiritualist society at that time, and a number of spiritualist groups at home and abroad – in Liverpool, Cardiff, Brixton, Dalston, Brussels, Budapest, and Madrid – had become allied with the BNAS.[76] Within a very few years, however, the Association was facing serious trouble on several fronts, which at length combined to close the office in Great Russel Street.

Part of the trouble was financial. It was costly to hire professional mediums for trial séances and to maintain headquarters in a suitable manner. Members, too, were negligent about paying their fees, and by 1878 real concern for the economic health of the Association colors the records of committee meetings.[77] Furthermore, the BNAS was not particularly comfortable with its position as a national center for British spiritualism. Although it welcomed the affiliation of other spiritualist societies and had given itself, in C. C. Massey's phrase, "a slightly magniloquent designation," it did not have the funds, or the desire, to work effectively on behalf of provincial spiritualism. In a letter to a leading Liverpool spiritualist, the secretary of the BNAS explained in June 1878 that, given the Association's limited resources, it was essential to the whole spiritualist movement, for "an honorable recognition and a better social standing," to promote "a strong and vigorous centre in London."[78] The genteel members of the BNAS were not altogether eager for prolonged contact with their provincial colleagues, and the ambivalence with which the Association approached its role as a national organization made it a less than galvanizing leader of British spiritualism.

The most severe problems arose, however, not over external relations, but over deep divisions concerning the internal government of the BNAS. The cumbersome council of between fifty and seventy officers was clearly not capable of functioning as the chief administrative authority, and real executive control passed to committees that could readily be managed by one or two diligent and efficient members. As the BNAS came to be run along ever more oligarchic lines, with Stainton Moses as chief oligarch, some members became increasingly uneasy over its departure from the democratic principles of spiritualism. W. H. Harrison led the opposition against Moses's preponderant influence, but with little result. Although a few other officers of the BNAS, including C. C. Massey, joined Harrison in resigning their positions, Moses and his supporters remained entrenched in control of the Association.[79]

Moses's victory was short-lived. The Association that he continued to dominate was in uncertain financial condition, riddled with ill will, and losing members. Many of those who nominally remained found other ways to spend their leisure time, and during the summer of 1881, a number of committee meetings had to be adjourned for lack of a quorum. Moses himself appears to have lost interest in this moribund group, and from about the same time his name disappears from the minutes of the General Purposes Committee that he had hitherto chaired so effectively. Thereafter, matters reached a predictable denouement. In 1882, in an attempt to rally its members and present a fresh face to the spiritualist community, the remaining officers of the BNAS reconstructed the society, drew up a revised constitution, and called themselves the Central Association of Spiritualists (CAS). E. D. Rogers, one of Moses's most loyal supporters, stayed with the CAS until the bitter end, with a handful of stalwarts like J. J. Morse and John S. Farmer, but Moses's name is conspicuously absent from the CAS minute books. By the autumn of 1883, with the membership continuing to shrink and an estimated deficit of over £100, even Rogers could see no point in trying to keep the CAS alive any longer.[80]

Stainton Moses was not, however, inactive during this gloomy period for his former colleagues, and in the spring of 1884, he brought a number of them together in a new society that he had been contemplating in the light of developments at the BNAS. As Rogers later recorded, the London Spiritualist Alliance (LSA) was launched by Moses who "proposed that an Alliance should be promoted, with no governing body at all, other than a few persons nominated by himself, as he had been disgusted with the experiences of the British National Association."[81] Moses remained the president, and autocrat, of the LSA until his death in 1892, and among the "few persons" whom he appointed vice-presidents and council members were E. D. Rogers, J. S. Farmer, C. C. Massey, Alaric Alfred Watts (Howitt's son-in-law), Major-General Alfred W. Drayson, Moses's close friend Dr. Stanhope Speer, and the Hon. Percy Wyndham, a Conservative MP for twenty-five years.

After Moses's death, Rogers assumed the presidency of the Alliance and, finding an autocratic body anachronistic in the 1890s, helped to establish the LSA as a legally constituted limited company. Although no longer vexed by questions of geographic commitment – its source of strength was clearly London – it attracted some members from beyond the confines of the capital and boasted such prominent outsiders as Alfred Russel Wallace of Dorset and the Earl of Radnor of Longford Castle.[82] Membership expanded under Rogers's benign leadership, and the LSA entered the twentieth century with every apparent likelihood of success. Like the Marylebone Association, the LSA has withstood two World Wars, depression, and inflation to reside today at 16 Queensberry Place, as the College of Psychic Studies.

Another survivor from the late Victorian age is the Society for Psychical Research. From its headquarters in Adam and Eve Mews, off Kensington High Street, it can look back on a century of investigation, experiment, and controversy. It has moved about London at various times in its history, and its very first meetings, early in 1882, were held at 38 Great Russell Street, in rooms provided by the BNAS, for a number of convinced spiritualists figured among the founders of the SPR.[83] The Society's *Journal* and *Proceedings* represented a substantial departure from the existing spiritualist press; its membership included a constellation of professors, MPs, JPs, and Fellows of the Royal Society unmatched by any spiritualist association. Yet the origins and early years of the SPR were nonetheless intimately involved with the world of London spiritualism, and that early intimacy influenced the Society's history for years to come.

PART II
A surrogate faith

Apart from the belief that the dead can and do communicate with the living, spiritualism, as it emerged during the second half of the nineteenth century, developed no single creed embraced by all species of British spiritualists. Many highly diverse people espoused it for highly different reasons, and most spiritualists thoroughly savored the variety of opinions that were free to take shelter under their capacious umbrella. But many, likewise, would have shared the view of J. S. Farmer, editor of *Light*, when he wrote about spiritualism that "standing midway between the opposing schools [of faith and science], it gives to the one a scientific basis for the divine things of old, whilst it restores to the other the much needed evidence of its expressed faith in the duality and continuity of life."[1] It was a matter of great pride and satisfaction to its enthusiasts that spiritualism appeared to solve that most agonizing of Victorian problems: how to synthesize modern scientific knowledge and time-honored religious traditions concerning man, God, and the universe.

The emergence of spiritualism in a period of religious perplexity and tension was not, of course, coincidental. There is no need to argue here that the Victorian years were times of spiritual disorientation. Volumes of diaries, letters, memoirs, essays, poetry, and fiction written in these decades attest to "the note of unsettlement in belief" that countless troubled Christians sounded during the course of their lives as they responded to accumulated evidence from a range of sources.[2] Not the least potent response was the revulsion of sensitive consciences against the harshness, the apparent indifference to human suffering that had accompanied the spread of Christianity throughout its history, and the vengeful, punitive zeal that formed a central part of Christian dogma. This ethical revolt against a faith that seemed entirely out of step with the progressive and humanitarian sentiments of the times offered a strong foundation for the doubts that were subsequently aroused by discoveries in geology, the life sciences, and biblical criticism.[3] As scientific discoveries gave ever greater authority to naturalistic interpretations of the universe and reduced the credibility of the Christian saga of man's fall and redemption, likewise close textual studies of the Old and New Testaments undermined the theory of divine authorship. They revealed the Bible to be a work of human artistry, but an unreliable guide to human history.

It did not, however, require scientific learning nor a professorial familiarity with the Bible to understand what was happening to Christianity, buffeted by a torrent of subversive theories and arguments drawn from ethics, history, and science alike. All that Christians had for centuries found most awesome, and most reassuring, in their faith was jeopardized: the holy authority of Scripture, the divinity of Jesus, the immutable design and purpose behind God's creation, humanity's unique place in the universe, the promise of eternal life after death. One did not have to grasp in full detail the process of biological change through the natural selection of chance variations; one did not need to read Hebrew and Greek, nor participate in debates over the classification of fossils, to fear that the flood of new knowledge and interpretations would shatter traditional Christian concepts of cosmic order. In place of the universe perfectly constructed for all time by the Master Builder, who designed each infinitesimal part to serve a specific and ultimately beneficent function, the Victorians glimpsed a universe both chaotic and hostile, the product of random, accidental juxtapositions. Instead of an essential harmony and morality in nature, many a Victorian came to detect with dread nothing but universal conflict in which all but the fittest succumbed. Worst of all, man turned out to be an integral part of this natural savagery, with a history inseparable from that of animal life on earth, and the terrible implication of that connection was clear. With man's mind as vulnerable to gradual evolutionary change as his toes, the divine image in which humanity for centuries believed itself created dissolved into primeval slime. So did the certainty of an immortal soul.

"Materialism," wrote the Scottish theologian John Tulloch in 1885, "fights with bolder and more far-reaching weapons than it has ever before done." Tulloch saw forces at work that were threatening to make untenable "any Divine theory of the world,"[4] and, like his contemporaries, he fully realized that facile explanations concerning divine intervention in the earth's history could no longer prevail against the more cogent explanations that science offered in purely naturalistic terms. Indeed, materialism was widely perceived as the archvillain of the age. With its partner atheism, the alarmists moaned, it uprooted churches, made a mockery of morality, undermined social sanctions, and sapped the very foundations of western society and culture. If even T. H. Huxley, that most outspoken advocate and publicist of science, shied away from the ultimate, devastating implications of materialism, it is understandable why others in this period should exert themselves to oppose the materialist position. That, in fact, became the special task of the spiritualists in Victorian and Edwardian Britain: to deplore and combat the materialism that they perceived as all too rampant in their time.

In proclaiming the existence and activity of spirit agencies throughout the universe, spiritualists were articulating a fundamentally religious point of view. No matter how much scientific terminology they employed, spiritualists could not conceal the metaphysical implications of their pronouncements. Even the simplest definition of religion as a belief in some sort of spiritual, or nonmaterial, beings – a definition suggested by the Victorian anthropologist E. B. Tylor[5] – labels spiritualism a kind of religious mentality. People were drawn to it, not just on the tantalizing chance of establishing contact with departed friends and relatives, but because spiritualism answered a fundamental need. The *British Quarterly Review* sagely observed in October 1862:

> There is no form of belief so deeply rooted in man's nature, so widely spread over his entire history in time and space, so apparently necessary to his very being, as a conviction of the existence of an unknown and invisible world, capable of signalizing its presence by becoming at certain times visible and palpable.[6]

Spiritualism, combining this deeply rooted belief with the reassurance of human survival after death, exerted a natural attraction. The magistrate who presided over Henry Slade's trial in October 1876 was not speaking lightly when he remarked that spiritualism "may be called almost a new religion"; nor was the Reverend Davies when, after "a lengthened examination," he found "that, to a very large number of persons indeed, Spiritualism is in the most solemn and serious sense a religion."[7] It would be a mistake, however, to assume that, in elevating spiritualism to the status of a surrogate faith, British spiritualists were desperate people, willing to swallow any medicine in order to feel well again. They were not simply embracing *any* source of reassurance about the human condition, nor giving their allegiance to *any* faith in their eagerness to believe something other than the ubiquity of matter. They were turning to the one set of beliefs which, they trusted, could meet the specific demands of their day by satisfying a religious need in language, and through procedures, acceptable to science. Spiritualism, boasted Stainton Moses, was "that platform on which alone religion and science can meet."[8]

To dub spiritualism a surrogate faith in Britain is not to say that British spiritualism was a fully developed religious alternative in the second half of the nineteenth century. Its status among the categories dear to sociologists, such as church, denomination, and sect, is very much open to debate, as the practice of spiritualism in different communities, and among different social groups within the same community, varied substantially. In some cases, spiritualism and Christianity made not so strange bedfellows, while in others, spiritualism accompanied a virulent hostility to virtually all forms of the Christian religion. Sometimes spiritualist societies arranged services for themselves that bore a close resemblance to more traditional church services; in other instances, all formal trappings of religious observance were carefully eschewed. Freethinkers might embrace spiritualism, and so might church-going Anglicans, Nonconformists, and even, occasionally, Roman Catholics. In Victorian and Edwardian Britain, the way in which spiritualism became entangled with religion, organized and disorganized, exoteric and esoteric, is not easily summarized; to argue that it functioned as a surrogate religion for an outspoken secularist or unabashed atheist needs fuller substantiation than to claim that it reassured slightly worried Christians on the all-important question of immortality. Yet spiritualism filled both roles in the late nineteenth century, offering a path away from uncompromising materialism when secularist zeal faltered as well as a buttress for faith when Christian optimism sagged.

3

Spiritualism and Christianity

British spiritualists from the 1850s on shared the conviction that they lived at a climactic moment in human affairs. "We have arrived," one of their most prolific spokesmen cautioned, "at a grave crisis in relation to that which is the very life and soul of all religious faith and hope – belief in a Future Life – or as I prefer to state it, belief in life unbroken by the incident we call 'death'."[1] Spiritualists did not, of course, have any monopoly on the tone of urgency that frequently colored their writing. In 1876, Archbishop Tait warned the country: "When I was young we were told there was no such thing as an atheist in the world, but all that is changed . . . A materialistic atheism is in the air." And Gladstone, that ever vigilant observer of the religious state of the nation, informed his sister at the end of 1872: "The battle is to be fought in the region of thought, and the issue is belief or disbelief in the unseen world, and in its Guardian, the Creator-Lord and Deliverer of Man."[2] Spiritualists, however, had a distinct advantage in battle discussions of this nature, for they believed that they possessed a secret weapon of unbeatable force. Free of the doubts that assailed other combatants, they were confident that, in spiritualism, they had the means to annihilate once and for all the unholy alliance of atheism and materialism.

The secret weapon was demonstration or proof. Knowing how great a premium science placed on empirical evidence, British spiritualists, in endless chorus, asserted the evidential strengths of spiritualism. Samuel Carter Hall, for example, writing of spiritualism in 1884, explained: "I believe that as it *now* exists, it has mainly but one purpose – TO CONFUTE AND DESTROY MATERIALISM, by supplying sure and certain and *palpable* evidence that to every human being God gives a soul which he ordains shall not perish when the body dies." Newton Crosland, London wine merchant and spiritualist, was certain that "the miracles of Spiritualism, acted out in our presence, furnish us exactly with the demonstration we require to overwhelm the reasoning of the unbeliever, and to baffle his stern logic with a triumphant success that no polemics ever yet achieved." Thomas Shorter likewise boasted that spiritualism demonstrated immortality "as it can be demonstrated in no other way."[3] Spiritualists, in short, were convinced that their case was watertight, that they had "met the men of science on their own ground."[4] Spiritualism, they felt confident, could lessen the pain of personal bereavement and soothe individual fears about man's future state by providing the definitive answer to those cancerous philosophies that lurked so menacingly in the modern consciousness.

It was easy enough to discount such optimistic claims. Scientists could do so on the grounds of slipshod method, philosophers for faulty logic. Spiritualists were accused of everything from improper observation, through naiveté, to downright mendacity. What must have bothered devout spiritualists most, however, since they saw themselves firmly on the side of the angels, was the occasional accusation that spiritualism was an instrument of the devil. Clearly, those who voiced these accusations were not of the unconscious cerebration school that viewed séance phenomena as purely psychological in origin and nature. The men and women who cried "Diabolic agency!" in response to reports of spiritualist manifestations undoubtedly believed in spirits, but argued that what occurred at séances was caused by the "actions of evil spirits or devils, personifying who or what they please, in order to undermine Christianity and ruin men's souls."[5]

Given this perceived threat to Christianity, it is no wonder that clergymen were among the most insistent spokesmen for the satanic interpretation of spiritualism. Podmore has quoted entertainingly from a half-dozen or so pamphlets written during the 1850s by assorted provoked reverends, warning their readers against the snares that Satan set for the unsuspecting around the séance table. The major clerical alarms on this issue were sounded in the opening years of modern British spiritualism, but 'dread of diabolic machinations can still be found in the writings of clergymen as late as the 1870s. The Congregationalist preacher John Jones of Liverpool emphasized the satanic origins of spiritualism in a pamphlet asssertively entitled *Spiritualism the Work of Demons*, which he published in 1871, and another Nonconformist minister of Liverpool, Thomas Worrall of the Free Church of England, echoed his sentiments in a sermon issued three years later.[6] Neither pamphlet makes for particularly stimulating reading; the argument, predictably, is hammered home by repetition rather than logic. But subtlety and sophistication of argument were of little use to writers like Jones and Worrall. They were fulminating against what they perceived to be dire apostasy from Christian truth, and indeed Worrall was one of a few clergymen who specifically looked to spiritualism as evidence of the predicted apostasy of the last times.[7]

Clergymen as well as laymen, who approached spiritualism from this angle reflected a continuing millennial restlessness, a fascination with apocalyptic prophecies and signs, that had remained just below the surface of Christianity in Britain, particularly among the lower classes, for centuries. During the first half of the nineteenth century, this mood had stimulated the formation of a handful of enthusiastic sects, such as the Southcottians, Irvingites, and Plymouth Brethren. Bolstered by the influence of Owen's thought, it lingered on into the second half of the century, largely in evangelical circles, and far more often in Nonconformist than Anglican ones. Nor was clerical opposition to mediums and séances without earlier parallels. The Catholic priest of the Middle Ages, like the Protestant minister of the Reformation, had keenly felt the rivalry of the village wizard as an alternative purveyor of magical remedies, and the church had for centuries unsuccessfully attempted to monopolize all access routes to the spirit world.[8] What was involved was not merely the salvation of souls, but the very foundation of ecclesiastical authority and power on a popular basis throughout the countryside. Clerical hostility to the mediumistic profession, when it emerged after 1850 in Britain, was therefore predictable. After all, some mediums claimed to be in contact with exalted members of the heavenly host, and it is more than likely that the clergy who cautioned against traffic with the devil were also, at base, warning mediums not to trespass on ecclesiastical territory.

The spiritualist response to the charge of diabolic agency in the nineteenth century was not quite so simple as might be expected. Outright denial was, naturally, one line of defense that could be adopted with vigor. "Talk of these discourses coming from Satanic Spirits," exploded one contributor to the *British Spiritual Telegraph* in 1857. "Why man, such Spirits were worthy to have discoursed with Adam in Paradise in his state of innocence. Satanic!" Stainton Moses, himself an Anglican minister, smugly noted:

> On the orthodox hypothesis that Satan transforms himself into an Angel of light for the purpose of luring me to ruin, I must say he has been a very long time in revealing the cloven hoof, and meantime he and his emissaries have been very consistently truthful, pure, and good in the only sense I can attach to those terms.[9]

The more typical spiritualist position was, however, less cocksure. Asking "Is Spiritualism Satanic?" in the columns of the *Christian Spiritualist* in 1871, a contributor carefully explained that "Spirits are, as men and women are, of all grades and shades of holiness and of sin, from the glorious Archangel, . . . to the wailing denizens of the pit." But to acknowledge that spirits might communicate from hell, the writer assured her readers, was in no way to doubt that Christ had triumphed over the ways of Satan. Newton Crosland came right out and announced that good and evil spirits flocked with equal zeal to séances, and that constant vigilance was required lest the latter "creep in, seize the message almost in the middle of a word and finish it with a Satanic colouring, or render it ridiculous." William Howitt was similarly frank in conceding that "undoubtedly the devil takes care to have a finger in this matter, as he does in everything on earth." Howitt's recognition of Satan's earthly omnipresence did not undermine his belief that spiritualists, through prayer and faith in Christ, would ultimately learn to trample the devil under foot, but it did suggest his awareness that spiritualists could no more pick and choose the spirits they contacted than select their fellow passengers in a railroad carriage.[10] If, however, the forces of evil had as ready access to the channels of communication between the living and the dead as the forces of good, then it was all the more incumbent upon spiritualists to distinguish between the two, discouraging the former while promoting and publicizing the messages of the latter through every possible means.

The issue of satanic influence, doggedly though it was pursued in the 1850s and gravely though it was raised from time to time thereafter, was not the focal point for most orthodox Christians as they articulated their attitudes towards spiritualism. Even for many men of the cloth, old Beelzebub was not an immensely real or menacing figure in the nineteenth century. It was, rather, the sheer impiety of spiritualism that distressed the author of an article on "Spiritual Manifestations" that appeared in *Blackwood's* in May 1853. He deplored the fact that spiritualists believed themselves to be "favoured with spiritual revelations through means which the divine word has denounced." A correspondent to the *Sussex Daily News* likewise ignored the role of the devil, but faulted spiritualism on more general grounds for being devoid of religious content. "Utterly destitute of all the ennobling ideas of Christianity," he thundered, "Spiritualism has at best revealed a state of existence, if such it be, which is far more a loathsome parody of the Christian heaven than is the most degraded type of monkey of the 'human form divine'."[11] Yet whatever the specific basis for criticizing spiritualism, a broader issue lay behind most of the diatribes. The critically important consideration, as far as numerous clergymen and their devout parishioners were concerned, was whether they perceived spiritualism as supporting a Christian world view and cast of mind, or whether they saw spiritualism as a threat to Christianity, turning men's loyalties away from the faith of their fathers.

In an attempt to come to terms with the complex relationship between spiritualism and organized Christianity in nineteenth-century

Britain, contemporary observers, as well as scholars investigating the subject subsequently, have often divided British spiritualists into two mutually hostile factions: the Christian and the anti-Christian spiritualists. A rough distinction can be readily drawn between spiritualism in London and in the provinces during the Victorian and Edwardian decades. The former – dominated by middle-class professionals, intellectuals, and people of some means and social standing – was largely Christian in emphasis (despite James Burns). Outside the metropolis, particularly in northern England, there was strong anti-Christian sentiment among the lower-middle-class and better educated working-class members who formed the bulk of provincial spiritualism. Some of these had moved to spiritualism from a position of aggressive free thought; in embracing spiritualism, they may have been attempting to satisfy an unquenchable religious longing, but they were emphatically not returning to the Christianity that they had rejected, often after strenuous study. This categorization of pro- and anti-Christian spiritualism is valid enough, as far as it goes, but like so many attempts to classify elusive attitudes, it merely highlights the black and white, leaving obscure the shades of gray. It will be helpful for a time to use this distinction while examining the place of spiritualism in Victorian religion, but, after a while, the distinguishing features become blurred, and questions need to be raised that the simple dichotomy ignores.

CHRISTIAN SPIRITUALISM

Christian spiritualism is no contradiction in terms. The men and women who proudly called themselves both Christians and spiritualists discounted scriptural exhortations against communion with spirits and saw nothing in their activities opposed to Christian doctrine. Let more conventional Christians shake their heads over the spiritualists' apparently obsessive fascination with the hereafter; Christian spiritualists regarded it as a perfectly legitimate interest in the light of Christ's promise of eternal life. There was no doubt in their minds that spiritualism strengthened Christianity, not only proving once and for all the reality of life after death, but furthermore, in doing so, supporting the entire structure of Christian dogma. The phenomena produced around the séance table verified all that had hitherto seemed miraculous and therefore, in the modern scientific age, suspect in Christianity. S. C. Hall spoke for many of his fellow Christian spiritualists when he rejoiced that spiritualism had confirmed his Christian faith.

> I humbly and fervently thank God it has removed all my doubts. I can and do believe all the Bible teaches me, – in the efficacy and indescribable happiness of prayer, in the power of faith to save, in

the perpetual superintendence of Providence, in salvation by the sacrifice of the Saviour, in the mediation of the Redeemer: in a word, I am a CHRISTIAN.[12]

How were the churches of nineteenth-century Britain to deal with spiritualism? Could they afford to dismiss and repudiate so potentially effective an ally at a time when the encroaching waves of modern science and scholarship were visibly eroding the rock of Christ's church? Could they afford to deny the enormous religious comfort which spiritualism offered to its eager believers in an age when religious comfort was painfully hard to find? In the case of the Anglican church, the answers to these questions which it tacitly adopted offer a fascinating glimpse of the Establishment's ongoing pursuit of a *via media*.

Officially, the Church of England – in accordance with its longstanding opposition to the conjuring of spirits or any other kind of unconsecrated magic – had no choice but to turn its back on spiritualism as doctrinally unacceptable, morally hazardous, and socially demeaning. In 1893, in the first issue of Stead's *Borderland*, the Archbishop of Canterbury, Edward White Benson, set forth at some length his objections to spiritualism, calling its manifestations "phenomena of a class which appears mostly in uncivilised states of society, and . . . in persons of little elevation of intellect." Explaining these phenomena in terms of thought transference and other automatic activities of the brain, the Archbishop nonetheless thought it advisable to warn:

> If it were really believed that the impression was produced by spiritual beings of a bad or mean or foolish order, it would be at least as unworthy a course to seek their intimacy as to seek that of degraded human creatures, and . . . more likely to be "dangerous," so to speak, to a struggling moral nature.

He concluded with the cautionary reminder that "the higher spiritual phenomena are approachable by the way of faith and intercourse with God."

Interestingly enough, as an undergraduate at Cambridge in the early 1850s, Benson had helped to launch a Ghost Society to collect and investigate ghostly tales, but such youthful pursuits must have appeared trivial indeed from the perspective of Lambeth Palace. The Right Reverend Brooke Foss Westcott, bishop of Durham, likewise discounted his participation in the same society[13] when he expressed his opinions in the columns of *Borderland*:

> It appears to me that in this, as in all spiritual questions, Holy Scripture is our supreme guide . . . I cannot, therefore, but regard every voluntary approach to beings such as those who are supposed to hold communication with men through mediums as unlawful and perilous.

I find in the fact of the Incarnation all that man (so far as I can see) requires for life and hope.

Similarly Bishop Temple of London, soon to succeed Benson as primate, saw little use for spiritualist investigations. "Hitherto," he explained, "the only result of the investigations has been, in my judgment, to show the extreme probability that the investigators will be self-deluded, and tempted to consciously delude others. To this temptation many of them have yielded."[14]

The public statements of leading Anglican prelates did not, however, dictate Church of England policy toward Anglican clergymen who chose to explore for themselves the borderland between life and death. Those who did so, whether out of mere curiosity or in the genuine hope of strengthening their ministry, continued to hold church office and to exercise parish duties. Reverend William Williamson Newbould, for example, who added half a dozen species to the classification of British flora, was doubtless a greater botanist than theologian, but his interest in spiritualism in no way compromised his pastoral position. A member of the Scientific Research Committee of the BNAS during the 1870s, he continued to take part in church services, at Kew and Petersham, until his death in 1886.[15] The careers of Thomas Colley, Hugh Reginald Haweis, and Charles Maurice Davies all illustrate, even more strikingly, the willingness of the Anglican church to accommodate spiritualism. The fact that the accommodation was unenthusiastic is less significant than the accommodation itself.

On 11 August 1876, the *Spiritualist Newspaper* ran a not-so-discreet advertisement:

> The Rev. Thomas Colley, curate of Portsmouth, so well known to many of the readers of these pages, is about to leave his present charge, St. Mary's, of which he has been sole minister for the past two years, and much appreciated, and is seeking an incumbency or curacy where he may have the same liberty of opinion on tabooed subjects which he has enjoyed under the liberal vicar of the parish that now loses him. Is there no wealthy or influential Spiritualist who can further his interest in this respect?

Clearly, Colley saw no reason to conceal his spiritualist beliefs; he was still broadcasting them thirty years later, and still serving as a minister of the Anglican church.

Much of Colley's spiritualist energy was occupied in the defense of Francis Ward Monck, a Baptist preacher turned medium, who was imprisoned for three months as a rogue and vagabond after "conjuring apparatus [was] found in his room" at Huddersfield in 1876.[16] Colley was as certain of Monck's innocence as Massey was of Slade's. He and his wife had numerous séances with Monck, under conditions that Colley confidently believed made imposture absolutely out of the question. Time after time, Monck produced messages on closed slates for the enthusiastic clergyman,[17] and, what was far more extraordinary, Colley claimed to have witnessed, along with several other astounded spectators, full-form materializations growing out of the medium's left side. Throughout the autumn and winter of 1877–8, Colley reported, in prose apparently addled by excitement, that he had repeatedly observed the spectacular process.

> When in expectation of a materialisation . . . there was seen steaming, as from a kettle spout, through the texture and substance of the medium's black coat, a little below the left breast, toward the side, a vaporous filament . . . When it grew in density to a cloudy something, from which (and apparently using up which for the quick evolving of much white raiment) there would then stand . . . to companion with us . . . our frequent psychic visitors . . . and, exhaling again to invisibility in a cloud (sucked back within his body) were they withdrawn from us.

Not only did these materialized forms separate themselves entirely from the medium; they walked around the room, wrote messages, lifted Colley off the ground, and ate baked apples. On one memorable occasion, in December 1877, a second materialized spirit emerged from the first, which had already materialized from the entranced Monck. If Colley was experiencing hallucinations, as Podmore suggested, he was certainly getting his money's worth.[18]

Colley, of course, scornfully rejected Podmore's hint and went to great lengths to confirm his faith in the genuineness of Monck's materializations. Not only did Colley claim that the damning apparatus, found in the medium's room in 1876, was Colley's own, borrowed by Monck to show how fraud might be committed, but Colley was even willing to wager £1,000 that no one could reproduce the medium's materialization of spirit bodies. J. N. Maskelyne jumped at the offer and managed to imitate the performance to an impressive degree; Colley, however, always the true believer where Monck was concerned, denied that the magician had exactly reproduced the medium's phenomena and refused to pay a farthing. When the case came to court in 1907, the venerable A. R. Wallace testified on behalf of Monck and Colley, and Maskelyne never collected his prize money.[19]

Nor was Monck the only spiritualist cause that Reverend Colley championed. Early in the twentieth century, he became the ardent admirer of the alleged spirit photographer William Hope of Crewe and helped to publicize Hope's psychic talents. Colley, now rector of Stockton, Warwickshire, organized a "domestic Photo-Circle" at

Crewe to encourage the supernatural photographer and in 1908 was thrilled to recognize his own deceased parents, spiritually captured on film, in two murky little snapshots. These likenesses were reproduced in a flyer that also included excerpts from Colley's letter to his diocesan, the bishop of Worcester. "Can anyone suppose," he asked the bishop, "that after my many years' experience of amateur photography, I would befool myself in such a sacred matter of family affection, or being a Beneficed Clergyman of the Church of England, would be utterly lost to all sense of honour and truth as to enact so base a fraud?"[20] The issue, however, was not whether Colley had enacted a base fraud, but whether Hope had done so, and little in his career as spirit photographer suggested resolute integrity. Oliver Lodge had doubts about him at once, as he later explained to William Crookes:

> I confess I have been extremely sceptical about that man Hope . . .
> My only experience was some years ago, in Colley's lifetime, when I supplied him with some specially wrapped-up plates, to be taken to Crewe and dealt with as he liked. They came back after a few months, the man Hope bringing them, and there were definite signs that they had been tampered with.[21]

One more gullible spiritualist would not be of tremendous significance except for the fact that he remained, as he proclaimed in capital letters, a "Beneficed Clergyman of the Church of England." He argued, as early as 1878, that his ardent spiritualist convictions had "imperilled [his] clerical position and prospects," but he nonetheless continued to find employment within the tolerant bosom of the establishment. After leaving Portsmouth in 1876, he was appointed chaplain on H. M. S. *Malabar* for a year and shortly thereafter departed for Natal, where he served as archdeacon until 1888.[22] Upon his return home, he obtained a living in Warwickshire, and it was from the Stockton rectory, in September 1905, that he wrote to the bishop of Salisbury, president of the upcoming Church Congress at Weymouth, requesting the opportunity during the congress to state his belief in the reality of spiritualist phenomena. His request was ignored, but he went to Weymouth in any case and delivered his lecture, though not as part of the congress agenda.[23] Despite his unorthodoxy, Colley remained comfortably settled at Stockton and continued to promote spiritualism on rectory letterhead.

The lectures and sermons of Reverend Hugh Reginald Haweis were equally likely to unsettle his more conservative clerical colleagues. Described as an "Eclectic Broad Churchman" and "incomparably the best qualified to represent advanced thought in the Church of England at the present moment,"[24] Haweis supported his progressive theology with a dynamic style of presentation that made him one of the most effective and popular preachers in London during the second half of the nineteenth century. A tiny man with a limp – a childhood hip disease had left him "almost a dwarf"[25] – this third-generation Anglican clergyman managed, at least figuratively, to tower above his pulpit. For the thirty-five years between 1866 and 1901 during which he was incumbent of St. James's, in Westmoreland Street, Marylebone, his parishioners grew accustomed to their pastor's flare for the dramatic. In the great controversy over ritual that shook the Church of England in the third quarter of the century, Haweis came down squarely in favor of a richer church service. A great lover of music and a fine amateur violinist himself, he did not hesitate to introduce a full program of music into the regular services at St. James's, and even drew upon the visual arts, in the form of sacred pictures, to enhance his verbal message.

Haweis's departure from Anglican orthodoxy took a variety of forms, the most persistent of which was his explicit and reiterated repudiation of the doctrine of eternal punishment. Throughout his career in the ministry, he denounced clergymen who delighted in "preaching hellfire and frightening the poor children into fits and sending timid women into lunatic asylums." For Haweis, Jesus was speaking metaphorically, in terms of moral corruption not physical punishment, when He spoke of eternal torment. Haweis wanted no part of "the capricious, fanciful, irrational kind of God who is supposed to judge His creatures in a way that would be a disgrace to a common magistrate, without intelligence, pity, sympathy, or knowledge; such a God as has revolted so many sensible religious people." Speaking in 1871 about the Article on predestination in the Anglican creed, he found that it reflected "a state of civilization which we have almost entirely outgrown. Its propositions are not easy for us to discuss at all, for they lie outside all our modern modes of thought, and its value, as far as we are concerned, is rather historical than doctrinal."[26]

Rejecting vehemently as he did all dogma that consigned a human being to everlasting damnation, Haweis was particularly receptive to the spiritualist belief in the "unending progress" of the soul after death. He was, in fact, sympathetic to spiritualism in general and became an outspoken advocate of the complete compatibility of spiritualism and Christianity. In a sermon on "The Character of Christianity," delivered in June 1870, he took a cautious, but conciliatory stand:

> I am propounding no theory about spiritualism. I hardly know what it means, or why it is called spiritualism. I merely affirm that occurrences which cannot be confounded with conjuring tricks . . . seem to me to occur, and they certainly seem to me still to await some adequate explanation. I will commit myself to no theory. I have none.

> I merely aspire to be honest enough to admit what I believe – that a class of phenomena are daily occurring in our midst which have not been explained; . . .
>
> But whatever truth or untruth there may be in these opinions, one thing is tolerably evident to my mind, and it is this – that if you accept the Christian miracles you cannot reject all others.[27]

By the turn of the century, Haweis's endorsement of spiritualism had grown far bolder. He attended séances in the 1870s at the Mayfair home of Mrs. Makdougall Gregory, widow of William Gregory, professor of chemistry at the University of Edinburgh. He joined the Society for Psychical Research as an associate member in the early 1880s. He told *Borderland* in 1893 that "Occultism is not only *a* question; it is *the* question of the day." He regularly scanned the pages of *Light* and clearly moved beyond his earlier posture of open-minded ignorance.[28] When he addressed the London Spiritualist Alliance in April 1900, he did so as an unabashed partisan. In the first place, he told his audience, spiritualism was not "in the least degree contrary to what he believed to be true Christianity. Indeed, Spiritualism fitted very nicely into Christianity; it seemed to be a legitimate development, not a contradiction – not an antagonist." Furthermore, he argued that Christian clergymen were actually indebted to spiritualism, because

> faith in and reverence for the Bible was dying out, in consequence of the growing doubts of people regarding the miraculous parts of the Bible . . . But now the whole thing had been reversed. People now believed in the Bible because of Spiritualism; they did not believe in Spiritualism because of the Bible . . .
>
> The phenomena of Spiritualism are reliable and happen over and over again, under test conditions, in the presence of witnesses; and . . . similar phenomena are recorded in the Bible, which is written for our learning. It is not an opinion, not a theory, but a fact . . . this is what has rehabilitated the Bible. The clergy ought to be very grateful to Spiritualism for this, for they could not have done it themselves. They tried, but they failed.

Nor did Haweis forget to remind his fellow clergymen that they "ought to be grateful to Spiritualism for giving them a philosophic basis for the immortality of the soul," since spiritualism "*demonstrated* . . . that mind actually does exist apart from brain and nervous system."[29]

Haweis's unorthodoxy, combined with financial irresponsibility and long absences from his parish on international lecture tours, did not endear him to his superiors in the Church of England. By the mid-1890s, he had "received a hint from high quarters that he could look for no more preferment."[30] Yet he served as an Anglican representative to the Parliament of Religions held in Chicago in 1893 and retained his parish post right up to his death in January 1901. The fact that he did so suggests, as his wife's biographer points out, that Anglican officialdom dreaded losing the services of a highly persuasive and powerful public speaker – one whom Dean Stanley had invited "to preach at a course of 'services for the people' in Westminster Abbey."[31] "His church was nearly always full," the *Times* reported in his obituary, and he was probably right in thinking that many of those who flocked to hear his sermons would not have been drawn to more traditional Anglican services.[32] If Haweis's skills as an orator were valuable for the channels of communication that they opened up to disaffected and bored parishioners, the men and women whom Christianity was on the verge of losing, then surely spiritualism was not the least effective of his enticements to faith.

Like Haweis, Charles Maurice Davies was an Anglican clergyman who firmly believed that spiritualism could peacefully coexist with Christianity. "As to the fruits of Spiritualism," he wrote in *Mystic London* in 1875,

> I can only say that I have never witnessed any of these anti-Christianizing effects which some persons say arise from a belief in Spiritualism. They simply have not come within the sphere of my observation, nor do I see any tendency towards them in the tenets of Spiritualism – rather the reverse.

Although the Reverend Davies fully acknowledged his duty carefully to scrutinize the principles of spiritualism, he was comfortable in his conviction, as he told a conference of spiritualists in 1874, *"that they readily assimilate with those of my own Church."*[33]

But what, precisely, was Davies's own church? In *The Great Secret*, a volume that he wrote toward the end of his life, Davies compared himself to Haweis and concluded: "If I had to put the matter epigrammatically I would say that whereas Mr. Haweis was a Church of England clergyman first and a spiritualist afterwards, I reversed the process; I was a spiritualist first and a Church of England clergyman afterwards." Certainly Davies never made a career out of the Anglican ministry alone. After being ordained in 1852, he held several curacies, in Somerset and London, but his major energies were devoted to education and journalism. Between 1861 and 1868, his time was largely occupied with the West London Collegiate School of which he was headmaster; thereafter he became a regular contributor to the *Daily Telegraph*, wrote leaders for the National Press Agency, and also worked on the *Western Morning News*.[34] It was as a series of articles for the *Daily Telegraph* that Davies began his studies of religion in the metropolis, which led to *Unorthodox London, Orthodox London, Het-*

erodox London, and *Mystic London*. He also turned out a series of
novels on religious themes, with such forbidding titles as *Broad
Church: A Novel, Philip Paternoster: A Tractarian Love Story*, and
'Verts; or, The Three Creeds. He is, furthermore, credited with authorship of the tongue-in-cheek novel *Maud Blount, Medium. A Story
of Modern Spiritualism*.[35]

Davies initially encountered spiritualism in Paris during the mid-
1850s, when D. D. Home set the Tuileries agog. Almost from the first,
Davies believed that there was "something in it," but it was not until
1865, when his young son died of scarlet fever, that he eagerly committed himself to thorough spiritualist inquiry. He was overjoyed when
his wife, writing automatically, received a reassuring message from the
guardian spirit into whose care their little boy had passed. Spiritualism,
Davies had suddenly discovered, "left room for hope – hope at a time
when we are mostly hopeless." For more than fifteen years, Davies
pursued his research into spiritualism, growing ever more eager "to
prove unbroken continuity between the life in this world and the life
beyond." Benjamin Coleman, the prominent London spiritualist, took
him under his wing, as did S. C. Hall through whom Davies was privileged to witness the levitation of Home. Davies attended countless
séances, many of them at the home of Mrs. Gregory, where over a
period of several years he saw "pretty well everything and everybody
there was to see in connection with the Modern Mystery, and in its
most pleasant form," thanks to the unstinting hospitality of this "high
priestess of spiritualism." Davies participated in the spiritualist inquiries of the London Dialectical Society and served on the council of
the BNAS. He observed "Katie King" in action and had some strong
suspicions of a liaison between Crookes and Florence Cook.[36]

Davies's approach to spiritualism in the 1870s appears as a strange
blend of good-humored amusement and that most compelling of motives, a firm will to believe. On the one hand he could write with gentle
sarcasm of the patently unspiritual physical manifestations that he repeatedly witnessed, and on the other he could assert flatly: "I felt
almost from the first – certainly from an early stage of my occult studies
– that occultism was either a religion, or it was, for me, nothing." One
half of his brain, evidently, did not want to know what the other was
thinking, and for a while the believing half dominated. Early in the
1880s, he determined to put his spiritualist religion into a more formal
context, and accordingly he organized a small group of followers into
the Guild of the Holy Spirit. The Guild rented two rooms at 38 Great
Russell Street and "fitted one up as an oratory," the other as a séance
room. An oratory service preceded the séance and was designed to
introduce what Davies considered the necessary "ecclesiastical treat-

ment" of spiritualism. These services, at which Davies "officiated exactly as though [he] had been in church," were held twice a week, on Monday afternoons and Thursday evenings.[37]

The format of these exercises in ecclesiastical occultism varied slightly over the years. For a couple of months, Davies also tried a public Sunday evening service, mixing elements of Anglicanism and spiritualism in a novel potpourri. Readings from the Church of England Prayer Book, addresses on "The Enfranchised Spirit" or "An Hour's Communion with the Dead," excerpts from Cardinal Newman's remarks on the ministry of angels, hymns ancient and modern – the eclecticism of Reverend Davies's service is striking. When the public services failed to attract a permanent audience, Davies moved his endeavors to the London suburbs, where the twice-weekly services resumed, in a room adorned like a chapel, with altar, organ, pulpit, and picture of the Madonna and Child.[38] Davies presided over this spiritualist temple without relinquishing his claim to function as an Anglican minister. In one address to his small audience, he proudly described himself as "an ordained clergyman of the English branch of Christ's Catholic Church," who had that very morning "officiated at a Church of England service." He vowed to his followers that he would "always retain the chapel and the chapel service," explaining that the "key to [his] whole position" was precisely "the combination of mysticism with the Christian services of the Church of England."[39]

But Davies did abandon the mystic oratory when his one-year lease on the suburban house expired. The great truths had not unfolded; the departed had not flocked to fill the chairs "left vacant for them." Davies emerged from the experience disheartened, if not disillusioned, and during the last twenty years of his life, until his death in 1910, he was largely absent from spiritualist circles and inquiries. He was, similarly, absent from clerical duties, spending much of his time as general editor of a series of translations, commissioned by Cecil Rhodes, of Gibbon's sources for *The Decline and Fall of the Roman Empire*.[40] *Crockford's Clerical Directory* loses track of him entirely, and it is altogether possible that he left holy orders toward the end of his life. For a good two decades, nonetheless, he was totally committed to his efforts to make Anglicanism and spiritualism congenial companions. Although neither partner proved as amenable to the union as Davies had hoped, the effort was not negligible. It reflected his conviction that Christian spiritualism was a viable response to the anxieties of life in the nineteenth century, a response "prevalent in all the various religious bodies, not even excepting the Roman Catholic, but claiming particular notice . . . as spreading widely over the Established Church."[41]

Certainly the minister of the established church who gained the widest renown among British spiritualists was Stainton Moses. In addition to his involvement with the BNAS and the London Spiritualist Alliance, he served on the council both of Cox's Psychological Society and of the Society for Psychical Research. In 1882 he helped to launch the Ghost Club, a monthly dining club whose members had to "provide an original Ghost Story, or some psychological experience of interest or instruction, once during the year."[42] Nor were Moses's gifts confined to the establishment and organization of clubs and societies. Shortly after embarking on an intense study of spiritualism in 1872, he began to develop extraordinary powers as a medium himself, and over the next decade a highly impressive array of physical manifestations supposedly occurred through his mediumship. These ranged from raps, spirit lights, scents, and musical sounds, to apports of objects, sometimes through locked doors, levitations both of furniture and medium, and the materialization of spirit hands.[43] Such phenomena, appearing in his presence, were all the more readily accepted in spiritualist circles because of the medium's unimpeachable character. Davies, who met the Reverend Moses frequently at Mrs. Gregory's, explained that "Stainton Moses' testimony is *so* respectable, and he gave it to you in such a transparently sincere way over a cigar, that it could not fail to carry weight." Ada Goodrich Freer of *Borderland* concurred. "It may be said, once for all," she announced, "that it is unnecessary to insist on the absolute sincerity of Mr. Stainton Moses. It is a point which has never been so much as raised. His life has been of a kind not to be called in question."[44]

Both the phenomena of direct writing and of automatic writing were likewise associated with his mediumship. In direct writing, the medium was said to play no role at all, and spirits (visible or invisible) controlled the writing instrument; sometimes, by even more mysterious means, the written message simply emerged on paper or slate, apparently without the use of any instrument whatsoever. In the case of automatic writing, for which Moses was particularly known, the medium clearly held the implement that wrote at length, but he was entranced when he did so, or busy reading a book, or "reciting some passages of Virgil . . . in order to eliminate the disturbing element of [his] own mental action."[45] Whatever his state of mind, he consistently argued that it did not, in any way, influence or direct the automatic script that filled more than a score of notebooks. *Spirit Teachings*, a collection of these writings published in 1883, became one of the most influential texts of British spiritualism; indeed, it has been dubbed the "Bible of British Spiritualism."[46]

The lofty tone that this comparison with Holy Scripture implies characterized all of Stainton Moses's spirit communications. His were no frivolous messages, but carried words of the most solemn import, didactically expounding questions of philosophy, ethics, and theology. Throughout the 1870s, his inspirational writings circulated in the spiritualist press and appeared in volumes of spiritualist literature, under the pseudonym "M. A., Oxon." His growing stature among British spiritualists loomed even larger as rumor spread that his "controls," the spirits who communicated through him and gave him access to the "Other Side," were no ordinary spirits. Although he was reticent about their identity during his lifetime, research after his death revealed that among the more than eighty spirits that manifested themselves to Stainton Moses were the prophets Malachi, Elijah, Ezekiel, and Daniel, St. John the Baptist, Plato and Aristotle, Mendelssohn and Beethoven, Swedenborg, Benjamin Franklin, Napoleon III, President Garfield, and Bishop Samuel Wilberforce.[47]

Given this roster, it is understandable that some psychical researchers have subsequently categorized Stainton Moses as "a deeply dissociated personality," and a fit subject for the student of "morbid psychology."[48] Of intentional deception there seems little likelihood, but Moses's mediumship is nonetheless fraught with disturbing questions. Not only is it impossible to prove that the medium's mind, conscious or unconscious, played no role in producing his celebrated automatic writing, but problems of a more general nature cast a shadow over all the phenomena produced through his powers. Most of these phenomena manifested themselves to a very restricted circle of close and trusted friends, limited usually to Dr. Stanhope Templeman Speer and his wife, but sometimes augmented to include their son Charlton and another good friend, F. W. Percival. Dr. Speer, a retired physician, had ministered to Moses's never vigorous health as early as 1869, and Moses served as tutor to the Speers' son during the 1870s, filling "that office in a way which attached to him both parents and pupil more closely than ever." Moses's burgeoning interest in spiritualism in 1872 and 1873 was shared, and encouraged, by the Speers. Clearly they were not the people to examine his mediumship with a watchful and critical eye. It does not help Moses's reputation as a medium – with critics, at least – that the records of his mediumship were kept, in the first place, by Moses himself and, in the second, by Mrs. Speer, with briefer corroboratory accounts by Dr. Speer and Percival.[49]

As in the case of D. D. Home, it is easy to point out that Moses was performing for a rapt and eager audience, willing to believe without question the validity of the phenomena produced by an admired family friend. Again as with Home, the argument of collective hypnosis might help to explain Moses's influence over his sitters. Alternatively, one can fall back on the hypothesis that Moses, despite external appearances, was not of sound mind.[50] He certainly suffered much physical illness, starting with a nervous breakdown from overwork while an Oxford undergraduate, and including congestion of the liver, catarrh, bronchitis, whooping cough, "an affection of the throat," suppressed gout, and, toward the end of his life, bouts of influenza, difficulties with his eyes, "extreme depression and nervous prostration, and severe neuralgic pains." What finally caused his death in 1892, while still in his early fifties, was apparently Bright's disease.[51] Such a record of ill health, strewn as it was with neurological disorders, may provide the perfect background for an extended exercise in self-deception and involuntary misrepresentation. Moses's mediumship, nevertheless, is not readily explained away. A. W. Trethewy was not the only psychical researcher who argued, after careful inquiry, that there seemed "less difficulty in supporting the claims of the 'controls' to be truthful discarnate entities than in trying to make any other explanation fit the facts."[52]

If Stainton Moses was engaged, inadvertently or not, in a prolonged deception of himself and his friends, his motives for doing so may well have been a growing dissatisfaction with the Church of England. Although he served conscientiously and effectively as curate from 1863 to 1870, calling himself at the time "a sound High Churchman," there is good evidence that he felt increasingly constricted by the formal doctrines of his faith. He frankly admitted that his introduction to spiritualism led him toward "a development of thought amounting to nothing short of spiritual regeneration."[53] It is true, as Podmore points out, that most of the ideas that broadened Moses's theology appeared to him, and to the world, in the context of his automatic script – a context that enabled Moses to deny responsibility for any unorthodoxy. Yet Moses did not hesitate to express his revised religious views even when writing consciously and purposefully. As early as March 1874, he praised and justified the "religious sentiments put forward" by his spirit teachers; "if they contravene the teachings of Orthodox Creeds as accepted in this corner of God's universe," he announced, "they at any rate teach a nobler theology and preach a Diviner God."[54] In 1876, in an impassioned letter to the editor of the *Spiritualist Newspaper*, "M. A., Oxon." denounced religious intolerance that committed to eternal hellfire any man whose "mind is of a different complexion to another man's." "What!" Moses thundered,

> has God shut up all avenues of inquiry? Has He placed pitfalls before the doors that lead to new truth? . . . The days, thank God! are fast passing away when any such belief will be tolerated. Dogma and ana-

thema, fire and brimstone, stereotyped beliefs, and Athanasian creeds
are fast dying a richly merited death.

And in 1887, in a presidential address to the London Spiritual Alliance,
Moses informed his audience: "We are receiving new developments
of truth now as certainly as in olden days it was revealed to our fore-
fathers."[55]

For Stainton Moses, spiritualism became a new form of "personal
religion," one which provided the much desired proof of human immortality, and which turned articles of Christian faith "into logical
deductions from experience."[56] He would have liked to make available
to them, he told members of the LSA, a "form of religious service
expressive of our faith," but, perhaps influenced by Davies's disappointments, he had concluded by 1885 that "the time was not yet come
for such a step."[57] Clearly, if Moses was not any the less a Christian
for his spiritualism, he was much the less an Anglican. Although he
might invoke the "Ever Blessed Trinity" to exorcise an evil spirit that
interrupted one of Florence Cook's séances in 1873,[58] he did not continue to function as an Anglican minister. Whether sickness or inti-
mations of theological unorthodoxy compelled him to give up the ministry in 1870, Moses did not wholeheartedly commit his time and
attention to spiritualism until he had abandoned the clerical collar.[59]
In this respect, his experience was notably unlike that of Colley, Haweis, or Davies. When Moses became enmeshed in the politics of London
spiritualism and the throes of mediumship, he was an English master
at University College School, not a practicing clergyman. Perhaps for
this reason he appears less illustrative of that blind eye which the Anglican church was willing to turn toward its wayward spiritualist priests.
Yet Moses nonetheless continued to style himself "Reverend" for
many years after 1870, and it is certain that the public who knew of
him associated the medium with that title and position.[60]

In 1881, at the annual Church of England Congress, held that year
in Newcastle-on-Tyne, "The Duty of the Church in Respect of the
Prevalence of Spiritualism" was the subject of discussion during one
evening session. The points reiterated by several of the speakers suggest some degree of orthodox recognition that spiritualism might perform invaluable service in the struggle against materialism. While conceding that certain aspects of spiritualism scarcely conformed to the
official Anglican creed, Dr. Robinson Thornton, vicar of St. John's,
Notting Hill, London, concluded: "Let us thankfully acknowledge the
truths of Spiritualist teaching, as weapons which we are too glad to
wield against Positivism, and Secularism, and all the anti-Christianisms
of this age of godless thought." Reverend Basil Wilberforce, younger

son of Samuel and subsequently canon, then archdeacon, of Westminster, pointed to the "potent influence" of spiritualism "upon the
religious beliefs of thousands"; he, too, urged his fellow Anglicans to
adopt "a conciliatory rather than hostile or dogmatic attitude towards
believers."[61] Theirs was a pragmatic understanding of the perilous environment in which they had to protect and nurture the increasingly
vulnerable Church of England.

Henry Sidgwick, one of the founders of the Society for Psychical
Research, may have had their speeches in mind when he observed the
following year, in the first presidential address to the Society, that

> the attitude of the clergy has sensibly altered. A generation ago the
> investigator of the phenomena of Spiritualism was in danger of being
> assailed by a formidable alliance of scientific orthodoxy and religious
> orthodoxy; but I think that this alliance is now harder to bring about.
> Several of the more enlightened clergy and laity who attend to the
> state of religious evidences have come to feel that the general prin-
> ciples on which incredulous science explains off-hand the evidence
> for these modern marvels are at least equally cogent against the rec-
> ords of ancient miracles, that the two bodies of evidence must *primâ
> facie* stand or fall together, or at least must be dealt with by the same
> methods.[62]

The marvels of spiritualism, in short, might be just what was needed
for the wasting sickness of the Anglican church. They might provide,
as Haweis suspected, the excitement, the sense of spiritual immediacy
and efficacy, needed to draw Anglicans back inside their houses of
worship. A few Anglican ministers, in any case, determined that the
aid proffered by spiritualism should not be ignored, and they were not
drummed out of the establishment for their determination.

Peaceful coexistence can be a useful policy in religious diplomacy,
and the Anglican church doubtless adopted it with regard to spiritualism
for obvious and urgent reasons. It remained, however, an implicit
working policy, not an official acknowledgment of that church's pressing need for support. If the Church of England was hoping to enlist
the assistance of spiritualism in its campaign to broaden the appeal of
the national church, establishment spokesmen were not willing to concede this policy publicly for too many compromises with orthodoxy
were clearly involved. Psychical research, however, with its attempt
to maintain neutrality on the crucial theological issues at stake, did not
involve such compromises; the Society for Psychical Research, with
its prestigious social and intellectual status, was the perfect home for
Anglicans who were uneasy over spiritualism, but deeply concerned
to learn if there was, after all, "something in it." The membership lists
of the SPR were liberally sprinkled with the names of Anglican curates,

vicars, deans, canons, and even a few bishops. Although Archbishop
Benson did not himself join the Society – despite a greater interest "in
psychical phenomena than he cared to admit" – Mrs. Benson did from
the first.[63] Two Anglican bishops, Carlisle and Ripon, served as Society
vice-presidents during the 1880s, and the latter, William Boyd Carpenter, became SPR president in 1912. Although he had retired from
his diocese a year before assuming the presidency, Carpenter obviously
found psychical research and Anglican commitments perfectly compatible. Indeed, he commended the SPR for performing exactly the
same sort of Christian rescue work for which less cautious Anglicans
gave credit to spiritualism. "It cannot be denied," he announced in
his 1912 presidential address to the Society,

> that in the revolt against the low-levelled materialism which threat-
> ened to rob mankind of its noblest prerogatives, its most splendid
> inheritances, and its most inspiring outlook, this Society has played
> a quiet, modest and most effective part in recalling men's thoughts
> to the need of a more careful study of their own innermost nature.[64]

Throughout the better part of the nineteenth century, concerned Anglicans struggled, in their various ways, to revitalize their church, to
make it a viable religious institution in an age of urbanization, democ-
ratization, and secularization. Their efforts included the heroic dedi-
cation of slum pastors, the attempt to make peace with the new universe
revealed by science, and acts of Parliament designed to correct abuses
and irregularities within the establishment. New churches were built,
and others were redecorated. Some Anglicans went very high in their
search for religious truth, others very low, and still others found in
spiritualism or psychical research the grounds for renewed optimism
about the human condition. These latter Anglicans did not face the
difficult choice of pursuing their unorthodox inquiries or remaining
within the Church of England. Both options were left open to them,
and the first became a means of strengthening their resolve to follow
the second as well. The author of Haweis's obituary in the *Times* might
have been describing any number of the reverend's fellow Anglicans
when he found in Haweis's career "an indication of the vast compre-
hensiveness of the Anglican Church."[65]

Anglican ministers were not alone in their spiritualist sympathies,
and Davies claimed that he could "name clergymen of all denominations who hold Spiritualistic views."[66] If his boast was a trifle hyperbolic, there did exist clergymen of virtually all Protestant denominations who were at least willing to approach the subject with an open
mind and who hoped that spiritualism would prove an effective weapon
against materialism. It goes without saying that materialism and athe-

ism loomed as threateningly to Nonconformists as to Anglicans, and
that the former bewailed the secular character of the age with as much
conviction as the latter. If, nonetheless, the Nonconformists felt less
urgently the need for propping up their churches than did the defenders
of the establishment, that was perhaps because the growing strength
of Nonconformity, both in numbers and in political organization, gave
Dissent a sense of vitality that the Anglican church lacked. It is, moreover, far harder to make valid generalizations about Nonconformist
attitudes toward spiritualism than about Anglican views on the same
subject. With the varieties of religious dissent in Victorian England,
particularly with those independent congregations that imbibed distrust
of central organization and authority through their historic roots, the
search for a definitive position on spiritualism is, in fact, finally unre-
warding. Sympathetic, and even enthusiastic, individual clergymen
emerge from the records, but never a coordinated policy, not even
within a single denomination. Although it seems fair to suggest that
the Nonconformist mentality, characteristically intenser than the Anglican, jumped more readily to the satanic interpretation of spiritualism,
plenty of Nonconformist ministers were no more deterred by warnings
of demonism than were their Anglican counterparts. The *Nonconformist* quite firmly declined to "sneer at and ridicule the beliefs of the
Spiritualists, however much we may sometime deprecate and doubt
the wisdom of the form and manner of the revelations."[67]

Some of the most fervent Christian spiritualists served as clergy in
sects that occupied the outskirts of Nonconformity. Frederick Rowland
Young, for one, was editor and proprietor of the *Christian Spiritualist*
in the early 1870s, and minister of the Free Christian Church in Swindon, Wiltshire. Although the *Westminster Review* in 1862 described
the theological foundations of the Free Christian Church as "half Unitarian and half Swedenborgian" – two faiths whose relationships with
Christianity were decidedly equivocal – Reverend Young had no
doubts whatsoever about his Christian identification and allegiance. In
the columns of his paper, he offered the clearest, the most confident
justification of Christian spiritualism that appeared in any of the spiritualist serials of this period. What Young lacked in subtlety of perception or theological sophistication, he made up for in earnest conviction. In the first issue of the *Christian Spiritualist*, in January 1871,
he characterized his fellow Christian spiritualists as men and women

> who profess and call themselves Spiritualists, but who at the same
> time feel that their highest and first allegiance is due to the Lord Jesus
> Christ, as the Son of God, the Saviour of the world, and the Word
> made flesh Who dwelt amongst us, . . . They are Spiritualists, simply
> because they cannot help being so; but they are Christians also, and

they believe that there is perfect unity between what they understand by Spiritualism and the Christianity of Jesus Christ.

Switching to the first person plural, he explained:

> *We* have one thing to do and one only, to be loyal to our Master Christ, while we show, as we believe we can do, that Spiritualism is not the enemy but the friend of Christ's religion, and that we can speak of ourselves as Christian Spiritualists, without being justly liable to the imputation of trying to mingle things which are essentially opposed to each other.[68]

The Catholic church, of course, continued to maintain the fundamental opposition between spiritualism and Christianity. The international Roman Catholic hierarchy spoke with customary authority and certainty when it condemned spiritualistic practices, as it had for centuries condemned supernatural magic outside its own bailiwick. The Vatican expelled Home early in 1864 on a charge of sorcery; in the United States, the Plenary Council of Baltimore ruled in 1866 that "some, at least, of these [spiritualist] phenomena must be attributed to diabolic agency"; in England, Cardinal Manning fulminated against spiritualism as very much to the devil's purpose.[69] The Catholic bishop of Nottingham, Dr. Edward Gilpin Bagshawe, summed up the Catholic position neatly when he warned W. T. Stead in 1893:

> The intelligence which uses your hand, and of which you are not conscious, is no other than the Devil, and if you continue such unlawful intercourse with the unseen you will necessarily be misled to your ruin by the enemy of God, the murderer of souls, and the liar from the beginning.[70]

Determined spiritualists who also happened to be Roman Catholics did not, however, allow dire predictions of future perdition to stand in their way. Florence Marryat and Marie Sinclair, Countess of Caithness after 1872, were two Catholics who doubted not at all the religious significance of spiritualism, nor denied themselves the pleasure of frequent séances.

Christian spiritualists, whether Anglican, Nonconformist, or Catholic, shared one fundamental point of view, well expressed in 1881 by the sympathetic clerical observer, Reverend Wilberforce. "Those who are following Spiritualism as a means and not an end," he explained to the Church Congress at Newcastle,

> contend warmly that it does not seek to undermine religion, or to render obsolete the teachings of Christ: that, on the other hand, it furnishes illustrations and rational proof of them, such as can be gained from no other source; that its manifestations will supply deists and atheists with positive demonstration of a life after death.[71]

This is the argument that runs, refrain-like, through the writings of Christian spiritualists. It was, as we have seen, S. C. Hall's basic contention, as it was William Howitt's. Alaric Watts echoed the sentiment, together with J. S. Farmer and Thomas Shorter. J. H. Powell, editor of the *Spiritual Times*, exulted that spiritualism was "destined to take the scales from the eyes of scepticism, and make Immortality and Christ ever living spiritual realities." Newton Crosland went so far in underscoring his continuing Christian allegiance as to invoke Christ at the start of séances.[72] Indeed, many Christian spiritualists sought to emphasize the close compatibility between their two systems of belief by giving their séances a specifically Christian coloring. Invoking Christ was not the only way to do so; as Davies had found, a similar effect could be achieved through prayers, Bible reading, and the singing of Christian hymns before, during, or after the séance.

During the Victorian and Edwardian years, Christian spiritualism escaped the confines of sectarian partisanship. While staunchly maintaining their Christianity and even, in a number of cases, continuing to attend their traditional churches, Christian spiritualists frequently found that spiritualist beliefs blunted the edge of denominational differences and led them to abandon the specific doctrines or practices that distinguished one Christian sect from another. For example, spiritualist views of the afterlife – emphasizing progress from heavenly sphere to sphere, with each individual spirit gradually losing its earthly imperfections – made irrelevant the often bitter controversies among diverse Christian churches concerning questions of salvation and resurrection.[73] A middle-class London society like the LSA was necessarily Christian, but it was nondenominational and would have endorsed the message communicated spiritually to Florence Marryat: "Creeds are some of the wickedest things on earth. They have caused more bloodshed, and wrangling, and hatred, and wrong, than all the wicked people put together. Do you suppose there will be any creeds in the spheres? . . . Not a bit of it!" John Jones, the Christian spiritualist who was later known as J. Enmore Jones, wrote with similar emphasis that "God is not confined to a sect or ism," and reminded his readers that all Christendom shared "one leading Faith," which he reduced to three fundamental components: belief in God, the Trinity, and life everlasting.[74] If some Christian spiritualists would have argued with Jones's capsule summation of the meaning of Christianity, few could have persuasively denied that Christian spiritualism was moving toward a notion of Christian fellowship that minimized divisions among the sects and denominations into which Christendom had splintered.

ANTI-CHRISTIAN SPIRITUALISM

In startling contrast to the pious homilies of the Christian spiritualists, the rallying cry of James Burns and the anti-Christian spiritualists rang out raucous and bold. "Let us," Burns urged his fellow progressive spiritualists,

> . . . with all the power we possess, oppose every effort to Christianise, Mormonise, Mohammedanise, or otherwise pollute Spiritualism. . . . Spiritualists! surely we may call our souls our own? Let us resist as traitors and dangerous foes those who would enthral our minds by their personal opinions under the term of "Christian Spiritualism," or any other authoritarian bondage whatever.[75]

Burns was, indeed, the principal spokesman for anti-Christian spiritualism in Britain throughout the second half of the nineteenth century, but despite his zealous efforts, his point of view remained comparatively obscure. Unlike Christian spiritualism, it lacked prominent publicists and flourished most luxuriantly among the less influential ranks of Victorian society, particularly the lower middle and upper working classes. Even if it is not true that the majority of English spiritualists were Christian, as is sometimes assumed for lack of accurate statistics, such certainly appeared to be the case. Most of the spiritualist message that was communicated to the public at large was the Christian spiritualist message, and it was communicated by men and women who, for the most part, were fully entitled to respect, as writers, teachers, lawyers, doctors, or even clergymen. Anti-Christian spiritualism, by contrast, could not even obtain a hearing in many a Victorian periodical, and its advocates had no claim whatsoever to a polite audience. On the contrary, from many an anti-Christian spiritualist, particularly from the working classes, society only expected trouble.

Working-class spiritualism aroused uneasiness in the hearts of society's guardians not simply because it was predominantly anti-Christian, but because it was firmly linked with several other unsettling movements. Many a working-class convert to progressive spiritualism came directly from atheism, as an author in the *Spiritual Magazine* pointed out in 1872. Begging the magazine's Christian spiritualist readers not to condemn too harshly the anti-Christian version of their faith, the writer observed that anti-Christian spiritualist missionaries had "converted within the last 20 years hundreds of thousands, and possibly several millions of unbelievers or so-called Atheists."[76] The number of converts in this estimate was grossly inflated, and the nature of the conversion is open to question, but the fact that numerous working-class secularists and freethinkers ended up as spiritualists is beyond doubt. One Mr. Johnson, for example, a trance medium who conducted spiritualist services on Sunday in the Temperance Hall, Grosvenor Street, Manchester, and who was a compositor by profession, "had been a Methodist, but had lapsed into Atheism previous to his conversion to Spiritualism." Joseph Gutteridge, a skilled Coventry weaver, began to hold séances in his home as early as 1849, and was gradually persuaded by the phenomena that he observed to embrace spiritualism in place of free thought. And somewhat later, around the turn of the century, V. S. Pritchett's great-uncle Arthur, a cabinet-maker in York, migrated from atheism to spiritualism.[77] The pattern followed by these examples, among many, is typical of Victorian working-class spiritualism. These men, all skilled workers, had received little or no formal education, but they possessed independent minds, relished study, and were eager to tackle weighty theological questions for themselves. They had all espoused atheism at some time in their lives, but, finding it ultimately an uninhabitable region, moved on to spiritualism. As one contemporary succinctly observed: "Among the lower orders it is not the church or chapel-goer who makes the readiest convert to Spiritualism. It is the secularist or positivist."[78]

Not every working-class secularist who turned to spiritualism embraced the anti-Christian variety. Quite a few, like Thomas Shorter, experienced conversions that took them all the way to Christian spiritualism. George Sexton, for another, born in a Norfolk cottage to a family of small tenant farmers, so sharply veered away from the secularism that he endorsed in the 1850s that he developed mediumistic powers and became editor of the *Spiritual Magazine* in the 1870s, when it merged with the short-lived *Christian Spiritualist*.[79] The greater number of lower-class spiritualists, however, seem to have resisted the call of Christianity, even when hitched to the spiritualist wagon; for them, the lure of free thought was perhaps too powerful to abandon completely, and anti-Christian spiritualism offered its adherents the chance to emphasize the allegedly scientific, purely rational underpinnings of their beliefs, while minimizing the religious aspects. Free thought had deep roots in working-class culture, particularly among radical, self-educated skilled workers for whom Tom Paine, Shelley, and the *philosophes* had once been obligatory reading. It was a heritage that such segments of the population would not willingly relinquish for it had helped to mold their identity and self-esteem. Even in the 1850s, the "*Yorkshire Spiritual Telegraph* was conducted by men who had been accustomed to look up to Paine and Voltaire as Biblical critics, and to see in the Baron d'Holbach's 'System of Nature,' an authoritative text-book of theology."[80] For numbers of working-class converts, anti-Christian spiritualism was not, therefore, simply a successor to free thought, but also a sort of supplement, not entirely replacing its point of view, but enriching it with the reassurance of immortality.

If the writings of Voltaire and d'Holbach were inaccessible to all but the most diligent of self-taught workers, there were plenty of pamphleteers eager to popularize and vulgarize the ideas of the Enlightenment for working-class consumption. And Robert Owen, the most famous secularist-cum-spiritualist, likewise provided for his working-class disciples a direct link between eighteenth-century rationalism and nineteenth-century spiritualism. Yet the impact of Owenism was complex and ambivalent. As much as rationalism formed a major strand of Owenite thought, so did millenarian convictions. These convictions found their way into anti-Christian spiritualism just as surely as did the rationalist legacy, and they particularly left their mark on the progressive spiritualist vision of life after death. There, in the "Summerland" that awaited all spiritualists after physical demise, the transformation of social relations, so long obstructed on earth, would come to fruition, as individuals learned to master those shortcomings that stood in the way of universal brotherhood and the triumph of the communal over the personal.[81] Millenarian expectations also crept into working-class spiritualism from the Church of the New Jerusalem, for Swedenborgianism attracted artisans of an intellectual bent and could readily function as an antechamber to spiritualism.[82]

Owen's influence on anti-Christian spiritualism inevitably, if somewhat vaguely, also affected attitudes toward social reform here and now. It is true that social crusades were never a coherent aspect of British working-class spiritualism, but they figured nonetheless in the literature of the movement and gave more than a simply religious meaning to its pervasive anti-establishment attitudes. A range of problems besetting British society came under the scrutiny of progressive spiritualists, from an excess of alcoholic beverages to a deficiency of social equality. The rights of women found a staunch champion in the British Association of Progressive Spiritualists,[83] while a number of ties connected progressive spiritualism with socialism in general, and specific socialist organizations in particular. In the early days of British spiritualism, these ties were largely forged by Owenism, as at Keighley, where the active spiritualist society imbibed its socialism from Owenite thought. Also illustrative of the early Owenite connection between social reform and spiritualism were the experiences of James Smith, a Scottish mystic of highly idiosyncratic theological views who was associated with Owen in the leadership of the Grand National Consolidated Trades Union; twenty years later, in the 1850s, he brought to his new spiritualist beliefs the "idea that a man ought to belong to his Union."[84] Later, at the turn of the century, several spiritualist associations in the north of England – in Bolton, Bradford, Burnley, Rochdale, and Preston, for example – shared members with the Independent Labour Party and, more rarely, with the Social Democratic Federation.[85] Free thought, Owenism, feminism, trade unionism, socialism:

here was a quintet of disturbing movements guaranteed to confer on the anti-Christian spiritualism with which they were sometimes associated a dangerous reputation indeed. Given the democratic nature of anti-Christian spiritualism, with its impatient rejection of all authoritarian figures and hierarchies, it is not difficult to understand why this brand of spiritualism attracted most of its followers from the working and lower middle classes, rather than from those more affluent layers of society which were generally comfortable with the status quo.

Nor is it surprising that a set of ideas opposed to Christianity proved immensely attractive to thousands of Great Britain's less fortunate citizens. Hostility to Christianity and an often passionate anticlericalism had become virtually traditional in many a working-class community by the latter half of the nineteenth century,[86] and it was a tradition supported by social antipathies as well as doctrinal scruples. The Christian churches – preeminently, of course, the Church of England – seemed too ensconced in existing social relations for many a working-class conscience, too willing to accept without question the appalling excuse for poverty that the poor are always with us. In the Anglican church particularly, the time-honored bonds between the church, the landed classes, and conservative politics gave the poor little reason to feel much loyalty to the establishment. Inadequate church facilities in the overcrowded industrial centers further undermined any possible Anglican claim to the allegiance of the masses. Nor did the Nonconformist churches necessarily offer an appealing alternative. While their social identity was usually not as class-bound as that of the Anglican faith, their theology might be even more repulsive to the questioning minds of working men and women. A harsh and punitive attitude toward sin, combined with a still vivid portrayal of hellfire and brimstone, made the Baptists and Primitive Methodists, for example, uncongenial company for a doubting Christian with a mind of his own.

It is even questionable whether the label "doubting Christian" can apply to large segments of Britain's lower classes in the nineteenth century. As the famous Religious Census of 1851 so amply revealed, church attendance in Great Britain left much to be desired, and it was widely acknowledged that the greatest rate of absenteeism occurred overwhelmingly among the working classes. Probably the majority of these Sunday shirkers were utterly indifferent to religion. Christianity had never figured significantly in their uneducated, undernourished, and exploited lives, and, no doubt, their Sundays were spent "ignoring religion more than denying it."[87] But there were others whose absence from church represented a deliberate decision, the expression of an intellectual, as well as emotional, repudiation of all that Christianity meant in nineteenth-century Britain. It was from the ranks of these freethinkers that anti-Christian spiritualism recruited its most articulate working-class enthusiasts, and part of their enthusiasm arose from the fact that anti-Christian spiritualism was as deliberate a challenge to the established order as Owenism and secularism.[88]

But the pleasure of thumbing one's nose at the ideologies of self-proclaimed social superiors cannot alone explain why substantial numbers of the upper working and lower middle classes were drawn to anti-Christian spiritualism. Atheism was as effective an affront to the prevailing value system of Victorian Britain as could be devised; anti-Christian spiritualism was no improvement on that insult. On the contrary, in practice it usually represented a softening of the secularist position, a kind of mellowing with regard to the religious cast of mind. Instead of dismissing the need for, and validity of, all religious systems of thought, those secularists who turned to anti-Christain spiritualism more often than not sought an alternative to the Christian religion. For all their pride in the independent posture of the freethinker, they were characteristically unable to sustain the endless negativity required of the atheist. They may well have continued to perceive themselves as inveterate opponents of religion, but in embracing anti-Christian spiritualism many were, in fact, guilty of backsliding.

Why secularists and atheists, whose antipathy to formal religion often emerged after great expenditure of time and thought, should come to endorse anti-Christian spiritualism is still a question that can only be answered by generalizations, despite information about several such converts. Mr. Cogman, who presided over a developing circle for budding mediums in London's East End, was converted from unbelief when his teenaged daughter became a medium and explained religious questions to him with entranced sagacity. Johnson, of Temperance Hall, Grosvenor Street, Manchester, was persuaded to abandon atheism by the experiences of mediumship that he himself underwent. Gutteridge was likewise convinced by the phenomena that he witnessed at séances after the death of his first wife.[89] These are, however, only partial explanations; they do not really shed light on the disposition of these men to invest séance phenomena and trance utterances with profound spiritual significance. To understand that predisposition, one can only turn, like the contributor to the *Westminster Review* in 1862, to such fair, but lame, observations that these surprising champions of spiritualism "evidently retain[ed] religious susceptibilities."[90] They may have been more deeply marked than they realized by the Christianity imbibed in their youth and could not permanently shake themselves free of a religious outlook. And for those working-class spiritualists who had never received any Christian teaching, formal or informal, and therefore could not blame lingering Christian influences for their spiritual yearnings, the hope of immortality was a powerful magnet drawing men and women toward the comforts of religious belief. The ability to face the prospect of total annihilation with equanimity is given to few. Anti-Christian spiritualism offered, in an "accepted scientific way" and on the basis of "well-accredited facts," the reassurance of "a continued existence for the spirit of man after the cessation of the physical life," while not subjecting the believer to the constraints and prejudices of the Christian churches.[91] If anti-Christian spiritualism was, after all, a religion, it was a religion rationalized, modernized, and brought firmly into line with advanced thought, both social and scientific. This, at least, its adherents could assert.

In trying to conclude this imprecise line of reasoning, it is valid to say that, outside of London, spiritualism flourished most vigorously in northern England, particularly in the industrial communities of Lancashire and Yorkshire.[92] It was predominantly lower-class spiritualism, and it was strongly anti-Christian. A number of threads of intellectual and social history combined to impose this pattern on British spiritualism. On the one hand, working-class skepticism, and even atheism, figured noticeably in the north, augmented both by indigenous developments and by the influx of workers from other parts of the country whose ties to organized religion had been severed, if they ever existed. The full extent of the bonds between free thought, Owenism, and spiritualism are still uncertain, but surely it was no coincidence that the areas with the largest number of major provincial free thought societies between the 1830s and the mid-1860s – the Midlands, Yorkshire, and Lancashire – were also the areas which saw the greatest activity among provincial spiritualists after 1850. And not only was there regional similarity, but many of the same towns and cities figure in lists of free thought and spiritualist groups from this period: Manchester, Oldham, Rochdale, Blackburn, Lancaster, and Liverpool, for example. The fact that the majority of the free thought organizations in these regions were specifically Owenite strengthens the suspicion that there was a direct, and fairly common, continuity between the earlier secularist form of infidelity and the later spiritualist one.[93]

On the other hand, however, there was also a potential for intense religious feeling among working-class spiritualists in northern England. Primitive Methodism, a distinctly lower-class Methodist splinter group whose meetings became renowned for their fervent revivalist atmosphere, was popular in the north and may have provided anti-Christian spiritualism with recruits who were recoiling from too much Christianity rather than reflecting too little. Much northern spiritualism, even when explicitly anti-Christian, exuded an aura of religiosity, an aura that might well have been inherited from the more passionate fringes of Nonconformity. The evidence from the spiritualist press and contemporary observers implies that most working-class spiritualists were drawn from the Nonconformist churches, especially Methodism,[94] and although this information is based on a certain amount of guesswork and assumption, adequate personal records remain to support the basic contention. What seems to have been occurring in northern England in the second half of the century, therefore, was a convergence of two streams of anti-Christian sentiment that involved men and women from the lower middle and upper working classes. One stream emerged from a secularist tradition that provided too meager emotional nourishment for some adherents; the other escaped from an excessively zealous Christianity that offended the moral principles and intellectual integrity of certain of its followers. Both dissatisfied groups found refuge in anti-Christian spiritualism that gave the former a warm dose of religious comfort and the latter a milder, less threatening, and infinitely more palatable form of religious belief.[95]

In their rejection of Christianity, anti-Christian spiritualists were refusing to endorse a handful of specific Christian doctrines. It is a tricky business attempting to pigeonhole anti-Christian spiritualist beliefs, because spiritualism was a highly personal faith; one can never be quite sure that individual spiritualists subscribed to all the positions set forth in their own publications.[96] But since we are dealing with segments of the working and lower middle classes that were for the most part literate, and even in many cases highly articulate, there is some reassurance that they did not suffer their views to go unpublicized, nor allow them to be distorted. The constant reiteration of a few basic arguments in the literature of the progressive spiritualists, furthermore, supports the belief that these were absolutely central to their repudiation of orthodoxy.

Above every other Christian teaching, the anti-Christian spiritualists abhorred the doctrine of eternal punishment. It was at once unreasonable and unjust to consign fallible human beings to endless damnation, particularly for no better reason than the misfortune not to be selected to eternal salvation. E. W. Wallis, for one, remembered going to a Baptist Sunday school and chapel as a boy in Twickenham and "being horrified at the thought of the dreadful Hell pictured" there. As a grown man, Wallis vehemently lambasted Christianity, in part expressly because "*Jesus taught the eternity of hell torments*, and the powerlessness of those in Abraham's bosom to help the sufferers!" J. J. Morse, too, recorded that his "reason revolted against the dogmas of eternal torment hereafter," and Burns dismissed "Hell torments" as one of the several forms of ancient "idolatry or Paganism" that disfigured Christianity.[97]

Anti-Christian spiritualists were not the only scathing critics of a religion based on threats of hellfire. Christian spiritualists like Haweis and Moses shared their opinion, and, indeed, the doctrine of eternal punishment was becoming increasingly distasteful to a wide range of Christians, from all social classes in Victorian Britain. Among its critics was no less a theologian than F. D. Maurice, and "of all the articles of accepted Christian orthodoxy that troubled the consciences of Victorian churchmen, none caused more anxiety than the everlasting punishment of the wicked." While, doubtless, the orthodox belief in the torments of hell continued to command the allegiance of the majority of clergymen, both within and outside the establishment, not a few concerned Christians eagerly studied their Bibles to find whether or not this distressing doctrine was actually scriptural; as always, biblical texts proved elastic enough to satisfy almost every inquirer.[98] The inquiry was in deadly earnest, for it was widely assumed that traditional views of hell were reducing the numbers of Christians in the country. An alternative attitude, of course, was to scoff at hell, to treat it breezily as a relic from the superstitious past that had become utterly irrelevant in the enlightened nineteenth century, and there is likewise evidence of this response at all levels of British society, from the irreverent working classes to the otherwise reverent Queen. (Princess Louise once told a Scottish minister that her mother did "not altogether believe in the Devil.")[99] It was one thing, however, to ridicule outmoded images of devils and damnation; it was quite another to build an alternative religion around their absence. The anti-Christian spiritualists set to that task with enthusiasm.

Hell, and consequently predestination, were not the sole Christian doctrines that the progressive spiritualists jettisoned. The vision of man's utter depravity and innate sinfulness similarly vanished, together with its necessary accompaniment, the doctrine of Christ's vicarious atonement. Both were anathema to anti-Christian spiritualists, for they reduced to nought the area of meaningful human activity and annihilated all hope of genuine freedom for humanity. In an address to the second convention of the British Association of Progressive Spiritualists, held in a temperance hotel at Newcastle in 1866, one speaker, Dr. Hugh McLeod, was specific and bitter on this point. "Our predecessors preferred darkness rather than light," he declaimed,

> – they went into willing and stupid bondage, and, necessarily, took posterity with them. This is the lamentable result of that pernicious system which declares man's total degradation . . . It is this system of mighty falsehood which I find fault with and would contend against. A system of lies inaugurated by fools and conspirators against man's ETERNAL LIBERTY.

J. J. Morse was as sickened by "the doctrines of original sin and total depravity" as by the thought of eternal punishment; E. D. Rogers, raised by a devout Wesleyan Methodist mother, and hardly an outspoken member of the progressive spiritualist camp, nonetheless could not accept the Wesleyan doctrine of atonement, "namely, that Christ died to appease the wrath of the Father and reconcile God to man." Such a doctrine, he recorded, was "repulsive" to his reason.[100]

Original sin and Christ's vicarious sacrifice for mankind were unacceptable, indeed hateful, doctrines from the anti-Christian spiritualist point of view because they diminished man to a mere puppet, incapable of taking moral responsibility for his own actions, and ultimately incapable of fundamental growth and improvement, except through the outside intervention of divine grace. Man's moral responsibility and the possibility of an eternally progressive future were central to anti-Christian spiritualist theology; they provided the basis for that vision of life after death that contrasted so sharply with the traditional Christian one. When anti-Christian spiritualists wrote of life "beyond the veil," there was neither pearly gates nor agonies of hell. St. George Stock, an Oxonian essayist and uncharacteristically intellectual figure among anti-Christian spiritualists, explained:

> There is no eternal torment, no heaven of ecstatic bliss. What Spiritualism does bring to light is the prospect of a progressive future for human beings – no sudden break, no violent transformation – death but the birth into another sphere of existence, a sphere in which every human being is exactly that which himself and society have made him, . . . There, as here, are all grades and varieties of being, and it is the work of the higher to lead up the lower.

The anti-Christian spiritualist afterlife was "but a prolongation, on a higher plane, of the present life, with its human aims and interests." It was "the continuance of the same scheme of development which we see in operation around us, only under more favourable conditions." Indeed, it was "the apotheosis of evolution."[101] The Spiritualists National Federation explicitly incorporated these views of life after death into a Declaration of Principles issued in 1901, which set forth as fundamental: "The Fatherhood of God, the Brotherhood of Man, the continuous existence of the human soul, personal responsibility, compensation and retribution hereafter for all the good or evil done on earth, and eternal progress open to every human soul."[102]

One did not get off scot-free in the anti-Christian spiritualist afterlife. Sins of omission and commission demanded atonement, but it was an atonement that the individual sinner could perform for himself, shouldering the blame for all that he had done during his physical life. The consequent punishment was finite, commensurate with the "evil done on earth," not a sentence without end. Once endured, the possibility of improvement was "open to every human soul," an eternity of moral and spiritual progress, with salvation available to all who would make the effort to obtain it. The only hell that figured in anti-Christian spiritualist thought was "such as the sinner may make for himself in this world as in the next."[103] Here was a scenario to tempt even the most obdurate of unbelievers, one that drew upon dreams of the millennium nurtured in working-class Owenism, theories of evolution borrowed from contemporary science, and the generally forward-looking, optimistic assumptions that molded the cultural and social stereotypes of the day. It was certainly a more appropriate vision of life after death for an age devoted to progress than anything the Christian churches could offer. As E. W. Wallis boasted, in underscoring the superiority of spiritualism over Christianity: "Spiritualism demonstrates that Law reigns supreme, and is beneficent; that progress here and hereafter is the Law of life; instead of starting with a failure (the Fall) and ending with a mistake (endless and useless torment)."[104]

It would be misleading to imply that there was absolute unanimity about life after death among anti-Christian spiritualists. There was never absolute unanimity among spiritualists about anything, except the possibility of communicating with spirits of the dead. Anti-Christian, as well as Christian, spiritualists expended considerable energy debating the difficult question of exactly what vestige of individual personality survived death.[105] Fortunately the arguments over a material body, quasi-material body, or purely ethereal spirit need not occupy much space here. In no way did they alter the anti-Christian spiritualist vision of ongoing progress in the "Summerland," and it is easy to understand why adherents of that viewpoint hailed it as a major moral advance over Christian teachings about heaven and hell. Not only did it bestow upon man both moral freedom and responsibility, but it also banished from the skies the angry, vengeful God whose threats of divine wrath effectively undermined the intermittent promises of divine mercy scattered through the Bible. It was the doctrine of predestination, and the orthodox portrayal of hell, that made God appear vindictive and cruelly punitive – a god who did not deserve to be the object of human worship. "Any God who demands the worship of fear," observed Gerald Massey, "is unworthy the service of love," and other anti-Christian spiritualists repeatedly expressed similar opinions. For Burns, "the angry God" was merely another remnant of ancient paganism, to be swept aside by the truly modern religion of progressive spiritualism.[106]

In tossing out the concept of a harsh, unforgiving Jehovah, the anti-Christian spiritualists virtually rid themselves of a personal deity altogether. With a vision of life, both before and after death, that stressed human responsibility and progress, any role left for God to play was minimal.[107] That was why atheists and secularists could comfortably espouse anti-Christian spiritualism, for it allowed them virtually to ignore God and, if they chose, to keep the element of worship entirely absent from their spiritualist beliefs. They might go so far as to speak of some overarching abstract principle of goodness, justice, or progress, or of some equally impersonal universal spirit, but they acknowledged nothing more definite by way of a deity. Other anti-Christian spiritualists, unwilling to abandon the Supreme Being altogether, but likewise vague about the essence and appearance of divinity, turned to pantheism as one possible solution to their quandary; it proved an attractive concept to spiritualists, both Christian and progressive, although they nearly always tempered it, illogically, with belief in the enduring uniqueness of the individual personality.[108] St. George Stock, not quite a full-fledged pantheist, adopted the evasive, but incontrovertible, position that God is "incomprehensible, the something that transcends all knowledge, but underlies all existence." James Burns expounded a concept of the Deity that so enhanced the stature of man as virtually to eclipse God. Burns upheld the essential divinity of humanity, arguing that "God is within the innermost of every man."[109]

God was thus subordinated to the service of man, not vice versa, and the anti-Christian spiritualists could be equally hard on Jesus. Presumably the Son had as little work to do as the Father, if human beings were washed clean of original sin and capable of infinite moral progress on their own. Anti-Christian spiritualists took the business of salvation entirely out of divine hands, with Jesus accordingly reduced in glory. Nor had they any use for the Trinity, or the doctrine of the immaculate conception (both figured on Burns's list of pagan vestiges), because these Christian teachings separated Jesus from humanity. Progressive spiritualists stressed, instead, what bound him to mankind, and their typical portrayal of Jesus depicted him as a first-class medium. Stock phrased it elegantly when he asked: "What was Christ but an incarnate type of man's higher nature? It was as such he lived and acted, unconsciously to himself, under the impulse of a power not his own." Burns was more blunt about Jesus' powers: "He was psychometric, clairvoyant, could heal the sick even at a distance, exercised wonderful biological influence over mind and matter, . . . We see in this portrait a man with remarkable psychological endowments, and moreover, a medium for superior influences."[110] While continental writers of advanced religious views were offering biographical portraits of Jesus as an extraordinary human being, often emphasizing the moral impact of his message, the anti-Christian spiritualists in Britain were underscoring his humanity in their own particular fashion. His miracles, his great acts of healing, his claim to communicate with God were not substantially different, they implied, from the feats of modern mediums – men and women who were often perfectly commonplace, simple people. Burns argued that "the Spiritualist is doing the same work now that Jesus did in his day," and he spoke of "spiritual power delegated" to Jesus by "the Father" as if that were the title taken by some prestigious spirit control, but nothing more. The Old Testament prophets, needless to say, received even less consideration at the hands of the anti-Christian spiritualists, who frequently allowed their readers to assume that the words "prophet" and "medium" were utterly synonymous.[111]

With the prophets severely diminished in significance and Jesus' divinity more than doubtful, it is no wonder that anti-Christian spiritualists also challenged the authority of both the Old and New Testaments. Like large numbers of working-class freethinkers, progressive spiritualists scorned bibliolatry, because they saw the Bible as yet one more "thrall" and "fetter upon man's intellect and conscience."[112] Not only did anti-Christian spiritualists refuse to acknowledge the Bible as an infallible, holy text, but they believed that dependence on Scripture dangerously undermined man's capacity to make his own moral decisions. Although recognizing that parts of the Bible were "grand and sublime," they lamented that, for centuries, "slavish subjection to the Bible" had blocked the exercise of man's reason. They were confident that humanity would progress substantially, if only the Bible's "hold on the human mind" could be weakened.[113]

In the anti-Christian spiritualist indictment, Christianity came under a two-pronged attack. In the first place, the anti-Christian spiritualists excoriated a set of doctrines that had fueled centuries of cruelty, bigotry, and bloodshed, or so they contended. Secondly, they were impatient with a religion that they deemed inadequate to the demands of the modern age. E. W. Wallis incorporated both aspects of the anti-Christian spiritualist critique in his free-wheeling assault on Christianity, the "Open Letter to Christian Opponents of Spiritualism." Among the ten or so reasons that he cited to explain why "Spiritualism is superior to Christianity," Wallis argued that spiritualism, unlike Christianity, had not "left a blood-red trail of persecution and gross inhumanities, of Crusades, Star Chambers, Smithfield fires, Spanish maidens; nor has it tortured and murdered hundreds of thousands of people to establish an authoritarian system." On the contrary, the record of modern spiritualism was an immensely constructive and uplifting one, for "Spiritualism affords to the heart-hungry, demonstrations of the continued personal existence of their deceased loved ones, whereas Christianity has no *evidence* to offer, only *hope* and faith." Spiritualism, in short, fought with the weapons of the nineteenth century – good, hard facts[114] – while Christianity's weapons, hope and faith, had grown so dulled and blunted over the centuries as to become useless.

Whether faulting Christianity for its doctrines of persecution or for its alleged failure in the modern era, anti-Christian spiritualists were offering their own beliefs as a religion of humanity. It was the amelioration of the human condition that they sought, through liberation from the fear and passivity that they claimed Christianity inculcated, and through the greater comfort to the "heart-hungry," which they believed spiritualism provided. Their "central inspiration," the Progressive Spiritualists announced at their first convention in 1865, was "a Love of Truth and Humanity, with an undying determination to discover the former, and apply it to the development and happiness of the latter."[115] Unlike their Christian counterparts, progressive spiritualists, as we have seen, adopted a distinctly reformist posture, often more striking in announcements than activities; nonetheless, the concern for the betterment of humanity during this life was, in theory, as important to anti-Christian spiritualists as eternal improvement in the next. Their brand of spiritualism was outspokenly progressive on both counts. For its followers, it was nothing less than "that glorious Gospel of Emancipation"[116] from centuries of inhibition and intimidation forced upon man under the guise of Christianity.

In celebration of their emancipation, progressive spiritualists spurned all authority and dogma. The emphasis on individuality that colored every aspect of spiritualism left anti-Christian spiritualists with a strong distaste for religious organization. Their dislike was based on something more than regret over the sectarian divisions among Christians; it was a repudiation of "priestcraft," of all interfering mediators between the individual and his principles, whether religious or purely ethical. "No power, either in heaven or earth, had a right to come between a man's soul and his sense of right and duty," Burns warned, as he explained that the great task facing spiritualists was to avoid contamination "by the selfishness of priesthoods." Robert Cooper, another progressive spiritualist with little good to say about Christianity, reduced it to a mere "system that has grown out of the creeds and dogmas of priestcraft." Spiritualism, St. George Stock explained, "is a revelation that disclaims authority," and even mediums, essential though they were for the transmission of spiritualist truths, were not endowed with any priestly powers. "We are warned," he continued, "that the utterances of mediums are to be no substitute for the individual judgment and conscience."[117]

Individual responsibility was always the anti-Christian spiritualist rallying cry. Priests, sectarian creeds, religious rules and restrictions were all eschewed because they allowed the individual to resign that responsibility, to leave crucial personal decisions to outside authorities. Louisa Lowe, active in the BNAS, told a conference of spiritualists in 1877 that they should strive "to overthrow all external authority in matters of thought: to free mankind from religious dogma and the trammels of priestcraft – in a word, to teach the individual to make his own reason an ultimate court of appeal in all matters of personal concernment."[118] The anti-Christian spiritualists in the audience must have applauded, for they prided themselves that theirs was a religion expressly based on reason. They did not have to accept without question the validity of miraculous events outside nature's order, events that insulted their good sense and demanded a suspension of all critical thought processes. In their reiterated emphasis on the rationality of anti-Christian spiritualism, progressive spiritualists further betrayed their debt to the eighteenth century for, unwittingly or not, they were echoing the claims put forward on behalf of deism, that earlier religion of reason.

Anti-Christian spiritualists, then, insisted that their system of beliefs was readily comprehensible and utterly grounded within the context of plausible natural phenomena. Rejecting altogether the category of the supernatural, they argued that the phenomena that they chronicled could be plainly explained in the language of science, for all to understand who cared to. Yet the language of science never fully satisfied them, and one of the most fascinating aspects of anti-Christian spiritualism is the degree to which it ultimately reverted to a form of church organization. The reversion began with the decision, on the part of numerous progressive spiritualist groups in the late 1860s and early 1870s, to hold Sunday services, as a rational alternative to orthodox Christian worship. By the time Emma Hardinge Britten published *Nineteenth Century Miracles* in 1884, she could describe such services in spiritualist communities throughout Britain – concentrated in Yorkshire and Lancashire, strong also in Northumberland and the Midlands, and much less evident in southern and western England. Three years later, *Two Worlds* listed more than one hundred available spiritualist "Services for Sunday" in Temperance Halls, Co-operative Halls, Mechanics' Institutes, and other assorted meeting rooms of urban Britain. Anti-Christian spiritualists in London enjoyed a wide choice of Sunday services, and, needless to say, Burns figured busily in this enterprise. Lectures, prayers, and sermons, delivered by trance mediums, were the major features of these services, with a liberal dash of hymn-singing interspersed among the inspirational talk. There were even special hymnals for spiritualists, such as *Hymns of Faith and Progress*, used at the Sunday meetings in Temperance Hall, Grosvenor Street,

Manchester, and arranged by the Unitarian minister John Page Hopps.[119]

Although the anti-Christian spiritualists firmly believed that their trance discourses far surpassed the orthodox Christian liturgy in enlightened rationality, it seems clear that those who flocked on Sunday to spiritualist services were seeking the comfort and sense of worthy activity provided by Sunday church going. In all likelihood, they also sought the sense of communal experience and group solidarity that has always been one of the rewards of church attendance. The very term "church," by which many anti-Christian spiritualist groups designated themselves and the halls where they met, suggests the need that both buildings and services were designed to fill. "In Bradford, Yorkshire," Mrs. Hardinge Britten reported, "a large and zealous society of working men and women have combined to hire a good hall, which they entitle the Walton Street Church." "A special hall," she continued, "has also been built and devoted to the Spiritual Sunday services, at Sowerby Bridge, Yorkshire, where a fine and well-trained choir of young people adds the charm of excellent singing to the elevating influences which pervade the place."[120] Here were some of the garnitures of organized religion, without the specifically Christian context, and the compromise seemed eminently suited to the progressive spiritualists. Beyond the metropolis, their societies acknowledged the religious impulse with growing readiness, so that by the turn of the century, most of the anti-Christian provincial spiritualist organizations had emerged as churches. In doing so, they were merely giving substance to that "odd air of religion which accompanied so much antichurch and antichapel feeling" during the Victorian era[121] and which progressive spiritualism had exuded for years. It was part of the same tendency that lay behind the proliferation of Sunday observances, complete with choir and hymnal, in numerous secularist societies and that prompted the foundation of Labour Churches in the 1890s.[122]

Like many a secularist community, progressive spiritualists in the second half of the nineteenth century also turned their attention to the provision of Sunday schools. A movement that aspires to longevity must have at its disposal the means to train the children of its members, to prepare the rising generation of believers for active participation in its programs and deep commitment to its doctrines. At this very period, Nonconformist churches were also placing more emphasis on the role of Sunday schools as centers of recruitment and incubators of denominational loyalty,[123] but their schools, naturally, were antipathetic to the views and goals of progressive spiritualists. Accordingly, the latter groups established Sunday schools of their own. That these institutions, known as lyceums, emerged at precisely the same time that anti-Christian spiritualists started to adopt church services is hardly surprising, for the two developments were entirely complementary.

The spiritualist lyceum movement in Britain derived its inspiration from the famous American trance medium, Andrew Jackson Davis. Davis, dubbed the "Poughkeepsie Seer," had visions of life beyond death in which he studied the education provided for spirit children – that is, children who, although dead to earthly life, were continuing to grow and progress in the spirit world. It became Davis's purpose

> to have similar schools on earth, where little children could be taught how to grow up to be wise, good, honest, truthful, and useful men and women; understanding their bodies, and knowing how to take care of them. And also to learn something of the wonderful things of this world.[124]

A group of disciples in the United States set out to realize Davis's dream in the early 1860s, while on the other side of the Atlantic, Burns publicized Davis's writings and promoted the establishment of spiritualist Sunday schools along the lines of the American model. As Burns's active participation indicates, the British lyceum movement was from the first clearly affiliated with anti-Christian spiritualism.

A few Children's Progressive Lyceums were founded in Britain during the late 1860s and 1870s. The earliest one, at Nottingham, emphasized "what we might call *drill*, as marching and movements to music or singing have much to do with the work of the School."[125] The lyceum at Keighley was launched by a local grocer, David Weatherhead, who also subsidized the publication of the *Yorkshire Spiritual Telegraph* and its successor, the *British Spiritual Telegraph*, in the late 1850s.[126] By the late 1870s, following the efforts of Davis's provincial disciples in anti-Christian spiritualist communities, the idea of separate spiritualist Sunday schools was taking root. Their further development in Britain was largely stimulated by the labors of a Yorkshire coal miner named Alfred Kitson.

Kitson had traveled to spiritualism along a route well-worn by working-class, anti-Christian spiritualists. The self-educated son of a miner and local preacher in Gawthorpe, Yorkshire, Kitson had been raised as a Primitive Methodist until his father's conversion to spiritualism in 1867. Following the paternal example, Kitson joined the Gawthorpe spiritualist society and began to broaden his knowledge of spiritualism in northern England. By the early 1880s, he had become an outspoken advocate of lyceum expansion, an advocacy which was shared by *Two Worlds* when it was founded at the end of the decade. Its first editor, Emma Hardinge Britten, joined Kitson and Harry A. Kersey, a Newcastle spiritualist, in compiling a manual that set forth recitations, songs, marches, calisthenics, and lessons for use in the lyceums around the country. The suggested routine commenced with a hymn and invocation, and closed with another hymn and benediction – a necessary reminder that it was designed to serve, not the Boy Scouts, but a Sunday school of sorts.[127]

Kitson set forth his philosophy of spiritualism in general, and lyceums in particular, in *Spiritualism for the Young*, published in Keighley in 1889. In a comparatively brief space, the volume expounded all the main tenets of the progressive spiritualist faith and mixed Andrew Jackson Davis's organizational blueprint with a healthy dose of English working-class consciousness. It included the standard anti-Christian spiritualist demotion of Jesus to the ranks of specially gifted mediums, a discussion of the Bible's many errors and contradictions, the emphasis on man's divine nature coupled with an angry denunciation of "that pernicious doctrine of total depravity which stultifies his reason, cripples his best efforts, and keeps down his highest aspirations." Kitson rejected the teachings inculcated in Christian Sunday schools as "bigot-creating and soul-blighting in a high degree." He scoffed at the doctrine of vicarious atonement and the threat of an "angry God, and still more terrible Devil." He rejoiced that, in the lyceums, "each child is dealt with as being a part of the Divine Father. It is pure, spotless, and without blemish, planting its feet on the shores of time to gain an experience that will form a basis for its immortal nature."[128]

Kitson heartily endorsed the diverse physical fitness programs that figured from time to time in the progressive spiritualist literature of this period. Lambasting liquor and tobacco, he urged his readers to be moderate in all their pleasures, for he was convinced that "the basis of all intellectual and moral greatness is health and strength." The progressive spiritualists who endorsed Kitson's plans for the lyceums shared his high estimate of "health, sunlight, and fresh air, and cleanliness"; they agreed that marching and calisthenics were perfectly appropriate to a Sunday school, because they valued exercise as a means of stimulating both body and mind. In their elevation of man above the fallen, sinful creature of Christian tradition, the anti-Christian spiritualists celebrated more than his immortal and essentially divine spirit. They also appreciated the need to care for his material body and sought the harmony of all parts, the smooth integration of physical and spiritual in the best interests of the whole person. Harmony was, indeed, a central concept in the lyceum program. Each lyceum, for example, was urged to form a choir and to give the children ample opportunity to sing, in order to expose them to the "wonderful, harmonising power" of music. The lyceum, in short, according to *The Manual*, was "the school of a liberal and harmonious education."[129]

Kitson's design for the Children's Progressive Lyceums in Britain followed Davis in terms of the internal arrangements and the classification of children into age groups that bore such names as "Fountain," "River," "Ocean," "Beacon," "Excelsior," and "Liberty." The subjects that Kitson recommended as "Lessons Suitable for the Lyceum" were a mixed assortment, ranging from personal hygiene, through physiology, botany, geology, astronomy, phrenology, and the spirit world, to the evils of swearing, gambling, drinking, smoking, and "keeping bad company." But there was nothing borrowed or hybrid in Kitson's devotion to the working classes. In a chapter not entirely relevant to the rest of the volume, he wrote confidently of the ultimate triumph of the workers, the industrious segment of humanity. Adopting a mildly socialist tone, he observed:

> The remuneration of the various classes of workers is, unfortunately, out of proportion with the actual work done. The heads of our large industries claim far more than their just share of the proceeds of united labour. But as the workers have won in the past in the aggregate, . . . just so surely will they ultimately win in the other. It is only a question of time.[130]

Indeed Kitson, with his strong social biases and equally strong anti-Christian sentiments, is the near perfect exemplar of working-class progressive spiritualism in Victorian and Edwardian Britain.

His active affiliation with hard, physical labor in the mines continued until 1904, when he became a full-time salaried officer of the British Spiritualists' Lyceum Union. By then, lyceums had gone the way of most British spiritualist associations and become organized. The Union was launched in 1890, and gradually many of the independent local lyceums joined, bringing the total membership to nearly two hundred on the eve of World War I. Not surprisingly, lyceums and spiritualist societies in the same geographic areas did not always cooperate amiably; there were the usual cases of jealousy and rivalry that plagued British spiritualism in all its aspects. The lyceum movement, following the broad divisions of Christian and anti-Christian spiritualism across the country, was strong in the north, but weak in London and the south. Nevertheless, despite only partial national participation, and despite occasional clashes for the allegiance of local spiritualists, the lyceums, like the spiritualist societies, became a permanent feature of British spiritualism in the twentieth century.[131]

A COMMON GROUND

The distinctions between Christian and anti-Christian spiritualists in Great Britain have general validity, in terms of geography and theology alike. There was often bitter hostility between the two groups, as the Christian spiritualists accused the anti-Christians of abandoning the faith of their fathers and luring innumerable souls away from the safe harbor of Christianity onto the perilous waters of blasphemy and infidelity. That some anti-Christian spiritualist writing was vociferously blasphemous is undeniable. It is equally certain that numbers of anti-Christian spiritualists were indeed attracted to progressive spiritualism from one or another of the Christian denominations. As far as the evidence can show, however, most of these converts went through an intermediate freethinking stage, and it is probable that an equally large number embraced anti-Christian spiritualism without any previous involvement whatsoever in a Christian church. Rather than drawing its members from orthodox Christianity, anti-Christian spiritualism recruited the great majority of them from positions of profound skepticism or outright atheism. There is no compelling evidence to suggest that the progressive spiritualists would have been good Christians, or at least good Christian spiritualists, without the fatal beguilement of anti-Christian spiritualism.

Nor is there persuasive evidence to indicate that communications between Christian and anti-Christian spiritualists were characterized by animosity alone. On the contrary, the rigid line between the two positions apparently softened time and again; it seems, in this case as in so many others, that firm categories, while useful for generalizations, fail to take into account the complexity of individual men and women. Beneath the rhetoric of Christian and progressive spiritualists lay a substantial degree of cooperation. Both groups, after all, had to face reiterated denunciations by diverse spokesmen for orthodox Christianity; both were subjected to ridicule by much of the scientific profession, the press, and the public at large; both were unalterably committed to the reality of spirit communication. That anti-Christian and Christian spiritualism sometimes found mutual support to be their best policy is hardly astounding.

Such support usually took a practical turn. In the publication of spiritualist writings, for example, the important point was to broadcast the spiritualist message, not to be fussy about the beliefs of the publisher. Thus James Burns published the writings of several Christian spiritualists, including pamphlets by Thomas Shorter and Reverend Colley. Stainton Moses had no qualms about contributing articles to Burns's *Human Nature*, nor did he hesitate to make himself at home at 15 Southampton Row, attending séances at the Spiritual Institution and using its library. When Burns took over the publication of the *Spiritual Magazine* in 1869, it continued to print essays by Shorter and William Howitt for several years, and, on the other side of the coin, the Christian spiritualist J. H. Powell was likewise perfectly willing to print the proceedings of the British Association of Progressive Spiritualists.[132] Business was business, and other examples of reciprocal assistance abound in what spiritualists of all varieties may well have seen as a publicize or perish situation.

Cooperation between Christian and progressive spiritualists blossomed, too, within their associations. Although Christian spiritualist pressure had compelled the BNAS to abandon its early statement of religious neutrality, the BNAS was by no means closed to progressive spiritualists. Louisa Lowe, for one, served on the General Purposes Committee side by side with W. S. Moses and C. C. Massey, Christian spiritualists both.[133] J. J. Morse and E. D. Rogers, despite their rejection of fundamental Christian doctrines, were welcome at 38 Great Russell Street, where they loyally remained, even during the organization's last painful year as the Central Association of Spiritualists. Rogers reappeared prominently in the London Spiritualist Alliance; under his presidency, Morse was coopted onto the council and E. W. Wallis, outspoken anti-Christian though he was, served as secretary. Furthermore, while Rogers led the Alliance, it maintained cordial relations with the representatives of provincial, and anti-Christian, spiritualism. In 1897, for example, when its council was preparing for an international congress of spiritualists in the following year, it included the Spiritualists National Federation and the Lyceum Union among a list of societies to be consulted. The LSA agreed that the Children's Lyceums ought to be discussed at the congress and appointed Morse to serve as the Alliance spokesman on the subject of educating young spiritualists. There was, no doubt, much nodding of heads among the members of the LSA when the president of the Spiritualists' National Union, addressing them in 1903, "urged the desirability of the Alliance entering into association with it for the good of Spiritualism generally."[134]

The relationship between Christian and anti-Christian spiritualism, then, did not follow a straightforward path of mutual hostility. It was a complicated relationship, as complicated as the impossibly difficult question that it raises: what, precisely, does the term "Christian" signify? During the centuries that the question has stirred controversy, some have argued that the fundamental belief of Christianity is enshrined in the doctrine of Christ's divinity, whereas others have centered their faith in the conviction that Christ's body rose from the dead. Some consider Christianity nothing less than the entire structure of the church, its organization, ritual, Scripture, and dogma; others simply look to the inspiring person and teachings of Jesus. That the anti-Christian spiritualists of the Victorian and Edwardian decades wanted no part of formal, institutionalized Christianity is beyond dispute. Their attitudes toward the figure of Jesus, however, revealed profound ambivalence, and, despite their attack on his divinity, some progressive spiritualist writers betrayed a note of reverence for the founder of Christianity. Robert Cooper, for one, made a great point of distinguishing the Christianity that had sprouted out of priestcraft from the religion "that Christ taught on earth." The former, he asserted, was "no more like the religion of Christ, as taught and practised by Himself, than a player is, in reality, like the king he personifies."[135] In Liverpool's Meyerbeer Hall, during a progressive spiritualist Sunday lecture on "What does Spiritualism offer Superior to Christianity?" the speaker, Mrs. F. A. Nosworthy, "observed that if by 'Christianity' was meant the sublime teachings offered in the Sermon on the Mount, and in the life and example of Jesus Christ, she held that Spiritualism presented nothing higher." A subsequent speaker proceeded to read a portion of that Sermon[136] – surely an odd choice for a Sunday service designed to offer a rational form of worship in sharp contrast to Christianity.

Yet the choice would not be so peculiar if the anti-Christian spiritualists, or at least a portion of them, were seeking to indict only that part of Christianity that they believed had lost its pristine purity. Some anti-Christian spiritualists, much like certain Chartists before them and early socialist Labour leaders after them, were capable of criticizing "the Churches because they were unrepresentative of Christianity, . . . because they failed to follow Christ, not because they sought to follow him."[137] To their way of thinking, the Christian churches had deviated egregiously from the original teachings of Jesus. They accused the churches of distorting and corrupting Christ's gospel, the same charges that religious reformers have levied against Christian institutions for centuries. The anti-Christian spiritualists were not, of course, trained theologians; their attacks lacked critical acumen and scholarly perceptions. They were more likely to dissect Christianity with a bludgeon than a scalpel. Yet not all progressive spiritualists were hopelessly heavy-handed. There were those who wanted to retain a role for Christ, while vehemently rejecting Christianity. Indeed, in their fashion, they may be seen as part of what one astute scholar has called "the great Victorian enterprise, the attempt to separate the soul of Christianity from its body."[138]

Preserving as they did a deep respect for and even, in some cases, genuine devotion to Jesus, many progressive spiritualists seem inappropriately labeled as "anti-Christian." That label, in fact, becomes increasingly unsatisfactory the more the contours of progressive spiritualism are explored, and yet "non-Christian," as a neutral alternative, proves to be no more satisfactory. Certainly the progressive spiritualists saw no need for a closely organized, carefully hierarchical church structure, but Congregationalists had long espoused similar views of local independence without jeopardizing their Christian status.[139] Progressive spiritualists clearly denied the doctrine of eternal damnation, but so did a number of their Christian colleagues. What was distinctive and definitive about the anti-Christianity of progressive spiritualism was not a single belief, or even a cluster of beliefs, but rather the overwhelming conviction that the Christian churches had become forces of oppression instead of freedom, their concerns more earthly than spiritual, their message one of fear and despair. Yet, in the late nineteenth and early twentieth century, this conviction remained compatible with a strong attachment to Jesus as the greatest teacher mankind had ever known, an unrivaled moral guide, and the man whose true message was the promise of life eternal in the light of God's love.

Today, in the 1980s, for countless churchgoers such an attachment is the most religious feeling that they can muster, and it passes for an acceptable expression of faith among many Christian modernists. In the nineteenth century, few orthodox Christians could have accepted so slight a criterion as a legitimate basis for faith, and so the problem of adequate nomenclature remains. Neither "anti-" nor "non-Christian" will fully explain the progressive spiritualist position, and "anti-religious" is, of course, even further from the mark. "Anti-church" seems promising until one remembers the progressive spiritualist penchant for gathering themselves into their own "churches." Perhaps the "anti" route leads to no permanently acceptable conclusions, although no doubt there were some progressive spiritualists who could live entirely without religious gratification, who were solely interested in the scientific implications of spiritualism, and whose attitudes toward religion were, indeed, anti-everything. Such a frame of mind, however, was far from characteristic of progressive spiritualism, and a less negative approach to that movement may illuminate its motives more sharply.

The most fruitful approach suggests synthesis in place of dichotomy, an emphasis, not on the differences that separated the two branches of nineteenth-century spiritualism, but on the profound similarities uniting them.[140] In addition to the common problems of ridicule and denunciation, they shared important attitudes toward religion, improbable though such agreement may initially appear. For one thing, both heartily deplored sectarianism. For another, numerous Christian spiritualists called for the revivification of Christianity along lines that the progressive spiritualists could warmly endorse. The Countess of Caithness repeatedly spoke of the dawning of "the new Dispensation" and predicted that

> The established religions of the world are on the eve of great changes; they will be thoroughly sifted, the fine gold of good they contain will be preserved, and the dusty cobwebs of theological lore that have so accumulated as almost to exclude the growing light, will be swept away.

Thomas Shorter, too, was given to pointing out "the dust and cobwebs" clinging to the Christian churches, while J. S. Farmer bluntly observed: "That a vitalising and purifying influence is needed is plain enough; of something radically wrong, the Church stands self-convicted, self-condemned." W. S. Moses, whose own relations with Christian orthodoxy were troubled and ambivalent, moved beyond criticism to prophesy the resolution that spiritualists of all viewpoints awaited. "One mind," he explained,

> rebounds from a narrow cramping Christianity miscalled orthodoxy, to a broad and rather shadowy Theism, or to a still more shadowy Pantheism; while another rests in the familiar by-paths of the creed of its childhood.
> . . . As time rolls by these views will harmonise, and out of their fusion will come what I think I can dimly discern in bold outline, looming through the mists that hang around me – the Religion of the Future.[141]

The religion of the future, the new dispensation – these were more than mere figures of speech for British spiritualists whose dissatisfaction with denominational schisms prompted them to envision a faith that, far from fragmenting the religious consciousness of humanity, emerged from the common elements in all religions, in all times and places.

The desire to discover and probe the universal religious impulse linking together all human societies and cultures was certainly not new to the second half of the nineteenth century. But the quest for universality that underlay much of Victorian spiritualism probably derived particular support from contemporary developments in both religious and social thought. The emerging study of the anthropology of religion, as well as Max Müller's work to introduce the sacred Sanskrit texts to the British public, could not have failed to make an impression on the better educated spiritualists;[142] the yearning to realize the universal brotherhood of man, so fervently expressed in socialist literature in this period, must have left its mark on those whose spiritualism had Owenite roots. In either case, when British spiritualists spoke of a universal religion, they meant something distinct from Universalism, the belief in ultimate salvation for one and all. While they did indeed believe that every human being was capable of infinite progress in the hereafter, the religious universality that many of them sought was on the earthly side of paradise. They wanted to find the fundamental substratum on which all the religions of mankind had been built and on that base to establish, not just Christian fellowship, but, beyond that, the single faith that could serve the needs of all humanity.

Although spiritualists of different perspectives were inclined to argue among themselves whether spiritualism was actually a new religion, a sect, or a common religious principle, there was no doubt in their minds that its application was universal. That was why Elizabeth Barrett Browning, "out of sympathy with institutionalized religion in general," and eager to find "the Church Universal," clung with such interest to spiritualism. That was why Louisa Lowe, almost vitriolic in her critique of the social and emotional repressions perpetrated in the name of Christianity, could hail spiritualism as a "grand new charter," not only of "free thought" and "boundless toleration," but also of "universal brotherhood."[143] It is true that many a spiritualist used the word "universal" rather loosely. James Smith, for example, having abandoned the Scottish ministry, the Southcottians, and the Owenites on his travels toward spiritualism and mysticism, wrote happily about the "coming age of universality" without bothering over too many details. Similarly, Burns and Wallis were inclined to state flatly that spiritualism was "the essence and form of all religion," without laboring long to substantiate their claims. But perhaps substantiation was not what mattered, for the word "universal" triggered potent feelings of pride and optimism that did not demand exhaustive evidence. To spiritualists, it was virtually self-evident, as St. George Stock observed, that "every religion is founded upon spirit-manifestation."[144]

With this certainty firmly ensconced in the attitudes of British spiritualists, their writings at times exuded an air of complacency. They were absolutely convinced that theirs was the faith that united all faiths, that reconciled religion and science, and gave man the facts to prove his immortality. Whether they honored spiritualism as the foundation of every religion in human history, including Christianity, or whether they hailed it as the new "revelation suited to the needs of the time,"[145] their beliefs gave them a hopefulness about the future that could not be daunted. In some respects, the Victorian spiritualists were a caricature of their society, harboring those traits that were supposed to have exemplified the British attitude toward the world in general during the second half of the nineteenth century. They were confident; they could lapse into smugness; they took for their canvas the entire globe and proclaimed a system of beliefs that transcended barriers of geography and culture. If Victorian society and its values were a great deal more than complacent, confident, and expansionist in outlook, spiritualism likewise had its complexities and apparent contradictions. Most of its followers, however, were not attracted to spiritualism for any subtle reason. They embraced it, rather, as an unequivocal statement about the human condition, and one that, like religions throughout history, allayed their most fundamental fears of death and loss.

4

Psychical research and agnosticism

The distance between a James Burns, E. W. Wallis, or J. J. Morse, and Henry Sidgwick must be measured in intellectual light-years. Sidgwick, Knightbridge Professor of Moral Philosophy at Cambridge, possessed a mind incapable of complacency or smugness; for him, it was far easier to see all sides of a question than to adopt one position with confidence. Yet as a founding father and first president of the Society for Psychical Research, his concerns were closely allied to those of Burns, Wallis, and Morse. There was as much groping after religion in his agnosticism as in their hostility to Christianity. As John Maynard Keynes observed about Sidgwick in 1906: "He never did anything but wonder whether Christianity was true and prove that it wasn't and hope that it was."[1] If Sidgwick spent his life doing nothing, he had much to show for it at the end, including Newnham College, Cambridge, which he helped to establish. Keynes's witticism was not, however, altogether hyperbolic, for the central concern of Sidgwick's intellectual inquiry was always religion – not just the truths of Christianity, in particular, but the human propensity for religious faith in general. Given Sidgwick's prominence in the SPR during its first eighteen years, and the important role played in its affairs by a number of his close, like-minded friends, it is not at all surprising that many of the Society's investigations before World War I were tacitly predicated upon religious questions. Indeed, it would not be an exaggeration to say that the early leaders of the SPR zealously explored the terra incognita of telepathy with the aim, whether purposeful or subconscious, of providing new, unassailable foundations for religious beliefs.[2] If it were true, after all, that men were capable of communicating in ways far beyond the range of normal sensory powers, then there was surely more to this world than the materialists could fathom, including a possibly active role for soul or spirit.

Sidgwick, like so many of his fellow doubters, passed his childhood and adolescence peacefully in the bosom of the Anglican church. Although his father, the Reverend William Sidgwick, died in 1841, when

Henry was only three, other members of the family were on hand to maintain the ecclesiastical tradition. Of particular importance was Edward White Benson, Henry's older second cousin and assistant master at Rugby where Henry was sent in 1852. Benson, who married Sidgwick's sister Mary in 1859, provided an inspiring model on which Henry for some years sought to pattern his own life.[3] Benson's son recalled that Sidgwick

> was brought up in orthodox Christianity; he was a serious and convinced Christian as a boy; he had a more or less definite intention of taking Orders. These tendencies were fostered both in his own home, where his mother was a devout High Churchwoman, of the old-fashioned type, and still more by my father, whose influence over Henry Sidgwick at an impressionable time was very great.[4]

Like Benson, Sidgwick pursued his university education at Trinity College, Cambridge, which he entered in 1855 and where he, too, had a distinguished undergraduate career. Thereafter, however, the parallel lines of their early lives began to diverge. While Benson rose to the highest position in the Church of England, Sidgwick eventually abandoned the Anglican faith. In 1860, he was already entertaining doubts about a clerical career and had definitely rejected that course of action by 1861. The following year he confessed that his "old theological trains of thought and sentiment" had not been discarded so much as paralyzed by "the scientific atmosphere," and, in 1869, after much painful probing of his conscience, he resigned the Trinity Fellowship, which by law presupposed his subscription to the Thirty-Nine Articles. As Sidgwick told Benson at the time, "I could not accept the dogmatic obligation of the Apostles' Creed, which *primâ facie* I have bound myself (in confirmation) to accept."[5] Other Fellows did not take the responsibility so seriously and managed to combine, without qualms, university fellowships and religious doubts. But Sidgwick's conscientious scruples made him an unusual sort of Fellow. Although he did not formally renounce membership in the Anglican church, he ceased, in effect, to be a member.

The resignation of his Fellowship and assistant tutorship did not terminate Sidgwick's affiliation with Cambridge. So promising a don was not lightly tossed aside, and both Trinity and the university found positions for him over the years, as Lecturer in Moral Sciences, Praelector in Moral and Political Philosophy, and, in 1883, as Knightbridge Professor of Moral Philosophy. The professional sacrifice that he made in 1869 helped to expedite the final elimination of religious tests at the university, and for the rest of his life Sidgwick continued to participate actively in a variety of campaigns to reform the ancient regulations that governed Cambridge administration and teaching. He became a fixture of university life during the forty-five years that he spent at Cambridge, a man widely admired for subtlety of intellect and integrity of character, and deeply loved for his charm, kindness, and capacity for enduring friendship. Lady Jebb evoked the quality of Sidgwick's personal relationships when she wrote in the biography of her husband, Sir Richard Claverhouse Jebb, Regius Professor of Greek at Cambridge and Conservative MP for the university: "On August 28th [1900] Professor Henry Sidgwick died, and thus there was broken for Jebb an almost life-long friendship. Cambridge never seemed quite the same to him again, when it was no longer possible 'to consult Sidgwick,' or to be cheered by that delightful mind."[6]

That mind did not, however, always pursue delightful thoughts. The loss of religious certainty permanently denied Sidgwick any genuine peace of mind, and while he was too committed to open-minded inquiry to wallow in despair, he was too honest to allow himself the luxury of abiding hope. During the 1860s, he examined with merciless rigor the host of arguments raised by science, history, biblical criticism, studies in Arabic and Hebrew, and not least of all philosophy.[7] He came to perceive clearly how fragile were the underpinnings of Christian thought and directed his keenest attention to the ethical implications of that fragility. What compulsion toward moral conduct, he wondered, would mold human behavior if religious sanctions crumbled? It was, of course, an old question and one that had been much discussed in the eighteenth century. With increased urgency, it troubled numbers of Sidgwick's contemporaries and informed not only his magnum opus, *The Methods of Ethics* (1874), but also figured in many of the well over one hundred articles that he contributed to the periodical press between 1860 and 1900.

Sidgwick indulged in a rich diet of philosophical nutriment in the 1860s, while he sampled what utilitarianism, positivism, empiricism, and intuitionism had to offer. Yet none of these systems of thought, and none of their prophets, adequately solved for him the problem of duty, or moral responsibility, in a world without omnipotent authority. He could not confidently assume, like some Victorians, that morality is an innate human sense, capable of flourishing outside a religious context. Nor did the Broad Church theology of F. D. Maurice ultimately satisfy him; its apparently infinite flexibility could not attract a man of Sidgwick's mental discipline for very long.[8] By the time that he resolved to resign his Fellowship, he had freed himself from the grasp of all ready-made philosophies and had struck out on his own to create a new metaphysical recipe.

The ingredients of that recipe – the proportions of the mixture of intuitionism and utilitarianism – occupied Sidgwick until his death at the turn of the century. Although he first enunciated his principal arguments in 1874, in *The Methods of Ethics*, he constantly refined his work thereafter, in subsequent editions, in other books, and in articles and essays by the dozen. Always at the center of his inquiries was his acute awareness of the fatal flaw lurking in the utilitarianism propounded by Bentham, James Mill, and, with modifications, J. S. Mill. The Utilitarians argued that the interests of self and society were one and the same, that in serving self, the individual furthered social goals as well. Sidgwick, however, could find no evidence of such convenient harmony of personal and public ends. On the contrary, he recognized their fundamental incompatibility and pointed out that the greatest happiness of the individual was not necessarily, nor even probably, identical with the greatest happiness of the community. The former, in fact, was often dramatically opposed to the latter. In the absence of an omnipotent judge, Sidgwick saw no guarantee that people would behave altruistically, if such behavior served no immediate, personal purpose, and he sought to discover what other motives, besides self-interest, might prompt people to perform acts of duty, even of self-sacrifice.[9]

Duty and service were two concepts that ranked high both in Sidgwick's own pantheon of values and in his published philosophy. Viscount Bryce, a friend and colleague in the campaign for educational reform, recollected shortly after Sidgwick's death that "the love of truth and the sense of duty guided his life as well as his pen."[10] At Rugby, he had been deeply impressed by the ideal of service to his fellow man, and he never lost the desire, as his biographers explained, "to be of some service to humanity," to do "some practically useful work." It comforted him to think that, if the life of an academic philosopher would not quite fill the bill, that of a crusader for female education would.[11]

What was the source of Sidgwick's profound sense of duty? To a superficial degree, it was the Rugby created by Thomas Arnold, but some deeper, far more potent motivation was also at work, and after the early 1860s, Sidgwick could not look to traditional Christian precepts for the answer. Neither the exhortations of the Gospels, he was convinced, nor the whisperings of selfish egoism underlay his efforts to contribute to the public good. Why, then, did the dictates of duty still appear so imperative to him? A partial explanation might lie outside Sidgwick himself, in the cultural milieu in which he was raised, for by the 1840s and 1850s the evangelical revival had deeply colored English Protestantism, Anglican and Nonconformist alike. The evangelical tradition, with its unrelieved emphasis on personal probity and constant self-examination, and with its distinguished record of public service as well, influenced the moral climate of England for the rest of the century.

Even those men and women who came to reject the specific Protestant theology on which evangelical morality was raised felt the influence of that morality throughout their lives and tended to measure themselves against its yardstick. Perhaps the potency of the evangelical influence in England helps to explain why some British intellectuals exerted themselves so doggedly – apparently more than either their European or American counterparts – to reconcile science and faith. When the fabric of national life seemed to be woven on the warp of Christian morality, whatever challenged religious dogma threatened the entire existing social pattern. Loss of faith could, accordingly, manifest itself not merely as a personal crisis to be endured, but as a national catastrophe to be forestalled.

These speculations may or may not apply to Henry Sidgwick. What remains certain, however, is his obsession with the springs of altruistic behavior in the absence of any compulsion to drive people toward altruism. He devoted a large portion of his life to unraveling the enigma, and the effort not only helped him to articulate his ethical assumptions, but also led him to a highly tentative theological compromise. In a frequently quoted and justly famous letter, written to a former Rugby schoolfellow in 1880, Sidgwick explained his delicately balanced religious attitude.

> Frankly, then, I must first draw a distinction, in order to explain my position between Theism and Christianity. It is now a long time since I could even imagine myself believing in Christianity after the orthodox fashion; . . .
>
> But as regards Theism the case is different. Though here my answer will doubtless surprise you. For if I am asked whether I believe in a God, I should really have to say that I do not know – that is, I do not know whether I *believe* or merely *hope* that there is a moral order in this universe that we know, a supreme principle of Wisdom and Benevolence, guiding all things to good ends, and to the happiness of the good. I certainly *hope* that this is so, but I do not think it capable of being *proved*.
>
> . . . Duty is to me as real a thing as the physical world, though it is not apprehended in the same way; but all my apparent knowledge of duty falls into chaos if my belief in the moral government of the world is conceived to be withdrawn.
>
> Well, I cannot resign myself to disbelief in duty; in fact, if I did, I should feel that the last barrier between me and complete philosophical scepticism, or disbelief in truth altogether, was broken down. Therefore I sometimes say to myself "I believe in God"; while sometimes again I can say no more than "I *hope* this belief is true, and I must and will act as if it was."[12]

Without such a belief – whether labeled God, or universal reason, or "a supreme principle of Wisdom and Benevolence" – Sidgwick could, as he said, find no foundation for a system of ethics and, consequently, no rational basis for human society. The final sentence of *The Methods of Ethics*, in its first edition, poignantly summarized Sidgwick's point of view. Without some conviction of perfect moral order, he wrote, "the Cosmos of Duty is thus really reduced to a Chaos: and the prolonged effort of the human intellect to frame a perfect ideal of rational conduct is seen to have been foredoomed to inevitable failure."[13]

Here was no proof of the existence of God, either as a traditional deity or an abstract metaphysical concept. All that Sidgwick could show, as he readily admitted, was how much he wanted to prove the existence of God. He could only argue that "beliefs in God and in immortality are vital to human well-being." Acknowledging the paramount social function of religion and, like the non-Christian spiritualists, hailing the life of Jesus as an inspiring example for mankind, he nonetheless wrote sadly in his journal, at the end of 1884: "The deepest truth I have to tell is by no means 'good tidings'." The young man who wrote in 1862 to H. G. Dakyns, another schoolfriend from Rugby, "You see, I still hunger and thirst after orthodoxy,"[14] had, by the end of his life, spent nearly four decades searching in vain for food and drink. The search had not, however, been lacking in drama; it had offered Sidgwick high adventure, as well as low chicanery, in the byways and back rooms of psychical research.

For Sidgwick, psychical research was a logical extension of his ethical concerns. All too fully aware that human sanctions could not suffice to make men behave selflessly, he was reluctant to abandon altogether the sanctions of a future life. Although the great traditional incentives for altruistic behavior – hope of heaven's rewards and fear of hell's torments – had lost much of their impact by the third quarter of the nineteenth century, some salvaging operations might yet restore the concept of human immortality. Stripped of both its lurid trappings and the elements of injustice that dismayed Christian consciences, belief in life after death could continue to provide precisely the motive necessary to reconcile the wishes of the individual and the needs of society. The survival after death of the human personality kept alive the possibility that individual pleasures foregone in this life, for the common good, would be entered as credit in the life to come.[15] If man's immortality alone could give meaning to human society and to the otherwise seemingly random motions of the cosmos, it was certainly worth every effort to prove human immortality. However many thousands of hours Sidgwick accumulated at the séance table during forty years of psychical research were hours devoted to an estimable cause. He was attempting to demonstrate that the human personality survives physical death and thereby to establish, irrefutably once and for all, the single basis for a system of ethics that he could endorse. The significance of the endeavor did not escape him, and in a bleak frame of mind, he wrote in his journal, early in 1887:

> I have been facing the fact that I am drifting steadily to the conclusion . . . that we have not, and are never likely to have, empirical evidence of the existence of the individual after death. Soon, therefore, it will probably be my duty as a reasonable being – and especially as a professional philosopher – to consider on what basis the human individual ought to construct his life under these circumstances.[16]

Human immortality, however, was not the sole hypothesis needed to preserve from chaos the cosmos of duty. If life after death existed, it had to be perceived as a life ruled by justice and wisdom, where men and women could reap the delayed rewards for altruistic sacrifices performed on earth. The afterlife, in brief, needed an ordering hand or mind, imbued with perfect morality.[17] Yet, if life after death proved hard to verify empirically, the existence of God, as Sidgwick knew well, was utterly unprovable by any late nineteenth-century standards of empirical inquiry. He had boxed himself into a philosophical corner, with very limited means of graceful exit. It was always possible to confess the insolubility of the ethical dilemma that Sidgwick had posed, and he was too candid not to say as much from time to time, especially in his private correspondence. He chose, however, to follow a different escape route, one that led him to the limits of empiricism and to a growing acceptance of nonempirical, nonrational modes of knowledge. He never lapsed into a muddled irrationalism – far from it – but he argued, most vigorously from the late 1870s, that even scientific theories allegedly based on the strictest of empirical methods were, in fact, predicated on nonempirical assumptions, such as the presumed uniformity of natural laws. Just as a comprehensive theory of cognition must take into consideration intuitions not perceptible by the senses, so, he came to contend, nonempirical postulates might also figure as valid components of an ethical theory.[18] Although Sidgwick could never place immortality and the existence of God among those facts marked "scientifically authenticated," he did not, consequently, have to expel them as useless from his bag of working hypotheses. If he was playing metaphysical games with himself, psychical researchers must nonetheless have been grateful that he bent his rules of intellectual inquiry enough to turn his formidable attention to supernormal phenomena.

So precise was Sidgwick in describing his loss of orthodox faith and his ongoing quest for a philosophically acceptable substitute that he has become almost a shorthand symbol for the Victorian religious crisis.[19] The generation of scholars and intellectuals who came of age in the 1860s and 1870s was characterized by a salient lack of theological convictions. Whether their loss of faith was part of the ethical revolt against the punitive rigidity of Christian dogma, a response to the persuasiveness of evidence from science and biblical scholarship, or a reaction against an orthodoxy that seemed at first unwilling to listen to the voices of modernity, the final result was the rejection of formal religion. Some, like Leslie Stephen and John Morley, not only embraced agnosticism without regret, but actually experienced a considerable sense of freedom.[20] A few found comfort, and even inspiration, in the Positivists' Religion of Humanity. Others, like W. K. Clifford, professor of applied mathematics at University College, London, also pinned their hopes on man, accepted the loss of God, but acknowledged that His absence left a certain gap in the universe. "It cannot be doubted," Clifford wrote in 1877,

> that theistic belief is a comfort and a solace to those who hold it, . . . It cannot be doubted, at least, by many of us in this generation, who either profess it now, or received it in our childhood and have parted from it since with such searching trouble as only cradle-faiths can cause. We have seen the spring sun shine out of an empty heaven, to light up a soulless earth; we have felt with utter loneliness that the Great Companion is dead.[21]

For another group, however, those "reluctant doubters"[22] who could not dispense with the solace of theism, who refused to endorse scientific naturalism as the sole key to unlocking the mysteries of the universe, the loss of Christian faith was merely the opening bout in a determined struggle to find alternatives to the materialistic vision that they rejected. Not a few pinned their hesitant hopes on psychical research.

It has become commonplace to say that droves of educated Englishmen embraced agnosticism during the late Victorian and Edwardian years. "At that time," writes one historian, ". . . to be an intellectual was to be agnostic." Another notes: "In some parts of the community, and notably among the intelligentsia, religion had either been replaced by an active agnosticism or had simply withered away."[23] But the agnostic embraces a peculiarly elusive "ism," one that is too complex for a single label adequately to describe. Principal Tulloch of St. Andrews understood the curious constitution of the "reluctant doubters" when he discussed that "class of minds who, while repelling the old solutions and the ecclesiastical connections identified with them, are yet restlessly impelled to new solutions. They are unable to leave religion aside."[24] Such minds, incapable of worship at the cult of science, were unprepared to settle down to any permanent set of attitudes about religious issues. They demanded instead the right to ask questions and seek answers concerning any and all aspects of the human condition, dead or alive. It may not be profitable to argue whether they are best classified as agnostics, deists, or would-be theists. A label they would have all agreed to share, however, was that bestowed on E. R. Dodds, a future Regius Professor of Greek at Oxford, member of the SPR from before World War I, and its president in the 1960s, whose "persistent sceptical curiosity earned [him] the nickname of 'the Universal Question Mark.' "[25]

The SPR has had its share of Universal Question Marks throughout its history. In the early years, Sidgwick was in good company, with Frederic William Henry Myers, Edmund Gurney, Frank Podmore, and Walter Leaf, among others. Although not all these colleagues at the SPR represented to the same degree as Sidgwick "the defection of the educated" from the church,[26] they were agreed in the cosmic scope of the problems that interested them and in their willingness to expend endless effort in the application of psychical research to those problems. They shared, too, a religious background similar to Sidgwick's. Myers, Gurney, and Podmore were all sons of Anglican clergymen, and although Leaf's father was a successful London businessman, Walter was raised in an unquestioning evangelical faith. It was his mother's fondest wish that he take Holy Orders and devote his life "to Evangelical propaganda"; although she failed to inspire him with a kindred fervor, he was as zealous as Sidgwick in the desire to use his talents "in the service of [his] fellow men."[27] For each of these psychical researchers, the confident religious convictions of their parents were impossible. Yet none could rest peacefully in "the camp of negation."[28]

Shortly before his marriage in 1894, Leaf wrote to his fiancée, Charlotte Symonds, the daughter of John Addington Symonds: "For me, reason refuses to be satisfied with Christianity – or at least with Christian formulas – and gropes, however blindly, for something more. I think it always will."[29] Reason demanded satisfaction from the likes of Sidgwick, Myers, Gurney, Podmore, and Leaf. They were men who had garnered an impressive array of honors at Cambridge and Oxford; their minds were trained to be disciplined and discriminating. Fully cognizant though they were of the limitations of scientific inquiry, they nonetheless thought and wrote at a time when the interpretation of "all natural and human phenomena in positivist terms"[30] appeared ever more authoritative, even irrefutable. They, too, were impressed by the rigors of the scientific method, and, in launching the SPR, they insisted that the highest intellectual standards, the same scrupulous attention to evidence found in the scientist's laboratory, would characterize their own inquiries in the shadowy world of psychical phenomena. In practice, they did not always succeed in reproducing the methods of scientific inquiry, nor in maintaining its lofty standards. At times, in fact, they failed to a quite embarrassing degree. They remained faithful enough to their initial intention, however, to endow the publications of the SPR with a stature unrivalled among the periodicals of British spiritualism.

In so intellectual a circle of investigators, the grosser forms of spiritualist materializations, redolent with fraud, tended to have less appeal than the mental phenomena of mediumship.[31] Automatic utterances, oral or written, the possibility of telepathic communications between two or more people, the relationship between hypnotism and telepathy were all questions deemed worthy of intensive study for what they might reveal about the workings of the human mind. Comparatively little space in the early volumes of the SPR *Proceedings* was devoted to physical manifestations, but virtually every volume, right down to World War I, contained papers on thought transference, automatic writing, trance speech, or Myers's pioneering work on subliminal consciousness. All of the articles that Gurney contributed to the *Proceedings* concerned mesmerism, hypnotism, hallucinations, or telepathy, so that, like Myers, the line between Gurney's contributions to psychical research and to the emerging study of psychology is often difficult to draw. In the first two years of the Society's existence, furthermore, its Committee on Thought-Transference produced four reports, more than any of the other five committees into which the initial work of the SPR was channeled. Indeed the entire first issue of the *Proceedings*, apart from the obligatory "Objects of the Society" and "First Presidential Address," dealt with nothing but thought reading. Within the SPR was gathered a corps of diligent workers who, through psychical research, wanted to reach conclusions about humanity equally as demonstrable as the conclusions of geologists, biologists, and physicists about the natural world.[32] For them, the study of the mind was the avenue along which to seek definitive insights about the human species. The distinction between mind and brain, the functioning of both, the nature of consciousness, and the definition of intelligence were all topics of profound importance to them, an importance that pertained neither to physiology nor to metaphysics alone. They were at once neurological, philosophical, ethical, and absolutely essential questions to pose – and, if possible, to resolve – if psychical research was to broaden the basis of self-knowledge.

One of the most diligent of the SPR workers, who fully shared the founders' concern for strict attention to detail as well as their preference for mental phenomena, was Sidgwick's wife, Eleanor Mildred

Balfour, sister of the prime minister. A skilled mathematician in her own right, she was also on occasion the able assistant of her brother-in-law, John William Strutt, third Baron Rayleigh, who was named Cavendish Professor of Experimental Physics at Cambridge late in 1879. For a period in the early 1880s, she was a principal aid in the redetermination of the electrical units of absolute measurement, the major project that Rayleigh undertook at the Cavendish Laboratory. The painstaking tasks demanded of her, the careful reading of instruments and the recording of figures, the complicated computations and verifications, were all work for which she was splendidly equipped.[33] When, a few years later, she became actively involved in the work of the SPR she saw no reason why the same set of skills should not be brought to bear on psychical research.

In summarizing the fifty years of time and service that Eleanor Sidgwick gave to the SPR, almost up to her death in 1936, a member of the Society wrote in 1958 that "it was in the mental phenomena of psychical research that Mrs. Sidgwick's main interest lay."[34] While she harbored the eminently practical capabilities needed as Principal of Newnham from 1892 through 1910, and as its Treasurer for forty years, she evidently found her greatest pleasure in abstract mental exercises.

> She told a friend, Miss Johnson, that mathematics especially appealed to her in early youth because she thought a future life would be much more worth living if it included intellectual pursuits. "I imagine," says Miss Johnson, "the abstract nature of pure mathematics seemed to her specially adapted to a disembodied existence."[35]

"Disembodied" is surely the key word here. Whatever afterlife Mrs. Sidgwick was able to imagine, it was an existence of pure intellect or mind, and her attention in psychical research was, accordingly, concentrated on whatever evidence could be marshaled for the independent operation of mind. The possibility "that the mind of one living person [could affect] the mind of another otherwise than through the recognised channels of sense" continued to intrigue her throughout her long life, and she came to treat mental telepathy as close to a proven fact.[36]

Years before the founding of the SPR, in his controversial Bampton Lectures at Oxford in 1858, Professor Henry Mansel had argued that man lacked the ability to ponder the unknowable, particularly the qualities of divinity and infinity. "The office of Philosophy," the final lecture warned in conclusion, "is not to give us a knowledge of the absolute nature of God, but to teach us to know ourselves and the limits of our faculties."[37] The Sidgwicks and their like-minded friends in the

SPR were all in favor of tracing the limits of man's faculties, but they were, at the core of their work, opposed to the view that such exploratory studies were irrelevant to cosmic issues. For all their important endeavors to strengthen psychology in its infancy, they were not fundamentally challenged by the idea of playing postnatal physicians to a new science. What aroused their curiosity and abiding commitment with regard to psychological inquiry was the hope that they might find, in certain hitherto unrecognized attributes of mind, the answer to their religious perplexities. Gurney, for one, was convinced that "it was useless to speculate" about life after death

> until many aspects of personality which, like hypnotism, had been insufficiently explored by science, or, like paranormal faculties in general, entirely neglected by it, had been submitted to a thorough, systematic investigation. This might *possibly* provide a solution of 'the Controversy of Life' more firmly based on fact than the traditional formulae, more satisfying to human emotions than the substitutes for religion then being offered.

Sidgwick, too, recognized the close links between inquiry into mental states and the theological questions so common in his circle of acquaintances. It was patent to him that psychical and psychological research could have a profound bearing on ethical and metaphysical problems, and he clung all his life to "the hope of a final reconcilement of spiritual needs with intellectual principles."[38]

So did Leaf, who had lost the evangelical faith of his youth when he arrived at Trinity College, Cambridge, in 1870, and began "to think and talk freely upon religion." He became "at best a theoretical Agnostic," he wrote many years later, "but with a strong natural bent to Theism of a rather intimate sort." He nurtured that strong natural bent with

> the purely provisional attitude that I had an instinct, which I could not justify, but which told me that true reality was to be found not in Matter, but in Mind; that the Universal Mind, which we called God, was imminent in all of us, and that the voice of conscience which we call Duty, was the voice of God.

As an undergraduate, he was unable to transform his provisional hypothesis into any emotional certainty; he remained torn between the agnosticism whose logic he could not deny and the pantheism that spoke to his "inmost heart." It is no surprise that, when Sidgwick introduced the talented young classicist to psychical research in the mid-1870s, Leaf turned to it with eager interest.[39] He fully agreed with the Sidgwicks, Myers, and Gurney that there, at last, was a subject that might challenge the cold logic of agnosticism and reassure its students both that human life had an intelligible purpose and the universe an ultimate meaning.

FORERUNNERS OF THE SPR

The Society for Psychical Research was not a novel organization when it was founded in 1882. On the contrary, it bore some resemblance to several ancestors, including the many spiritualist societies that had waxed and waned in Britain since the 1850s and some of whose members joined the SPR at its inception. Despite the declared intention of Henry Sidgwick, the first president, that the Society planned to pursue its investigations "with a single-minded desire to ascertain the facts, and without any foregone conclusion as to their nature,"[40] a number of spiritualists saw no qualitative difference between the SPR and a spiritualist association like the BNAS. Some changed their minds fairly quickly, however, and heatedly repudiated any kinship with the SPR. A less troubled parentage could be ascribed to earlier associations for psychical inquiry at both Cambridge and Oxford, particularly the Ghost Society, which E. W. Benson assisted in founding around 1850, and the Oxford Phasmatological Society, which originated among a group of University College undergraduates in 1879 and lasted until 1885. When still an undergraduate himself, Sidgwick had joined the Ghost Society, more formally known as the Cambridge Association for Spiritual Inquiry, and a youthful enthusiasm for ghost stories was apparently one route that led him to serious psychical research.[41]

He began to study spiritualist phenomena more systematically in the 1860s, but it was not until the following decade that he and a few close friends joined together in an informal study group to pursue the subject intensively. Their association was very much a precursor of the SPR, and the lessons that its members learned from their inquiries during the 1870s were eventually applied to the SPR in the 1880s. In the 1860s, however, Sidgwick's lessons were largely exercises in frustration and suspended judgment. As early as 1860, he wrote hopefully to his mother from London: "What do you think? To-night I am going to witness some spirit-rapping. I do not know the least what phenomena I shall see, but I intend to have as absolute proof as possible whether the whole thing be imposture or not." As he continued to investigate rappings and automatic writing in the following years, he came to realize that this was one area of study where absolute proof was virtually impossible to obtain. Although some mediums were doubtless charlatans, one could not therefore condemn the validity of all spiritualist phenomena, nor besmirch the good name of other mediums who were not easily dismissed as frauds. The most definite comment that Sidgwick could bring himself to make about his psychical research by 1867 was that "it gives life an additional interest having a problem of such magnitude still to solve."[42]

Sidgwick was notoriously unlucky as a psychical researcher. As William Crookes commented, several years after the philosopher's death: "There are some people so constituted that nothing psychic will take place in their presence. Prof. Sidgwick was one. In spite of repeated trials he never witnessed anything."[43] In the face of such a disheartening record, it is conceivable that by the mid-1870s Sidgwick's ardor for psychical inquiry might have cooled, but for a few sources of external encouragement that confirmed his determination to keep probing the strange world of séances. One such incentive came from Crookes himself – from the articles that he published in the *Quarterly Journal of Science* and other periodicals, including the *Spiritualist Newspaper*, in the early 1870s. While Crookes's endorsement of Florence Cook may have dismayed his scientific colleagues, the very fact that a highly respected scientist and FRS was willing to investigate the subject without predisposition to sneer seemed promising to Sidgwick. "No one," he informed his mother in July 1874, "who has not read Crookes's articles in the *Quarterly Journal of Science*, or some similar statement, has any idea of the weight of the evidence in favour of the phenomena." A. R. Wallace's defense of spiritualism in the *Fortnightly* in the same year further reassured Sidgwick that at least a couple of reputable scientists were beginning to recognize the importance of inquiry into allegedly spiritualist phenomena.[44]

Of more immediate importance to Sidgwick in stimulating his activity in that field of endeavor, however, was the enthusiasm of Frederic Myers. Sidgwick had privately tutored him in classics when Myers first went up to Trinity in the autumn of 1860,[45] but their friendship took the better part of the decade to blossom. Myers had to undergo a variety of emotional crises before he could appreciate Sidgwick's capacity for friendship without flamboyance and counsel without dogmatism. He had to pass from an overripe Hellenism, to an equally fervent Christian zeal, to end in a gradual disenchantment. "This came to me," he reported in his autobiographical sketch, "as to many others, from increased knowledge of history and of science, from a wider outlook on the world. Sad it was, and slow; a recognition of insufficiency of evidence, fraught with growing pain. Insensibly the celestial vision faded."

By the late 1860s, Myers had no fixed beliefs on which to anchor himself. Fleeting "moods of philosophical or emotional hope" yielded inexorably to "an agnosticism or virtual materialism which sometimes was a dull pain borne with joyless doggedness, sometimes flashed into a horror of reality that made the world spin before one's eyes." Although Sidgwick could face the possibility of a meaningless universe with a graceful courage that covered his profound uneasiness, the concept of cosmic void was intolerable for Myers. It was then, while a Trinity Fellow and College Lecturer in Classics, that Myers learned to value Sidgwick's guidance and advice, and turned to him as the man who seemed "to know in every problem where the possible answers lay." He asked Sidgwick if spiritualist phenomena might hold any answers to the difficulties that troubled both men, and, typically, "Sidgwick replied with modified encouragement."[46] In the early 1870s, Myers initiated his own study of the subject and threw himself into the work with a characteristic gusto that did not abate even after he joined the Education Department as a school inspector. A meeting with Stainton Moses in 1874 put the finishing touches on Myers's newest conversion. The following year Mrs. Jebb wrote tartly to her sister:

> Human nature is certainly a very credulous thing, and eager after novelty. Fred Myers is a complete convert to the existence of spirits able to materialize themselves through the presence of a medium, and he now spends all his time in sitting in these séances, most of which are failures.[47]

It was Myers who, in May 1874, urged Sidgwick to form the group that became the forerunner of the SPR.[48] This was a casual association with no fixed membership, but over the next six years or so it included, in addition to Sidgwick and Myers, a few of Trinity's brightest students. Edmund Gurney, who won great distinction in classics as an undergraduate in 1870–1 and was elected to a Fellowship in 1872, joined the endeavor shortly after it was launched. Walter Leaf had just received his degree, with a handful of classical honors to his credit, and was studying for the Trinity Fellowship examination early in 1875, when Sidgwick and Myers invited him to assist their investigations.[49] Arthur James Balfour, Gurney's undergraduate contemporary at Trinity, had a less lustrous university career, but Sidgwick, under whom Balfour read for the Moral Sciences Tripos and who deeply influenced Balfour's intellectual development, had no doubts about the versatility and independence of his student's mind. He was glad to have Balfour's collaboration in psychical research, perhaps not least of all because the séances held at Balfour's London home in the mid-1870s gave the professor a chance to meet Eleanor Balfour, who frequently participated. In 1871, another Balfour sister, Evelyn, had married John Strutt (soon to be Lord Rayleigh), and they, too, occasionally gave a hand to the proceedings. Sidgwick, Myers, and Strutt had all held Trinity Fellowships; Gurney and Leaf still did; Balfour, if not so traditionally promising a student, was nonetheless a graduate of whom his college could expect great things. Together they formed a select group of Trinity luminaries.

It is curious to imagine these bright lights of Cambridge intellect flickering in the murky company of Mary Showers, Annie Eva Fay, and Henry Slade. Yet these were only several of the mediums with whom Sidgwick and his colleagues experimented at séances between 1874 and the end of the decade. They met with Kate Fox Jencken and the much publicized Eglinton, among others, and not a little of their time, energy, and hope was expended on Annie Fairlamb and Catherine Wood, a pair of young Newcastle materialization mediums with whom varying members of the group sat in Newcastle, London, and Cambridge between 1875 and 1877. Sidgwick was initially enthusiastic about these young women, but his enthusiasm rapidly declined, as the "probability of fraud became painfully heavy." His wife summed up their mature opinion of Wood with devastating finality in her terse, somewhat impersonal prose: "The indications of deception were palpable and sufficient." She treated Fairlamb only slightly more charitably when she observed, "all that occurred was within the power of the medium." Even Leaf, an utter novice in psychical research, found their materializations highly suspicious. Myers retained his faith in the deceptive duo far longer than his partners in the inquiry,[50] but he, too, realized in the end that Fairlamb and Wood were not providing any useful clues for solving the riddle of human immortality. In fact, no such clues were forthcoming from any of the mediums, professional or private, with whom the group sat in the 1870s. At the turn of the century, shortly before his death, Myers recalled those early labors and pronounced them "tiresome and distasteful," yielding "unsatisfactory" results, "so contradictory, so perplexing, that we could neither feel sure that there was nothing discoverable, nor yet that any valid discovery had in fact been made."[51] What the Cambridge Trinitarians had ascertained, however, was that the principles that they espoused for their inquiries were valid in practice and worthy of being incorporated into a procedural method for psychical researchers. They had confirmed that strict testing was the *sine qua non* of legitimate investigations; they had learned to distrust the evidence of human observation and perception, and therefore sought to devise tests that did not require continuous observation of the medium or of the medium's séance paraphernalia.[52] Above all, with the exception of Myers, they had realized that skepticism was reasonable and optimism generally misplaced in a field where error, self-delusion, and potent motives for fraud all flourished abundantly.

The SPR was not, however, merely the descendant of older spiritualist and psychical research associations. It was, to some degree, also indebted for its identity to the tradition of discussion and debating clubs in British intellectual circles. That most exclusive of university clubs for exceptional young men, the Apostles, had been an outstanding feature of Sidgwick's formative years at Cambridge, as it was of Leaf's. The idea of a group of men meeting regularly to discuss, with utter frankness and without restrictions, questions of religious, philosophical, and ethical import remained immensely appealing, and both men recalled their Apostolic days with nostalgia bordering on veneration.[53] Sidgwick, in fact, seems to have been particularly attracted to societies of this nature – he was a member of one or another for most of his adult life. In the 1860s, he participated in the "Grote Club," an informal and "speculative gathering" of young Cambridge dons who met, initially, at the home of John Grote, Knightbridge Professor of Moral Philosophy between 1855 and 1866, and later in college rooms, in the company of Grote's successor, F. D. Maurice.[54] When the star-studded Metaphysical Society, most famous of all Victorian discussion clubs, was founded in 1869, Sidgwick soon joined and, for almost a dozen years, he remained one of the most devoted attendants at its meetings. The Metaphysical Society met "once a month, usually at an hotel, where, after dining together, a paper was read by some member, and afterwards discussed." Discussion among the likes of Gladstone, Tennyson, Ruskin, Maurice, Bagehot, Huxley, Morley, Clifford, J. A. Froude, Frederic Harrison, Archbishop (later Cardinal) Manning, James Martineau, A. P. Stanley, John Tyndall, and James Fitzjames and Leslie Stephen was an intense and serious matter. The Society, at its outset, gave itself no less an assignment than "to collect, arrange, and diffuse knowledge (whether objective or subjective) of mental and moral phenomena."[55] Sidgwick contributed six of his own essays to the proceedings and clearly relished his affiliation with an association that investigated "what those of every shade of faith and lack of it really did believe, and to what extent this left them with a common moral understanding and culture."[56]

Toward the end of his life, Sidgwick could not resist the attraction of yet another gathering of intellectuals, this time called the Synthetic Society, founded in 1896. Unlike the Metaphysical Society, the members of the Synthetic were all favorably disposed toward religion, if not by conviction at least by aspiration, and their goal was not merely the discussion and elucidation of their beliefs, but also the purposeful construction of "a new synthesis of religious positions and convictions." Sidgwick, clearly, belonged in a society that sought "a philosophical basis for religious belief," for that search had preoccupied him throughout his life. He had been fortunate always to have stimulating companions on the quest, and in the Synthetic these included A. J. Balfour, James Martineau, Albert Venn Dicey, James Bryce, R. B. Haldane, Oliver Lodge, Lord Rayleigh, Myers, R. C. Jebb, and G. K. Chesterton. Some were veterans of the Metaphysical Society; a few were Sidgwick's co-workers in the SPR. The latter helped to make the SPR a peculiarly hybrid creation: In part, it shared the concerns of traditional spiritualist organizations, testing mediums and pursuing ghosts; yet it also aimed, in part, to serve as a central depot for information about psychic phenomena, collecting, arranging, and diffusing that information with all the earnestness and sense of high purpose that had characterized the Metaphysical Society.[57]

These qualities also characterized Richard Holt Hutton, a fine example of a man whose interests led him to both the Metaphysical Society and the SPR. A founding member of the former, who presented no less than seven papers to its meetings,[58] he served, too, as one of the first vice-presidents of the latter. By profession, he was a journalist, and an extremely successful one during his long tenure as coproprietor and literary editor of the *Spectator* between 1861 and his death in 1897. By avocation, however, he was a theologian, and theological questions always held paramount interest for Hutton. Despite his demanding work as an editor, he managed to compose an impressive array of articles and essays on religious subjects, many of them for the *Spectator*, and drew from his own religious development the insights for much of what he wrote. Raised as a Unitarian in a family where both father and grandfather were Unitarian ministers, Hutton himself had studied for the ministry under James Martineau. His talents, however, did not lie in clerical directions, and he went to work for his religion in a different capacity, helping to write and edit two Unitarian publications, the *Inquirer* and *Prospective Review*, in the early 1850s. During the decade, however, he became increasingly estranged from Unitarianism as he suffered a crisis of faith brought on by the death of his young wife.[59]

His religious progress thereafter followed steadily along a path toward formal, authoritative religion. Eased out of Unitarianism into the Church of England with the transitional help of F. D. Maurice's theology, he remained an Anglican for the rest of his life, appreciating the fullness of the church's liturgy and embracing the doctrine of the Trinity as "the central truth of Christian Revelation."[60] So high church had he become by the end of his life that rumor, wrongly, had him converting to Roman Catholicism. There is no doubt, however, that he came to feel deep respect for Catholicism, recognizing in its "dogmatic impregnability . . . one of the surest guarantees of the survival of Christian theism."[61]

What drew Hutton to psychical research was not, as with a number of colleagues in the SPR, the longing to believe in some transcendental, universal, immaterial guiding hand. His belief in God, after the crisis of the 1850s, was steadfast, and in the debates of the Metaphysical Society, "he bore more of the burden of defending the theistic postulate than many of the churchmen." Hutton's peace of mind also extended to human immortality, about which many of his fellow psychical researchers sought reassurance.[62] If, however, he did not turn to psychical phenomena for religious comfort, they nonetheless served his religious interests in a fundamental way. Despite the spiritual serenity of his later years, he still pondered the age-old difficult questions about the identity of mind and spirit, the distinction between body and soul, and the ability of intellect truly to fathom the working of natural laws by empirical evidence alone. If any line of inquiry could suggest answers to these puzzles, he reasoned, it would be inquiry into the powers and scope of the human mind, such as the SPR pledged to undertake.

Like Sidgwick and Hutton, A. J. Balfour enjoyed pondering things theological and metaphysical in congenial intellectual company. He joined the Metaphysical Society just as its meetings were coming to an end in 1880, but he was a founder of the Synthetic[63] and, with Hutton, figured among the first vice-presidents of the SPR. He served as president of the latter in 1893-4, and if he was conspicuously absent from his duties, he could at least legitimately point to other demands on his time: As Conservative Leader in the House of Commons, and subsequently deputy prime minister to his uncle, the Marquis of Salisbury, he was in training for the top job himself, a position he achieved in the summer of 1902. Psychical research was nonetheless an abiding and genuine interest of his,[64] for politics never managed to occupy the whole of Balfour's attention, nor to satisfy his apparently insatiable appetite for abstract speculation. His niece and biographer observed fairly: "No one can understand Arthur Balfour, who forgets that interest in speculative thinking was part of the fabric of his everyday existence, wherever he was, whatever he was doing."[65] Psychical research came much closer to that aspect of his personality than legislative maneuvers.

Psychical research was especially germane when Balfour's thoughts took a religious turn. The fervent evangelical faith of his mother, Lady Blanche Gascoigne Cecil, had deeply colored his childhood, particularly after the early death of his father in 1856, and he was brought up in an atmosphere "saturated" with "religious convictions."[66] Although doctrinal disputes held little attraction for him, the broad philosophical and ethical issues at the root of religious faith never lost their appeal. Nevertheless, his was not the confident belief that his mother possessed, and Mrs. Jebb, when she first met him at Cambridge in 1880, was struck by his air of melancholy. She explained to her sister:

> I think the sadness arises partly from the fact that the spirit of the age prevents him, a naturally religious man, from being religious, except on the humanitarian side . . .
> All the Balfour family take hold of the end of religion they can be sure of, the helping of people here. Their mother, Lady Blanche, belonging to a different generation, was absorbed in dogmatic religion, and they inherit her unworldly nature, without the power of her unquestioning faith, so they miss her happiness.[67]

Arthur Balfour was not, however, quite so passive in his spiritual malaise as Mrs. Jebb implied. He wielded his own carefully crafted rapier on behalf of religion, and if it lacked the blunt thrust of the Christian crusader's sword, it was as sharp as his subtle mind could make it. His principal contributions to the literature of science *contra* faith – all, of course, weighted toward the latter – were *A Defence of Philosophic Doubt* (1879), *The Foundations of Belief* (1895), and the Gifford Lectures delivered at the University of Glasgow in 1914 and 1922–3. His volumes of essays and addresses are also liberally sprinkled with many briefer excursions into theological fields. Throughout his work, Balfour consistently refused to acknowledge that science and religion could be at cross-purposes, that the former could fatally undermine the latter. They were, he argued in the *Defence of Philosophic Doubt* and in numerous later writings, completely separate systems of belief, neither of which could claim to be totally verifiable by empirical evidence or the dictates of reason. They should, and could, be complementary, he claimed, not locked into "immutable and perpetual antagonism."[68] Despite the misleading title of the earliest major attempt to define his position, he never defended doubts about the existence of God and human immortality. He defended, instead, the validity of doubting that scientific methodology provided the only legitimate way to make inquiries about man and the universe. With Sidgwick, he questioned whether the empirical approach to knowledge, at its most fundamental level, was more firmly based on facts than the supposedly nonrational assumptions of religious faith.

For Balfour, belief in a personal God was the pivot around which religious convictions turned. In an address to the Church Congress in 1888, he propounded the advantages of a God-centered religion over a man-centered one,[69] and in the first set of Gifford Lectures, he explained precisely what the concept of God meant to him. When "I speak of God," he told his audience,

> I mean something other than an Identity wherein all differences vanish, or a Unity which includes but does not transcend the differences which it somehow holds in solution. I mean a God whom men can love, a God to whom men can pray, who takes sides, who has purposes and preferences, whose attributes, howsoever conceived, leave unimpaired the possibility of a personal relation between Himself and those whom He has created.

Balfour, like Hutton, was incapable of basing his religious principles on the love of "a mere law of any kind."[70] No deity was worthy of the name God, he maintained, unless the faithful could enter into some kind of direct communication with Him.

Nor could Balfour dispense with the doctrine of human immortality. In language that again clearly underscores Sidgwick's influence on the development of his ethical theory, Balfour argued that only the conviction of "another phase of existence in direct moral relation with this one" could reconcile the occasional, but inevitable, conflict between rational self-love and "the disinterested love of man."[71] His trust that life on earth was not everything predated the horrors of World War I by many years, but it was that catastrophe that prompted the simplest expression of his faith. When Lady Desborough lost two sons in the fighting, he wrote to her in August 1915:

> For myself, I entertain no doubt whatever about a future life. I deem it at least as certain as any of the hundred-and-one truths of the framework of the world, as I conceive the world. It is no mere theological accretion, which I am prepared to accept in some moods and reject in others. I am as sure that those I love and have lost are living today, as I am that yesterday they were fighting heroically in the trenches.

Apparently, then, Balfour did not need the SPR to prop up a sagging faith, nor to afford the evidence without which he could enjoy no peace of mind. He would have welcomed irrefutable proof of immortality, but doubted that the results of any psychical inquiries and experiments ever afforded that proof.[72] Besides, the philosophy that he constructed did not demand irrefutable demonstration of religious truths; indeed, it recognized the impossibility of obtaining such evidence.

Balfour's theology was grounded, not on sublime certainty, but rather on the conviction of man's spiritual needs. Again and again, his arguments reduced themselves to this: Human life was meaningless and valueless without religious faith. Religion was worth fighting for because it was an indubitable "benefit" to mankind. For the sake of that benefit, Balfour could deny that biblical scholarship had invalidated portions of Scripture as holy writ and could contend instead that it had made the Bible "far more a living record of the Revelation of

God to mankind," "a more valuable source of spiritual life now than it could ever have been in the precritical days." With no belief in God, Balfour warned, humanity was left "face to face with the unthinking energies of nature which gave us birth, and into which, if supernatural religion be indeed a dream, we must after a few fruitless struggles be again resolved." He could not accept such a postscript to human endeavor on earth and, seeking to avoid a "hopeless pessimism," asserted that he could not "conceive human society permanently deprived of the religious element."[73]

Considering the alternatives, Balfour purposefully chose to believe in God and life after death, but he never forgot that a choice was possible. Although he did not rely on the SPR to justify his own choice, he also did not hesitate to impute great importance to its work. The investigations of the SPR, he suggested in his presidential address early in 1894, offered "the best starting point from which to reconsider, should it be necessary, our general view, I will not say of the material universe, but of the universe of phenomena in space and time." As an undergraduate at Trinity, he had already noted how "absurd" it was "for any philosopher to start with the assumption that the 'positive' or naturalistic view of the universe was the best to which we could rationally aspire"; he was, accordingly, overjoyed that the inquiries of the SPR, particularly into mental telepathy, were revealing "that there are things in heaven and earth not hitherto dreamed of in our scientific philosophy."[74]

Among the strangest things undreamed of in scientific philosophy were the automatic writings, cross correspondences, and trance pronouncements collected primarily from four automatists over a period of about thirty years, commencing in 1901. These materials bore directly on Balfour, since one of the alleged spirit communicators was Mary Catherine Lyttelton, a young woman whom Balfour deeply loved, who had died of typhus on Palm Sunday, 1875, before they had become officially engaged. Balfour himself was not made aware of her posthumous attempts to communicate with him until 1916, and he was not immediately persuaded of the authenticity of the transmitted messages. He was, however, apparently convinced before he died in 1930; toward the end of his life, he even tried to send a message in reverse back to Mary through Mrs. Winifred Coombe-Tennant, one of the principal automatists, who wrote under the name Mrs. Willett. The "Palm Sunday" scripts, nonetheless, did not dramatically alter Balfour's religious outlook. He had believed in the immortality of the soul independently of concrete proof and did not rush to embrace the evidence that Mrs. Willett offered. One assumes that the alleged messages from

Mary brought him great solace, but they were not needed to effect a spiritual conversion.

The significance of the Palm Sunday case is perhaps greater for psychical research in general than it was for Arthur Balfour in particular. That some three thousand scripts, written over three decades by several different automatists, should, when sifted and studied in minute detail, seem to develop in conjunction, using a similar and elaborate symbolism, is on the face of it absolutely extraordinary. The four automatists were not all acquainted (although two were closely related, Margaret Verrall and her daughter Helen). Even more important, none of them knew of the highly secret romance between Mary Lyttelton and Arthur Balfour at least until 1911. Certainly some of the later scripts, after Mrs. Willett had become friendly with Balfour's younger brother, Gerald, may have been influenced by her growing knowledge of the family. The Verralls, too, through their association with the SPR, had known Sidgwick, Myers, and Gurney, all of whom were said to participate as spirit communicators. The Verralls had also heard of another Balfour brother, Francis Maitland, killed in 1882 in an Alpine climbing accident, who was likewise identified as a major communicator. The fourth important automatist, Mrs. Alice Fleming, who took the pseudonym Mrs. Holland, was the sister of Rudyard Kipling and early in the twentieth century lived in India with her husband, an army officer. She was acquainted with Gerald Balfour, but had little intimate knowledge of his family.

To emphasize the links, however tenuous, between the automatists and the hidden subject of their scripts is not to offer any explanation for the startling fact that those writings showed an astonishing interconnectedness and consistency of theme and imagery. Taken together, they dealt with a number of topics, including the founding of a better world order, but they centered preeminently on the attempts of Mary Lyttelton, helped by a band of sympathetic spirit intelligences, to convince Arthur Balfour of her continuing love for him. Any theory of the automatists' subconscious minds at work has to be stretched literally to the breaking point to cover the Palm Sunday case, particularly for the first decade when the four women, by all accounts, knew nothing whatsoever of the Lyttelton-Balfour relationship. Nor is the likelihood of collaboration in the production of so massive a collection of intricate scripts very strong, especially given Mrs. Holland's vast geographic distance from the other writers. The only remaining loophole to explore is the possibility that the men and women, including Gerald Balfour and Eleanor Sidgwick, who interpreted and collated the scripts were subconsciously compelled by their own knowledge of the story to put the fragmentary pieces of the puzzle together in a particular way. But in 1912, when they first began to try to make sense out of the scripts, what would lead them to remember a romance nearly forty years old? And did even they know all its details, since their brother guarded his affections so closely? It is possible that collusion, fraud, and self-delusion played their part in the Palm Sunday case, as in so many other incidents in psychical research, but, for once, it is highly improbable, given the number of people, the sheer volume of material, and the span of time involved.[75]

The implications of the cross correspondences cannot be lightly dismissed, although they are still a long way from proving the reality of communications from the dead. Gerald Balfour was one psychical researcher, however, who had no problem accepting the idea that Mary Lyttelton was sending messages to his brother. He discovered the fascination of psychical research later in life than Arthur had, but it became for Gerald a far more preoccupying interest than it ever was for the prime minister. Although a member of the SPR almost from the start, for years Gerald played no active role whatsoever in its work, as he confessed when he became its president in 1906. He nonetheless possessed the most sterling credentials: A fine classical scholar, he had attended Trinity where he was elected to the Apostles and to a Fellowship in 1878. His mind, clearly, ran along philosophical grooves, as Mrs. Jebb reported in 1879: "He is full of earnest thoughts, works hard for hours every day at Mental Philosophy and Metaphysics." Walter Leaf recollected that when he visited Gerald's Florentine villa in 1883, his host "after breakfast . . . used to retire to his thinking room, trying to work out his system of ultimate categories of thought."[76] Although Gerald abandoned this awesome task and, like Arthur, immersed himself in Conservative politics for a while, he always remained strongly attracted to metaphysical inquiry. After serving as MP for Leeds for twenty years, Chief Secretary for Ireland, and President of the Board of Trade, he retired from politics in 1906, and thereafter his interests in psychical research blossomed. Much of his time was devoted to the study of Mrs. Willett's mediumship and to the deciphering of the Palm Sunday scripts. He gained at second hand the insights that the Sidgwicks had learned through long, tedious, and disillusioning séances, and like them he placed little credence in the physical manifestations of spiritualism, concentrating instead on the philosophical and psychological implications of automatic writing and speaking. He made his most extensive contributions to the literature of the SPR after 1914, but even before the outbreak of World War I he had articulated his strong supposition that "direct telepathic action between mind and mind" was widespread. In subsequent years, after he had studied the baffling cross correspondences of the Palm Sunday scripts and worked out their meaning to his own satisfaction, he did not hesitate to suggest "some intelligence or intelligences not in the body" as possible participants in the telepathic process, nor to derive from his studies the confidence that individual consciousness continues after death.[77]

EARLY TENSIONS WITHIN THE SPR

Whether or not they came to the SPR from serious intellectual discussion clubs, the group of men who presided over the early history of the Society were, by any measure, distinguished academically, politically, and socially. Henry Sidgwick, one of the leading British philosophers of the nineteenth century, was its first president in 1882 and, except for 1885–7, remained in office through 1892. The man who filled his shoes in the brief interim was Balfour Stewart, FRS, astronomer and professor of physics as Owens College, Manchester. Before World War I, other presidents of the Society included the Balfour brothers, William James, Eleanor Sidgwick, Andrew Lang, W. Boyd Carpenter, and Henri Bergson. Among the first eight vice-presidents were two Members of Parliament, Arthur Balfour and John Robert Hollond, the Liberal MP for Brighton who had been at Trinity with Myers. In addition to R. H. Hutton, the earliest group of vice-presidents also included William F. Barrett, professor of physics at the Royal College of Science, Dublin, and the Hon. Roden Noel, son of the Earl of Gainsborough, Sidgwick's close friend at Trinity, amateur philosopher, minor poet, founding member of the Metaphysical Society, and Groom of the Privy Chamber to the Queen between 1867 and 1871.[78] Gurney and Myers participated on the Society's first council, and in addition to army officers, clergymen, and JPs, initial membership attracted Charles Lutwidge Dodgson (Lewis Carroll), Lord Houghton (formerly Richard Monckton Milnes), Francis Burdett Money-Coutts, and Sir Baldwyn Leighton, Bt., Conservative MP for South Shropshire.

In the following years, its clientele became even more illustrious as membership in the Society expanded – the number of members, honorary, corresponding, associate, and regular, surpassed nine hundred by 1895. The honorary members in 1887, for example, included Gladstone, Ruskin, and Lord Tennyson, with the President of the Royal Academy, Sir Frederic Leighton, as a later addition.[79] The membership list for 1895 bore such names as Mary Sidgwick Benson, wife of the Archbishop of Canterbury; Lady Brooke, Ranee of Sarawak; the Marquis of Bute; the Earl of Carnarvon; the Earl of Crawford and Balcarres; Mrs. Leopold de Rothschild; Henry Morton Stanley, famous for his encounter with Livingstone in darkest Africa; and J. J. Thomson, FRS, Rayleigh's successor as Cavendish Professor of Experimental Physics at Cambridge.[80]

One could go on dropping SPR names indefinitely, but the point has already been sufficiently stressed: The SPR was no ordinary spiritualist association, but had a social and intellectual cachet all its own. Certainly, the snowball effect was at work to produce for the Society such a collection of professional and hereditary titles. Once Sidgwick had agreed to serve as president, his name and connections in church and state helped to attract other influential figures, like Arthur Balfour, to whom Sidgwick wrote in February 1882, just as the SPR was being launched, delicately urging him to accept a vice-presidency in the Society in order to encourage others to join.[81] It was not merely the presence of eminent academics and rising politicians, however, that gave the SPR its special appeal. Someone like Lord Rayleigh or Professor Thomson would not have been willing to serve on the Society's council in the 1890s if its intention to follow scientific procedures had not been firmly enunciated, and the SPR made every effort to preclude doubts on that critical point. In setting forth its aims in the first issue of the *Proceedings*, the SPR specifically called itself "a scientific society"; Sidgwick was careful, in all his early presidential addresses, to emphasize the need for the most painstaking, rigorous accumulation and analysis of evidence, urging his audience to employ "methods as analogous as circumstances allow to those by which scientific progress has been made in other departments." In private, he admitted to Myers his conviction "that if, where religion and philosophy had failed in establishing certainty, Science were to fail also, 'the human race . . . had better henceforth think about these matters' (the basis of morals, the government of the universe) 'as little as they possibly can.'" But he did not allow these fears to darken the public face of the SPR, and Bishop Carpenter was far from the only member whose interest was first "aroused by the studiously scientific spirit which it claimed."[82]

Given the intellectual achievements of many of the earliest members of the SPR and the scientific training of a few, its claims to acceptable scientific methodology must be treated more respectfully than similar claims by scores of spiritualist associations across the country. It is nonetheless important to recall that such claims were standard practice in the spiritualist movement of the late nineteenth century, and the fact that the SPR was founded with similar pronouncements and fanfare hardly seemed much of an obstacle to spiritualist membership in the new society. On the contrary, convinced spiritualists were involved in the SPR from the moment of its conception. This occurred late in 1881 when William F. Barrett was visiting the Finchley home of E. D. Rogers. For the past several years, Barrett had been carrying on experiments in thought transference, with results that, by 1881, seemed promising even to so obdurate a doubter as Sidgwick. In a late night conversation at Finchley, however, Barrett confided to Rogers his disappointment that "scientific and literary men of influence" had hitherto shown so little interest in any systematic study of thought transference, or of any other question related to "psychic phenomena and so-called Spiritualistic manifestations." Rogers, in reply, "suggested that a society should be started on lines which would be likely to attract some of the best minds which had hitherto held aloof from the pursuit of the inquiry." Barrett jumped at the proposal and thereafter tended to assume that the idea had been his own, although, on Rogers's death, he did acknowledge that "the original impulse . . . to the foundation of the Society for Psychical Research" had come from Rogers.[83]

With the energy of initial enthusiasm, Barrett dispatched a number of invitations to attend a planning conference on 5 and 6 January 1882, to be held at the BNAS headquarters in London.[84] One meeting led to another, and in February the SPR was formally constituted, with a thorough mix of committed spiritualists and cautious psychical researchers to assist its birth pangs. It had not been easy to marshal the support of the leaders of each camp. Sidgwick was not desperately eager to return to the tedium of the séance table, but Professor Barrett's work had given him some new grounds for hope, and he accepted the presidency, believing that there was no work "of equal moment to be done." With Sidgwick at the helm, Myers and Gurney joined as well.[85] But Stainton Moses proved a harder catch, not because he disapproved of the new society's goals, but, oddly enough, because he could not conceive of BNAS members amalgamating smoothly with the Sidgwickians. In a letter to Barrett, dated 12 January 1882 and understandably marked "*Private*," he had nothing but scorn for the BNAS, an organization that he was, at that very moment, abandoning to its inglorious demise. "When you invited me to attend a Conference my first impulse was to decline," he informed Barrett.

> Socially, I thought that such men as you and I wd desire to see governing a Society, & directing its aims, wd not find themselves able to act with the average BNAS man. And I felt pretty sure that their members wd either feel uncomfortable and retire, or wd be found impracticable.

Moses, furthermore, doubted that so heterogeneous an organization could publish a journal to satisfy all members.

> Men who associate themselves with you will not print wht they may have to say in a paper wh., like *Light*, wd frequently contain matter

> distasteful to them: . . . Such papers must cater for their readers; a
> class quite distinct fr. those you care to reach.

"I believe," he concluded, "this strongly marked intellectual and social distinction will be found to be insuperable. If we try our best to
use existing tools we shall fail." How or why Moses finally overcame
his scruples is not clear, but it is very likely that Rogers was mainly
responsible for changing the mind of his spiritualist colleague.[86] In any
case, Moses became one of the first vice-presidents and council members of the SPR.

A number of other confirmed spiritualists joined them, both as officers and members. The elderly Hensleigh Wedgwood, Darwin's cousin, became a vice-president after participating actively in the prelim-
inary planning of the Society. His great interest and life's work had
been the study of philology – his pioneering *Dictionary of English Et-
ymology* was published in 1857 – but in the 1870s he had also become
absorbed in spiritualism and had joined the BNAS.[87] C. C. Massey
and E. D. Rogers were on the Society's council in the company of
other BNAS veterans, like Dr. George Wyld, F. W. Percival, Alexander Calder, and Morell Theobald, a London accountant whose manifest credulity is exhibited on every page of his *Spirit Workers in the
Home Circle: An Autobiographic Narrative of Psychic Phenomena
in Family Daily Life*.[88] Other familiar spiritualist names, such as Lady
Mount-Temple, Dr. Stanhope Templeman Speer, St. George Stock,
Alaric Alfred Watts, and the Hon. Percy Wyndham, MP, appear on
early membership lists.[89] Moses's fear of an obvious social distinction
between the two groups within the SPR appears largely unfounded.
The spiritualists who joined were, with few exceptions, from the middle
and upper classes, and had been well educated.

Nor were the differences in perspective, of which Moses warned,
insuperable barriers to communication between spiritualists and psychical researchers in the SPR. Mrs. Sidgwick's biographer described the
Society as "planned and constituted on lines at once very wide and
very strict,"[90] and it was the width that first impressed the spiritualists.
In announcing the creation of the SPR on 25 February 1882, *Light*
commented:

> It will be seen that, while the new Association seems intended to
> embrace a field of inquiry almost bewildering in its extent, it will in
> no way interfere with, or supersede, the BNAS, or any of the existing
> societies for the pursuit of Spiritualism. Indeed, it is calculated to
> help them, by encouraging and systematising the work of a large num-
> ber of students and inquirers into cognate subjects, the investigation
> of which we, as Spiritualists, believe is calculated to lead the searcher
> after knowledge to an acceptance of Spiritualism.[91]

A year later, *Light* was still confident that the SPR was performing yeoman service on behalf of spiritualism. "Spiritualists cannot doubt what the end will be," announced an article on 3 February 1883, " – they cannot doubt that, as time goes on, the SPR will afford as clear and unquestionable proofs of clairvoyance, of spirit writing, of spiritual appearances, and of the various forms of physical phenomena as they have so successfully afforded of thought-reading."[92]

But somewhere between 1883 and 1886 *Light*'s prediction went awry. The scenario that unfolded in those years did not bring spiritualists and psychical researchers into ever greater unanimity of purpose and outlook, but brought into sharper focus the possibilities for misunderstanding that existed between them. It was then that the strictness of the SPR became more apparent to the spiritualists than its breadth of interests, and some spiritualists found the Society altogether too confining. Already in 1883, *Light* had lamented that the SPR was "studying the mere bones and muscles, and [had] not yet penetrated to the heart and soul" of spiritualism, but the weekly still held out hope of real conversion for the Society.[93] By 1886, that hope had faded, following the SPR investigation of William Eglinton, one of the most renowned physical mediums of the late nineteenth century.

By 1886, Eglinton's mediumship had passed through its more flamboyant and dramatic phase of the previous decade, when spirit figures obligingly materialized at his séances, objects flew through the air with the greatest of ease, and the medium allegedly levitated in the best tradition of D. D. Home. Unfortunately his reputation, unlike Home's, was slightly blemished, and there were incidents in Eglinton's earlier career that suggested an inclination "on his part to pass off conjuring performances as occult phenomena."[94] It was his skill as a slate writer, however, that members of the SPR undertook to examine; they did so partly in response to pressure from the Society's spiritualists who were dissatisfied with the group's emphasis on mental phenomena. Reports of Eglinton's success in obtaining slate writing under the most improbable conditions easily rivaled reports of Slade's accomplishments ten years before, and Eglinton's talents seemed a likely subject for investigation. Mrs. Sidgwick pursued the topic with her usual thoroughness and with the able assistance of Richard Hodgson. A young Australian with a doctorate of law from Melbourne University, Hodgson had decided on further education at Cambridge. He entered St. John's in 1878 and became Sidgwick's student in moral sciences. His astute powers of observation made him a valuable, if sometimes abrasive, member of the SPR, which he joined in 1882.[95] Hodgson, in turn, received invaluable help from S. John Davey, an associate of the SPR, whose early admiration for Eglinton waned as Davey himself learned to produce the same effects by conjuring. A spate of articles in the SPR *Journal* and *Proceedings* during 1886 and 1887 punctured Eglinton's reputation, as Mrs. Sidgwick, Hodgson, and Davey all stressed the fallibility of human observation and memory in order to explain Eglinton's success in diverting attention from his sleight of hand expertise. Even brazen substitution of slates (whereby clean, locked slates were whisked away and replaced by others already bearing long messages in fine handwriting) was possible if the medium maintained "a constant stream of chatter," which kept the sitter's eyes on his face instead of his hands.[96]

Eglinton, naturally, took personal offense and urged his admirers in the SPR – including Massey, Moses, Noel, Wedgwood, Wyld, and Wyndham – to abandon the Society forthwith.[97] Some kind of break appeared inevitable, and Mrs. Sidgwick, for one, had no regrets. With little interest in the physical phenomena on which the spiritualists pinned their fervent hopes, she was inclined to share "the contempt with which scientific men generally" treated those phenomena. It is hardly surprising, therefore, to find her writing privately to her husband in 1886:

> I really think the spiritualists had better go. It seems to me that if there be truth in spirit[m] their attitude and state of mind distinctly hinder its being found out . . . we are better and stronger without them, so that if they wish to go I should not like to hinder it . . . and people who fly into rages are such a bore . . . Their spirit is theological not scientific, and it is so difficult to run theology and science in harness together.[98]

Perhaps to Mrs. Sidgwick's disappointment, however, there was no mass exodus of spiritualists from the SPR in protest against the harsh treatment of Eglinton. What did occur was the departure of Stainton Moses with a handful of loyalists, like Dr. Speer and A. A. Watts. Already in 1885, Moses had expressed his impatience with all the SPR fuss over telepathy and the unconscious,[99] and the Eglinton exposure may simply have provided the convenient issue over which to break with an organization that no longer offered Moses a congenial atmosphere for the exercise of his abilities. He was never much of a team player in any case, and, by the time he left the SPR, his own London Spiritualist Alliance gave him the chance for undisputed leadership, which the SPR would not afford him. There is a nice little irony to the fact that the author of that snobbish and supercilious letter, written to Barrett in January 1882, was among the first of the former BNAS members to find association with the SPR "impracticable."

Even Moses's exit, however, did not provoke a spiritualist stampede from the SPR. Dawson Rogers, although he "withdrew from the Council in consequence of the attitude which [he] thought the new society evidently desired to take up in reference to Spiritualism,''[100] retained his affiliation with the SPR. Wyndham remained a member, as did Massey, Lady Mount-Temple, Hensleigh Wedgwood, and A. R. Wallace, among others. Arthur Conan Doyle and W. T. Stead joined in the 1890s. There is, in short, nothing to suggest that the SPR became an organization overtly hostile to spiritualists. All that happened was that leadership of the Society by the Cambridge friends became ever more strongly confirmed in the absence of any substantial challenge to their control. Even after the death of Sidgwick and Myers at the turn of the century, Mrs. Sidgwick continued to keep her hand in SPR affairs, showing a flair for politics worthy of a Balfour.[101]

INCONCLUSIVE INQUIRIES

It may well be that spiritualists and psychical researchers could sustain their uneasy alliance within the SPR because the Society never reached any definitive conclusions about the subjects that its members investigated before World War I. If the SPR reached no final conclusions on these critical issues, however, it was certainly not through any lack of trying. Its records attest to an astonishing expenditure of time and effort; its *Proceedings* and *Journal*, published from 1882 and 1884 respectively, reveal the thoroughness with which it undertook the task of psychical research. At the Society's foundation, six committees were established to share the work at hand; they dealt with thought reading, mesmerism, Reichenbach's experiments (the work of the German investigator who invented the theory of an odic force), apparitions and haunted houses, physical phenomena, and the accumulation of information about the history and incidence of psychical occurrences. The latter job, assumed by the Literary Committee, proved the most overwhelming, or perhaps it was that this committee's honorary secretaries, Myers and Gurney, had an insatiable appetite for work. Letters flowed freely from the committee – over ten thousand in 1883[102] – as its members amassed quantities of material concerning cases relevant to the Society's inquiries.

Hundreds of those cases were categorized and analyzed in the famous study by Gurney, Myers, and Podmore, *Phantasms of the Living*, published in 1886 under SPR auspices. The book was an epic attempt to examine

> all classes of cases where there is reason to suppose that the mind of one human being has affected the mind of another, without speech uttered, or word written, or sign made; – has affected it, that is to say, by other means than through the recognised channels of sense.

It was, indeed, an attempt to prove that telepathy, a word which Myers coined in 1882, was "a fact in Nature." The authors particularly focused on apparitions of dying people, perceived by geographically distant friends and relatives who had no knowledge of the impending demise. They argued that such apparitions occurred with "a frequency which mere chance cannot explain"; instead, the authors sought to elucidate these apparitions as hallucinatory images produced through the varying operations of telepathy. The *Times*, although it devoted a leader to the volume on its publication, was critical of the argument and expressed a common frustration when it observed that to explain the phenomena "by telepathy [was] surely, in the present condition of our knowledge, to beg the whole question."[103]

Groundbreakers, however, rarely have an easy time, and the weaknesses of *Phantasms of the Living* do not invalidate the volume's claim to pioneering status in the study of mental telepathy. "Since Gurney's time," Dr. Gauld has observed, "every serious discussion of crisis apparitions has taken its start from his classification and arrangement of them."[104] The work *was* primarily Gurney's. Although a number of psychical researchers, preeminently Podmore, helped him to interview perceivers of apparitions and to collate the mass of collected materials, and although Myers composed the introduction to the study, Gurney wrote and researched most of the two-volume enterprise. By 1883, Gurney had abandoned his successive attempts to establish a career in music, medicine, and the law, and had begun to dedicate the full scope of his wide-ranging talents to the SPR. Other members of the Society gave generously of their time and money; the SPR, in fact, might not have survived but for its independently affluent members, like Sidgwick and Myers, who could afford to subsidize a variety of its endeavors. Gurney, however, was the only one of them for whom the SPR became a professional commitment, his only source of employment, albeit unpaid. Sidgwick remained a university professor, Myers a school inspector, Leaf a banker, Balfour a politician, but Gurney became a psychical researcher first and last. For five years he was at the center of SPR activities, editing publications, superintending investigative projects, and serving as the Society's honorary secretary. Coming from a family of some means – his grandfather was a Baron of the Exchequer, his uncle Recorder of London[105] – Gurney was able to lead an unsalaried existence, pursuing the subjects that aroused his interest.

His interests were catholic, but by the 1880s they were tending to focus, directly and indirectly, on psychological questions. When he reluctantly realized that he lacked the abilities to excel either as a composer or a performer of music, he could at least theorize about it, in *The Power of Sound*, published in 1880, and in numerous articles and essays, such as "The Psychology of Music." When, in 1881, he ended his medical studies because he found the clinical training unendurable, he could, once again, turn his experiences into speculative channels and write "A Chapter in the Ethics of Pain" and "An Epilogue on Vivisection."[106] As these titles suggest, the presence of suffering in the world was, for Gurney, an acute concern. In his alternating moods of joyous enthusiasm and bleak melancholy, he never lost "the profound sympathy for human pain, the imaginative grasp of sorrows not his own, which made the very basis and groundwork of his spiritual being."[107] If he could not mitigate human suffering through beautiful music or skillful medicine, he could at least try to establish, on philosophical and psychological grounds, reasons for not finding the universe wholly evil.

The only reasons that seemed at all cogent to him centered on the possibility of a future life, the chance that the "oppressed" human spirit could feel with some assurance "that this is *not* all." Gurney, although lacking any "strong personal craving for a future life" himself, well understood how some hope of immortality could make a painful existence bearable.[108] To defend those grounds, to fortify them against the attacks of the positivist and scientific materialist, seemed to Gurney an uncontestably worthwhile task. He could readily justify devoting the rest of his life to such an effort, because it catered, not to his own personal needs, but to the profoundest needs of his fellow man.[109]

Gurney's contributions to the Society's investigations during the 1880s concentrated particularly on hypnotism and telepathy. He was attracted to these subjects because he believed they could be tested in a way that the highly suspect and erratic physical phenomena of spiritualism could not. With his capacities for analysis, "intellectual insight, penetrating criticism, dialectic subtlety," Gurney needed a subject that appeared capable of precise, repeatable experimentation and verification. At the same time, however, he also needed one that could satisfy his broad philosophic interests, that allowed him to speculate and theorize while he tested and calculated. The human mind, then, was the perfect subject for Gurney, and he made "the borderland between physiology and psychology" his particular domain. With the theological or metaphysical import of his work never intrusive, but always implicit, he devised several series of experiments in telepathy, both with and without the use of hypnotism. Sometimes the experiments featured the transference of what are now called "target" drawings or diagrams from an agent to a percipient; at other times numbers, colors, or names were tested in telepathic transfer.[110]

The agencies that worked to affect the mind remained the focus of Gurney's studies throughout the 1880s. He sought to render them less elusive to the psychical researcher, not only through the exploration of telepathy and hypnotism, but also through his study of hallucinations and his inquiry into the operation of memory, which he believed to be the key to personality.[111] Much of his work on hallucinations appeared in *Phantasms of the Living*, where Gurney set forth his intricate and "question-begging" theory of collective hallucination. Together with a paper on the subject published in the SPR *Proceedings* and another read at a Society meeting, his contributions amounted to "an elaborate survey of the psychology of hallucination which has an independent value."[112] "I believe," wrote Mrs. Sidgwick about Gurney, "he knew more about hypnotism than anyone else in England. Also about subjective hallucination. And his knowledge of Physiology and Psychology specially qualified him for the work."[113]

The work ended abruptly with Gurney's sudden death in June 1888, at the Royal Albion Hotel, Brighton, at the age of forty-one. When his body was found in his hotel room on 23 June, all the evidence pointed to death by an overdose of chloroform. What the evidence did not indicate was whether or not the overdose was accidental, and much turbulence was generated in SPR circles in the 1960s when Trevor Hall published *The Strange Case of Edmund Gurney*, a book that more than suggested suicide. Hall belittled reports that Gurney had inhaled the anaesthetic to relieve the pain of facial neuralgia and to induce sleep; he argued instead that Gurney intentionally gave himself an overdose when he discovered egregious deceit on the part of two men who had played central roles in his 1882–3 experiments in thought transference. Of these two, one – Douglas Blackburn – was a Brighton journalist with apparently no redeeming personal characteristics. The other, however, was George Albert Smith, a young resort entertainer whose performances at Brighton had featured mesmerism and alleged mind reading, but who had subsequently become Gurney's private secretary. Hall contended that Blackburn and Smith systematically tricked Gurney and Myers in 1882 and 1883, during a series of experiments held both in Brighton and at the SPR headquarters in Dean's Yard, Westminster. In June 1888, Hall's argument continued, Gurney learned that the men had used a code to simulate the operation of thought transference; he was devastated by the discovery because it not only destroyed his confidence in Smith, a valued co-worker, but, more catastrophically, it also destroyed all Gurney's work in which Smith had assisted. Believing that his most important endeavors had been rendered meaningless and even ridiculous, Gurney committed suicide, leaving a wife and small daughter.[114]

It is certainly true that Blackburn, in 1908 and 1911, announced in the press that he and Smith had successfully deceived the SPR back in the 1880s. Smith vehemently denied his former partner's statements,[115] but their truth or falsity is not really the central issue in Hall's argument. What one needs to know – and what Hall cannot say with certainty – is whether Gurney unearthed any trickery before his death, and, if he did, whether the revelation would have prompted so drastic a response as suicide. Gurney's personality, which today would probably be labeled manic-depressive, harbored traits that suggest he was perfectly capable of taking his own life, but Hall's reconstruction of events falls far short of proof.[116] The mystery is not likely to be resolved almost a century later, and it remains to be seen whether Gurney's tireless activities should be judged ultimately fruitful or futile. He was assuredly motivated by the noblest desires to contribute to a science of the human mind, but those desires may have led him sadly astray, prompting him to trust the untrustworthy.

After his death, his colleagues at the SPR carried on a number of projects and investigations with which Gurney had been identified. Mrs. Sidgwick continued his inquiries into the relationship between hypnosis and telepathy, using none other than G. A. Smith as hypnotist. She was also involved in the Census of Hallucinations that the SPR undertook between 1889 and 1894 in order to strengthen the statistical foundations of the conclusions reached in *Phantasms of the Living*. Her assistants in this undertaking included Frank Podmore who, with Myers, became joint honorary secretary of the SPR after June 1888. Podmore's affiliation with the Society dated from its origins. Still only in his twenties when he served on its first council, he was one of the Society's youngest active members. He remained on the council for more than two decades and managed to combine time-consuming SPR investigations and committee work with his responsibilities as a senior clerk in the Secretary's department of the Post Office. His youthful attraction to spiritualism, which had led him to the BNAS, gave way to an ever more critical outlook, until he became the SPR "'sceptic-in-chief' concerning spirit agency, and the official *advocatus diaboli* when the society undertook to adjudicate on the claim to authenticity of spiritualistic phenomena."[117] Although his university education bore the stamp of Pembroke College, Oxford, he had much in common with Sidgwick, Myers, and Gurney, particularly in that he outgrew the Anglican faith of his clerical father. Like them, his search for an alternative set of intellectually and emotionally satisfying beliefs led him to the SPR. It also inspired him to become a founding member of the Fabian Society.

Before his involvement with the Fabians, Podmore participated in two earlier associations that had tried to mix moral renewal with social and political reform. The first, the Progressive Association established late in 1882, was a gathering of well-educated young men who had abandoned Christianity and who met Sunday evenings in Islington to listen to "ethical sermons, political speeches and secular hymns." Havelock Ellis also joined; so did Edward R. Pease, a lapsed Quaker, a young stockbroker with a social conscience, and, like Podmore, a member of the infant SPR. Podmore and Pease were both temporarily impressed by Thomas Davidson, a peripatetic and pretentious scholar of humble Scottish birth, who collected disciples as he roamed over Europe and America creating his own synthesis of various philosophic systems. In the wake of one of his periodic visits to London, Podmore and Pease participated in discussions to form a second group, the Fellowship of the New Life, which was designed, in some unspecified way, to realize Davidson's philosophy in a social utopian setting. It was a hopeless endeavor. The several planners in the autumn of 1883 could not agree on the necessary components of utopia, although most of them concurred in rejecting Davidson's proposal for a monastic order of like-minded disciples. Pease and Podmore favored an organization devoted, not just to spiritual communion among a chosen few, but to political activity and social reconstruction as well. It was Podmore who sponsored the proposals, aired in January 1884, that led to the foundation of yet another society, separate from the Fellowship, yet sharing some of its members for a time. It was Podmore, a classical scholar at Pembroke, who suggested the name Fabian for the new society, based on "a dubious political reference to the Roman general Fabius Cunctator" and his famous delaying tactics against Hannibal.[118]

Yet Podmore's break with Davidson's devoted followers was not at first decisive. On 31 December 1883, he wrote an almost wistful letter to Davidson explaining his reasons for not joining the Fellowship and for contemplating a splinter society.

> I have not subscribed to the articles of the Fellowship because I am not yet ready to do so . . . I feel so uncertain what is meant by religion, that I don't like to use the word – at present. I wish to learn: & I have not yet learnt enough to enable me to sympathise with the creed of the Fellowship heart & soul . . . I am afraid of binding myself in a moment of enthusiasm to what I only half-feel, & half believe . . . I am not at all sure that I feel religion yet: & I think, therefore, that my aspirations shall remain unexpressed, kept for myself alone, until I am more assured of them.

A year later, when he reported to Davidson on the activities of the Fabian Society, he reminded the philosopher that "We look upon you as our founder."[119]

While he busied himself with the affairs of the Fabian Society, serving on its executive and coauthoring with Sidney Webb its tract on *Government Organisation of Unemployed Labour*, Podmore did not forget those unexpressed aspirations. For a time, he seems to have associated them with Davidson; perhaps he stressed Davidson's connection with the Fabians in order to persuade himself that his creation had spiritual values underlying its advocacy of social, economic, and political reorganization. But as the Fabian Society fell under the dominance of Webb, Bernard Shaw, Sydney Olivier, and Graham Wallas, Podmore devoted more and more of his spare time to the SPR. With his appointment as its joint honorary secretary in 1888, the realignment of Podmore's concerns was confirmed. Henceforth psychical research was to be his major preoccupation.[120]

Unlike Gurney or Myers, Podmore was not a pioneering expounder of provocative theories about psychical phenomena; his skepticism, strongest during the 1890s, kept him from playing such a role. It did not, however, inhibit him from writing widely in the field of psychical studies, and for a broader public than the SPR's rather technical *Proceedings* enjoyed. His *Modern Spiritualism* is still among the best surveys of the movement up to the turn of the century. In another volume, *Studies in Psychical Research* (1897), he characteristically offered a number of natural explanations – including fraud, coincidence, poor observation, and weak memory – for allegedly supernatural occurrences. The victims of his criticism could at least derive comfort from the fact that he was scrupulously fair: He seemed to distrust the observational skills of witnesses as much as the integrity of mediums.

Podmore's caution regarding the physical manifestations of spiritualism amounted very nearly to an obsession, but he shared with other SPR leaders the conviction that mental telepathy was a genuine phenomenon. In an earlier work, *Apparitions and Thought-Transference*, he presented "a selection of the evidence upon which the hypothesis of thought-transference, or telepathy, is based,"[121] and he returned to the subject repeatedly in his subsequent work. He toyed with the idea that telepathy was one of several "long lost but once serviceable faculties" which man, ages ago, used consciously, but had since "learned to acquiesce in as beyond his guidance."[122] Although he conceded that psychical researchers had not yet satisfactorily explained how telepathy worked, he was hopeful that they would do so, and the overarching goal of his own psychical research could be summarized in the title of yet another of his books, *The Naturalisation of the Supernatural*.[123]

During the final ten years or so of his life, Podmore was deeply impressed with the SPR investigations of the mental mediums Mrs. Piper and Mrs. Thompson, as well as with the early cross-correspondence scripts of Mrs. Verrall, Mrs. Holland, and others. In his last book, published the year of his death, he clearly distinguished the "revela-

tions made in trance and automatic writing" from the physical phenomena of spiritualism, which he persisted in assigning entirely to
fraud. Although he still believed that the feats of mental mediumship
could be ascribed to telepathy between the living, and although he
continued to reject the possibility of communication from the dead, he
came to exclude at least a few mental mediums from his scornful dismissal of their profession in general. Indeed, he came to the conclusion
that no "imaginable exercise of fraudulent ingenuity, supplemented by
whatever opportuneness of coincidence and laxness on the part of the
investigators, could conceivably explain the whole of these [trance and
automatic] communications."[124]

Podmore was fully aware that the concept of telepathy was being
stretched ever more tautly to explain the otherwise inexplicable. If
one ruled out the spirits of the dead as agents in the work of mental
mediums, he argued, the only possible remaining cause was "the
agency which has been provisionally named telepathy, but which no
one has yet ventured to define in other than negative terms, as communication apart from the recognised sensory channels." Podmore
himself could provide no further definition, although he offered, in *The
Newer Spiritualism*, a gloss on the subject that strikes a curious note
in the body of his writing. Describing a telepathic experiment with
cards, in which the percipient's senses were blocked from normal functions, Podmore commented: "If, under such conditions, he correctly
names the concealed card, the result may be due to reading our
thoughts, it may be due to communion with the world soul, or to the
direct interposition of the Deity. In any event, we have agreed provisionally to call the result telepathy."[125] Podmore was not given to
such phrases as "the world soul"; nor did he pepper his prose with
references to the Deity. It is possible, of course, that he was indulging
in a little tongue-in-cheekery when he wrote those sentences, but nothing else in the book's concluding chapter suggests a similarly light
touch. It is tempting to speculate that Podmore, to an extent which he
could not articulate, ultimately invested in the concept of telepathy
some of the hopes he had once entertained of contributing to the ren-
ovation of human society. In telepathy, perhaps, he came to perceive
a chance for deeper understanding and sympathy among people than
utopian communities, Owenite or Davidsonian, could ever hope to in-
augurate.

Just before Henry Sidgwick died of cancer in August 1900, Podmore
wrote him a letter in which he paid tribute to the beneficial influence
of Sidgwick's life and character on his own development. "I am not
sure now that I very much care," Podmore concluded, "whether or
not there is a personal, individual immortality. But I have at bottom

some kind of inarticulate assurance that there is a unity and a purpose in the Cosmos: that our lives, our own conscious force, have some permanent value – and persist in some form after death." Maybe this was as close as Podmore ever came to feeling religion, as close as his spiritual aspirations came to the expression he had withheld from Davidson more than fifteen years earlier. In 1900, in any case, he sounded optimistic about the ultimate meaning of human existence.[126] Ten years later, he also seemed optimistic that his work to probe the functioning of mental telepathy might number among those "many crusades" that prove "worth the pains, not for those who took part in them, but for the later generations which have entered into the fruit of their labours."[127] His concern for the inheritance of later generations had impelled him since young adulthood, although it had hovered ambivalently between concrete social and abstract moral goals. It had led him, on the one hand, to draft a statement on unemployment and to study Robert Owen's blueprints for New Harmony. It also drove him, on the other, to examine the interaction of mind with mind and to probe the hidden processes by which people might communicate with one another.

When Podmore's body was discovered on 19 August 1910, drowned in the New Pool, Malvern, the evidence for suicide was at least as strong as the case for accidental death. As with Gurney, the circumstances of Podmore's demise are open to rich conjecture, and Trevor Hall joined the fray with insinuations of homosexuality and financial ruin.[128] If the suicide verdict is the correct one, it is possible – although his last book hardly suggests it – that telepathy's elusiveness made him despair of ever attaining certainty in his inquiries. Perhaps his vision of a unified cosmos faded, together with his confidence that telepathy held the key to discoveries immensely fruitful for humanity. If he quite purposefully drowned himself in the New Pool, perhaps he had also acknowledged that his own criteria for valid investigative procedures ruled out the emotional commitment which, for years, he had been tacitly bringing to his psychical research.

Although telepathy and hypnosis remained the preferred subjects of investigation for several leading SPR members before World War I, the Society did not ignore other kinds of psychical phenomena. There was, for example, an inquiry into second sight in the Scottish Highlands, suggested and subsidized by the Marquis of Bute.[129] There were frequent visits to haunted houses to interview witnesses and to catch a ghostly glimpse – without any verifiable success. And there were sittings with Eusapia Palladino, the renowned Neapolitan medium, who unsettled the Society periodically between 1894 and 1910. Palladino, a lusty, illiterate Italian peasant, seems to have stimulated the prejudices latent in the Sidgwick group, not only against physical mediums, but also against mediums whose background was not socially acceptable.[130] Nevertheless, the SPR continued to be intrigued by Palladino and returned to investigate her questionable physical mediumship on several occasions. That the Society did so indicates its determination to leave no psychical stone unturned.

The SPR connection with Palladino began in 1894, when Myers, Lodge, and the Sidgwicks held séances with her in southern France, as guests of Charles Richet, professor of physiology in the Paris Faculty of Medicine. Myers and Lodge were certain that they were witnessing the real thing, at least some of the time. The Sidgwicks were, as ever, guarded and noncommittal, but not unwilling to have another round with Palladino. She was, accordingly, invited to Cambridge for a long visit in the summer of 1895.[131]

The phenomena that allegedly occurred in her entranced presence were not spectacular by spiritualist standards: Loud raps sounded, tables levitated, furniture and small objects moved about, sitters were touched, seized, or poked, and musical instruments played under their own power. Less frequently, curtains billowed when the windows behind them were closed, and a few sitters reported seeing odd, limblike growths protruding from the medium's body. Fundamentally, however, Palladino produced the classic repertoire of spiritualist mediums: strange sounds and mobile furnishings. The fame that she was gaining by 1895 arose, not from the originality of her performance, but from the fact that distinguished scientists, doctors of medicine, and budding psychologists from across Europe had been unable to prove, once and for all, that her manifestations were merely conjuring tricks. Evidence that she freely used her hands and feet to help the show along, despite the best efforts of sitters assigned to restrain her energetic limbs, did not always seem relevant. There were times when even the Sidgwicks were reluctantly compelled to admit that she baffled them. In this quandary, during the 1895 Cambridge sittings, they sent for Richard Hodgson, who was in Boston serving as executive secretary of the American Society for Psychical Research, affiliated with the British SPR at the time. With Myers and Sidgwick paying for his transportation, Hodgson crossed the ocean in August, in time to render judgment on Palladino. Her talents, he ruled, included nothing more noteworthy than the ability, through a variety of deceptive movements, to wriggle hands and feet free from the control of sitters. That agility, Hodgson was convinced, explained all her so-called spiritualist phenomena, and as far as the Sidgwicks and Podmore were concerned, Hodgson's was the last word on Eusapia Palladino. Their worst suspicions had been confirmed; there was nothing further to add.[132] At a meeting of the SPR

in October 1895, Sidgwick pointed out that "she had attempted to produce results by fraudulent means. On this they were all agreed. There
was some difference of opinion as to the extent to which previous
experiments were vitiated." Edward T. Bennett, secretary to the SPR,
describing the meeting to Lodge, continued: "For [Sidgwick's] own
part he considered (at least so I understood him to say) *all* previous
results were in his estimation so far vitiated, that the small portion
remaining unexplainable was practically valueless."

Myers and Lodge, however, could not abandon Eusapia and the
hopes that she had aroused. Bennett reported that Myers also spoke
at the same October meeting, "but in so low a voice that he was only
indistinctly heard at the bottom of the Room. He seemed to *feel* the
position more than anyone else; and after the Mtg several spoke to me
of *sympathy* with him."[133] Myers could rarely remain downcast for
long, and discouragement gave way to his more customary optimism
once again in 1898, when early in December Richet invited him to Paris
to resume sittings with Palladino. "Phenomena last night *absolutely
convincing*. Not a shadow of doubt," Myers wrote ecstatically to
Lodge, and subsequently added, "The second séance was even better
than the first." Lodge's heart may have warmed to the news, but Sidgwick and Hodgson remained unmoved. The former reminded Myers
that the evidence of Palladino's trickery in previous investigations was
"overwhelming";[134] Hodgson, then editor of the SPR publications,
barred Myers from publishing any lengthy discussion of his recent sittings and thoroughly supported Sidgwick in the "attempt to put that
vulgar cheat Eusapia beyond the pale."[135] There she stayed as far as
the SPR was concerned until 1908.

She might have stayed there much longer had not Sidgwick and
Hodgson both died in the next few years.[136] In 1908, the SPR council
allowed itself to be persuaded that the time had come to reopen the
Palladino file, since her twentieth-century séances, with investigators
in Genoa, Naples, and Paris, had once again raised the tantalizing possibility of genuine physical phenomena. Under SPR auspices, three
men were dispatched to Naples: the Hon. Everard Feilding, son of the
Earl of Denbigh and the Society's honorary secretary; W. W. Baggally,
also a member of the SPR council, as well as an amateur conjuror; and
the American psychical researcher Hereward Carrington. They were
capable and experienced investigators, without predisposition to believe in spiritualist marvels, and they understood fully the array of
trickery at the medium's disposal. Their favorable impression of Palladino in 1908 was, therefore, significant, and their verdict that many
of her phenomena were not attributable to trickery caused no slight
annoyance in London. Apparently Alice Johnson, the SPR research

officer, was especially insistent that deceit and faulty observation were at work, because Feilding finally exploded in a letter to her from Naples: "I wish to goodness you had come out when I wired so that instead of sniffing at us when we return you might be sniffed at yourself by Podmore."[137] Unfortunately for Feilding, Palladino soon afterward embarked on a trip to the United States where, under less than ideal conditions for serious investigation, her capacities for cheating were again exposed. In 1910, Feilding tried once more to obtain from her the manifestations that had so impressed him two years earlier, but instead he watched as she methodically tried to trick him.[138] Palladino made life very hard indeed for her admirers, and none could ever argue that her mediumship was beyond reproach. The best they could maintain was that, on occasion, genuine phenomena managed to appear in her presence, while at other times chicanery triumphed. Despite the obviously suspect nature of many Eusapian manifestations, her mediumship remains nearly as puzzling as Home's.

THE PREAMBLE OF ALL RELIGIONS

The men and women in the SPR who fancied themselves applying the rigors of science to the protean queries of psychical research might not have appreciated their inclusion in this study under the heading of surrogate faith. Yet religion was at the root of their inquiries, for religious yearnings – sometimes no more than a vague spiritual malaise – had played a role in bringing them together in the study of psychical phenomena. More than one intellectual historian has noted the "strong religious sensibility" inherent in mid-Victorian agnosticism,[139] and Sidgwick, Myers, Gurney, Podmore, Leaf, and Hodgson were no exceptions. Although repudiating orthodox Christianity, they longed to find some other basis for the ethical precepts they cherished and some reassurance that all human suffering was not utterly devoid of purpose. Implicitly they sought to use science to disclose the inadequacies of a materialist world view and to suggest how much of cosmic significance scientific naturalism failed to explain.[140]

As befitted people whose formative years had been molded in many ways by the tensions between science and religion, the Sidgwick group in the SPR was keenly alive to both the cogency of the materialist arguments and the urgent necessity to counter them. The resolution toward which its members groped was a masterful attempt to have and eat their cake simultaneously. They were quick to convey the shortcomings of materialism, but could not quite bring themselves to jettison the interpretative framework that scientific naturalism had imposed on their world. Rather, they aimed to naturalize the supernatural by inserting into that framework some nonmaterial phenomena, preeminently mental telepathy, thereby challenging the sole sufficiency of physical agents in the universe. They persisted in believing that they could operate as scientists and could preserve for themselves and their fellow human beings, on strictly scientific grounds, legitimate reason to endow man's life with a significance far transcending blood, bones, and tissue. Yet the desire to so endow human life was not, at base, a scientific one.

None worked harder at this effort, nor ultimately with greater conviction of success than Myers. For him, agnosticism became merely a passing state of mind, a period of "apathy in the brief interspace of religions," while the articulation of his mature religious beliefs occupied the last twenty-five years of his life. At the time of his death, he was still making alterations in the structure of his "final faith,"[141] but a paper that he presented to the Synthetic Society, in March 1899, at least suggested the sources from which he was constructing it. The foundation stone or fundamental concept, he explained, was "the coexistence and interpenetration of a real or spiritual with this material or phenomenal world." He found such a belief in pantheism, Platonism, and mysticism, as well as in Christianity, for that faith, too, in the resurrection of Christ, celebrated the triumph of the spirit world over matter. Myers's admiration for Buddhism was likewise evident throughout the paper and contributed to the wide-ranging eclecticism of his synthesis. Contacts with the Theosophical Society in the preceding two decades had, furthermore, exposed Myers to an array of occult theories. Absorbing all these influences thirstily, he exclaimed that "man cannot be too religious."[142]

Myers is an easy target for ridicule. He was so unguarded in his enthusiasms, so willing to emphasize the positive and minimize the negative findings in all his investigations. If any psychical researcher could be accused of succumbing too eagerly to the "will to believe," it must be Myers. Especially after the suicide in 1876 of a woman whom he adored, but could not marry – his cousin's wife, Annie Marshall – he was desperately anxious for convincing evidence that she was getting in touch with him from beyond the tomb.[143] Yet at the same time, he was not utterly incapable of exercising his critical judgment. The year following Mrs. Marshall's death, for example, he wrote to Barrett about an alleged clairvoyante whom he had visited: "I think she also is disposed to fish for hints and pretend to see accordingly." He knew that to believe in the face of evidence was to be guilty of "resolute credulity" and to ignore the canons of scientific methodology that he highly valued.[144] If he dreaded a meaningless universe, he was, in that respect, no different from countless numbers of his contemporaries whose Christian faith had been shattered. But he did differ from most of them in that he devoted years to designing a new religious faith with what he thought were the sturdiest building blocks available: the fruits of ancient wisdom and the findings of modern psychology. Although he distorted and misinterpreted the latter in the interests of upholding the former, his goal was always to win a secure niche for the human soul in the world of natural science.

The cement that held together the scientific, metaphysical, and mystical elements of his final faith was telepathy, a word of his own invention and a concept to which he had granted the status of scientific law by the mid-1890s. It is important for an appreciation of Myers's thought to realize that he did not cite the workings of telepathy in order to refute scientific materialism, but rather to contend that science still had much to learn about the cosmos. The expansion of intellectual horizons that he desired was to take place within the context of a scientific world view, not in bitter opposition to it. There was no reason, he argued, why scientists could not fundamentally enhance religious beliefs as they went about making their discoveries. He, for one, wanted psychology to serve as "a science of the soul," and the very juxtaposition of the words "science" and "soul" suggests how thoroughly Myers came to ignore the lines of demarcation between science and faith established during the course of the nineteenth century. He hoped to see the day when "the old opposition, even the old distinction, between Science and Religion [would] melt away," and he did his very best to hasten the melting process.[145] Religion, according to Myers, was nothing more than "the sane and normal response of the human spirit to all that we know of cosmic law," and telepathy was part of that law, demonstrated empirically in the best scientific fashion and leading to profound religious insights. After all, he observed, "that same direct influence of mind on mind," which telepathy revealed "*in minimis* would, if supposed operative *in maximis*, be a form of stating the efficacy of prayer, the communion of saints, or even the operation of a Divine Spirit."[146] Whereas adherence to restricting religious orthodoxies, including Christianity, had failed the human race, Myers believed that a natural religion based on cosmic law could answer humanity's most fundamental spiritual needs. Such a faith, in fact, made science and religion into the two diverse sides of one and the same coin.[147]

Science always remained a highly malleable discipline for Myers. With him, as Lodge observed in a memorial address to the SPR following his colleague's death in 1901, "the word science meant something much larger, much more comprehensive: it meant a science and a philosophy and a religion combined. It meant, as it meant to Newton, an attempt at a true cosmic scheme."[148] Indeed, Myers's religious temperament always pulled him toward the universal, where the splendor of his vision frequently obscured the clarity of his thought. Few scientists, outside the SPR, could have bothered to read his theories of cosmic integrity, which during the 1890s brought him close to Eastern faiths and neo-Platonic aspirations in his glimpses of the individual spirit merging "ineffably in the impersonal All."[149] One wonders how much even his sympathetic scientific co-workers within the SPR could tolerate. Although he was given to lofty discussions of matter and energy, Myers failed to relate them in any adequate way to the component parts of his cosmic scheme. His mind may have imagined itself at home in a scientist's laboratory, but Myers's heart always yearned for a church. In a sense, the universe became his place of worship, a vast cathedral where material and spiritual were intertwined and where Myers thrilled to recognize that "the Primal Life has . . . woven our lesser lives into one shining fabric."[150]

Myers, in his most exalted moments, completely abandoned the language of the scientist. His colleagues in the inner circles of the SPR rarely allowed themselves such loss of self-control and never came as close to embracing outright spiritualism as Myers did. Indeed, Dawson Rogers assumed that Myers *was* a spiritualist by the end of his life,[151] and others were similarly persuaded. Sidgwick, by contrast, clung to the "attitude of sceptical neutrality" which he claimed for himself in the mid-1890s, because he saw no compelling empirical evidence to clinch the case for human immortality.[152] Sidgwick was not even absolutely certain about telepathy between the living, although it seemed surer to him than any other psychical phenomenon. As he wrote in his journal, in July 1888: "I have not much hope of our getting out positive results in any other department of our inquiry, but I am not yet hopeless of establishing telepathy."[153] It is true that during the 1890s, through his work on the Census of Hallucinations, he edged closer to endorsing the existence of telepathy, but he was never convinced that spirits of the deceased were necessarily involved in the telepathic process. Chronic skepticism was not a state of mind that Sidgwick particularly relished. He did not find that serving reason offered much emotional satisfaction and was "conscious of hankerings after Optimism," which, if indulged, he observed in a fascinating confession to J. R. Mozley, would lead him to Roman Catholicism.[154] Needless to say, Sidgwick never indulged his hankerings that far. Attractive though it might occasionally have seemed to drown all his doubts in a doctrine that commanded adherence, his was not the temperament nor the intellect of Cardinal Newman.

Yet despite the fact that Sidgwick "reached a purely negative conclusion" in his attempt to provide persuasive proof of immortality,[155] there remained a fundamental buoyancy in his outlook. Reverend

Charles Gore, canon of Westminster and a prominent member of the Synthetic Society, recalled after his death that Sidgwick

> always seemed to expect that some new turn of argument, some new phase of thought, might arise and put a new aspect upon the intellectual scenery, or give a new weight in the balance of argument . . . The quality of his mind was profoundly different from ordinary scepticism: for it was inspired by a fundamental belief in the attainableness of positive truth.

As he liberated himself from the clutches of an impossibly rigorous empiricism, he came to accept the multiplicity of avenues to truth. Toward the end of his life, Sidgwick could commend the work of the late Lord Tennyson as a man who let science shape his thoughts about the physical world, but who would not accept that world without God. That the first Tennysonian attitude may have undermined the second no longer seemed to disturb Sidgwick. In 1898, he could announce, with only one qualifying phrase, that since sociology was incapable of answering man's "deepest questions," "the need of Theism – or at least some doctrine establishing the moral order of the world – seems to me clear."[156]

For Sidgwick, the leap between recognizing the need of theism and claiming that telepathy definitively rescued theism from materialism was never absolutely out of the question. He devoutly hoped for such a rescue and warmly welcomed the evidence that seemed to support the hypothesis of telepathy between the living, but by the time of his death he was no wiser than he had ever been about what awaited him in the future. By contrast, his wife came to anticipate life after death, in part thanks to the intricate puzzle of the cross correspondences. Mrs. Sidgwick's tough-mindedness, so irritating to a number of spiritualists in the 1880s, softened after the turn of the century. In her presidential address to the SPR in May 1908, as she praised the Society's investigations of telepathy, she singled out for comment "the automatic writing of Mrs. Verrall, Mrs. 'Holland,' and others, and the trance writing of Mrs. Piper." The evidence from these automatists, she warned, "needs careful and critical study," but she was certain of its significance.

> Those . . . who follow the work of the Society carefully will, I think, perceive that in these scripts we have at least material for extending our knowledge of telepathy. They will probably be disposed further to admit that the form and matter of the cross-correspondences that occur between the different scripts (produced at a distance from one another) afford considerable ground for supposing the intervention behind the automatists of another mind independent of them. If this be so the question what mind this is becomes of extreme interest and

importance. Can it be a mind still in the body? or have we got into relation with minds which have survived bodily death and are endeavouring by means of the cross-correspondences to produce evidence of their operation? If this last hypothesis be the true one it would mean that intelligent cooperation between other than embodied human minds and our own, in experiments of a new kind intended to prove continued existence, has become possible, and we should be justified in feeling that we are entering on a new and very important stage of the Society's work.[157]

Four years before her death in 1936, the SPR celebrated its fiftieth anniversary, and Mrs. Sidgwick, a venerable lady of eighty-seven, was named president of honor for the occasion. In the address that she wrote to mark the Society's jubilee – but which her brother, Gerald Balfour, read for her – she allowed herself to go so far as to observe, when discussing the cross correspondences: "The general effect produced by study of these scripts is that some intelligence behind the communications is acting by design." Her language was still careful, reserved, and unemotional, but she did permit Lord Balfour to make a dramatic announcement to the SPR audience. "I have Mrs. Sidgwick's assurance," he told his listeners," – an assurance which I am permitted to convey to the meeting – that, upon the evidence before her, she herself is a firm believer both in survival and in the reality of communication between the living and the dead."[158]

Back in 1886, Mrs. Sidgwick had complained of the difficulties inherent in trying "to run theology and science in harness together," but the SPR in fact continued the attempt right down to World War I. Some of the leading members assumed that science firmly had the upper hand, that the experimental method that they copied utterly precluded all occasion for "fraud, self-deception or incompleteness of data."[159] Others recognized the importance of allowing science to behave as the dominant partner, but trusted that it was a genuine partnership nonetheless. Gerald Balfour described the service that many SPR activists wanted science to perform for theology when he observed in 1906: "There is probably no subject which more keenly interests a majority of our members than that of the rational grounds for a belief in immortality."

The SPR never managed to provide those grounds – not, at least, to the satisfaction of so honest a participant in the effort as Henry Sidgwick. Those of his friends and family who thought otherwise did so for reasons that were intensely personal. What prompted Gerald Balfour, in the same speech of 1906, to wax lyrical about telepathy as "the manifestation in the world of spirits of the supreme unity of the Divine Mind" was not devotion to scientific procedure.[160] When Richard Hodgson became convinced of human survival, thanks to the American trance medium Mrs. Piper, the language that he used to convey his deep satisfaction with the universe echoed, not the scholarly scientist, but the intense Nonconformist he had once been before his crisis of faith in young adulthood. In a letter to an unnamed correspondent, written in 1901, one can hear him "giving his testimony at Methodist meetings" in earlier times. "Be of good courage whatever happens," he wrote,

> and pray continually, and let peace come into your soul . . . . Everything, absolutely everything, – from a spot of ink to all the stars, – every faintest thought we think of to the contemplation of the highest intelligence in the cosmos, are all in and part of the infinite Goodness. Rest in that Divine Love.[161]

Indeed, for all his excesses of enthusiasm and credulity, Myers was not far from speaking for most of his colleagues in the SPR when he addressed them as the Society's president in 1900:

> *To prove the preamble of all religions*; to be able to say to theologian or to philosopher: "Thus and thus we demonstrate that a spiritual world exists – a world of independent and abiding realities, not a mere 'epiphenomenon' or transitory effect of the material world – . . . " This would indeed, in my view, be the weightiest service which any research could render to the deep disquiet of our time.[162]

It was not the voice of detached scientific inquiry that spoke through these men.

5

Theosophy and the occult

THE LURE OF THE EAST

Spiritualism and psychical research not only served as substitute religions for refugees from Christianity in the late nineteenth and early twentieth century; both also had to define their positions with respect to the range of occult and mystical sects that studded the spectrum of Victorian and Edwardian heterodoxy. The relationships, on all sides, were complex and shifting. In theory, spiritualists and psychical researchers alike rejected the secretive and ritualistic elements of the occult tradition, proclaiming instead their absolute devotion to the standards of open, rational, empirical inquiry set forth by modern science. "The Spiritualism which has soothed so many," wrote the Glasgow spiritualist James Robertson, "does not run into the realm of the mystical. The rational Spiritualist looks with clear, open eyes at what is presented."[1] E. R. Dodds articulated at least one psychical researcher's objections to the occult, when he explained:

> The occultist, as his name betokens, values the occult *qua* occult: that is for him its virtue, and the last thing he will thank you for is an explanation. He is an intellectual anarchist, a rebel against the concept of natural law, and his unconfessed aim is a destructive one: he would like, if he could, to undermine the whole arrogant structure of modern science and see it crash about our ears . . . The genuine psychical researcher may feel this fascination, as I have sometimes done, but he has disciplined himself to resist it . . . Far from wishing to pull down the lofty edifice of science, his highest ambition is to construct a modest annexe which will serve, at least provisionally, to house his new facts with the minimum of disturbance to the original plan of the building.[2]

Both spiritualists and psychical researchers insisted that their inquiries were part of the mainstream of modern thought, not remnants of magical mumbo jumbo from bygone ages.[3]

Such declarations were easy to make, but difficult to implement. At heart, although few spiritualists and psychical researchers would admit it, they were groping for a knowledge that was beyond the scope of physical science either to confirm or to deny. The questions they raised, the very language they used, often had close affinities with the concerns and the vocabulary of occultists. If spiritualists and psychical researchers deplored the exclusivity inherent in occult sects, or grew impatient with the dogged study of venerable texts supposed to embody the key to all wisdom, there were nonetheless fundamental bonds linking them to the occult. Many spiritualists came to embrace an essentially occult view of the universe, an animistic vision of closely interconnected parts all bearing the mark of cosmic soul, or world force, or ultimate spirit. For the occultist, there is no sharp distinction between matter and spirit, tangible and intangible, and that frame of mind proved highly tempting, particularly to spiritualists with inclinations toward mysticism.[4]

Indeed the occult enjoyed a striking popularity in the late Victorian and Edwardian decades, and by no means only for spiritualists and psychical researchers. Although there has never been a moment in human history when magic has not exerted its fascination, some periods are more noticeably marked than others as far as public interest in the occult is concerned. The late nineteenth and early twentieth century was such a time, as triumphant positivism sparked an international reaction against its restrictive world view. In England, it was the age of "Esoteric Buddhism," of the Rosicrucian revival, of cabalists, Hermeticists, and reincarnationists. In the late 1880s, the Hermetic Order of the Golden Dawn first saw the light of day in London, and during its stormy history, the Order lured into its arcane activities not only W. B. Yeats, but also the self-proclaimed magus Aleister Crowley. Its founders included the extraordinary Samuel Liddell (MacGregor) Mathers, who spent years of his life poring over manuscripts of sacred magic in Paris and London.[5] Palmists and astrologers abounded, while books on magic and the occult sold briskly. Without doubt, much of the attraction of these and related subjects depended on the dominant role that science had assumed in modern culture, for Dodds was right when he identified the occultist's destructive anger against "the whole arrogant structure of modern science." It was an anger shared by many men and women who would not have dreamed of calling themselves occultists, and yet who resented the confidence and certainty with which science reduced nature's majesty to measurable quantities.

Among those who did specifically ally themselves in this period with one or another occult sect, dissatisfaction with the value system imposed by western science was often a paramount motive. Hargrave Jennings, for one, began his reverent history of an old occult sect, the Rosicrucians, with an assault on the vanity, the "intolerant dogmatism," of modern science.[6] Alfred Percy Sinnett, for another, one of the most active and prolific members of a new occult organization in Britain, the Theosophical Society, characterized modern science by its fanatical thirst for measurement, and its neglect of all knowledge that was not quantifiable. "That which is commonly called science," he observed,

> is exclusively "physical" science. It works with instruments made of metal, glass and so on, and has accomplished work that may be fairly termed sublime in its examination of what I will venture to call the outsides of things, but it always stops short in groping after a comprehension of their innermost essence.
>
> Its failures are most obvious when we deal with any of the mysteries of Nature that are associated with life.

Life, Sinnett believed, was not accessible to quantitative measurement, and its puzzles could not be probed by the methods of physical science. He deeply regretted that contemporary western scientists summarily dismissed older scientific traditions, such as astrology, for being incompatible with their own approach to the natural world. Thanks to their narrowness of vision, he argued, they were barring themselves from examining "a multitude of Nature's most interesting mysteries." "A problem must come within the range of laboratory experiment to *be* a problem for modern science," he complained, and yet how many of life's profoundest questions, he wondered, could be analyzed in the laboratory?[7]

What Sinnett was fundamentally attacking, of course, was not just the measuring mentality, but the omnipresent threat of materialism. "Dense materialism" he called it, "which cannot conceive of consciousness as anything but a function of the flesh and blood."[8] His wife, a staunch partner in the study of Theosophical wisdom, set out to show that Theosophy could serve to counter the modern "intellectual bias in favour of Materialism," and other Theosophists were similarly inspired to oppose "a brave front to Materialism."[9] The physician and Wagnerian William Ashton Ellis, who, having once attended a sick Madame Blavatsky, knew something about Theosophy, linked that movement to spiritualism and psychical research because all three shared the same goal: "to shake off this great pall of gross matter that shuts men off into separate prison cells of personal egoism."[10] Theosophists, needless to say, were not alone among occultists in seeking to liberate themselves from matter and materialism. The lure of the occult, from the 1870s to World War I, lay precisely in its antipathy to the strictly rational, empiricist outlook that was increasingly perceived as the hallmark of Victorian thought. Involvement in occult studies provided one means of challenging and of discarding a frame of mind that seemed to glory only in the concrete, the factual, and the substantive.[11]

Because much of the discontent that underlay the resurgence of interest in the occult during the late Victorian and Edwardian years arose as a response to the hegemony of science in occidental culture, it comes as no surprise to see the East emerging, in the eyes of the disgruntled, as the repository of true wisdom. If Christianity had been hopelessly compromised by its concessions to science, the Hindu and Buddhist faiths might still be studied for their ageless spiritual teachings. If Christian clergymen could no longer pretend to speak authoritatively for their times, there were wise men in the East whose learning applied to the human condition at all times. If the Bible had been criticized so minutely as to leave scant room for divine afflatus, there was still inspiration aplenty to be found in the *Bhagavad-Gita*. The East, ever exotic, mysterious, alien, was an escape from and an alternative to the shallow, externally-oriented culture of the West. Western scientists might examine the outsides of things, as Sinnett noted, but Eastern sages looked inward where, in the realm of essence, eternal truth resided. The members of Britain's "counterculture" in this period created the East, if not exactly in their own image, then as a reflection of their discontent with their own society.[12] In most cases, however, unlike the counterculture of a century later, British representatives of the movement were not drop-outs from that society. The majority of them were utterly respectable, hard-working professionals. If they succumbed to what J. N. Maskelyne scornfully dubbed "the artificial glamour surrounding Indian mysticism,"[13] it was not because they wished to undermine the socioeconomic bases of their culture, but because they found that their cultural milieu catered so inadequately to the needs of the spirit.

AFFINITIES WITH SPIRITUALISM

No one did more to encourage an artificial glamor surrounding Eastern wisdom than Helena Petrovna Blavatsky; nor did any other occult group in this period receive more public attention and press commentary than the Theosophical Society over whose infant destinies she presided. Friedrich Max Müller, the German-born orientalist and professor of comparative philology at Oxford, considered the movement significant enough in 1893 to compose a twenty-page article debunking its claims to any genuine foundation in Eastern religions. The following year Gladstone felt called upon to denounce the attitude toward atonement held by Annie Besant, the woman who succeeded Madame Blavatsky as matriarch of Theosophy, after the latter's death in 1891.[14] Clearly, Theosophy could not be ignored.

Founded in New York in 1875 by Blavatsky and her American partner, Colonel Henry Steel Olcott, the Theosophical Society was at once a product of the late nineteenth century and the beneficiary of centuries of occult thought. Blavatsky herself stressed the roots of her teaching in the venerable texts of the Far East, but the very term "theosophy" conjured up a rich variety of associations with the cabalist, neo-Platonic, and Hermetic strands in western philosophic and religious thought. Meaning "divine wisdom," or "wisdom of the gods," theosophy was a familiar term in the vocabulary of the occult long before Madame Blavatsky stamped it with the mark of her own impressive personality. Belief in the existence of specially initiated adepts, or of secret documents that held, in coded signs and symbols, the key to understanding nature's deepest enigmas, had haunted the fringes of European thought for centuries, tantalizing susceptible minds with the possibility of attaining truly godlike power over the natural world. C. C. Massey dubbed the Jewish cabala "a system of theosophy," while Hargrave Jennings used the label "theosophists" to describe the Paracelsists of the sixteenth and seventeenth centuries. The links between the new, Blavatsky brand of Theosophy and the older tradition related to Hermetic teaching were nicely encapsulated in Annie Besant's claim to have been none other than Giordano Bruno himself in a previous incarnation.[15]

When Madame Blavatsky produced her own corpus of wisdom, beginning with *Isis Unveiled* in 1877, she was by no means averse to drawing on time-honored occult beliefs. In fact, one of her critics, the author of *Isis Very Much Unveiled*, concluded that her philosophical erudition consisted of nothing more than a "rehash of Neo-platonist and Kabbalistic mysticism with Buddhist terminology."[16] It was the Buddhist terminology that particularly appealed to Blavatsky. Her religious or metaphysical preferences lay in the Orient, especially in remote Tibet, where she claimed to have received instruction in ancient learning from the Mahatmas, or Masters, who, she insisted, guided her along the bizarre path of her life. Max Müller acknowledged the Eastern provenance of Blavatsky's doctrines, but was utterly scornful of her scholarship. In the so-called "primeval wisdom" of Theosophy, he could find nothing

> that cannot be traced back to generally accessible Brahmanic or Buddhistic sources, only everything is muddled or misunderstood. If I were asked what Madame Blavatsky's Esoteric Buddhism really is, I should say it was Buddhism misunderstood, distorted, caricatured. There is nothing in it beyond what was known already, chiefly from

books that are now antiquated. The most ordinary terms are misspelt and misinterpreted.[17]

Fortunately there is no need to attempt an analysis of Blavatsky's personality here, to probe her motives, to measure the extent to which she revitalized Western occultism with her contributions from the East, or to evaluate the degree of fraud that props up the foundations of modern Theosophy. Whether she was the "greatest impostor in history," "a fibbing, cheating, variety performer," or "the most successful creed-maker of the last three hundred years,"[18] is not the issue at hand. Possibly Max Müller's generous estimation of her character was apt; perhaps she was, initially, "dazzled by a glimmering of truth in various religions of the world" and did crave "a spiritual union with the Divine," only to end by deceiving herself and her disciples alike.[19] That she resorted to trickery much of the time, in order to produce assorted "miracles," that she fabricated tall tales about herself in order to heighten the mystery and glamor of her past, is beyond doubt. Equally certain, however, is her striking success in capitalizing upon, and further promoting, contemporary fascination with oriental systems of religious and philosophical thought. The teachings of her supposed Masters may be only a rehash, mishmash, or muddle of many texts, both sacred and occult, but her mixture has proved to exercise an enduring appeal.

In launching Theosophy in the 1870s, Blavatsky also effectively capitalized on modern spiritualism. In no way was Theosophy more a child of the late nineteenth century than in its affiliations with the spiritualist movement whose origins antedated Blavatsky's Society by some twenty-five years. Indeed, it is difficult to conceive of the Theosophical Society meeting with much success at all had it not followed some of the trails already blazed by the spiritualists.[20] Not only did Olcott and Blavatsky themselves first become acquainted through their mutual spiritualist interests, but "during the first year of its existence, the English Theosophical Society continued to be recruited almost entirely, if not solely, from the Spiritualist ranks."[21] Yet, after 1875, Madame repeatedly expressed her hostility toward spiritualism and announced, in no uncertain terms, that she rejected the "crude theories" of modern spiritualists.[22] Uncharitable critics might well assume that, having failed to establish herself in the first rank of mediums, she decided to adopt even bolder means to achieve world fame. More sympathetic students of her life and work accept the theory that, from the first, her goal was to convert the materialists to the reality of spirit phenomena and that she embraced spiritualism until her Masters taught her to see how far removed from profound occult wisdom the silly séance phenomena actually were.[23] In any case, the relationship between Theosophy and spiritualism provoked considerable debate in the last decades of the nineteenth century, as spiritualists and Theosophists scrutinized their respective tenets to determine how they merged or clashed with those of the other camp.[24]

The greatest clash arose over that most fundamental of spiritualist beliefs – the reality of communication with identifiable spirits of deceased people. Spiritualism was, of course, predicated on the proposition that, after death, a person's spirit could remain in close touch with the living and could relay messages to them with the help of a medium. Theosophical denial of this principle, and denunciation of séance practices, seemd to many an angered spiritualist an attempt to cut the very heart out of their faith. But Theosophists had learned from Madame Blavatsky the dangers that followed all attempts to commune with spirits around the séance table. Not only did such efforts try to force the "Ego of the departed . . . back into earthly conditions" when it should be allowed to progress to higher planes of spirit existence, but the lesser forms of spirit life that could be summoned to séances might be of a decidedly unsavory nature, prolonged contact with which regularly ruined the health and sanity of mediums.[25] The astral plane, which Theosophy taught was intertwined with the plane of daily life, was far from a realm of pure spirit, as spiritualists might assume. On the contrary, Theosophists warned, the astral plane was inhabited by nonhuman "spooks, elementaries and elementals," as well as "cast-off lower principles of former men and women, helped by certain elementals to utilize the vital forces of the medium, masquerading as the personalities of such departed friends as the persons assisting at the *séances* desired to invoke." Suspicious of the comforting assurances so important to spiritualists, Theosophists closed the door on dear friends and beloved relatives, except in the rarest of cases, because the souls of decent people, they believed, characteristically departed for more rarified realms of existence. In doing so, these souls abandoned the astral plane to a variety of nonhuman spirits, or, much worse, to the spirits of the most loathsome and reprobate human beings which, remaining earthbound, positively delighted in the chance to deceive credulous séance sitters.[26] In short, Theosophists expected no good to come from séances, and much evil. They boasted that the spiritualism that they endorsed was something far grander than slate writing, table tilting, or rappings on furniture. Its purpose was not contact with the dead, but " the cultivation of the inner life and the systematic sacrifice of the lower instincts of our nature to the higher law."[27]

If spiritualists and Theosophists had only had the veracity of spirit messages to discuss, the chances for cooperation between the two groups would have been slight. On a wide range of other issues, however, Theosophical convictions were scarcely distinguishable from spiritualist arguments. At base, the two groups were partners in the struggle to convince contemporaries that spirit existed independently of matter. If Blavatsky was tactless enough to distinguish the true spiritualism of her Theosophy from the false spiritualism of the spiritualists, at least she made common cause with them in the formidable task of combating materialism.[28] Furthermore, their position on religion in general, and Christianity in particular, made Theosophical teachings congenial to that large body of British spiritualists who had abandoned the Christian faith. In principle, the founders of Theosophy did not intend to launch a new religion; "There is no religion higher than truth" was their motto. From the first, Theosophists directed their gaze, according to Blavatsky's instructions, on the ancient "WISDOM-RELIGION" from whose trunk sprang all the religions of the world, like so many "shoots and branches."[29] Just as the spiritualists felt confident that their beliefs could furnish the lowest common denominator to reduce all theological differences, so Theosophists emphasized the universality of their own precepts and the unlimited nature of their tolerance. Mrs. Sinnett summarized the position of the Theosophical Society when she wrote in 1885:

> As Theosophy is not in itself a religion in the common acceptation of the word, hardly even a philosophy, it may and does include among its followers representatives of almost every form of religious belief in the world, as well as many who have no belief at all. It teaches people to search for the fundamental truth that is the basis equally of every creed, philosophy, and science, . . . and thus to lay bare the fact that one truth supports every religion, no matter how divergent they may now appear; that truth being the Divine wisdom of the ancients, discoverable alike in the symbolical writings of the Kabbala, the Book of Hermes, the Vedas, and other sacred books of the East, in the Talmud, the Koran, our own Bible, as well as in the teachings of Pythagoras, Socrates, and many of the more recent philosophers.[30]

Even though the Theosophists looked much more directly to the East for their inspiration than did the spiritualists, both movements professed a similar longing to reveal timeless religious truth. The Theosophical emphasis on universal brotherhood, through shared religious and philosophical beliefs, was likewise guaranteed to win friends among the spiritualists.[31]

Since the Theosophists, in company with progressive spiritualists, refused to elevate Christianity to special prominence among the world religions, they bestowed upon the historical figure of Jesus no special eminence among world religious leaders. As the journalist who interviewed Madame Blavatsky for the *Pall Mall Gazette*, in April 1884, reported, the founder of Theosophy admired Christ who "was, like Buddha and Zoroaster, a great Mahatma, versed in the occult science of which she at present is the chief authorized exponent." She believed, nonetheless, that greatest reverence was due "Gautama Buddha beyond all other Mahatmas, because he alone of all religious teachers has ordered his disciples to disbelieve even his own words if they conflicted with true reason." Blavatsky may have spoken respectfully about Christ, but she only had scorn for the hypocrisy of those who called themselves Christians, and, as her remarks about Buddha suggest, she detested all systems of clerical authority. "For clergymen as a body she felt hatred," Olcott recollected, "because, being themselves absolutely ignorant of the truths of the spirit, they assumed the right to lead the spiritually blind, . . . and to damn the heretic, who was often the sage, the illuminatus, the adept."[32] It is true that she came to impose a fairly rigid hierarchical structure on the Theosophical Society, as a means of leading its members through the necessary stages of initiation into the esoteric knowledge that she claimed to possess, and some spiritualists accordingly accused her of instituting a new sort of priesthood. Yet Blavatsky eschewed the ceremonial trappings of priesthood, and while she remained in control of the Theosophical Society, ritual played a minimal role at its meetings. Indeed, the reading of minutes and the pursuit of business according to Robert's Rules of Order might prove the highlight of a Theosophical gathering.[33] Such austerity, however, was no more permanently satisfying to Theosophists in search of religious enlightenment than it was to non-Christian spiritualists. Both groups experienced that "'afterglow'" of church-worship [that] ran through all the heretical societies in varying degrees," and the Theosophical Society, after Blavatsky's death, gradually took on a theological coloring that would have made the founder explode in one of her awesome temper tantrums.[34]

As expounded by Madame Blavatsky, therefore, Theosophy espoused a set of teachings fully as anthropocentric as were those of progressive spiritualism; it was the development of the human faculties to their profoundest capacities that Theosophists strove to achieve. In fact, despite the hierarchical organization of their society, Theosophists went even further than non-Christian spiritualists in their zeal to eliminate all vestiges of priesthood, all notions that only certain people, specially endowed or prepared, could mediate between humanity and the eternal spirit. Spiritualists, after all, needed mediums to communicate with the spirit world; each Theosophist, by contrast, was imbued with the conviction that his own Higher Self *was* the divine spirit, or God within him, in the Theosophical sense of the Deity as "the mysterious power of evolution and involution, the omnipresent, omnipotent, and even omniscient creative potentiality." Every Theosophist was entirely and solely responsible for the nurture of that Higher Self. No prayer or ritual observance could absolve him from the consequences of his own actions.[35] There was thus held out to the Theosophist the promise of unlimited evolution to a state of absolute spiritual perfection, but, at the same time, the warning that no divine mercy could wipe clean the record of his past sins. Charles Webster Leadbeater, a prominent Theosophical lecturer, former Anglican clergyman, and close friend of Annie Besant, underscored both aspects of Theosophical thought when he lectured in Buffalo, New York, at the turn of the century. Outlining the fundamental truths that Theosophists embraced, he explained, on the one hand, that "Man is immortal; that he is a creature who is ever evolving, and whose power and glory will in the future have no limit." On the other hand, however, Leadbeater advised that "as we sow, so shall we reap; that as we are reaping now, so we have sown in the past; that there is an eternal law of justice and of equilibrium which operates in just the same way on the higher and spiritual planes as it does down here on the physical plane."[36] How close these views were to the beliefs of the non-Christian spiritualists scarcely needs to be stressed. Certain teachings – the absence of a personal deity, the denigration of priesthoods, the conviction that each person creates his own future destiny, for example – were virtually interchangeable among Theosophists and progressive spiritualists. Except for the reference to "returning spirits," the outspoken Scots spiritualist James Robertson could have been writing as a good Theosophist when he observed: "That our earth deeds affect our future life is what all returning spirits keep telling us. Creeds do not count in the eternal court; it is not what we believe, or profess, but what we are: nothing avails there but the life lived."[37]

Spiritualists did not, for the most part, employ the term "Karma" to describe the force at work molding the individual's existence beyond the grave, although such a concept was implicit in much non-Christian spiritualist literature on the afterlife. It was, of course, explicit in Theosophical writings, in which the workings of Karma play the central role in the unfolding drama of every human life. An article in *Theosophical Siftings*, demonstrating alleged similarities between Theosophy and Buddhism, pointed out in 1889 that

> the Theosophist believes in the three essentials: – 'Maya,' or illusion; 'Karma,' or fate; and 'Nirvana,' the condition of rest, which is neither sleep nor death, but is the longed-for conclusion to all the chances and changes of Life. Coincident with this belief is that of 'Reincarnation,' by which each new life is but the entrance upon existence

of a spiritual entity which has passed through many other lives, and whose conduct in each of these – and in all of them – is, in fact, its 'Karma' self-created, the doom which it inaugurates and works out for itself – according as it is or is not in harmony with the Divine Will and the law of its own structure.[38]

The Theosophist's Karma, as the article made clear, was no arbitrary power imposed upon an unsuspecting and defenseless victim. Each person, Theosophists averred, constructed his own Karma out of the totality of actions, words, and thoughts that compose a lifetime. From reincarnation to reincarnation, the individual molded his fate, profiting or suffering in the next life from the good or evil performed in the preceding one, and aiming to progress along the spiritual evolutionary scale until his lower and higher selves finally merged. The soul, attaining at length its eternal rest, then finally liberated itself from Karma and from the necessity for rebirth in the flesh. In their condemnation of the Christian doctrine of eternal reward or punishment, and in their insistence on the superior virtue and justice of their own Karmic law, Theosophists forged yet another bond between themselves and the non-Christian spiritualists.[39]

For the Theosophist, the concept of Karma was as inextricably intertwined with the belief in successive reincarnations as the twin menaces of atheism and materialism were inseparable for the spiritualist. The length of a single life was simply not deemed enough time for the soul to work out its fate according to principles of absolute justice; it needed opportunities spanning hundreds, even thousands, of years, for too many forces beyond an individual's control might intervene, too many coincidences might occur in the course of one life, or even a few lives, to hinder its free and equitable growth.[40] Sinnett explained that Karma was particularly linked with "the teaching concerning Reincarnation," because Karma

> shows us that the bodily form to which the soul is drawn back is not selected at random . . . Governed by the all-sufficient discernment of Nature, the soul ripe for Re-incarnation finds its expression in a body which affords it the exact conditions of life which Karma – in this sense its desert – requires.[41]

Reincarnation was a doctrine perfectly suited to the diversified origins of Blavatsky's Theosophy, whether these were ancient occult traditions leading back to Pythagoras, the philosophic and religious teachings of the great Eastern faiths that she plundered freely, or the impact of evolution on contemporary thought. Whereas Darwinian theory subsumed the individual in the evolution of the species, however, the Theosophical belief in reincarnation focused sharply on the indi-

vidual. It is not necessary to investigate here the elaborate theories
that emerged in Theosophical thought to explain what happens to the
individual soul between incarnations (it rests and renews itself on the
heavenly plane known as Devachan) or the complicated Theosophical
vision of the human being as composed of numerous bodies (physical,
etheric, astral, mental) that peel off gradually after death.[42] What is
fundamental to the Theosophical doctrine of reincarnation is the absolute refusal to accept a meaningless universe. All the complicated
teachings of Blavatsky and her successors served one basic purpose:
to preach the message that human life does have meaning, that the
evolutionary process is not a random one, that there is a goal toward
which the individual progresses, and that moral justice guides every
step of the way. The Theosophists were very much products of their
era when they recoiled in dismay from the possibility of a chaotic cosmos without underlying moral values.

The great majority of British spiritualists in this period did not embrace
a belief in reincarnation. They held instead to a vision of eternal progress in the spirit world, without interruption by repeated return to assorted earthbound, physical frames. Their assumption that the individual retains his personal identity throughout spirit existence seemed
incompatible with the multiple identities implicit in the doctrine of reincarnation. Yet the differences between spiritualists and Theosophists
in this respect were more apparent than real, for they shared an immensely optimistic view of gradual human progress to spiritual perfection.[43] Hence it is not surprising that the idea of reincarnation proved
attractive to some British spiritualists who tried to incorporate it into
their faith, with or without formal membership in the Theosophical
Society. The "spiritists," for example, distinguished themselves from
other spiritualists by their adherence to the teachings of Hippolyte
Léon Denizard Rivail, a Frenchman whose theories about the spirit
world featured the doctrine of spiritual progress through reincarnation.
Known to the public as Allan Kardec, founder and editor of the *Revue
spirite* in Paris, he wrote numerous volumes of devotional and instruc-
tional literature concerning spirit communication. They circulated
widely on the European continent, and even in parts of Latin America,
but were little known or admired in Britain.[44]

Yet his work suggested that aspects of Theosophical, spiritualist,
and Christian beliefs could form a viable, though bizarre, *ménage à
trois*. If Christians were repelled from the idea of reincarnation by
Blavatsky's hostile tone, they could find in Kardec far more sympathy
for their own faith.[45] A striking example of such a tripartite synthesis
occurred in the life and writings of Marie, Countess of Caithness and
Duchesse de Pomar, Kardec's warm admirer, and a Christian spiri-

tualist who approached Theosophy through "spiritism." This betitled lady was one of the most assiduous séance attendants of her day, from the 1860s when she sat with D. D. Home until her death in 1895. Born in London of a Spanish father, who in time left her very wealthy in her own right, she was first married to the Condé de Medina Pomar, a title later raised to ducal status by papal order. Her second marriage, to the fourteenth Earl of Caithness, brought her to Scotland and gave her a sense of kinship with that country that deeply influenced the final, and most sumptuous, phase of her spiritualist career. After the Earl's death in 1881, his widow lived grandly in Paris, in an opulent mansion on the Avenue Wagram, and her home became the center of fashionable French spiritualism and Theosophy.[46]

Dressed magnificently, "in priceless lace, falling over head and shoulders, and a beautiful tiara of various coloured jewels arranged over the lace" – this attire for nothing fancier than afternoon tea – or, on statelier occasions, sporting "diamonds the size of pigeons' eggs," the Duchess presided in style over her salon. A visitor, the English-woman Katharine Bates, described how the Duchess received guests in her bedroom "quite after the manner of the French kings in the days of the old monarchy,"[47] and, indeed, she apparently identified herself very closely with one former monarch, Mary Queen of Scots. Whether the Duchess actually assumed that the unfortunate queen was rein-carnated in her own body, or rather believed that the queen's regal spirit was her special guardian angel, there is no doubt that Lady Caith-ness was obsessed with Mary of Scotland, for the Queen Mary cult dominated the Caithness-Pomar household.[48] The reception rooms of the Avenue Wagram mansion "were arranged and furnished in close imitation of Holyrood Palace." More than fifty miniatures and other paintings of the Scottish sovereign decorated Lady Caithness's bed-room. There was "a sort of chapel," built according to instructions from the Queen, used for séances and featuring "a full-length, life-sized portrait of Mary herself."[49] Lady Caithness did not take her convictions lightly.

Chief among those convictions was the doctrine of reincarnation, "the SUCCESSION OF EXISTENCES, or earth-lives, as the estab-lished means of purification and progression for the spirit, . . ." She evidently first derived this belief from Kardec, rather than Blavatsky, because her treatise on the subject, *Old Truths in a New Light*, was published in 1876 before Theosophy had had the chance to make an impact in Europe. The volume paid copious tribute to Kardec, and its author attempted to distinguish the grosser spiritualism, with its "mor-bid craving for the wonderful," from the purer, more philosophic "spi-ritism," or the study of "the soul of things," to borrow Kardec's own definition.[50] It was a cast of mind that made her highly susceptible to the doctrines of Theosophy, and she soon became a prominent Theosophist on both sides of the Channel. When Madame Blavatsky descended on Paris in the spring of 1884, the Duchesse de Pomar undertook to act as official hostess and opened her home to everyone of suitable social or intellectual pedigree who wished to meet the high priestess. On the occasion of that visit, Blavatsky gave her stamp of approval to the first Theosophical society in France, a "Société théosophique d'Orient et d'Occident," which the Duchess had already founded. When Blavatsky died in 1891, the Duchess fancied herself a possible contender for the leadership of the international Theosophical movement, but events proved otherwise.[51] In the end, she turned to the SPR for help in setting up an Anglo-French Société de Psychologie in 1894, with Professor Richet as its first president and Myers its initial vice-president. When she died the following year, the infant society came into a substantial inheritance.[52]

For Lady Caithness, the ultimate purpose of psychical research and spiritualist inquiry was "moral and religious elevation." For her, it was a "holy cause," which, she believed, she had "received the mission to promulgate." Certainly the religious significance of the subject attracted her far more than its potential scientific contribution, although she frequently paid lip service to that aspect as well.[53] It would be difficult, however, to pin a theological label on her. "Eclectic" would be accurate, but not particularly helpful. Unorthodox she assuredly was, but Christian she nonetheless considered herself. She was a Theosophist, but having graduated to Theosophy from spiritism, her fundamentally favorable attitude toward Christianity seems to have successfully withstood the anti-Christian outbursts of the Theosophical Masters.

Membership in Theosophical associations in Britain from the late 1870s suggests that Theosophy remained doctrinally flexible, or vague, enough to draw spiritualists from both the Christian and progressive camps. Whether they were attracted by the Eastern allure of Theosophy, by its emphasis on each person's sole responsibility for his own fate, by the echoes of old occult lore, or by the hope of experiencing reincarnation, numerous British spiritualists moved into Theosophical circles during the 1880s. They did not, in all cases, stay there for very long, but they were at least tempted to familiarize themselves with Theosophy. Not only were its central doctrines frequently to their liking, but ancillary elements might also lure spiritualists and put them at ease in a Theosophical lodge.

The early miracles associated with Madame Blavatsky could not fail to rivet spiritualist attention upon her, for she seemed to be in touch with the spirit world in a remarkably direct and palpable way. Having in the past tried her hand at mediumship, she could obtain standard spiritualist rappings when the occasion required, and she could produce missing objects as well. More impressive, however, were the times when she caused mail to be delivered in an extraordinary fashion; it was nothing for a folded note to come floating down from the ceiling in her presence – a note which, when opened, proved to bear some message or instruction from her Mahatmas.[54] In 1883, Madame contrived to have a special cabinet, or "shrine," created at the Theosophical Society's Indian headquarters in Madras, into which such letters tumbled abundantly. Nor was it beyond the powers of the Mahatmas to mend broken objects, such as the china tray which crashed off a shelf in the shrine in August 1883. It was restored to perfect wholeness after the broken pieces were returned to the shrine for a mere five minutes, and not so much as "a trace of the breakage [could] be found on it!"[55] After her supposed miracles had been exposed, largely thanks to the SPR, Blavatsky spoke contemptuously of them, pretending that they had never played an important part in the unfolding of Theosophical truth. But she was equivocating; the physical phenomena of Theosophy were as critically important to that movement in its early days as were the physical manifestations of spiritualism. The Russian author and occultist V. S. Solovyoff, who became well acquainted with Blavatsky after 1884, rightly commented, with regard to these phenomena:

> It was with their help that H. P. Blavatsky founded her Theosophical Society, they were her panoply when she appeared in Europe to disseminate her doctrine, by them she advertised herself and gathered about her those who for one purpose or another wished to see them. It was these phenomena *only* which interested and brought into her circle of acquaintance such men as Crookes, . . . and the English savants who had established the London Society for Psychical Research.[56]

These phenomena, no doubt, first caused British spiritualists to take note of Theosophy.

Other Theosophical practices also underscored its affinities with spiritualism. Automatic writing, for example, was elevated to near sacred status in Theosophy, since Blavatsky claimed that the great founding text, *Isis Unveiled*, was dictated to her by the Mahatmas. A decade later, when she was struggling to produce another magnum opus, *The Secret Doctrine*, the Mahatmas aided her again, this time with "precipitated messages," not unlike the written exhortations and encouragements mysteriously received by spiritualists from loved ones beyond the veil. Often Blavatsky would find, on returning to her writing table in the morning, "a piece of paper with unfamiliar characters traced on it in red ink," and she would have her marching orders for the day. The Masters not only told her how to proceed with her literary task, but they also informed her when a previous day's effort deserved to be "consigned to the flames."[57] Furthermore, certain assumptions about the possibility of spirit photography were common to Theosophists and spiritualists, and likewise facilitated the cross-fertilization that occurred between the two groups toward the close of the nineteenth century.[58] Even as late as 1910, the London Spiritualist Alliance hailed as a "welcome sign of the times" "the growing friendliness between Theosophists and Spiritualists."[59]

H. P. BLAVATSKY AND THE SPR

Nor were spiritualists alone in greeting Theosophy with interest when it migrated to England, first in the late 1870s, but more successfully in the early 1880s. As Solovyoff noted, the SPR was by no means indifferent to Blavatsky's miracles. Before her first extended visit to London in the spring of 1884, the British branch of the Theosophical Society had already established "cordial relations" with the SPR. C. C. Massey was a member of both groups, as were Myers, Sinnett, and a few others, including Dr. George Wyld, president of the British Theosophical Society and active in the BNAS. When Barrett was planning the inaugural conference of the SPR late in 1881, he had invited Wyld to attend, and Wyld had replied with enthusiasm.[60] Even Blavatsky's explosive presence in 1884 did not immediately threaten the good feelings between Theosophists and psychical researchers. Believing that Madame and her claims were well worth investigating, the SPR established a committee to interview the Theosophical delegation to London that year, and in August Blavatsky and her entourage were invited to Cambridge. Sidgwick's reaction to Blavatsky's visit indicates the fundamentally friendly attitude that the SPR initially held toward Theosophy's founder. "On the whole," he wrote in his journal for 9 August 1884,

> I was favourably impressed with Mme. B. No doubt the *stuff* of her answers [to questions posed by Sidgwick and Myers] resembled *Isis Unveiled* in some of its worst characteristics; but her manner was certainly frank and straightforward – it was hard to imagine her the elaborate impostor that she must be if the whole thing is a trick.[61]

The SPR was inclined to give Theosophy the benefit of the doubt because psychical research, too, shared with it certain assumptions and attitudes. The hidden powers of the human race, its unexplored capacities for growth and development, were a fundamental concern of the Theosophists. Their Society's formal statement of purpose included the goal: "To investigate the unexplained laws of nature and the psychical powers latent in man."[62] Few members of the SPR would have taken issue with that aim. Nor would they have protested against the social biases of British Theosophy, at least as articulated by Mr. and Mrs. Sinnett, in whose home London Theosophists gathered in the mid-1880s. "We were at home, always, on Tuesday afternoons," recalled Sinnett,

> and my wife's Diary is filled every week with long lists of our Tuesday visitors. The movement in this way spread at first in what may in a broad sense be called the upper levels of society, and it appeared to me desirable that it should take root that way to begin with, its influence being left to filter downwards with social authority behind it, instead of beginning on lower levels and trusted to filter upwards if it could.

Although some psychical researchers may have been amused by Sinnett's social pretensions, the SPR in general would have applauded his desire to place "social authority" squarely behind Theosophy; the SPR was doing no less for psychical research.[63]

The community of interests, and even of personnel, which maintained amicable ties between the SPR and Blavatsky for a couple of years, was shattered in 1885, when the former issued a devastating report about her movement. The report, described as "easily the most dramatic and entertaining bit of work that the Society has ever published," resulted from Hodgson's detective research. When the SPR committee that interviewed Blavatsky, Olcott, and other Theosophists in the spring and summer of 1884 decided that further, more extensive, and on-the-spot inquiry was needed, Sidgwick paid for Hodgson to undertake an Indian adventure.[64] Having completed his Cambridge education in 1881, with an honours degree in moral sciences, Hodgson had obtained an academic position as University Extension Lecturer, but he abandoned his teaching post in 1884 to track down Theosophical mysteries in Madras. After a brief subsequent stint as extramural lecturer in philosophy at Cambridge, he became secretary of the American Society for Psychical Research in 1887 and thereafter remained a fulltime psychical researcher for the rest of his life.

Despite the harshness of its final report, the SPR inquiry into Theosophy began with no preconceived critical opinions. As Podmore, who was a member of the investigative committee, later explained:

> When . . . we found ourselves confronted with evidence for occurrences in India, analogous in some respects to those which had already formed the subject of our inquiries in England, and when we found that some of these occurrences were vouched for by witnesses of good

repute and good intelligence in other matters, we held that we should not be justified in summarily dismissing their evidence.

If anything, the SPR was startled by the vehemence with which Hodgson accused Blavatsky of wholesale fraud. As late as March 1885, when Hodgson had been in India for three months and was sending to the SPR investigative committee weekly letters in which he conveyed his increasingly dim views of Theosophy, Sidgwick could nonetheless write in his journal that there were "still some things difficult to explain on the theory of fraud."[65] Even those few things, however, vanished when Hodgson returned to England in April with the full array of damning evidence. His findings, preceded by a short supporting endorsement from the committee, were published in the third volume of SPR *Proceedings*.

Hodgson's debunking task had been facilitated by the publication, in September and October 1884, of letters allegedly sent from Blavatsky to M. and Mme. Coulomb, formerly two of her chief assistants at the Theosophical Society headquarters in Madras. These letters revealed, in ample detail, how Blavatsky arranged for various miracles to occur, even in her absence: how the holy shrine was manipulated by the Coulombs on Blavatsky's instructions, with a dummy head and shoulders atop a disguised confederate to provide the faithful with glimpses of a Mahatma's "astral" projection; how telegrams from the Mahatmas were carefully arranged in advance to be dispatched by the Coulombs at specific times in order to create a maximum effect on the lucky recipient. The tale of the broken and instantly mended china tray lost its miraculous aura when it emerged that the shrine had a sliding back panel that leaned against the wall of what was called the Occult Room. At that very place in the wall, a window had formerly existed. Although almost entirely covered with bricks and plaster, it still allowed narrow access to the shrine from the room (Blavatsky's bedroom) on the other side of the wall. The mystery of the shrine was not specifically revealed in the Blavatsky–Coulomb correspondence; that Hodgson had to piece together for himself when he arrived in Madras in December 1884, for the shrine had already been destroyed, following the publication of the Coulomb letters. Because those letters were published in the *Madras Christian College Magazine*, Theosophists tried to dismiss them as forgeries, produced by Blavatsky's Christian enemies who hated her for rivaling their own missionary efforts among the Indians. One of Hodgson's principal responsibilities, accordingly, was to determine the authenticity of the documents, and he sent a selection of them to two handwriting experts in England, one of whom was employed by the British Museum. Comparing the Coulomb letters with documents known to have been written by Blavatsky, each expert, independently of the other, pronounced them to be the handiwork of one and the same author.

Hodgson's investigative zeal, together with the explanations willingly provided by the treacherous Coulombs, also uncovered the means by which Theosophical letters might suddenly flutter down from ordinary looking ceilings. Thanks to a construction method that filled spaces between ceiling beams with blocks of wood and mortar, it was possible to scrape away some of the mortar, leaving space for a letter to rest on top of one of the beams. If a fine thread, the color of the ceiling, were loosely twined around the letter, and held by a confederate standing just outside the room, it could be pulled at a designated moment; the letter would make its startling impression as it floated downward, while the confederate, having drawn back all the thread, strolled away.

The letters, like those precipitated into the shrine or sent to particularly favored followers, were signed by one of two Mahatmas who took a special interest in the Theosophical Society: Koot Hoomi and Morya. Hodgson painstakingly ascertained, with the help of the same handwriting experts, that here, too, Blavatsky's pen had been busy. He announced, without hesitation, that these treasured Theosophical documents emanated from no loftier source than Madame herself, or from one trusted Indian disciple who learned to imitate the Koot Hoomi handwriting which Blavatsky had created.[66] This was, of course, an even graver charge than plagiarism, for which the Mahatmas themselves had been blamed ever since they, supposedly, had first used Blavatsky as their amanuensis. It was certainly insulting to accuse the Masters of lifting their profound wisdom from the pages of numerous, readily accessible sources, such as Ennemoser's *History of Magic*, or to point out glaring errors in their scholarship, but their very existence was not necessarily thrown into doubt as a result. Such, however, was the conclusion to which Hodgson's report inexorably led. Not only was the written proof of their involvement with Theosophy denied any but the most humdrum human status, but the leading witnesses to their special guardianship of the Theosophical Society all turned out to be more than a little suspect – either because of deliberate falsehood on their part or because, being so patently gullible, they were clearly putty in Blavatsky's hands.[67]

If the Mahatmas were fictitious, as the SPR report unequivocally stated, the whole structure of Theosophy was assuredly perched on very "sandy foundations."[68] All that was central to Theosophical belief, its claim to authority as a universal philosophy of life and death, rested on the contention that its teachings derived from superhuman sources transcending time and space. Reveal the falsity of that contention, and Theosophy collapsed into just another cult revolving around its crank founder. The SPR investigative committee might conclude its brief summary of Hodgson's findings with the somewhat condescending observation that Blavatsky had "achieved a title to permanent remembrance as one of the most accomplished, ingenious, and interesting impostors in history,"[69] but that was a devastating comedown from the rank of inspired prophetess to which she had pretensions. It goes without saying that her staunchest disciples refused to concede that the report damaged Blavatsky's reputation. They accused Hodgson of undertaking his Indian inquiries, not in a mood of impartial research, but as prosecutor, judge, and jury all at once. The integrity of the Coulombs was, with justice, assailed, and Blavatsky complained that she had never even been shown the incriminating letters which, she insisted, were largely fabrication. Sinnett accused the SPR of pandering to public opinion in its denigration of Theosophy and triumphantly concluded that Hodgson's logic served no purpose, because Blavatsky's complex character was not explicable "by any commonplace process of reasoning."[70] It would take more than Hodgson's detective work to drive true believing Theosophists from their convictions, just as repeated exposures of materialization mediums did nothing to shatter the confidence of true believing spiritualists. Hodgson was quite right when he told Sidgwick that "Theosophy will go on," for an organization like the SPR had no power to halt a movement like Theosophy. The best Hodgson and Sidgwick could hope for was to assist in preventing "people of education from being further duped."[71]

PROMINENT BRITISH THEOSOPHISTS

People of education were, in some cases, willing to have their eyes opened, and departures from Blavatsky's circle occurred, both before and after 1885, whenever common sense prevailed over occult impulses. Even W. S. Moses, who apparently for some years enjoyed a mutual admiration with Blavatsky, grew disgusted by the lack of originality manifested in her Mahatmas' writings. C. M. Davies confessed to having "sat at the feet of Madame Blavatsky," but found that Theosophy offered him little besides her "Turkey cigarettes."[72] In the case of C. C. Massey, however, the break with Blavatsky came after he had been closely affiliated with, and deeply committed to, the Theosophical movement in Britain.

For Massey, Theosophy represented an important phase in the ongoing metaphysical and theological explorations that filled the better part of his adult life. Already a convert to spiritualism when he met Olcott and Blavatsky on a trip to the United States in 1875, Massey eagerly embraced the doctrine of reincarnation and placed it at the center of the Christian mysticism with which he became identified. With Massey, as with Lady Caithness, seemingly incompatible doctrines merged effortlessly under the magic wand of an entirely earnest, undogmatic, and eclectic seeker after truth. So impressed was he with the Theosophical Society, whose birth he had witnessed in New York in the autumn of 1875, that shortly after his return home he organized in London the first European branch of the Society. Significantly, early meetings of the British Theosophical Society, in 1878 and subsequently, were held at the Great Russell Street home of the BNAS; among the first members of the British Theosophical Society, in addition to Massey and Wyld, were other BNAS participants, such as Dr. C. Carter Blake of Westminster Hospital, and Emily Kislingbury, onetime secretary of the BNAS who was also chosen to be the first secretary of the British Theosophical Society.

Blavatsky appreciated Massey's efforts on behalf of Theosophy, and in 1879 she saw fit to bestow on him a special mark of favor – a Mahatma letter addressed to him personally. It appeared mysteriously in the minute book of the British Theosophical Society and came in answer to Massey's urgent request for empirical evidence that the Masters existed. For a while, Massey was satisfied with the authenticity of the document and believed it to have been genuinely precipitated into the minute book by occult agencies. A few years later, however, he was shown other letters that made him realize that the event had been staged by Blavatsky, then in India, with the help of a member of the London Society who was a medium. When he wrote to Blavatsky, asking for an explanation, she equivocated, admitting that she had forwarded the letter to the medium in question, but insisting that the communication from the Master was absolutely genuine. Massey, whose credulity had its limits, resigned from the British Theosophical Society in the summer of 1884.[73]

The physician George Wyld left the ranks of British Theosophy even earlier than Massey. Although Wyld served briefly as the Society's second president, between 1880 and 1882, he was apparently never as impressed by Blavatsky and her teachings as Massey once was.[74] On first acquaintance, Wyld found Madame rather similar in appearance and manner to "a worn-out actress from some suburban theatre in Paris." Being a convinced spiritualist, he could not help noticing "her undoubtedly mediumistic powers," and he joined her Society out of a mixture of curiosity, interest, and "belief in her promises." Yet he could never accept her "coarse and rude behavior in public," and ultimately it seemed to him "a marvellous thing how any refined and thoughtful man or woman could *continue* to believe in this queer woman who smoked so incessantly, as an inspired expounder of the highest spiritual secrets of the human race."

It was not, however, Blavatsky's abuse of tobacco (mixed, according to speculation, with hashish) that especially offended Wyld; it was her scornful treatment of sacred subjects. He winced when, on being asked her views of Christ's nature, "she replied, '. . . I have not the honour of the gentleman's acquaintance.'" He was outraged when he read her assertion, in the May 1882 issue of *The Theosophist*: "There is no God, personal or impersonal." Indeed, that bold statement convinced Wyld to part company with her, for he reasoned that if there were no God, there could be no God-wisdom, or Theosophy. As it turned out, nevertheless, Wyld abandoned only Blavatsky, and not theosophy. Like Massey, he took what appealed to him of Eastern occult wisdom, blended it with his "intense belief in the life, teachings, and works of Jesus Christ," and produced his own hybrid faith which he appropriately dubbed Christo-Theosophy.[75]

A. P. Sinnett's rupture with Blavatsky came more gradually than Wyld's or Massey's and was engendered largely by a clash of personalities. For years, Sinnett was able to accept the deceit and chicanery that evidently formed an integral part of Blavatsky's character, and he could comment, without the slightest irony or sarcasm:

> Madame Blavatsky's shortcomings or defects of character did not alter the fact that through her intermediation the Veil had been lifted (more or less) from the Occult World previously so totally concealed from view – so far as the world at large was concerned. My wife and I had long been alive to her strangely diversified nature but had attained to a condition of mind and knowledge that enabled us to look behind her at those [the Mahatmas] who for want of a better agent had accepted her with all her disqualifications, as the intermediary who should make their existence known to us.[76]

Sinnett first met Madame Blavatsky in India, late in 1879. He was an influential journalist at the time, editor of the Anglo-Indian daily, *The Pioneer*, and an important conquest for Blavatsky. Sinnett, too, came to Theosophy from spiritualism, and he did not prove a difficult convert. During long visits at his homes in Allahabad and Simla, Blavatsky dazzled the journalist with her "manifestations of occult power then freely given . . ." When Madame, beginning in 1880, delivered mail to him from Koot Hoomi and Morya, Sinnett was overwhelmed. These "Mahatma Letters" became the basis of Sinnett's two books, *The Occult World* (1881) and *Esoteric Buddhism* (1883), which effectively, in clear, sensible language, advertised abroad Theosophy's

claims to ancient and secret wisdom. In fact, Sinnett identified himself
so closely with the new movement, and campaigned so tirelessly as a
propagandist for Theosophy, that the proprietors of the *Pioneer* decided to dispense with his editorial services. It seemed a good time for
Sinnett and his wife to return to London, where they arrived in April
1883.[77]

Sinnett quickly made himself, and his home, a center of Theosophical
activity in London. The following year, he maneuvered out of office
the president of the London Theosophical Society, Dr. Anna Kingsford, who had dared to criticize Sinnett's *Esoteric Buddhism*, and
helped to install a more tractable officer in her place. Even after the
disastrous Hodgson Report began to take its toll on membership in
what had become known as the London Lodge, the Sinnetts kept it
alive, if not well. Their dominance of Theosophical circles in London
came to an abrupt and inevitable end, however, when Madame Blavatsky settled permanently in that city in the spring of 1887. Sinnett
had deplored the idea of such a move and had not hesitated to say so,
certain as he was that Blavatsky's presence could only have a dele-
terious effect on British Theosophy, not to mention his own control of
the London Lodge. Blavatsky was not easily disconcerted by human
opposition and, shortly after her arrival in England, formed a new
group, the Blavatsky Lodge of the Theosophical Society, as the unabashed rival of the London Lodge. It was an uneven competition. Not
only was the incomparable Blavatsky herself president of the new
lodge, but she further stacked the deck by setting up within it the
Esoteric Section, to tantalize prospective members with a "promise
of some mysterious teaching not given to the rank and file of the Theosophical Society." When she required applicants for membership in
the Esoteric Section to take pledges of obedience to her, Sinnett announced that such a policy "sinned . . . against fundamental principles
of Theosophy." Blavatsky retaliated by condemning certain sections
of *Esoteric Buddhism* in the pages of her *Secret Doctrine*, published
in 1888. By this time, however, the Sinnetts claimed that they were
receiving private messages directly from Koot Hoomi, without Blavatsky's intermediary services. They felt that they had their own, un-
sullied corner on truth, which they continued to communicate to the
few "faithful members of the old London Lodge" who still gathered
for evening meetings at the Sinnetts' home.[78]

After Madame's death in May 1891, the relations between the London and Blavatsky Lodges blew hot and cold, with Sinnett now leading
his group into outright secession from the Theosophical Society, now
working to bring the London Lodge back into the Theosophical fold.
In Sinnett's long and rocky relationship with Blavatsky and Theosophy

are evident the dangers latent in a movement where doctrine is articulated piecemeal, and where truth can always be reinterpreted by a
leader who claims to be above criticism. Questions of dogma were
manipulated to hide what amounted to little more than petty power
struggles. Nevertheless, a substantial difference between Sinnett's
Theosophy and Blavatsky's did in time emerge. Once Sinnett's refusal
to cooperate with Madame in launching the Esoteric Section had driven
the two former comrades apart, Sinnett discovered that some severe
doctrinal errors had crept into Theosophical teachings. When he realized that the initial Mahatma Letters, sent to him between 1880 and
1884, were not, as he first assumed, written by the Masters themselves,
but dictated to Blavatsky, he could freely criticize them for the passages with which, he asserted, she had manifestly tampered. Nowhere,
he argued, were Blavatsky's distortions more blatant than in the sections "on after-death conditions" where her "bitter detestation of spiritualism" prompted her to produce messages that were actually "a
travesty" of Koot Hoomi's intended meaning. Sinnett, in short, came
to challenge Blavatsky's scorn for spirit communications, and after his
wife's death in 1908, he was deeply consoled by the opportunity to
communicate with her. He was convinced of her spirit identity and had
no fear that some elemental was playing pranks on him. Although he
offered a Theosophical explanation for his wife's ability to convey genuine messages – even Blavatsky did not altogether rule out the possibility of such communications when the spirit in question was of the
purest and most elevated nature – it is difficult to avoid the conclusion
that Sinnett, toward the end of his life, found more comfort and encouragement in his early spiritualist beliefs than in his later Theosophical ones.[79]

Other spiritualists who were temporarily Blavatskyites similarly
found no permanent resting place at Madame's feet. Their reasons for
departure were varied. Emma Hardinge Britten, like Massey, participated in the earliest meetings of the Theosophical Society in New
York, but left soon afterward when she found that the doctrine of reincarnation was incompatible with her spiritualist beliefs, and when she
suspected that Blavatsky and Olcott were straying from "simple Spiritualism into the realms of dreamland."[80] Mabel Collins, private trance
medium and novelist, had a dramatic falling out with Blavatsky after
the two women worked together as joint editors of *Lucifer*, the Theosophical periodical which Madame launched in London in 1887. W. B.
Yeats gained entrance to the Esoteric Section of the Blavatsky Lodge,
but ran afoul of Theosophical authority and was asked to resign in 1890.
His thirst for experimental corroboration of certain occult claims struck
some of his colleagues as inappropriate and disturbing.[81]

Spiritualists and psychical researchers were certainly not the only ones to experience both Blavatsky's charm and wrath. After its first few years, the Theosophical Society began to attract men and women who were not affiliated with spiritualism and who, in some cases, were even repelled by aspects of spiritualist teaching. Dr. Anna Kingsford, for example, a London physician, vegetarian, vehement antivivisectionist, and feminist, was president of the Theosophical Society's London branch in 1883–4, despite her strong antipathy to spiritualism. An even more prominent convert to Theosophy was, of course, Annie Besant. Although, in her private withdrawal from secularism, Besant had been attending séances and investigating spiritualism, she was by no means ready to call herself a spiritualist. Her leap of faith from secularism to Theosophy in 1889 struck contemporaries as all the more inconceivable precisely because there were no visible intermediate jumps.[82] The young A. R. Orage, still an elementary schoolteacher in Leeds and not yet famous as editor of the *New Age*, pored over Blavatsky's writings and organized a Theosophy Group in the 1890s. Without previous exposure to spiritualism, he was fascinated by the avenues to Eastern religious literature that Theosophy opened to his wide-ranging speculative curiosity. Furthermore, he found in Theosophy, as he likewise found in socialism, an alternative to the "doctrines and discipline" of organized religion that he vigorously repudiated.[83] Primarily, however, Theosophy for Orage was one of the several systems of thought and modes of experience to which he turned throughout his life in the hopes of finding the hidden reality that could give meaning to a world without values. Plato, Blavatsky, the *Bhagavad-Gita* and *Mahabharata*, Nietzsche, diverse socialist prophets, and, most important of all, Gurdjieff – all provided milestones in the landscape of Orage's intellectual and emotional pilgrimage.

In the ferment of ideas and movements that animated the closing decades of the nineteenth century and the opening years of the twentieth, it was possible to perceive Theosophy as part of a vast liberation movement designed to topple the materialistic, patriarchal, capitalistic, and utterly philistine culture of the Victorian Age. Edward Carpenter, iconoclast par excellence and a man who proclaimed Walt Whitman and the *Bhagavad-Gita* as the two major influences on his own work, found a central place for Theosophy in his characterization of the "years from 1881 onward." "It was a fascinating and enthusiastic period," he recalled;

> The Socialist and Anarchist propaganda, the Feminist and Suffragist upheaval, the huge Trade-union growth, the Theosophic movement, the new currents in the Theatrical, Musical and Artistic world, the

torrent even of change in the Religious world – all constituted so many streams and headwaters converging, as it were, to a great river.[84]

Therein surely lay much of the enduring attraction of Theosophy. For all the disillusioned who dropped out of Blavatsky's orbit, there were others who stayed, undismayed by the SPR report, by Madame's personal shortcomings, by the squabbles and bickering among her followers. The early history of Theosophy contains the names of numerous disciples whose devotion to Blavatsky was nothing short of saintly, given the abuse that she tended to heap upon her most loyal followers. There was Countess Wachtmeister, for example, the widow of a Swedish diplomat, who took care of Blavatsky in Würzburg and Ostend, from the autumn of 1885 until Madame's move to London in 1887, and who continued her ministrations for a brief time in London as well. Dr. Archibald Keightley, a Cambridge-educated physician, together with his cousin Bertram, a Cambridge-educated lawyer, arranged to bring Blavatsky to England and actually had her living with them throughout the summer of 1887, "in a tiny cottage" in the London suburb of Norwood. Of all possible house guests, Madame must have been the least congenial for an extended visit in close quarters. Undaunted, the Keightleys established a Theosophical headquarters for her, largely at their own expense, at 17 Lansdowne Road, Holland Park.[85] What kept these people, and many others, attached to Madame and Theosophy, despite all the evidence of deceit in high places?

Theosophy survived the many attacks aimed against it in its vulnerable infancy because it was exceedingly elastic; its teachings, as they evolved over the years, could address surprisingly diverse viewpoints and satisfy seemingly contradictory needs. For those who wanted to rebel dramatically against the constraints of the Victorian ethos – however they perceived that elusive entity – the flavor of heresy must have been particularly alluring when concocted by so unabashed an outsider as H. P. Blavatsky. Yet Theosophy could also comfort the anxious heretic, the frightened exile from Christianity, while it likewise attracted men and women who kept the Christian label, but adorned it with assorted unorthodoxies. It could even complement the Christian mystic's urge to achieve the extinction of conscious personality, for the ultimate goal of Theosophy was the release of the higher self into the eternal world of universal spirit.

One of the most attractive aspects of Theosophy was that it offered much the same promise of special protection and care as has always figured among Christianity's most appealing elements. In place of Christ's guardianship and intercession (as well as those of the saints, in the case of Roman Catholicism), Theosophy offered the Mahatmas, the so-called "White Lodge" or "Great White Brotherhood," who took particular interest in Theosophists and showed deep concern for their welfare. From the mountains of Tibet where they resided, the Brothers, Mahatmas, Masters, or Adepts, by all of which terms they were designated, watched closely the development of their followers. These occult Brothers, including Morya and Koot Hoomi, were not deities, but rather "men of great learning, . . . and still greater holiness of life." They were men who, having passed through numerous reincarnations, had attained a state of being where the physical body ceased to restrict their movements. Their astral forms traveled freely around the world, appearing to startled disciples, conveying messages both written and oral, and surveying the international progress of Theosophy.[86] What was particularly encouraging about the Theosophical Masters – far more so than anything that Christianity had to offer – was the suggestion that each diligent student of the Masters' wisdom might, in lives to come, progress toward their lofty status.

The conviction that Theosophists are under the special, benevolent supervision of the Brothers is, however, only one reason why Theosophy has survived a century's vicissitudes. The idea of secret knowledge has always been irresistible to human nature. It is highly flattering to believe oneself the possessor of esoteric wisdom, revealed only to a select group of cognoscenti. Even when the bonds with Christianity proved too firm to sever altogether, as in the case of Massey, Wyld, and Lady Caithness, Christian Theosophists were drawn to the allegedly arcane nature of Blavatsky's revelation. The suggestion of magical powers played no small part in luring people to Theosophy. Wyld admitted as much when he observed that "all the theosophists that joined the society in his time, did so in the hope of mastering the secrets of magic. Each wished to be an Apollonius of Tyana."[87] Theosophy could give its devotees a sense of importance and potential power that Christianity, with its traditional emphasis on humility, passive acceptance of suffering, and impotence in the face of divine wrath, could rarely offer.

It is this sense of importance that suggests certain generalizations about the role of women in the Theosophical movement. Blavatsky is, naturally, the preeminent example, but Dr. Anna Kingsford and Annie Besant also come readily to mind. While Blavatsky's life history is so obscured by fabrication that her motives for launching Theosophy are anybody's guess,[88] some striking similarities in the less mythicized lives of Kingsford and Besant tempt speculation. Both were high-spirited, intelligent women who found intolerable the limitations of married life in the third quarter of the nineteenth century and who, it appears, ultimately found satisfaction in the role of religious prophet. Although, admittedly, conditions for women in Britain had changed since the seventeenth century, the same frustrating restrictions still largely prevailed during the lives of Kingsford and Besant as had prompted a few women in the earlier period to adopt the stance of inspired prophetess.[89] University education, access to the pulpit, political power – all were denied women in the seventeenth century, so that, almost of necessity, those who struggled to comment on social and political conditions in the country were driven to play a part that guaranteed them at least a temporary audience. There is no strict parallel here with the experiences of Kingsford and Besant, despite the continued paucity of higher educational and career opportunities open to women in the 1870s, for both women sought religious fulfillment as sincerely as political or social influence. Yet it was precisely the religious mantle, assumed after gaining prominence in the world of men, which both Kingsford and Besant apparently believed gave them greater authority than any of the others that they had donned during full and varied lives.[90]

When Anna Kingsford and Annie Besant rebelled against the limitations of married life in Victorian Britain, they were specifically rejecting the role of Anglican clergyman's wife. Mrs. Besant separated completely from her husband, Frank Besant, vicar of Sibsey, in the autumn of 1873, and, as is well known, traveled from religious skepticism to a position of outspoken prominence in the National Secular Society. Kingsford's religious scruples took her in a very different direction. While Annie Besant, in a reaction shared by many contemporaries, rejected the Christian doctrines of eternal punishment and vicarious atonement, and wondered how a loving God could have created a world full of misery and sin, Kingsford's response to the religion of her parents and husband was more intensely personal. She grew to loathe the "hardness, coldness, and meagreness" of the Anglican church and felt "its utter unrelatedness to her own spiritual needs, intellectual and emotional." She longed for a more satisfying church service and spiritual atmosphere and, not surprisingly, found them in Roman Catholicism, to which she had been beckoned by "an apparition purporting to be that of St. Mary Magdalen." She was formally received into the Catholic Church in 1870, under the name Mary Magdalen, and was confirmed two years later by Archbishop Manning. Kingsford's husband, Algernon Godfrey Kingsford, became vicar of Atcham, Shropshire, but, unlike Frank Besant, was willing to give his wife free rein. Throughout all the adventures and campaigns of her life, he remained very much a shadow figure in the background. The man who joined her in many of those campaigns and who accompanied her to Paris, where she studied medicine from 1874 to 1880, was not the

Reverend Kingsford, but her loyal collaborator and biographer, Edward Maitland.[91]

Defiance of convention became a way of life for Annie Besant and Anna Kingsford, but it was not empty defiance merely for its own sake. They were never rebels without a cause. Whether the struggle was for free thought, birth control, Fabian socialism, or Indian nationalism, Besant brought her intense zeal and energetic commitment to the effort and became a formidable public speaker in the process. Kingsford's social causes included higher education and suffrage for women, dietary reform, and, above all, antivivisection.[92] She told Maitland at their first meeting that she detested cruelty and injustice in all their forms, whether toward man, woman, or beast, and if her battle for social justice was somewhat impaired by her physical fragility, her devotion to the principles involved was always ardent.[93] Nor was she solely an armchair sympathizer. Her long efforts to obtain a medical degree were not only inspired by the desire to do something meaningful with her life; the professional education was particularly intended to help her argue against the doctors' claims that vivisection was necessary for the advancement of medical science. Finding that the doors to medical school in England were closed to her, she enrolled in the Paris Faculty of Medicine and received the M.D. degree in 1880.[94] Thereafter she set up medical practice in London and soon attracted a substantial number of female patients. Unfortunately, she had only a few years to pursue her profession, for early in 1888 she died of consumption. Annie Besant, by contrast, one year younger than Dr. Kingsford, flourished until 1933.

Both women, despite all their other interests and concerns, gave their deepest commitment to crusades of the spirit. For Besant, the involvement with Theosophy that began in 1889 lasted for the rest of her long life, although under her leadership the Theosophical Society followed some trails that Blavatsky had never blazed. For Kingsford, the practice of medicine in London during the 1880s could never compete with the pleasure that she derived from her occult studies. The religious restlessness that led her to Roman Catholicism as a young woman never left her; the last years of her life saw her assume first the presidency of the London Theosophical Society in 1883 and, after her clash with Sinnett, the leadership of the Hermetic Society, which she and Maitland launched in 1884. It was the Hermetic Society that allowed Kingsford complete freedom to pursue her own brand of religious occultism. Scorning to be anyone's disciple, during the 1880s she articulated an individual theology, drawn from Christianity, Renaissance magic, Eastern mysticism, and late Victorian feminism, that was resoundingly her own.[95]

188 A surrogate faith

Indeed all the causes that she passionately espoused found their way into her intoxicating religious brew. The ferocity with which she opposed vivisection, for example, becomes fully explained only when one learns that she and Maitland embraced a highly specific theory of the soul's upward movement through successive reincarnations, from plants through animals to man, with the elimination of the material body as the final goal of the journey. Her theology likewise bore the firm imprint of her fervent feminist sympathies; she announced that "the object of all sacred mysteries, whether of our Bible or other," was "to enable man anew so to develop the Soul, or Essential Woman, within him, as to become, through Her, a perfect reflection of the universal Soul, and made, therefore, in what, mystically, is called the image of God." She waxed lyrical about the feminine nature of the divine principle and warned men that they could never attain "to full intuition of God," until they exalted "the Woman in themselves." One scholar of the occult has even linked Kingsford's thought to a contemporary French visionary extravagance, the doctrine of the Woman-Messiah.[96] Furthermore, she venerated the Virgin, and one of the strongest attractions of the Catholic Church for her was "the worship of Our Blessed Lady."[97]

Although in a letter to her close friend Lady Caithness, Kingsford dismissed Blavatsky as "an occultist, not a mystic," Kingsford was no mean occultist herself. She, too, was fascinated by the possibility of magical powers and was eagerly receptive in 1886 when "a proposal to study occultism was made to her by a notable expert, who, being well versed in Hermetic and Kabalistic science, had attained his proficiency in the best schools."[98] In the previous year, Kingsford and Maitland had published their version of *The Hermetic Works*, and she had a highly particular purpose in mind when she determined to undertake a regular course of occult studies: In learning to obtain power "over the elemental forces," she intended to direct them "against some of the leading vivisectors, and especially M. Pasteur." Suffering acutely from her physical ailments, this frail, neurotic, but determined woman grasped at magic in a final, desperate effort to exercise power, to exert some significant control over her environment, to achieve fame and greatness. She convinced herself that her magical exercises were potent and that, by the sheer concentration of her will, she could eliminate her opponents. When Professor Paul Bert, prominent in Third Republic politics and science, and "among the most notorious of the vivisecting fraternity of Paris," died in November 1886, she gloated malevolently in her diary:

> For months I have been working to compass the death of Paul Bert, and have but just succeeded. But I *have* succeeded; the demonstration

of the power is complete. The will *can* and *does* kill, . . . Paul Bert
has wasted to death. Now only one remains on hand – Pasteur, who
is certainly doomed, and must, I should think, succumb in a few
months at the utmost. Oh, how I have longed for those words – "*Mort
de M. Paul Bert*"! And now – there they actually are, gazing at me
as it were in the first column of the *Figaro*, – complimenting, con-
gratulating, felicitating me. *I* have killed Paul Bert, . . . as I will kill
Louis Pasteur, and after him the whole tribe of vivisectors, if I live
long enough. Courage: it is a magnificent power to have, and one that
transcends all vulgar methods of dealing out justice to tyrants.[99]

It seems that Dr. Kingsford had neither sound mind nor sound body
by this point in her life.

Despite her certainty that she was infinitely superior to Theosophy's
founding mother, Kingsford set forth a theology that was similar to
Blavatsky's in many ways. The belief in reincarnation, astral bodies,
the concept of Karma, the certainty that "every man makes his own
fate," the description of God as "the Substance of humanity"[100] – all
help to explain why Kingsford initially found a ready welcome in London's Theosophical Society. In 1883, a year before their nasty row,
Sinnett warmly praised Kingsford's book, *The Perfect Way*, and
pointed out that its "inner inspirations" appeared identical with Theosophy's.[101] Yet, like Massey and Wyld, her sojourn in Theosophy was
troubled, not only by personality conflicts, but by an underlying disagreement over the place of esoteric wisdom in Christian thought. Blavatsky wanted no part of Christianity; Kingsford, by contrast, thought
of herself as a Christian reformer. Two days after the Hermetic Society
was founded in the spring of 1884, she wrote in her diary: "What we
really seek is to reform the Christian system and start a new Esoteric
Church." Massey concurred warmly and was one of her staunchest
supporters in the new venture; he wrote to Maitland in July that the
Hermetic Society ought to become a sort of "Speculative Church Reform Society."[102]

From the publication of *The Perfect Way* in 1882 until Kingsford's
death six years later, she reiterated the same reform message: Christianity, at its most fundamental core, incorporated the esoteric wisdom
of the pagan schools, particularly "the Hermetic 'mysteries' of Egyptian and Hellenic origin." Christian revelation, she and Maitland argued, was the "descendant and heir" of these schools, not the arch-
competitor. The members of the Hermetic Society should therefore
strip away the false accretions and distortions imposed on Christianity
over the centuries, in order to elucidate its "original esoteric and real
doctrine." Kingsford set the example by interpreting the Creed as a
Hermetic document, just as she and Maitland had already interpreted

the Bible, that supreme "depository replete with occult and mystic lore." Everywhere they looked, in fact – in neo-Platonism, Gnosticism, Sufism, in Pythagorean, Platonic, and Alexandrian texts, in the Eleusinian mysteries, the Cabala, and Hebrew legends – Kingsford and Maitland discovered ever more overwhelming evidence to convince themselves that the Christian religion was not created during the lifetime of Christ. Rather they saw it as a faith whose deep roots traveled back to an ancient, hidden truth, an esoteric knowledge far more fundamental than the teachings of the New Testament.[103] Their "Esoteric Christianity" may have borne as little resemblance to Christianity as Blavatsky's "Esoteric Buddhism" bore to Buddhism, but the distinction between their goals should, nonetheless, be clear. For Blavatsky, the revelation of occult knowledge, through her own agency, rendered Christianity irrelevant and exposed its absurdities. For Kingsford and Maitland, the revelation of esoteric wisdom underscored both the venerability and universality of Christianity and equipped it to participate fully in what they believed would be the future development of religious thought.

It is exceedingly difficult to know what, if any, were Blavatsky's spiritual longings, but with Kingsford, spiritual malaise seems to have been an inextricable part of a more general dissatisfaction with her life and an abiding sense of frustration. Her ceaseless groping for a set of beliefs to stimulate her emotionally, intellectually, and aesthetically led her to Roman Catholicism, Theosophy, and Hermeticism, not to mention the secular causes that she championed with fanatical fervor. She drank deeply from occult waters and embraced unorthodoxy in diverse guises. Yet she never quite managed to break the fine thread that bound her to the Christian faith.[104]

With Annie Besant, the thread appeared to be severed irrevocably as she made herself famous on secularist platforms around the country. Her rejection of Christianity seemed too resolute for backsliding. With the aid of Theosophy, however, she did slide back, at least to a position not unlike the Esoteric Christianity of Kingsford and Maitland. If it is true, as Besant's mother claimed, that her daughter "was always above all else pre-occupied with religion,"[105] then Theosophy helped to turn her attention away from the social crusades in which, for years, she had sought alternative outlets for her religious zeal. Theosophy, in time, also gave her an international stature that neither the National Secular Society nor the Fabian Socialists could ever have offered her. As always with personal motivation, the historian who ventures to assign reasons does so at substantial risk. Whether it was the drive for world fame and unquestioned authority that impelled Besant first to join and then to dominate the Theosophical Society, or whether she was driven forward by the religious problems that had plagued her for years, is not a resolvable question. What is part of the historical record, however, is the trajectory of her spiritual migrations from the high church fervor of her adolescence through free thought to the occult messianism of her last, long Indian phase.[106]

Whatever it was that Besant sought all along the way was not likely to turn up in the National Secular Society. When she became a Theosophist, she admitted that "the Materialism from which I hoped all has failed me," and indeed it is hard to imagine her ardent, emotional personality flourishing in the ranks of Victorian freethinkers.[107] Yet she had committed too much time and drained too much of her energy on behalf of materialism to dump it all in a single moment of Blavatskian bewitchery. In defending her move to Theosophy, she placed great emphasis on the Theosophical rejection of the supernatural. The first lesson taught to the Theosophical novice, she explained,

> is that every idea of the existence of the supernatural must be surrendered. Whatever forces may be latent in the Universe at large or in man in particular, they are wholly natural. *There is no such thing as miracle* . . . This repudiation of the supernatural lies at the very threshold of Theosophy: the supersensuous, the superhuman, Yes; the supernatural, No.

She denied that she had veered abruptly from materialism to a belief in the existence of pure spirit entities and complained that

> "Spirit" is a misleading word, for, historically, it connotes immateriality and a supernatural kind of existence, and the Theosophist believes neither in the one nor the other. With him all living things act in and through a material basis, and "matter" and "spirit" are not found dissociated. But he alleges that matter exists in states other than those at present known to science.[108]

Theosophy perfectly served Besant's needs. Consciously or not, she used it to abandon an intellectual position that no longer satisfied her, if it ever really had. Through Theosophy, she could resume her earlier quest for life's hidden purpose without appearing to succumb entirely to blind religiosity. That she was, nevertheless, moving back toward a more overtly religious frame of mind is beyond doubt, and in her renewed receptivity to spiritual aspirations, she looked sympathetically at Christianity once again. The result, at the turn of the century, was *Esoteric Christianity or The Lesser Mysteries*, a volume in which Besant assumed yet another pose for her readers: the revealer of fundamental Christian truths whose exposure could strengthen the Christian faith in the modern world. She seems to have derived the book's argument from a union of Blavatsky and Kingsford. After explaining that the religions of the past all claimed a hidden, esoteric element, Besant proceeded in great detail "to prove beyond the possibility of rational doubt" that Christianity, too, had an original Gnostic aspect. The atonement, resurrection and ascension, the Trinity and the sacraments, she insisted, could not be completely elucidated until they were seen as manifestations of the esoteric mysteries that formed the universal components of religious experience.

Until Christianity was thus fully illuminated, she warned, it would continue to suffer attrition and decline.

> Christianity, having lost its mystic and esoteric teaching, is losing its hold on a large number of the more highly educated, . . . It is patent to every student of the closing forty years of the last century, that crowds of thoughtful and moral people have slipped away from the churches, because the teachings they received there outraged their intelligence and shocked their moral sense.

It was all the fault of the Protestant Reformation, she implied, with its assertion that the gospel ought to be accessible to everyone, no matter how uneducated. The consequence was a crude simplification of Christian teaching that was abhorrent to tender consciences. Once admit, however, that events recorded in the Bible and "thought to be historical have the deeper significance of the mythical or mystical meaning," and Christianity would enjoy an immeasurable resurgence of strength. The student of the Christian religion would, she promised, find "with joy, that the pearl of great price shines with a purer, clearer lustre when the coating of ignorance is removed and its many colours are seen."[109]

In her search for the roots of Christianity, Besant came, if not exactly full circle, closer to the initial point of departure on her spiritual wanderings than her freethinking colleagues of the 1870s and 1880s would have ever dreamed possible. Her theology was far from orthodox Christianity, but her quest for the hidden wisdom of the world became steeped in a Christian imagery that suggests the strength of her attachment to that faith, even while she remained a leader of the Theosophical Society. She closed her exposition of Esoteric Christianity, for example, with a paean to the "Virgin of Eternal Truth," and she expressed yearning to look upon "the splendour of the Face of the divine Mother, and in Her arms the Child who is the very Truth." In the final paragraph of *Esoteric Christianity*, the Christ imagery predominated, as she wrote exultantly:

> Yet since in man abides His very Self, who shall forbid him to pass within the Veil, and to see with "open face the glory of the Lord"? From the Cave to highest Heaven; such was the pathway of the Word made Flesh, and known as the Way of the Cross. Those who share

the manhood share also the Divinity, and may tread where He has trodden. "What Thou art, That am I."[110]

Christ, truth, esoteric knowledge, and divinity all appear as synonymous concepts for Annie Besant by the turn of the century. Accordingly, after she officially became president of the Theosophical Society in 1907, it seemed to her entirely appropriate to make the search for a new Christ the central feature of a movement devoted to the study of divine wisdom.[111]

THEOSOPHICAL SCIENCE

Theosophy had yet another trait that endeared it to the late nineteenth century, a trait that illustrated its elasticity even more sharply than did its curious relation to Christianity: It claimed to be scientific. At first glimpse, the claim appears simply outlandish, coming from a movement many of whose members were specifically protesting against the entire occidental emphasis on scientific knowledge. Yet Mrs. Besant was not the only Theosophist who found Blavatsky's doctrines intellectually respectable because they seemed capable of scientific proof. One needs to ascertain, however, what she and her Theosophical colleagues meant by scientific proof.

Like so much in Theosophy, venerable traditions and modern developments merged to influence scientific pronouncements. Official statements on science, contained in Blavatsky's *Isis Unveiled* and *The Secret Doctrine*, obscure more than they unveil, but it is nonetheless evident that Blavatsky saw herself as the great synthesizer of science and metaphysics. *The Secret Doctrine* was, in fact, subtitled *The Synthesis of Science, Religion, and Philosophy*, and it bears some comparison with other works of occult scientists, such as Paracelsus and Swedenborg.[112] When Blavatsky spoke of nature's laws, with which Theosophists were supposed to become intimately familiar, she harked back to a neo-Platonic comprehension of scientific inquiry. For her, as for other occultists of the late nineteenth century, the scientist's role was still to explore connections, or correspondences, between the diverse parts of the universe, and the universe, in Blavatsky's gospel, was thoroughly permeated by spirit as a creative, causative agent.

Blavatsky first set forth her overarching philosophy of science in the preface to *Isis Unveiled: A Master-Key to the Mysteries of Ancient and Modern Science and Theology*, where she bemoaned the conflict of reason and religion and pointed the way to true knowledge of the natural world. Looking around at contemporary civilization, she saw

> On the one hand an unspiritual, dogmatic, too often debauched clergy; a host of sects, and three warring great religions; . . . On the other

hand, scientific hypotheses built on sand; no accord upon a single question; rancorous quarrels and jealousy; a general drift into materialism. A death-grapple of Science with Theology for infallibility – "a conflict of ages."

In the midst of this death-grapple, Blavatsky lamented, contemporaries had lost sight of "genuine science and religion," the "twin truths – so strong in their unity, so weak when divided." Just as the spiritualists claimed to have revealed the sole possible meeting place of science and faith, so Blavatsky believed that it was her task to discover a "middle ground," from which the sincere seeker after truth could dispassionately study both fields of endeavor. She found it, not surprisingly, in Platonic philosophy, although Plato might have been astonished to learn that he, "the greatest philosopher of the pre-Christian era," "mirrored faithfully in his works the spiritualism of the Vedic philosophers who lived thousands of years before himself, . . ." Blavatsky then pointed out, on the authority of "Porphyry, of the Neoplatonic School," that Plato's philosophy "was taught and illustrated in the MYSTERIES." Her lack of knowledge about the provenance of Gnostic, Alexandrian, and Hermetic writings was patent, but it was enough for her that she established ostensible links between Eastern religion, ancient philosophy, and occult practice, all of which were pressed into service as part of the great campaign to reconcile science and theology. With her authorities thus jumbled together, Blavatsky proceeded to expound her own hybrid philosophy of cosmic *nous*, or spirit, in which a broad range of additional experts, from Pythagoras to John Tyndall, of all people, were also summoned to support her world view.[113]

As Tyndall's name suggests, Blavatsky, despite her fondness for occult scientists of the past, also derived stimulation from the present, and her writings about science reflected current trends as well as the state of knowledge in bygone centuries. She had a clever eye and ear for what was in the scientific air and could weave seemingly prophetic announcements – concerning the atom's divisibility, for example – into her occult tapestry.[114] In *The Secret Doctrine*, she turned her attention to the evolution of the human race, disagreeing vehemently with the Darwinian account of that process, yet clearly aiming to capture her share of an audience that was avid to learn more about natural organic change. In her account of evolutionary growth, Blavatsky denied that mere physical alterations could have produced the human intellect and spirit, that man emerged from simian forebears, and, in fact, that man had ever existed in any form other than human. Her alternative explanation is of such dense complexity that the simplicity of Darwin's stands out all the more impressively. The central tenet in her version of human evolution concerns the five root races of mankind, whose cyclical and successive patterns of development unfolded independently of animal evolution.[115] The agent of change, in Blavatsky's cosmic scenario, came from outside the physical organism, whether botanical, bestial, or human; it descended from above as a kind of spirit influx, the fuel on which the entire machinery operated.

What is noteworthy about Blavatsky's so-called scientific theories is not their characteristic eccentricity, but her reiterated emphasis on their empirical foundation. Theosophy, she told the *Pall Mall Gazette* in 1884, is "an exact science, based, like any other science, upon the recorded result of centuries of experience."[116] What sort of experience she did not specify, but it could not have been experimentation in any systematic manner – even Yeats's simple experiments sufficed to get him expelled from the Esoteric Section. Given their antipathy to scientific materialism, Theosophists might well have been expected to eschew laboratory investigation; consequently, they might also have been expected to claim empirical corroboration with some restraint. On the contrary, however, Theosophical writers typically exuded complete confidence on that score. Wyld, for one, boasted that he "had obtained *scientific* demonstrations" that miracles frequently occurred, even in his own day. He, at least, was a medical doctor, but Sinnett, with far less justification, assumed scientific airs, scattering scientific terminology throughout his work, and posing as an informed source on new scientific theories. Annie Besant, in keeping with her careful disentanglement from secularism, asserted that "Theosophy accepts the *method* of Science – observation, experiment, arrangement of ascertained facts, induction, hypothesis, deduction, verification, assertion of the discovered truth."[117] She was the proud author of a pamphlet on *Occult Chemistry* and wrote with equal authority about planetary chains and other celestial phenomena.

Besant's approach to research, however, reveals the limits of her commitment to the modern "*method* of Science" and, as with Blavatsky, places her in the tradition of the neo-Platonists, for whom "chemistry took on a quasi-religious aura."[118] Tangible evidence, as such, interested her very little. She was only being candid when, on publishing an expanded edition of *Occult Chemistry*, she subtitled it *A Series of Clairvoyant Observations on the Chemical Elements*. The surprising insights that it contained were derived from exercises of will, not laboratory calculations.[119] In the chapter on "Theosophy as Science," which begins Besant's little text on Theosophy, published in 1912, she wrote at length about will, vitality, intellect, and mind. She described the creative activity of the monad, or eternal man, and traced the links between the physical body and its astral counterpart. She examined the mental sphere, including its "heaven-portion," the intuitional sphere, "in which the Christ-nature unfolds in the Man," and the spiritual sphere, and she concluded with a few brief comments about the significance of religious rites and ceremonies. Although she asserted that these subjects fell within the purview of Theosophical science, since Theosophy had undertaken vastly to increase the areas of knowledge open to scientific inquiry,[120] it was clear that theology completely eclipsed science, as any twentieth-century reader would have understood that term, in Besant's volume.

If Theosophical scientists could boldly identify causative agencies where the physical science of the day could only raise questions, that was because Theosophists had a decidedly unfair advantage over professional scientists. The former could range freely over a spirit world of their own creation to designate instruments of change in the material world, and they could do so in the name of a "Higher Science." As the Theosophist William Kingsland, later a biographer of Blavatsky, wrote in 1888:

> There exists . . . a Higher Science, which is also Religion in its truest sense, and which deals with the hidden forces in nature at which Physical Science stops short, but which are more than suspected by the majority of mankind, because every form of Religion whatsoever is an acknowledgment of a *something*, which underlies, and is superior to, the phenomena of Nature.

Massey, likewise, stressed the difference between merely superficial "information concerning things" and real knowledge of them which brought the inquirer face to face with the "Divine principle of Nature." Science, Massey argued, was grievously incomplete without religion.[121]

The confusion of theological speculation and scientific concepts, of inner vision and outward inquiry, sprang from the Theosophical belief in the close intermingling of matter and spirit. In some cases, no doubt, both the vocabulary and authority of science were quite purposefully misapplied, to bolster metaphysical or theological conjectures that needed to be "packaged" for Western audiences. But what was significant about Theosophy was not, of course, its claim to contribute to biology, chemistry, or physics, but its role as a religious alternative. Like spiritualism, it arose at a time when Christianity's many weaknesses had been starkly exposed, but when people still felt the acute pain of trying to live without some sense of meaning, purpose, design, and beneficence in the universe. The certainty with which Theosophists maintained that there existed beings higher than the earthbound specimens of humanity had scarcely any specific connection with current theories of biological evolution, or even with current theories of material and technological progress. It was the reflection of a religious need – the need both to aspire to some future loftier condition of existence and to feel, in this life, the protection and care of something superhuman.

The manipulation and misuse of scientific language for spiritual purposes was hardly unique to Theosophy in this period. In addition to spiritualism itself in its several varieties, such manipulation characterized the Christian Science of Mary Baker Eddy and the Anthroposophy of Rudolf Steiner, which grew out of Steiner's affiliation with Theosophy in Germany. Swedenborgianism, too, combined pseudo-scientific terminology with its mysticism, but that faith had first blossomed in a far different intellectual climate and was even more immediately inspired by Renaissance science than were the occult movements that emerged in the last quarter of the nineteenth century. Blavatsky was quick to point out Swedenborg's intellectual debt to "the Hermetic philosophers," Pythagoras, and the cabalists, and, in order further to minimize his scientific contributions, she dubbed him merely "a seer; . . . *not* an *adept*."[122] Yet it was not Blavatsky's modern scientific acumen that drew men and women to Theosophy. Fundamentally, they were looking for antidotes to science, alternative ways of interpreting the universe and human destiny, even though they might be impressed by an apparent familiarity with scientific theories. Since Blavatsky could sprinkle the jargon of science throughout her writing, it meant to many a devotee that she had mastered the limited knowledge that modern science could provide and had gone beyond it to offer the far profounder knowledge that they sought.

C. C. Massey observed in the late 1880s that Christian teachings on immortality were meager indeed when compared to "some Eastern systems of religious philosophy."[123] What science had not rendered incredible in Christian doctrine, biblical criticism had left suspect, or moral outrage had made unacceptable. Furthermore, the late nineteenth-century Christian churches were less concerned with enriching theology than with trying to secure for themselves a popular base through increased emphasis on the social role of institutionalized religion. As care and cure of souls came to play a less prominent part in their daily concerns, other sources of comfort developed to supplement, or completely replace, their services. What Christianity left imprecise about human survival became fully explicated, if along rather different lines, both in spiritualism and in Theosophy. Future joys, as well as trials, in the life after death were vividly depicted in spiritualist and Theosophical literature alike, with exhaustive details that rescued the fearful from the terrors of uncertainty.

Frederic William Henry Myers. (Photo by his wife, Eveleen Myers.)

Eleanor Balfour Sidgwick (Mrs. Henry Sidgwick). From Ethel Sidgwick, *Mrs. Henry Sidgwick* (London: Sidgwick & Jackson, 1938).

Henry Sidgwick. From *Popular Science Monthly*, 1899.

Richard Hodgson. From Ruth Brandon, *The Spiritualists* (London: Weidenfeld & Nicolson, 1983).

Edmund Gurney. From Trevor H. Hall, *The Strange Case of Edmund Gurney,* 2d ed. (London: Duckworth, 1980).

Frank Podmore. From Edward R. Pease, *The History of the Fabian Society,* 2d ed. (London: The Fabian Society and George Allen & Unwin, 1925).

William and Mary Howitt.

Edmund Dawson Rogers. From E. D. Rogers, *Life and Experiences of Edmund Dawson Rogers* (London: Office of *Light*, [1911]).

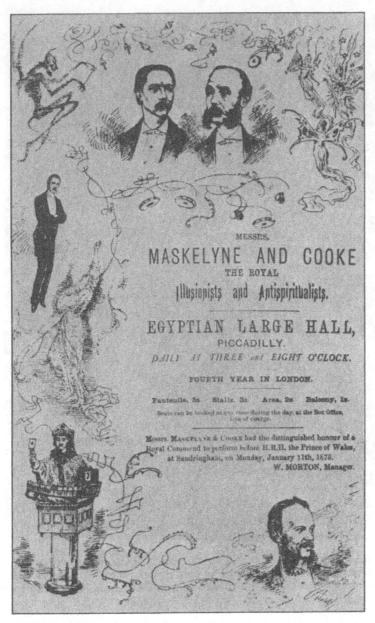

An advertisement for the performance given by Maskelyne and Cooke, conjurors, in the Egyptian Hall, Piccadilly, London.

The sequence of events depicted here chronicles Sir George Sitwell's exposure in 1879 of a medium masquerading as a materialized spirit. The occasion, very similar to Sitwell's capture of Florence Cook in 1880 (see p. 18), was illustrated in the *Graphic*.

"Katie King." From Richard Cavendish, *Man, Myth and Magic*, vol. 14 (Freeport, N.Y.: Marshall Cavendish Corp., 1970).

Florence Cook. From the Harry Price Library, University of London.

William Stainton Moses. From W. S. Moses, *Spirit Teachings through the Mediumship of William Stainton Moses,* 7th ed. (London: London Spiritualist Alliance, 1912).

Daniel Dunglas Home.

Helena Petrovna Blavatsky.

Kate Fox.

Eusapia Palladino. From *Hearst's International Cosmopolitan Magazine*, February 1910.

Hugh Reginald Haweis.

Anna Maria Hall (Mrs. Samuel Carter Hall).

Alfred Russell Wallace. From Karl Werckmeister, ed., *Das neunzehnte Jahrhundert in Bildnissen* (Berlin, 1898–1901).

Robert Chambers.
From Henrietta Ward,
*Memories of Ninety
Years* (London: Hutch-
inson, 1924).

Lord Rayleigh (John William Strutt,
third Baron Rayleigh). From *Popular
Science Monthly*, 1884.

Joseph John Thomson. From
Graphic, April 1908.

William Fletcher Barrett. From *Personality Survives Death. Messages from Sir William Barrett*, ed. Florence Barrett (London: Longmans, Green, 1937).

William Crookes.

Oliver Lodge.

PART III
A pseudoscience

Deeply as numberless spiritualists in Britain cherished spiritualism for the religious comfort that it offered, they tended to emphasize the purportedly scientific foundations of their beliefs when they urged the claims of spiritualism to public attention and respect. It was the reiterated contention of spiritualists during the latter half of the nineteenth century that spiritualist phenomena were verifiable by the empirical methods employed in the physical sciences. There was nothing in spiritualism, they maintained, that had not been seen, heard, smelled, or touched by witnesses in full possession of their senses. Phrases such as "firm scientific basis" and "careful scientific research" recur throughout the spiritualist literature of the period. In promoting a new spiritualist monthly, the *Spiritual Record*, for example, the publishers boasted in 1884 that it treated "the whole subject of Spiritual manifestations exactly as it would any branch of natural science,"[1] and spiritualists argued the affinities between their investigations and scientific experiments on every possible occasion. Their homage to science was far more sincere than the mere lip service paid by Theosophists. Whatever reservations spiritualists may have felt about the ultimate implications of scientific revelations, they did not regard scientific procedure with the cavalier license of Blavatsky and Besant.

Even the most credulous spiritualist doffed his hat, or her bonnet, to science in these decades and fully recognized the need to present spiritualist evidence with the apparent endorsement of its august authority. So much has been written about the Victorian veneration of science that it hardly seems necessary to emphasize the point. Not only were scientific discoveries greeted with acclaim and widely publicized, but the scientific method itself was hailed, almost reverently, as the surest means of attaining truth. Many Victorians shared the hope that this method could be adapted to all realms of intellectual inquiry, that all areas of knowledge would prove reducible to scientific laws, for physical science had "set up a new standard of what is antecedently credible."[2] It had established new criteria for the Victorian system of values, and scientists commanded unprecedented public admiration, for they appeared to hold in their collective grasp nothing less than the future of human civilization.

British spiritualists were scarcely unique in seeking the prestige and power associated with the methods of science, yet their veneration was not wholehearted. Although they busily posed as scientific investigators, they were in fact rejecting the grim conclusions concerning human life and universal disorder toward which science seemed inexorably directed. Their ambivalence about science was, of course, shared by countless other Victorians for whom it would be mistaken to assume an unquestioning and cheerful optimism about the social impact of scientific discoveries. For all the excitement engendered by such discoveries, for all the confidence that science would prove the source of future blessings to humanity, there remained a lingering awareness of its destructive force. Not a few Victorians watched with fascinated horror as the strength of scientific evidence and the power of scientific argument tore to shreds the comforts of traditional religious beliefs, without providing new shelters for man's spiritual needs.

It was the fond hope of British spiritualists that, through their faith, the constructive aspects of the scientific method might be harnessed to the search for philosophical or religious meaning in human existence, thereby mitigating the destructive impact of science. If the validity of spiritualist phenomena could be proven in acceptable scientific fashion, then science could become once again, as in past centuries, the defender and not the challenger of faith. If mediums could be subjected to laboratory experiments, the results, surely, would be as conclusive as any findings from a chemist's flask. If the possibility of communication with the dead could be thus indisputably affirmed, the promise of human immortality would cease to be mere metaphysical aspiration and would assume the imposing status of scientific fact. Clearly, the claim of spiritualism to offer an alternative source of religious reassurance depended heavily on its scientific pretensions.

Yet, paradoxically, central to those pretensions was the spiritualists' desire to redefine science in their own terms and to argue against the contemporary concept of positivistic science. British spiritualists repudiated current assumptions about nature and the extent of natural law, and they persistently sought to stretch the boundaries of the natural world beyond physical causes and effects into the realm of spirit.

That was Myers's constant goal, and C. M. Davies, likewise, provided something of a manifesto when he wrote that the tenets of spiritualism

> rest on the demonstrations of science – a science, however, which does not illogically stop short at the physical or intellectual, ignoring the spiritual portion of man's being, but applies its rigorous analysis to the domain of revelation hitherto disposed of in the wide category of the supernatural. Spiritualism has no such word as Supernatural. It substitutes the certainly less objectionable term supra-sensual – for who shall presume to define the limits of nature?[3]

When spiritualists used such language, they were not necessarily aspiring to a sort of Theosophical Higher Science, where any and all cosmic speculations might freely assume scientific airs. It is true that most British spiritualists ended, like Myers, by extending the laws of nature right into the errors of pseudoscience, but a few were fully conversant with the lessons of scientific pioneers in centuries past. They realized that, throughout history, the contours of natural science had time and again proved more elastic than orthodox defenders believed possible, and they took as their guiding examples the treatment of Galileo and Harvey. They asserted repeatedly that scientific methodology applied to the subjects that they explored because nature's rules extended over a far vaster range of phenomena than scientists had previously suspected.

In aiming to dissociate themselves from the supernatural, in reiterating that nothing miraculous occurred at séances, Victorian spiritualists underscored the eighteenth-century rationalist roots of their faith. Their efforts to present spiritualism in the guise of a rational, scientific pursuit, however, received little encouragement from the scientific professions, including medicine. The spiritualist assertion that the phenomena of the spirit world were accessible to the investigative methods employed in the physical realm tended to provoke amusement, scorn, or wrath among the members of Britain's scientific societies. Contempt and anger were largely justified, for spiritualists did not, in fact, demand of their investigators the true rigors of laboratory experimentation. They might solemnly pledge their devotion to the procedures of the physical sciences, but few had any substantial knowledge of the painstaking observation and analysis required of the scientist. They scarcely understood what constituted evidence in the scientific sense, nor were they capable of the detachment and impartiality needed for the successful completion of a genuinely open-ended inductive inquiry. Stainton Moses summarized a widespread assumption among his fellow believers when he glibly told the London Spiritualist Alliance in 1887: "I think I employed in my investigations what is known as 'the scientific method,' *i.e.*, I endeavoured to be sure of my facts and to appreciate their significance."[4] Such diminution of the scientific endeavor, whether unwitting or purposeful, had been characteristic of spiritualist writings in Britain from midcentury. As early as 1860, it had prompted G. H. Lewes to dismiss the mass of spiritualist believers as "so unacquainted with the principles of evidence and the ordinary methods of verification, that they willingly listen to the impudent excuses by which the impostors resist and evade inquiry."[5]

Given the notorious credulity of most spiritualists, even the more cautious psychical researchers aroused resentment in scientific circles and found it difficult to justify their inquiries to skeptical scientists. Albeit with distinguished exceptions, psychical researchers were also perceived as nonprofessional intruders in complex and exacting fields of inquiry. In this period, as British science emerged from the grip of gentleman amateurs to achieve a professional status both in academe and industry, trained scientists were likely to feel uneasy about a group like the SPR. In part, too, the scientific profession feared that spiritualism and psychical research threatened to reintroduce into modern science those links with magic and the occult from which it had only recently broken free. Perhaps, furthermore, professional scientists shrank from admitting into their ranks an element of popular vulgarization that could only reflect disreputably on their own inquiries. With good reason, they suspected that the spiritualists' concept of science had more to do with entertaining demonstrations by traveling performers than with laboratory apparatus. Spiritualists did indeed believe in a science that was demonstrable to the uninitiated public, and whose discoveries could be readily apprehended by the audience at a public lecture.[6] While British scientists in the second half of the century were taking pride in the increased opportunities for specialized scientific education in Britain, spiritualists were stoutly resisting the notion that science was the preserve of the magisterial few.

Accordingly, they never acknowledged that the scientific profession was justified in rejecting their credentials as scientific investigators. When the *Spiritualist Newspaper* assailed the arguments of scientists who belittled spiritualism in 1876, it described "English men of science" as "a standing example before the world, of men steeped in a kind of priestly bigotry founded upon physics."[7] Major-General A. W. Drayson of the London Spiritualist Alliance was only one of many outraged spiritualists who joined the counterattack. How, he asked in 1884, could scientists, after a cursory study of spiritualism lasting only a few weeks, presume to speak definitively on the subject, blithely castigating the work of people whose investigations were far more extensive and laborious than their own? "When," Drayson concluded a little too assertively, "we find that the mental condition of these individuals is such as to render them unwilling to collect or examine facts before they theorise, we are naturally disposed to question the competence of such minds to form conclusions on any branch of Science."[8]

The issue did not, however, reduce itself to a simple conflict of spiritualist versus scientist. The antagonists were no more readily consigned to fixed categories than were Christian and non-Christian spiritualists. Just as numerous spiritualists earnestly admired the achievements of professional scientists and wished to emulate them, so there were scientists, and very distinguished ones at that, who could not join in any easy repudiation of spiritualism. These men sought to maintain an open mind about psychical phenomena and séance manifestations. Some became convinced spiritualists themselves, to the dismay of their professional colleagues who tried to explain their apostasy as temporary lapses of judgment at best, and, at worst, as proof of an utterly uncritical "will to believe." Others never ventured beyond the role of psychical researcher. In both cases they initially approached the phenomena as part of their serious scientific investigations, hoping that these bizarre occurrences might reveal hitherto unknown scientific laws or hitherto unimagined powers of the human race. Yet what evidential basis could they find in perambulating furniture, in mysterious raps, voices, and handwriting, and in ectoplasmic materializations of spirit forms? Was it, in fact, possible to uphold the standards of scientific inquiry while pursuing the will-o'-the-wisps of psychical research?

6

Concepts of mind

Despite its advances, nineteenth-century science did not resolve all the puzzles that troubled natural philosophers of previous centuries. No question underscored the continuing metaphysical concerns of Victorian science more insistently than the age-old enigma of mind and body. The relationship of mental and physical was a fundamental issue underlying all studies of man, whether undertaken by inquirers in the developing fields of physiology, neurology, psychology, and psychiatry, by public health officials with practical concerns to address, or by students of ethics. The problem was equally significant to spiritualists and psychical researchers, for the independent existence of mind was, of course, an essential part of their argument against materialism. Whether dubbed mind, soul, spirit, or ego, a word gaining currency throughout the second half of the century, such an entity distinct from brain tissue was requisite to rescue man from a state of virtual automatism, a mere bundle of physical and chemical properties. That an entire religious world view and system of morality hung in the balance was patent to all concerned in the debate, and the controversy often proceeded within a philosophical or theological framework. It seemed increasingly clear, however, as the accomplishments of science demanded ever greater respect from the Victorian public, that the authoritative answers could only lie in the concrete explorations of the physical sciences. Thus spiritualists and psychical researchers alike found themselves drawn to the infant study of the human mind from a scientific perspective.

Yet definitive answers eluded the nineteenth-century students of the mind, as they had eluded their myriad predecessors in the past. James John Garth Wilkinson, a London medical doctor of unorthodox views, wrote in disgust to his American friend, Henry James, Sr., in 1850:

> When I look at the disconnection in science between man and his own body, I cease to wonder at the difficulty the great Emerson has in thinking of an incarnate God. Why, the philosophers have never yet got to think of man himself as incarnate. They admit either the

flesh without the spirit, or the spirit without the flesh. The thought
has yet to come which will combine the two.[1]

Many thoughts came in subsequent decades, from many different pens,
but flesh and spirit remained in their uneasy, ill-defined alliance. Des-
cartes's mind–body dualism obviously could not serve the needs of a
psychology that sought identification with the physical sciences, but
what was to replace the Cartesian solution to the problem remained
the subject of heated disputes. The growth of psychology, away from
its philosophical origins toward its biological destiny, did not carry with
it any overwhelmingly compelling new answers to the same repeated
question.[2] The answers which the psychologist William McDougall suggested in *Body and Mind*, published in 1911, were certainly no less
speculative for the passage of time than similarly titled books and articles that had appeared well before the turn of the century.

By about 1850, with the problem of dualism no longer being articulated in its Cartesian form, conscious mind was widely conceived as
interacting with unconscious matter.[3] Medical doctors, professors of
logic and mental philosophy, armchair metaphysicians, and natural scientists of varying expertise all had theories to propound concerning
the mutual influence of mind and body, and their diverse views came
together on at least one point: the conviction that to separate the one
from the other was to do violent theoretical injustice to the integrity
of the human organism. They agreed that mind and body had an inextricably intimate relationship, that "the ongoings of the one would be
often a clue to the ongoings of the other."[4] Nevertheless, to assert that
mind affects matter, and vice versa, is a long way from explaining the
impact of one on the other. Although, at some time in their lives, most
of the intellectual giants of the Victorian era undertook just such an
explanation, none ultimately succeeded in untying, or cutting, that Gor-
dian knot. One of the most influential theories posited a psychophysical
parallelism in which mental and bodily states were two different aspects
of the same reality. They were parallel, or corresponding, conditions
within a single context, not directly impinging on each other, as parallel
lines do not intersect, but both essential parts of the functioning organism, and indeed each responsive to the functioning of the other.
Another point of view sought to break away entirely from all vestiges
of dualistic thought and looked only to the world of physics, to the
inseparable connection between matter and force. Given that connection, the argument proceeded, the matter of the physical body had no
need of a distinct animating principle or spirit.

At one extreme, someone like the psychiatrist Henry Maudsley could
deny that separate categories of mind and body had any validity at all.

He upheld the inductive, positivist methods of physiology as the only way to establish psychology on solid foundations. By studying the brain in its manifold functions, by investigating all the interconnected parts of the nervous system, and by applying known physical laws to them, the psychophysiologists trusted that they were approaching nearer to mind than the old methods of introspection had ever allowed. Others, however, found it possible to question whether mind might not still slip through the physiological nets held out to catch it. If introspection seemed hopelessly outdated by the 1870s and totally at odds with the methodology of the natural sciences, some scientists nonetheless refused to accept the universal applicability of the "unified scheme of scientific explanation" that emerged during the third quarter of the nineteenth century.[5] Volition, emotion, imagination, and intellect seemed to demand more subtle analysis than that supplied by the apparently harsh reductionism of the psychophysiologists, whose arguments threatened to disperse the wholeness of the human organism among the fragments of its component parts, all interlocked in place with a finality that looked decidedly deterministic. There could be no hope of immortality for such a mechanistic creature, operating solely by sensory stimuli, reflex actions, and motor responses. Nor could there be any question of free will and personal moral responsibility. If carried to their ultimate conclusions, the views of the psychophysiologists, in short, deprived man of humanity's distinguishing traits, as traditionally perceived. At the other extreme of the mind–body controversy, the terrifying implications of these conclusions could lead, as in the case of T. H. Green, to the assertion that the methods of modern science were simply inappropriate and irrelevant to the study of the human mind.

More typically, however, scientists and philosophers alike sought the middle ground. They trusted that, as psychology developed, it would prove hospitable both to the experimental method and to the vague, yet often persuasive, insights offered by philosophical speculation about mind. Spiritualists and psychical researchers seized upon this tentative truce to interject their own concerns into the ongoing discussions. The pursuit of certainty about the human spirit led them far into the uncharted realms of psychology, where they brought to the work at hand their special commitments and aspirations.

PHRENOLOGY AND MESMERISM

Even before spiritualists added their contributions, two earlier movements had claimed to offer fresh insights into the ancient debate over thought and matter. Phrenology and mesmerism both commanded substantial attention in Britain during the first half of the nineteenth century, and both subsequently developed close ties with spiritualism. While appearing to advance a materialist, somatic approach to the study of the human organism, both in fact harbored mentalist tendencies, and their two-edged appeal made them, for several decades, central to the dispute over the nature of mind.

Phrenology's contribution to psychology did not lie in its division of the brain into a cluster of more than thirty propensities, sentiments, and intellectual faculties, each functioning as a separate unit within the cranium. Its claim to recognition by modern physiological psychology lay instead in phrenology's firm assertion that the brain alone is the organ of the mind – the center of thought and feeling alike – and in the affirmation that different parts of the brain affect human actions and behavior in varying ways. Franz Joseph Gall, the German physician who expounded the theories behind phrenology in the opening years of the nineteenth century, was not the first to assign to the brain sole responsibility for intellect, passion, reason, and all the other presumably nonphysical components of human life about which men had speculated for centuries. Nor was he the first to posit cerebral localization. What distinguished Gall's work from that of classical, medieval, and Renaissance metaphysicians, as well as anatomists of the more recent past, was his promotion of scientific methodology over philosophical and epistemological concerns. For him, the brain was as amenable to empirical observation and inductive reasoning as any physical organ.[6]

Gall's much emphasized empiricism was, in fact, heavily diluted with a priori assumptions about the links between human behavior and certain physical traits. He developed his complex organology by comparing the outsides of people's heads with their individual conduct, and he formulated his theory of the brain's structure independently of anatomical studies.[7] In time, evidence that the cranium did not strictly conform to the contours of the brain within, and that particular portions of brain tissue did not invariably correspond to predictable behavioral disorders or personality types, undermined phrenology's credibility. During the 1820s and 1830s, however, Gall's theories exerted considerable appeal in Britain, for they appeared to rescue mind from the vagaries of metaphysics by offering a persuasive physical explanation of its functions. With the organ of each feeling and faculty extended or limited in size as that feeling or faculty was more or less utilized, every human personality could be analyzed, according to phrenology, as the product of interacting cerebral organs in diverse states of dominance or weakness.

Contemporaries, needless to say, were quick to comment on the materialist implications of phrenology. "The objection, that phrenology leads to materialism, has been frequently urged against the science," observed the first issue of the *Phrenological Journal*, launched by the very active Edinburgh Phrenological Society.[8] The Edinburgh lawyer George Combe, a leading promoter of phrenology in Britain, was careful not to publicize the degree to which phrenology could undermine Christian faith, but he admitted as much in his private correspondence. "According to my perceptions," he wrote in 1847,

> man on earth knows absolutely nothing of spirit. What are called spiritual influences appear to me to be merely influences acting on the brain, and that the spirit which is felt within us is the action of the brain; which men mistake for spirit just because consciousness does not reveal the existence of organs.

Even kindly John Tulloch, writing many decades later and trying to depict Combe as "an earnest Christian theist," had to concede that he held "views of man's constitution and responsibility which seem constantly shading off into Materialism."[9]

Yet the mind is not easily reducible to clear interpretative categories. Even while phrenology was enticing and infuriating some contemporaries with its evident materialism, others were attracted by the apparent flexibility of its doctrines. Combe and other spokesmen for the new "science" refused publicly to adopt an unequivocally physical basis of mind and dismissed the question as one about which it was futile to speculate. "Of the mind, as a separate entity, we can know nothing whatever," commented Robert MacNish, another Scottish phrenological enthusiast, "and we must judge of it in the only way in which it comes under our cognizance."[10] Some practitioners and students, however, explicitly rejected the materialist interpretation of phrenology and maintained their Christian principles while avidly studying cranial contours. As late as the 1860s, James C. L. Carson, for one, published a volume that aggressively asserted *The Fundamental Principles of Phrenology are the Only Principles Capable of Being Reconciled with the Immateriality and Immortality of the Soul.*[11]

Carson was fully justified in maintaining that one could subscribe to basic phrenological beliefs without thereby having to endorse the material nature of mind. Within phrenological doctrine itself, there was leeway for some nonmaterial agent at work inside the brain. No phrenologist assumed that human character was merely the total of all the cerebral organs piled on top of one another. On the contrary, the individual personality emerged from the interaction of those organs which, when harmoniously balanced, produced rational conduct, but, when imbalanced, led to widely varied abnormal behavior. In short, phrenological theory acknowledged that the whole of the human personality was very much more than the sum of its component parts, and phrenologists never analyzed a single cerebral bump without taking the entire contour of the skull into consideration.

Early in the nineteenth century, phrenology both encouraged a clinical interest in cerebral functions within the expanding expertise of the medical profession and celebrated the uniqueness of each man, woman, and child as a moral being. It suggested to some that the scientific study of mankind, far from reducing humanity to mechanical automata operating according to strictly physical stimuli, might open vast new possibilities for the improvement of the species. If the organs of the mind that kept man tied to his lowest capacities and basest qualities could be reduced in potency, and if those organs that fostered man's higher nature could be nurtured, future human history might be glorious indeed. Such optimism faded with the declining authority of phrenology, but even at the end of the nineteenth century, phrenology's remaining devotees still embraced it as both scientific canon and article of faith. In designing the curriculum for the Progressive Lyceums in the 1880s, Alfred Kitson, for example, classified phrenology with the "laws of health," as a science to be taught to spiritualist children – but a "science which treats of the brain as being so many avenues or functions, by which the soul has to express itself."[12]

The nonmaterial strain, barely concealed by the more salient physical emphases of phrenology, assumed even greater significance in mesmerism, or animal magnetism. This self-proclaimed new science initially explained its extraordinary accomplishments in physical terms, but managed to survive the misconceptions of its founder only because scientists came to accept the role of imagination and suggestion in giving the mind dominance over the body. The close ties between mesmerism and faith healing, furthermore, and the occult ancestry of animal magnetism in the cosmic theories of Paracelsus, among others, gave the movement an aspect of magic and mysticism not to be found in the less flamboyant phrenology.[13]

In theory, animal magnetism depended upon a physical agent to explain the workings both of man and of cosmos. The Viennese doctor, Franz Anton Mesmer, who claimed to have discovered animal magnetism in the 1770s, attributed its effects to an invisible, superfine fluid existing in and around all objects of the universe.[14] It connected earth to sky, human to celestial bodies, and its proper or improper distribution within any individual explained that person's state of health or disease, for whenever the fluid could not flow freely through the body, illness ensued. The treatment necessary to restore the patient to harmony, with himself and with nature, had far more in common with the stroking, or laying-on of hands, practiced by faith healers and village wizards than with the standard eighteenth-century doctor's bag of frequently murderous medical tricks. Treating the body as a magnet, Mesmer and his disciples sought to activate the human magnetic poles (including nose and fingers) in an effort to send the fluid coursing properly around its bodily track.[15] Mesmer's first remarkable therapeutic successes were accomplished when he actually used metal magnets on suffering patients, but he soon discovered that magnets played no essential part in the curative process. By touching the patient with his hand, by rubbing the afflicted portions of the patient's body, or simply by making passes in the air above them, Mesmer could produce striking results – often too striking, as clients characteristically experienced a convulsive or hysterical crisis as part of the catharsis preceding recovery. Clearly, metallic magnetism could not explain the phenomena that Mesmer provoked. Yet he was so certain that he had revealed a force analogous to the effect produced by magnets that he persisted in labeling it animal magnetism.

The sensationalism surrounding Mesmer's cures should not obscure the commonplace quality of his theory, at least from the perspective of his contemporaries. When he moved to Paris in 1778, he established a fashionable clinic where attention focused on the paraphernalia of mesmerism. There was the mysterious *baquet*, or tub, around which numerous patients sat together, receiving the magnetic fluid from iron rods protruding at right angles through the lid of the tub. There was the rope that tied the patients to each other, and to the tub, in order to form a magnetic chain. Above all, there was the awesome figure of Mesmer himself, presiding as master of ceremonies, often in a magician's gown, and appearing as a veritable storehouse of magnetic force that he was willing to convey to the feeble and debilitated victims of the Parisian social round.[16] Here was enough excitement to capture and hold public interest without any need for an original cosmic vision to lend significance to the lively exhibitions. Indeed, by the measure of other late eighteenth-century cosmologies, there was nothing at all original about Mesmer's superfine, magnetic fluid. The eighteenth century was awash with fluids – universal, vitalistic, mechanistic, animistic, subtle, magnetized, and electrical. In the works of numerous writers, both for a scientific and a popular audience, electricity, light, fire, heat, and even gravity were liable to figure as universal fluids, or as the products of such fluids. It hardly required much effort to add the magnetic force to that list. As confidence in the validity of Aristotle's four basic elements waned with the early advances of modern chemistry, scientists and pseudoscientists alike sought alternative hypotheses to explain the composition of the cosmos. Mesmer's was only one among many theories that confidently proposed to offer its adherents a glimpse into the fundamental physical reality.[17]

The earliest prominent champion of animal magnetism in England was John Elliotson, professor of medicine at University College, London, Fellow of the Royal College of Physicians, and president of the Royal Medical and Chirurgical Society. Although several practitioners of the mesmeric art had visited Britain periodically since the 1780s, it was not until 1837, when Elliotson seriously turned his attention to the subject, that it became the center of a national controversy.[18] Controversy, in fact, characterized Elliotson's entire medical career. Among the first English physicians to use the stethoscope, an experimenter with acupuncture, and innovator in the prescription of extremely heavy doses of medicine, he devoted his professional life to battling the traditional, conservative practices of the English medical establishment. It was primarily his insistence that a school of medicine needed to have an affiliated training and research hospital that led to the founding of University College Hospital, where Elliotson was an effective and popular lecturer. He also maintained an extensive private practice and was a highly regarded diagnostician whose articles and books were widely read.[19]

Elliotson endorsed the therapeutic powers of animal magnetism after observing demonstrations by Dupotet, a French mesmerist performing in London in 1837. What intrigued Elliotson was both the apparent curative impact of the treatment and its potency as a form of anaesthesia far greater than any available drug. Fascinated as he had always been with questions posed by nervous disorders, he felt certain that mesmerism could provide essential clues to the neurological puzzles of the human organism. He began experimenting with mesmerism himself and was soon giving demonstrations at University College Hospital to crowded and eminent audiences. As always, Elliotson knew how to attract publicity, but the medical faculty of the college loathed public exhibitions that introduced a circus atmosphere into the hospital wards and associated their institution with a disreputable practice that seemed ready made for charlatans. The fact that Elliotson's mesmeric treatments frequently met with remarkable success in relieving the symptoms of illness did not soften the opposition of his hostile colleagues nor calm the ire of the hospital trustees. Given the choice of ceasing his demonstrations or resigning his professorship, he took the second option late in 1838. For the remaining thirty years of his life, he staunchly articulated his unswerving conviction that animal magnetism was an inestimably valuable addition to nineteenth-century medical practice. He continued to mesmerize his private patients, founded a mesmeric infirmary in London in 1849, and in 1843 began a journal, the *Zoist*, which both explained the theory behind mesmerism and publicized its therapeutic prowess.[20]

Elliotson's campaign on behalf of animal magnetism ran aground on the physical explanation that he persisted in defending. His belief in the existence of an invisible magnetic fluid exposed him to charges of scientific carelessness, to the point of irresponsibility. As early as 1784, a French royal commission had investigated animal magnetism and conclusively denied the existence of Mesmer's magnetic fluid. Elliotson's arguments in favor of such a substance seemed not merely naive and uncritical, but fatally untenable and ridiculously easy to disprove. Indeed most of the British medical profession in the 1830s and 1840s condemned animal magnetism for its lack of empirically demonstrable physical foundations, at a time when somatic theories of disease alone carried scientific credibility.[21]

Mesmerism, however, proved as flexible a set of beliefs as phrenology, as protean a bundle of ideas that could simultaneously serve contrasting systems of thought. Doctors might point to its inadequate, or nonexistent, material basis, but much nonmedical criticism was leveled against animal magnetism precisely for the materialism that it appeared to imply. Although Elliotson did not intentionally mold Mesmer's universal fluid into a weapon against Christianity, it could easily be perceived as replacing that independent, animating principle – spirit or soul – that was central to religious faith. The links between mesmerism and materialism were celebrated, for example, in *Letters on the Laws of Man's Nature and Development* (1851), written by Harriet Martineau and Henry George Atkinson, a dilettante of independent means with a taste for philosophy and mesmerism.[22] Yet, in the same year, William Gregory, chemistry professor at Edinburgh, could publish his *Letters to a Candid Inquirer on Animal Magnetism*, which dismissed all arguments that mesmerism led inexorably to materialism and infidelity. Earlier, an anonymous work by "a Beneficed Clergyman" was simply and bluntly entitled *Mesmerism the Gift of God*.[23] Mesmerism could be all things to all people because the mesmeric force remained utterly elusive. Despite the claims of Elliotson and his followers, it could not be observed and measured, nor reduced to chemical compounds, by doctors and scientists. Fundamentally, its properties were vitalistic, associated in the popular mind with a state of health and, at one remove, with the idea of life itself. For some enthusiasts, it embodied and disseminated the enigmatic life principle, a curative force in nature that never came within the reach of laboratory analysis, whether by microscope or telescope.

Although the equipment of the physical sciences proved inadequate to capture Mesmer's imaginary fluid, the mental sciences had different instruments at hand. If one hypothesized that the mesmeric power was not centered in the mesmerist, but in his subject – if one ceased to think in terms of magnetic fluid affecting the body of the patient and concentrated instead on the patient's mind – vast new interpretative possibilities arose. The role of the imagination in producing the so-called mesmeric phenomena had been recognized since the 1780s when the French royal commission attributed many of Mesmer's astonishing results to the patient's expectant imagination.[24] The study of mental disorders had not proceeded far enough in the late eighteenth century to pique the commissioners' curiosity about the ways in which imagination could provoke violent physical effects. It was enough for them to prove to their own satisfaction that no physical reality adhered to Mesmer's pretensions. Once they had denied his fluid material existence, they assumed that animal magnetism ceased to come within the scope of science. As that scope expanded in the nineteenth century, however, and particularly as the mental implications of animal magnetism began to attract the attention of the medical profession, mesmerism finally escaped its degrading associations with the occult and gained a respectable scientific status as hypnotism.

The first stages of the gradual transformation emerged from the work of James Braid, a Scottish surgeon living in Manchester. Braid became interested in animal magnetism when another French mesmeric performer, Lafontaine or La Fontaine, visited Manchester in 1841. Initially skeptical but with curiosity aroused, Braid undertook some mesmeric experiments of his own and became convinced that more than charlatanry was involved. He was also convinced that the phenomena he witnessed could not be explained by some strange influence, potency, magnetic fluid, or whatever, residing in the mesmerist. On the contrary, in his experiments Braid concentrated his attention on the mesmerized subject and sought to understand what internal agency enabled that person to fall into a profound somnambulistic trance. At first, Braid suggested a strictly physiological explanation for the patient's susceptibility. The mesmeric trance, he explained, could

> be accounted for on the principle of a derangement of the state of the cerebro-spinal centres, and of the circulatory, and respiratory, and muscular systems, induced . . . by a fixed stare, absolute repose of body, fixed attention, and suppressed respiration, concomitant with that fixity of attention.[25]

In order to distinguish the induced sleep that he studied from the fluidic theories of animal magnetism, which he found ludicrous, Braid dubbed his version neurohypnotism, from the Greek word for sleep, *hypnos*.

The more Braid studied neurohypnotism – a term later abbreviated to the form now in use – the more his interest focused on the patient's fixity of attention that was requisite to induce the hypnotic sleep. He became ever more certain that hypnosis resulted from a mental state, an intense mental concentration that excluded all external distractions and which he called "monoideism." As his certainty increased, his interpretation took a psychological turn. When at length he concluded that self-hypnosis was possible, he effectively eliminated all grounds for believing that trance sleep was imposed upon the sleeper from an outside source. Although Braid by no means denied the active role played by an effective hypnotist, he maintained that the subject had fully to participate in the endeavor. He did not believe that people could be hypnotized against their wills. "I contended," Braid wrote in the early 1850s,

> and endeavoured to prove that, by the patient concentrating his attention on any part of his body, the function of the part would, to a certain extent, be altered or modified, according to the predominant idea and faith which existed in his (the patient's) mind during the continuance of such fixed attention. I proved . . . that these ideas might arise in the minds of patients, and become operative on them, through their own unaided acts, or from the mere remembrance of past feelings, without any co-operation or act of a second party; and that, in certain subjects, they might also be excited by audible, visible, or tangible suggestions from another person, to any extent whatever – even before they passed into the state of sleep. My object, then, was to prove, by these results, the wonderful "power of the *mind*" of the *patient* over *his* OWN body.[26]

Finding enough to intrigue him within the human mind alone, Braid did not need to invent subtle fluids binding together all parts of the universe. No doubt that is why he, rather than Elliotson, has won acknowledgment as the man who inaugurated the scientific study of hypnotism in Britain.[27] Although the mesmerism that relied on fluids faded from the scene into its well-deserved obscurity,[28] the hypnotism explored by Braid and like-minded colleagues on the continent in time provided major contributions to psychological inquiry. Historians of psychology trace the connections between the Manchester surgeon of the 1840s and the Viennese psychoanalyst of the *fin de siècle*. Studies of spiritualism have merely to press the case that, by the mid-nineteenth century, mesmerism was no longer largely affiliated with materialism. It had, in fact, started to embrace a view of mind that could readily accommodate the spirits.

During the years of its development and transformation, mesmerism crossed phrenology's path and shared many of the same supporters. Long before his conversion to mesmerism, for example, Elliotson was a keen phrenologist. He was the founder of the London Phrenological Society in 1823 and, when he launched the *Zoist* in 1843, its subtitle – *A Journal of Cerebral Physiology and Mesmerism* – announced his conviction that phrenology and mesmerism were closely related subjects of inquiry. H. G. Atkinson, too, played an active part in the affairs of the London Phrenological Society, as the *Zoist* reported. Nor was the affiliation of mesmerism and phrenology found only among those who espoused the materialist interpretation of each. William Gregory devoted a chapter of his book on animal magnetism to the importance of phrenology for his studies, while John Forbes, physician and editor of the *British and Foreign Medical Review*, combined his phrenological interests with an acceptance of the psychological significance of mesmerism.[29]

Phrenomagnetism, the mongrel offspring of phrenology's union with mesmerism, flourished briefly in the 1840s and early 1850s. Under mesmeric treatment, it was argued, a patient would respond immediately to excitation of the phrenological organs. For instance, a contributor to *Blackwood's* in 1851 explained that

> by exciting (and that without touching it, but by waving the hand over it) the organ of acquisitiveness, a person would be induced to steal anything that came in the way, ... Then, by exciting in the same manner other organs, the thief would become a liar, a proud justifier of the deed, and a combative one; then . . . by altering the process, the same thief would become a highly moral character, and abhor theft.[30]

London audiences were not alone in their interested response to such exhibitions, for phrenomagnetism spread far beyond the metropolis in these years. Just as many spiritualist mediums earned their fame, if not fortune, on the provincial lecture circuit, so did the demonstrators of phrenomagnetism find a popular audience across the country. Indeed, mesmerism in general abounded with that class of performer whose work, at one extreme, merged with serious scientific lectures for popular consumption, but, at the other, collapsed into mere quackery.[31] Animal magnetism lent itself readily to the nonprofessional who could quickly master the technique and muster the verbiage needed to envelop mesmeric passes in impressive pseudoscientific jargon. Among the best known of these popular performers was Spencer Timothy Hall, a self-educated printer by profession, a staunch defender of phrenology, a firm believer in mesmeric cures, and, during 1843, editor of the short-lived monthly, *The Phreno-Magnet*.[32]

The eleven issues of the *Phreno-Magnet* provide a useful record of Spencer Hall's activities during 1843 and give a vivid picture of his public demonstrations. On a visit to Nottingham in January, for example, Hall mesmerized the daughter of a local worthy by drawing his hands from the front to the back of the head; in three minutes her eyes were closed, and in five her head dropped, and she was in a deep magnetic slumber. Under the influence of Veneration and Language, the young woman prayed in a low voice, and with Tune, sang a hymn in a rather louder tone. On the organ of Secretiveness being touched, when asked what she was thinking about, she replied, "I shall not tell you;" . . . Under the influence of Imitation, if the audience clapped, she clapped, and also imitated several other noises which caused considerable laughter.[33]

Such "lectures," surely closer to pure entertainment than adult education, were repeated in Sheffield, Manchester, York, Derby, Middlesbrough, Blackburn, Liverpool, and Birmingham, among other places. By July, Hall was boasting, no doubt hyperbolically, that there was "scarcely a single town of any importance in this country where, publicly or privately, a variety of Phreno-magnetic phenomena have not been exhibited."[34] The correspondence in the *Phreno-Magnet* indicates the extent to which local phrenomagnetists sprouted in the footsteps of prominent visiting lecturers, like Hall. Phrenomagnetic societies began to emerge as variant forms of older phrenological groups, so that the *Phreno-Magnet* could report on Northampton's Society for Investigating Phreno-Magnetic Phenomena and announce the formation of similar groups in London, Liverpool, Leicester, Coventry, Burnley, and Walsall.[35] The merger of mesmerism and phrenology is scarcely surprising. Both movements, on the fringe of science, perceived well-established adversaries in common. Both, furthermore, were willing to employ unorthodox medical practices on behalf of their commitment to the amelioration of human health and well-being. An eagerness to challenge orthodox scientific opinion and a passionate concern for human happiness brought spiritualism into the alliance after 1850 as a vigorous third partner.

PARTNERS IN UNORTHODOXY

"The subject of Mesmerism is so intimately connected with that of Spiritualism," wrote the *British Spiritual Telegraph* in June 1859, "that no apology is needed for occasional notices of that branch of phenomena." Similarities between mesmerism and spiritualism in terms of their external trappings were particularly striking, and they received special emphasis from nineteenth-century commentators.[36] The very vocabulary and practices of spiritualism after 1850 immediately called to mind the mesmeric sessions of earlier decades. In both cases, sitters gathered at meetings known as séances, while the participants often held hands in a linked circle, around *baquet* or table, to enhance the potency of the invisible forces at work. It takes little effort, furthermore, to see the magnetic trance as the precursor of the similar mediumistic state, especially since some mesmerized subjects became clairvoyant while entranced. Such clairvoyance seems to have been frequently employed for the diagnosis of disease. Elliotson, for example, claimed that one of his patients, Elizabeth Okey, could not only prophesy the future, but also predict the course of other patients' illnesses. These predictions might take a dramatic turn, as when, on one occasion, the entranced Okey announced that she saw the angel of death perched on a hospital bed.[37] The Reverend Davies was another of the many who believed in medical clairvoyance. He found that a governess whom he employed could, when magnetized, correctly diagnose his wife's serious illness.[38] Mesmeric clairvoyance was not, however, limited to medical purposes, but provided the same range of opportunities to the mesmerized subject as spiritualist mediums subsequently claimed for themselves. Sophia De Morgan reported that she mesmerized an eleven-year-old girl who "saw and reported things of which it was *impossible* for her to have obtained any knowledge in her normal state."[39] Mesmeric performers of particular talent were known to impart to their entranced stage partners the gift of reading, or even playing cards, blindfolded – virtually a preview of future spiritualist demonstrations among a certain genre of professional medium, and equally open to prearrangement or contrivance by the entertainers. Others, when mesmerized, could see in the dark, or visually pierce through walls and transcend intervening space to describe events happening elsewhere at the same time. In the popular mind, mesmerism became so closely associated with clairvoyant skills that a manual written in 1878, purporting to teach its readers *How to Magnetize*, was subtitled *Magnetism and Clairvoyance*. It was published, significantly, by a small firm whose principal business was phrenology.[40]

The summoning of invisible, imponderable fluids for explanatory purposes likewise provided an important thread connecting spiritualism with mesmerism. Since they looked primarily to spirit agency, spiritualists were, of course, less dependent than mesmerists on elusive liquids to elucidate séance phenomena. Nonetheless, many spiritualists embraced aspects of fluidist theory as subsidiary parts of their overall world view. W. S. Moses simply equated the Psychic Force with "the 'mesmeric fluid' of Mesmer; the odyle of Reichenbach; the nerve-aura of other investigators."[41] Such fluids were useful to the spiritualist, as to the psychical researcher, seeking to explain thought transference, for as it flowed from one person to another through the nervous system, the liquid might convey the thoughts of the first to the second. It might even, through obscure interactions between animate and inanimate, provoke rappings and movements of furniture. So influential were Reichenbach's theories among spiritualists and psychical researchers that, as late as 1882, the newly founded SPR established a special committee to undertake the "critical revision of Reichenbach's researches."[42]

In the early 1850s, just as modern spiritualism was finding its first audiences in Britain, the close relationship between spiritualism and mesmerism revealed itself repeatedly around gyrating tables. Not everyone who sat at those tables attributed their activity to spirits. An alternative explanation could be sought in animal magnetism, specifically in the motive power of the mesmeric fluid propelled outward from the sitters' hands. Given the long-standing analogies with both electricity and magnetism, it is understandable that not a few Victorians confused the pseudoscientific theories of mesmerism with the new theories of physics publicized at the same time. In 1853 Queen Victoria was not alone in assuming that table turning was a magnetic or electrical phenomenon; when the Manchester Athenaeum held a *conversazione* in June to try to determine the causes of table moving, the participants tested for an electrical impulse. It was indicative of the prevailing uncertainty over causative forces that the city's authority on animal magnetism, Dr. Braid, was called in to supervise the evening's experiments.[43]

Yet even while some magnetists found in table turning further proof of an all-pervasive mesmeric force, others could endorse the explanation proposed by spiritualists. If tables spun in circles and rapped out messages, they argued, spirit agents must be at work – perhaps connected to the living by mesmeric fluid. Such notions were scarcely new to Mesmer's theory at midcentury. As early as the 1780s, at his own Parisian headquarters, patients had claimed communion with spirits of the dead during deep magnetic slumber, while in Germany and Sweden, too, where Swedenborg's influence provided powerful encouragement, animal magnetism before 1800 quickly became an access route to the other world.[44] The role that mesmerism played in preparing the ground for spiritualism, in this respect, needs no further amplification. Emma Hardinge Britten, herself a somnambulist and "magnetic subject" as a child, succinctly summarized the situation when she wrote that "Mesmerism has been – humanly speaking – the cornerstone upon which the Temple of Spiritualism was upreared."[45]

In the first decades of modern British spiritualism, mesmerism frequently shared the spotlight at spiritualist séances, and vice versa, because the same individuals often proved susceptible to both influences. Podmore reported on a mesmeric "doctor," Hardinge by name, who presided over séances in the early 1850s, using as medium a magnetized "sickly young woman." Mrs. Guppy, a well-known medium of the 1860s and 1870s, had an earlier, brief career as a professional mesmerist before her marriage.[46] Burns's Progressive Library featured both mesmeric and spiritualist séances, and, even in the 1880s, it advertised demonstrations in "Medical Clairvoyance."[47] The dramatic potential in both spiritualism and mesmerism attracted performers eager to display their histrionic skills in public. To a great extent those skills were interchangeable. The gifts that made Emma Hardinge Britten a highly successful trance lecturer in later life were no doubt similar to the traits that enabled her as a child to receive direction from her mesmerist. Suggestibility, suspension of will, passivity – qualities necessary for certain kinds of mediumship – are very close indeed to those desirable in a mesmeric subject. Equally, where mediumship accompanied the power of mesmerizing others, both pursuits relied upon that dramatic flair and the ability to impress an audience that can be as important in mediumship as passive receptivity. Where charlatanry was involved on the mesmerist's part, the trickery learned at mesmeric séances could stand one in good stead at a spiritualist session. The Mademoiselle Prudence who advertised her show in London in 1852 with the promise of "Mesmerism, Transmission of Thought, Illusions, Clairvoyance, and Double Sight" would have had no trouble posing as a spiritualist medium.[48]

Nor was it only mediums who might make the easy transition from mesmerism to spiritualism. The annals of British spiritualism are filled with names of men and women for whom animal magnetism served as an introduction to their spiritualist beliefs. "I fancy," opined the Reverend Davies, "as far as any order is traceable in the somewhat erratic course of spiritualistic experiences, that most people arrive at spiritualism via mesmerism." Certainly in Davies's own case, his successful experiences as a mesmerist helped commit him to "the effort to prove unbroken continuity between the life in this world and the life beyond." *Blackwood's* similarly recognized the influence of mesmerism, although in negative terms, when it grimly commented in an early article on spiritualism: "The phenomena of animal magnetism have unsettled the minds, and, we fear, perverted the religious faith of thousands, both in the Old World and the New."[49]

Among the many people whose lives illustrate the accuracy of Davies's opinion, if not *Blackwood's* fears, the following are typical examples. The Coventry weaver Joseph Gutteridge, for one, "possessed some powers as a mesmerist" before embracing spiritualism. E. D. Rogers, who gained great proficiency as a mesmerist, explained that "Mesmerism has frequently proved to be a stepping-stone to Spiritualism" and insisted that his own "experiences as a mesmerist" helped prepare him to receive "the larger truths of the spiritual life." W. H. Harrison, another journalist, likewise reminisced that mesmerism had provided his introduction to spiritualism. The Howitts and Alfred Russel Wallace were indebted to Spencer Hall for their introduction to mesmerism in the 1840s, and Anna Mary Howitt Watts believed that both her father and Wallace were led from mesmeric knowledge directly to spiritualism.[50] Sophia De Morgan practiced mesmerism in the late 1840s, before her attention was riveted on spiritualist phenomena, and, like Wallace, she had also studied phrenology with avid interest. (Where Wallace was deeply influenced by phrenology in the formulation of his views on mind and body, however, Mrs. De Morgan ultimately found much amusement in the pseudoscience.) Similarly, Dr. George Wyld combined phrenology and mesmerism with his spiritualism. He joined the London Phrenological Society in 1844, a few years after he had first been attracted to mesmerism while still an adolescent. Wyld gained a thorough knowledge of mesmerism, although he never practiced it professionally, and it was through "an old mesmeric friend" that he first heard about D. D. Home. Immediately making a point to visit the young medium, Wyld was overwhelmed by what he witnessed and felt compelled "at once to believe in the phenomena as genuine."[51] Many of the literati who likewise found much in Home's artistry to admire had previously marveled at animal magnetism, although not all became permanent converts either to mesmeric or spiritualist theory.[52]

Even Elliotson, after much initial hostility to spiritualism, came to embrace spirit's real existence in the cosmos. John Ashburner, M.D., an enthusiastic phrenologist, supporter of Elliotson, and co-worker at his mesmeric infirmary, broke with Elliotson in the 1850s when Ashburner traveled further down the road to spiritualism, where Elliotson contemptuously refused to stray. Elliotson, Ashburner explained, preferred to follow paths "lighted according to his taste, with the flare of common gas. He is not up to the delight of a Spiritualist flame." Even in the early 1860s, William Howitt referred to Elliotson as "the martyr of Mesmerism turned persecutor of Spiritualism."[53] In the very year when Howitt's comment was published, however, Elliotson was converted to spiritualism by Home, to whom Mrs. Milner Gibson introduced him in 1863. Whether or not the seventy-two-year-old doctor was enduring a period of despondency that made him particularly susceptible to spiritualist claims, as has been suggested, he appears to have renounced all his former opposition and become a docile lamb in the spiritualist fold. Writing in 1867, a year before Elliotson's death, Ashburner changed his tune dramatically, praising the flexibility of

Elliotson's mind that had "not failed to land him on the domains of spiritual philosophy," and applauding Elliotson's "contribution to the history of Mesmerism, and its expansion into spiritualism."[54]

Mesmerism expanded effortlessly into spiritualism for a rich variety of reasons, not the least important of which was the combination of scientific, religious, and occult sources on which both movements drew. Each created a blend of theory and practice that could appeal strongly to a population wanting scientific authorization for its faith and the blessings of religion upon its scientific discoveries. Like phrenology, mesmerism could offer its followers the pretense of scientific methodology, with conclusions empirically tested and verified. Yet the demonstrable achievements of which mesmerists proudly boasted referred largely to medical cures that depended more on mental than on physical explanations. In both spiritualism and animal magnetism, the role of the medium or mesmerist, when involved in healing, assumed as much the priest's as the physician's function. The kind of medical recovery available through the services of medium or mesmerist had been offered for centuries in varying forms, by cunning men and wise women, village wizards, tribal shamans, famous strokers like Valentine Greatrakes in the seventeenth century, and notorious impostors in all ages. Inevitably, the religious implications of mesmeric cures led animal magnetism into "supernatural regions" where Mesmer himself had hesitated to wander.[55] By the mid-nineteenth century, however, mesmeric and spiritualist healing alike were widely equated in the public mind with the faith healing celebrated in the New Testament, whether mesmerists welcomed that equation or not.

"Among the Pentecostal charisms of the Church whereof I was an ordained minister," C. M. Davies observed, while describing his medical success as a magnetist, "this gift of healing by the laying on of hands stood pre-eminent." Others argued that the cures effected at spiritualist séances bore direct comparison with early Christian miracles. *Fraser's* reprinted the story of a séance where a woman, paralyzed from the waist down, "sprang on her feet" when directed by the planchette to do so "*in Jesus Christ's name!*" (The planchette had first inquired whether the woman believed in Christ's "invariable goodness and power," to which she had replied in earnest affirmative.) The Reverend F. R. Young, editor of the *Christian Spiritualist*, regularly offered healing services to his congregation of the Free Christian Church in Swindon. By the laying on of hands and "commanding the pain to depart in the name of God and Christ," Young was able to cure parishioners of "violent face-ache, lumbago, and paralysis." The American who introduced Young to faith healing, a Dr. James Newton of Newport, Rhode Island, visited London and gave public demonstrations at Burns's Progressive Library. Despite the setting, Newton, too, attributed his medical success to his deep faith in Christ. He claimed to have "thoroughly reproduced the 'Christ-Life'" and to receive frequent visitations from "the Nazarene." So strong was his faith, Newton explained, that it provided the force needed to heal sickness. Although he showed a slight lapse of taste when he remarked that "the Nazarene made some noise in Judaea, but nothing like what I'm going to make in London," he won the warm endorsement of the Baptist minister, Jabez Burns of Paddington.[56]

Mesmerists and spiritualist mediums who practiced healing were allied in providing relief from pain and disease, thanks largely to an intense relationship established between patient and healer. The elements in that relationship were manifold; they might include an intentional appeal to the patient's Christian beliefs, or, avoiding overt religious associations, build on the hope aroused by the patient's faith in, and dependence upon, the healer. They called for the active involvement of the patient's imagination in the healing process, the healer's belief in his own power, and often the supportive potency of an accompanying ritual.[57] During the nineteenth century, observers of mesmerism and spiritualism grasped much of this explanation, although it has been more fully elaborated with the growth of psychiatric, psychological, and neurological theory. Many of those who did not summarily reject mesmerism and spiritualism realized that both, with their profound appeal to faith of one sort or another, provided fertile areas of study for the expanding sciences of human life.

Yet the scientific and medical professions for the most part refused to take mesmerism and spiritualism seriously, except as manifestations of abnormality. In the Victorian period, science and medicine were seeking to anchor themselves firmly in the physical world, distinctly divorced from all disreputable thought systems whose insubstantial foundations could cast doubt on the weightiness of the scientific endeavor. Mesmerism, spiritualism, and even phrenology as well, had roots in a magical, occult past, where divination, humors, and conjured apparitions abounded. It was a past that nineteenth-century scientists were fully conscious of outgrowing, rapidly and with increasing disdain for the foolish misconceptions of their predecessors. Once the physiological bases of mesmerism and phrenology were successfully challenged by the 1840s, both appeared as suspect as spiritualism became shortly thereafter. That all three movements also stressed the untapped potential of the human mind – and telepathy was, in this respect, their natural successor[58] – did little to redeem them in the eyes of contemporary orthodox science, at least until the last quarter of the century. Mentalism, in any form, could find little place in the mainstream of

Victorian scientific activity, and no matter how vigorously some phrenologists and mesmerists might attempt to deny the nonmaterial foundations on which their beliefs rested, both phrenology and mesmerism
were indeed saturated with the conviction that mental processes were
much more than physiological interactions. With spiritualism, phrenology and mesmerism implicitly constructed their respective articles
of faith on the assumption that each human personality was a unique
and complex entity whose very individuality precluded any mechanistic
dismantling into diverse parts. Mind possessed a fundamental reality
for mesmerists, phrenologists, and spiritualists, and whether or not
they actually labeled it psyche, soul, or spirit, it was nonetheless associated with an intangible, imponderable element of life that successfully defied materialists throughout the century.[59]

The social implications of mesmerism and phrenology likewise underscored their close kinship with spiritualism, for they, too, appealed
broadly across the social spectrum from the working classes to the
aristocracy. Mesmerism had its artisan and lower-middle-class enthusiasts, who might pay as little as a few pence for admission to mesmeric
demonstrations, in London and the provinces. E. D. Rogers first encountered mesmerism while attending lectures on the subject at the
Mechanics' Institute in Wolverhampton, and similar talks, sometimes
varied with phrenomagnetic performances, enlivened the programs of
many such institutions in the 1840s. Equally, gentleman amateurs like
Atkinson were attracted to mesmerism and formed their own salons,
or private clubs, to pursue their studies amidst a select circle of social
equals. When they did venture forth to public exhibitions of the mesmeric art, they might pay as much as two shillings, six pence for a
reserved seat. A reporter from the Northampton Phreno-Magnetic
Society was apparently not guilty of undue flattery when he informed
S. T. Hall that "the subject of Mesmerism excited in the town of
Northampton, as at other places which have been favoured with your
presence, a vast degree of interest among all classes."[60]

The lofty and lowly followers of phrenology ranged from Prince Albert, who consulted George Combe about his eldest son's cranial
bumps, to George Jacob Holyoake, Owenite, Chartist, secularist, and
leader of the cooperative movement in Britain. In between was a full
quota of middle-class professionals, academics, and intellectuals, from
whose ranks the phrenological societies in London and Edinburgh were
filled. Phrenology, too, was taught at Mechanics' Institutes, as well as
provincial literary and philosophical societies.[61] Like mesmerism and
spiritualism, there was something in phrenology for everyone. If the
movement's masquerade as "a new scientific philosophy"[62] particularly appealed to middle-class societies with intellectual pretensions,

its progressive, reformist aspect, its deliberate challenge to old orthodoxies and vested interest groups, appealed mightily to the lower
eschelons of Britain's social hierarchy.

Although the progressive aspect of phrenology served as a connecting bond with the whole later spiritualist movement, it naturally
provided a particular link with non-Christian spiritualism, which derived much of its strength from working-class reformist zeal. Such theological views as lurked in Gall's craniology were part of the rationalist,
humanist inheritance from the eighteenth century that figured so prominently in the philosophy of Victorian Britain's self-taught artisans and
that likewise loomed so large in non-Christian spiritualism. There was
the same denial of original sin, or intrinsic moral failing, in the human
race, the same firm conviction of gradual improvement, as individuals
learned to identify, regulate, and master the temperament with which
each was endowed. Like the non-Christian spiritualists, phrenologists
repudiated the belief that man must meekly accept his destiny, and
they argued instead that each human being could play an active part
in molding his future, just as humanity collectively could participate
in a general moral advance.[63] As George Combe taught in his best-
selling volume, *The Constitution of Man*, "Intelligent beings exist, and
are capable of modifying their actions." To discover and obey the laws
under which nature operated, he asserted, was the key to enjoying life,
to experiencing the full portion of human happiness. Although he was
careful to sprinkle God, or the Creator, liberally through the pages of
his book, he was convinced that Holy Scripture held no important
instructions for man's sojourn on earth. Like the non-Christian spiritualists who later scorned bibliolatry, he argued that

> Revelation does not communicate complete or scientific information
> concerning the best mode of pursuing even our legitimate temporal
> interests, probably because faculties have been given to man to dis-
> cover arts, sciences, and the natural laws, and to adapt his conduct
> to them. The physical, moral, and intellectual nature of man is itself
> open to investigation by our natural faculties; and numerous practical
> duties resulting from our constitution are discoverable, which are not
> treated of in detail in the inspired volume.[64]

It is no wonder that Kitson found a place for phrenology in the Progressive Lyceum curriculum.

Attitudes such as these seemed particularly provocative in their social context. Phrenology and mesmerism have, in fact, been characterized by their historians as agents of confrontation with establishment
views and entrenched positions,[65] and both pseudosciences contained
social philosophies that found later echoes in non-Christian spiritualism. Mesmerists saw themselves as saviors of a suffering humanity,

alleviating man's terrible burden of pain and devoting their selfless efforts, despite opprobrium and ridicule, to the cause of human happiness. The *Zoist's* long subtitle explained that the journal not only intended to explore the subjects of phrenology and mesmerism, but would equally reveal *Their Applications to Human Welfare.* Furthermore Mesmer, and his disciples after him, were certain that the mesmeric force could be harnessed for man's good in wide-ranging and still unfathomed ways. Not only did personal health respond to the ministrations of the mesmerist, but somehow, they believed, human relations, social interaction, the functioning of government – in short the health of the nation at large – could all be ordered aright under the beneficent influence of mesmerism.[66]

Phrenology's challenge to the existing order came in its contention that all the institutions of society, particularly schools, prisons, and asylums, had to be redesigned according to the observable laws of human nature. Among phrenologists, there was much interest in Owen's attempts at social engineering; indeed, "Holyoake, who was studying phrenology in 1836, found it a natural corollary of Owenism,"[67] and he was not alone. Owen himself expressed interest in phrenological doctrine, and Combe, visiting New Lanark in the early 1820s, was impressed with much of what he saw.[68] It is true that phrenologists found Owenites utterly unrealistic in believing that fundamental changes in human nature, and the establishment of a new moral order, could occur within a mere generation or two – phrenologists knew that centuries were needed – but they nonetheless shared similar hopes and long-term goals. They both placed the amelioration of the human condition, individual and communal, within a thoroughly secular context, and both strenuously rejected the conservative concept of society as the fruit of gradual, organic growth. Phrenologists and Owenites alike were eager and willing to tamper with social institutions whenever these functioned as obstacles to human well-being. The concern of phrenologists for man's happiness on earth made them figure for decades among the unabashed critics of the Victorian status quo, regardless of their exact relation to Owenism or any other form of early socialist thought in Britain.[69]

The inherent radicalism of phrenology and mesmerism not only found occasional expression in the blueprints for social reform that their adherents sketched, but also, from time to time, in the professional policies that their followers pursued. Nowhere did phrenologists and mesmerists offer a more subversive threat to vested interests than when they challenged the medical profession of Victorian Britain. Much of that challenge came from outside the profession, from the amateur, and often lower-class, mesmeric healers and phrenological diagnosticians who, without the slightest training, claimed to do the physician's work with far more drama and much less expense than the doctors themselves. Spencer Hall, for example, son of a cobbler and a dairymaid, never received any medical education, but that did not stop the likes of Harriet Martineau and the Howitts from consulting him and singing his praises.[70] Countless other practitioners, much less prominent and with less noteworthy clients, gravely undermined the professional respectability of medicine and threatened to diminish medical earnings. What the non-Christian spiritualists later taught, particularly through the lyceum movement, the itinerant phrenological and mesmeric performers earlier implied through their demonstrations: Information about the functions of mind and body should be easily accessible to the public, not the exclusive knowledge of a closed and self-perpetuating elite. Phrenology, mesmerism, and spiritualism alike were thoroughly domesticated pursuits. They could be readily practiced in the home, apart from laboratory or hospital – those restricted preserves of the trained scientist and physician. James Burns spoke for all three movements when he declared during a mesmeric séance at his Progressive Library in the early 1870s that "we should be able to do without doctors as soon as the healing powers of animal magnetism were properly recognised and diffused."[71]

Yet the cluster of interests and beliefs represented by phrenology, animal magnetism, and spiritualism also attracted a few spokesmen within the medical profession itself. It is difficult to trace the precise relationship between phrenology and animal magnetism, on the one hand, and orthodox medicine, on the other, because both forms of unorthodoxy contained elements that eventually found acceptance within the profession: localization of cerebral function in the case of phrenology and the important role of the patient in the case of mesmeric, or hypnotic, treatment. There were even medical doctors who temporarily subscribed to Gall's map of the brain. Phrenology, nonetheless, never enjoyed a position in the mainstream of medical thought and lost all reputable medical support during the second half of the century. From the late 1830s, the theory of mesmeric fluid was systematically denigrated in the authoritative medical press, as it was in the serious, nonmedical quarterlies and reviews.[72] Even so, throughout the century a limited number of doctors continued to approach their patients with assumptions and methods derived from phrenology, mesmerism, or both. They did so, typically, as a form of protest against the standard practice of medicine in their own day. The fact that some of these doctors were also convinced spiritualists suggests that they viewed the interaction of mind and body in ways not countenanced by the leading medical texts of the time.

The medical doctors who incorporated spiritualism into their professional activities were reacting against what they perceived to be the increasing tendency of modern medicine to view the human body in mechanistic terms. They were attracted to an older vision of the physician as one who restored his patients to equilibrium, both within their own bodies and with the external natural world. In the past, they believed, doctors had treated their patients as totalities, as people whose physical complaints were accompanied by moral faults and emotional wounds. Doctors of an earlier generation had not hesitated to peer into their patients' souls, to ferret out the hidden causes of their distress, for it was the doctor's role to cure all that was amiss, with body and psyche alike. As the nineteenth century progressed, the spiritualistically inclined physicians were distressed to see their colleagues adopting more and more of the scientist's neutral stand. Sickness seemed to be increasingly viewed as an isolated problem within the organism, something to be treated by chemical agents, as the malfunctioning portion of a machine to be identified and repaired. In opposition to this apparently materialist approach to illness, the spiritualist doctors insisted that "vital activity cannot be explained by the laws of physics and chemistry."[73] They sought to offer a fuller picture of man, as a creature of rich intricacy, connected with the world of nature in ways that stethoscope, forceps, and lancet could never fully master. It was not that spiritualists rejected modern medicine out of hand; on the contrary, they were as willing as the next person to hail its triumphs. They contended, however, that orthodox medicine alone was incomplete. Since man possessed an immortal soul, since life did not spring from inert matter, they were confident that a greater breadth of vision was needed in medical practice. In fact, they advocated holistic medicine long before it was fashionable to do so – indeed, before it even had a name.

That mesmerism had a place in this alternative approach to medical treatment should be clear. Whether perceived as propounding vaguely vitalistic theories of the human organism or specifically animistic views predicated on the existence of the soul, it belonged squarely in Victorian counterculture medicine. A book reviewer in the *Athenaeum* certainly viewed it in that context. In denying a new study of mesmerism its self-vaunted status of medical panacea, he complained:

> The author gives us the usual collection of wonderful cures accomplished by his favourite agency, and at the same time deals largely in various other kinds of transcendentalism. Homoeopathy, the od-force, hydropathy, (we miss spirit-rapping) are all taken into commendably impartial favour; the appetite for one of these wonders being generally ready, we notice, for the whole group.[74]

The reviewer had a point. It was indeed common to find some combination of mesmerism, homeopathy, hydropathy, and spiritualism (subscription to Reichenbach's odylic force was not as frequent) among the beliefs and therapies advocated by doctors who challenged current medical assumptions in nineteenth-century Britain.

Homeopathy figured in this cluster of "medical heresies"[75] because it emphasized the patient's total organic equilibrium and relied heavily on the curative powers of the life force. Its central tenet, as propounded in the late eighteenth century by the German doctor Samuel Hahnemann, was that "like cures like." It was not a novel medical theory; it had appeared in ancient Greek and Islamic writings on medicine, and Paracelsus had toyed with the idea. Hahnemann's contribution was not only the name itself, but the attempt to systematize homeopathy and to make it a common medical practice in place of allopathy, the prevailing mode of treatment. The homeopathic doctor believed in helping the patient's body to heal itself, in supporting the efforts of the presumed life force to quell disease and recreate internal balance. Such medicine as he administered was designed to intensify the patient's physical reaction to the disease. The allopathic doctor, by contrast, sought to combat the disease directly with an arsenal of drugs, often in enormous doses. He wanted to induce a condition in the patient that would counteract the symptoms of disease. The homeopath, in short, "induces fever to fight fever, the allopath cools the patient."[76]

Homeopathic medicine was entirely constructed on vitalistic assumptions. By administering infinitesimally small doses of medicine, preferably simple, uncompounded substances from natural sources, the homeopath sought to stimulate his patient's life force. Hahnemann experimented extensively with diverse medicines of this sort until he produced in healthy people symptoms similar to those of any known disease. Thereafter he used the medicine in question to combat that disease, convinced that it assisted the natural response of the patient's body. He continued his experiments until he found medicines for a range of maladies. His followers considered his approach the best possible medical cure because it offered a form of treatment that was far less lethal than the allopath's megadoses and did not meddle obtrusively in the body's own therapeutic system. They were thoroughly aware of the limits to their medical skill and only hoped to foster a situation in which the patient's own vital strength could restore order to its troubled surroundings. For the homeopathic doctor, furthermore, each patient represented an entirely individual case, not to be handled routinely with a standard dose of medicine, but to be watched and cared for as a unique composition of parts, some well, some ill, but all needing treatment within the context of the whole organism. Just as phrenologists believed that the several organs of the brain could not be analyzed separately, so the homeopaths believed that disease was not an isolated occurrence within the body. Both saw human beings as more than agglomerations of tissue, and both celebrated the individual over stereotypes of sickness, whether physical or mental.

Homeopathy came to Britain in the 1830s, largely through the efforts of the socially prominent doctor, Frederic Hervey Foster Quin. It found disciples among the well-educated and well-to-do and aroused sufficient public interest to support the establishment of a homeopathic hospital in London at midcentury.[77] But it never enjoyed the support of the medical establishment, who saw it undermining the principles on which modern, scientific medicine was being built. It seemed to minimize the role of the doctor in the remedial process and to emphasize the importance of forces whose vague, elusive nature avoided precise analysis and appeared alarmingly reminiscent of medicine's magical, medieval past.[78] As it became increasingly clear, furthermore, that allopathic cures were more rapid than homeopathic and that, with the advance of pharmacology, a far wider array of impressive allopathic drugs was available, public support for homeopathy waned. Although, today, knowledge of the work of antibodies could provide scientific credibility for the hypothesis that like cures like, in the nineteenth century homeopathy had no authoritative scientific foundation from which to challenge allopathy. It remained on the periphery of medical practice in Britain, to attract independently minded physicians.

Hydropathy likewise stayed on the fringe of medical respectability, despite the popularity of hydropathic spas, and despite the eminence of some of the patients who tried the water cure – Bulwer-Lytton, Darwin, Ruskin, Alfred Tennyson, to name a few. There were spas at such places as Malvern, Cheltenham, and Leamington, which followed different regimes, but featured in common the frequent internal and external application of water. The patient might try sitz baths or foot baths; he might submit to being rubbed with wet towels or wrapped in damp sheets; he would drink tumbler after tumbler of water, all in the fond hope that this form of nature therapy would help to "restore an equilibrium of the bodily and mental systems."[79] Although hydropathic doctors might expound a learned-sounding theory to explain the efficacy of water cures, a theory that preached of nervous irritation soothed and inflammation relieved, there was really little more than common sense behind hydropathy. With the virtual elimination of mental exertion, and the encouragement to take fresh air and exercise with the quantities of water, the spa atmosphere was expressly designed to reduce tensions. The many middle-class professionals and intellectuals who frequented the hydros found them relaxing places where one could allow body and mind to uncoil from the pressures of busy lives.

Hydropathy was, indeed, intended to address the needs of both body and mind. Like other forms of nature cures that eschewed drugs of chemical composition, it advocated a form of therapy derived solely from nature as the surest way to rectify unnatural states of imbalance within the body. Doctors who relied on natural curative agents claimed, like homeopaths, to consider the entire state of their patients' health or ill-health, the mental and emotional condition as well as the purely physical. Since many of the customers who patronized hydropathic spas experienced nervous ailments of various sorts, the disinclination of their doctors to isolate a single cause of suffering is understandable. Such a practice, furthermore, would have been contrary to the hydropathic philosophy that recognized inextricable links between healthy body and healthy mind.

The career of George Wyld amply illustrates the manner in which medical heterodoxy could enhance a deep-seated repudiation of materialism. Mesmerist, phrenologist, theosophist, and spiritualist, he was also a leading advocate of homeopathy in Britain. He was first drawn to Hahnemann's theories while a medical student at Edinburgh, when the nervous tension of preparing his final examination in 1851 thoroughly upset his digestive system and led to repeated vomiting. When orthodox medicine failed to relieve Wyld's acute indigestion, he consulted a homeopathic doctor who prescribed "globules of Nux Vomica I^x and Byronia I^x," with splendid results. Within a few days, Wyld recovered health and appetite and was eager to spread the homeopathic word. In London, he joined the homeopathic hospital and, after studying the method of treatment, became a physician to the outpatients there. He wrote a pamphlet to publicize the virtues of homeopathy and in 1876 became the acting president of the British Homoeopathic Society.[80] In that capacity, he sought to reconcile the homeopathic school with what he termed the "Old School of Medicine" and was encouraged by the fact that, in recent years, the latter had gradually abandoned bloodletting, profuse blistering, and heroic doses of medicine. Among the new practices he was glad to see homeopathic doctors adopting were Turkish baths and "the *water cure*." As a homeopath, he stressed the practice of temperance, purity of diet and conduct alike, and "the action of a right mind" as "the chief factors in the sacred art and science of healing." Although he regretted that his homeopathic heresy "excluded [him] from all professional interchange of opinions and consultations with the leaders in medicine, and from all orthodox medical societies," he never doubted for a moment "the immense superiority of the homoeopathic as compared with the heroic treatment of acute disease . . ."[81]

Homeopathy was not the only cause for which Wyld believed that he had sacrificed standing and reputation within the medical profession. His ardent spiritualism likewise cost him clients, he recollected in his autobiography. His was not a private, domestic sort of spiritualism; he was a vice-president of the BNAS[82] and appeared at the Bow Street police court in October 1876 to defend Slade during the slate-writer's trial. In fact, he organized the collection of a spiritualists' defense fund to meet Slade's legal costs. For his efforts, he received the laughter of the medical profession and a decline in his medical practice. In an attempt to recover some authority among his professional colleagues, he became active in promoting a more potent form of smallpox vaccine derived directly from calf lymph.[83] What is interesting about his calf vaccine campaign is not only its similarity with the homeopathic concept of "like cures like," but also the manner in which he treated the subject in his autobiography. A single brief chapter, bearing the title "Slade and his Slate Writing: and Vaccination from the Calf Direct," suggests that, for Wyld, both crusades shared a similar significance and possessed the same level of reality.

Throughout Wyld's life, two disparate aspects of his personality proved surprisingly complementary. On the one hand, he was utterly preoccupied with spirit as "the fundamental substance of the Universe," and he exulted that the spiritualist movement of his own day "must greatly interfere with the widespread scepticism and materialism of our age."[84] On the other, Wyld was a man of practical concerns who wanted to see a variety of social improvements, for reasons of man's health and moral well-being. He played an active role in the London Smoke Abatement Society. He advocated the establishment of a "Society for the Simplification of Legal Proceedings," on the grounds that the administration of justice ought to be "as cheap and expeditious as possible," and he believed that legal advice ought to be free for the poor. He argued for health education in the schools, and for girls' instruction "by matrons" concerning the "laws of marriage." Wyld was certainly no socialist; he was not even mildly radical in politics and broke with Gladstone over Irish Home Rule.[85] He was, nonetheless, convinced that the individual must take a vigorous part in shaping his own environment, both the public environment of society in which the human species must live and the private environment of the body in which the immortal soul must flourish. His combination of concerns and career choices catered to both those realms.

So did the work of James John Garth Wilkinson, a self-proclaimed "pioneer in several heterodoxies."[86] As in Wyld's case, Wilkinson's medical practice was only one commitment in an immensely productive life. To many contemporaries, he was better known as the editor, translator, and biographer of Swedenborg than as a physician. Indeed, his private practice in Bedford Square, London, was slow to increase after he established himself there in 1835. Perhaps his lack of enthusiasm for the medical profession manifested itself to his patients. It was not Wilkinson's idea to become a doctor, but that of his father who was for many years judge of the county palatine of Durham.[87] Wilkinson's brother, William, likewise joined the legal profession – to become D. D. Home's attorney – but James was pushed into a line of work that was extremely distasteful to him. He hated the practice of "promiscuous drugging," and, from the very start of his medical career, Wilkinson was glad to spend more of his time poring over the writings of Swedenborg than dispensing prescriptions and selling drugs in his Bedford Square office. He was inevitably attracted to animal magnetism in the late 1830s and attended both Dupotet's and Elliotson's demonstrations of the subject. When he began contributing occasional articles to the *Monthly Magazine* around this time, the two topics of particular interest to him were mesmerism and Swedenborgianism.[88]

Mesmerism appealed to Wilkinson, not in its occult aspects, but in its challenge to the medical establishment.[89] Although he was a member of the Royal College of Surgeons and licentiate of the Society of Apothecaries, his was a lifelong battle against medical authority in all its guises, to the extent that he wanted to undo the regulation and control that was being introduced into the British medical profession in his time. "Perhaps," mused his biographer,

> there was never a practitioner of medicine who was less a member of the profession at heart. Almost from the first he would have dischartered that profession and have thrown the treatment of the sick open to all who chose to engage in it, holding them equally responsible whether formally qualified or not. Degrees and diplomas would have conveyed no privilege, had Wilkinson been allowed his way.[90]

Wilkinson loathed the concentration of power in the hands of a few elite bodies within the medical profession, and he waxed almost obsessively wrathful when he saw the government strengthening that inequitable exertion of authority by the few over the many. He became the champion of those whom he considered the victims of oppression by the combined forces of medicine and bureaucracy. He denounced the Contagious Diseases Prevention Acts of the 1860s as "an outrage on womanhood at the instance of the medical profession who, through the Royal Colleges of Physicians and Surgeons, had memorialized Government in favour of extension of the areas to be affected."[91] He was even more bitterly aroused against the compulsory vaccination of children and expended much energy during the 1870s speaking out, in print and before the House of Commons Committee on Vaccination, against the practice, which he called "blood assassination" and "the homicidal insanity of a whole profession."[92]

Wilkinson's intense dislike of his profession might have driven him ever more exclusively into literary pursuits had he not come upon homeopathy in the 1840s. Henry James, Sr., with whom Wilkinson maintained an active correspondence and warm friendship, called Hahnemann's doctrine specifically to his attention, although Wilkinson had been generally aware of considerable controversy over homeopathy since the late 1830s. He read whatever books he could find on the topic, experimented with homeopathic treatment, and found that he suddenly loved his work. By 1850, he considered himself a staunch homeopath. As with the other causes which he espoused, Wilkinson soon went into print to proclaim his new allegiance; in 1854 he wrote an open letter in which public concern over cholera and the Crimean War served as background for the contention that homeopathy was, beyond doubt, the medicine of the future.[93] He was not, in fact, an entirely orthodox follower of Hahnemann – he would, for example, prescribe herbal medicines that had not been tested homeopathically – but he nonetheless deserved the title, "champion of Hahnemann," which Ralph Waldo Emerson bestowed upon him.[94]

Wilkinson was attracted to natural remedies as alternatives to what he considered the drug-addicted practices of the medical profession. Not surprisingly, he also found hydropathy extremely helpful and wrote a correspondent in 1849: "For a length of time now I have been hydropathizing, and with huge benefit to mind and body. 'Tis astonishing how much more work I can do, . . . under the *wishy-washy* régime." He seemed incurably drawn to methods of treatment and forms of medicine that did not enjoy the blessings of the medical establishment, particularly if they lacked an apparent explanation in materialist terms. "Wilkinson held strong views as to the limitations of science," his biographer explained; "his tendency was to mistrust chemical and physical bases for vital processes." The tendency befitted a confirmed Swedenborgian and illustrated, in a different way, what Wilkinson meant when he told the London Dialectical Society in 1869: "I have been a believer in the spiritual world, and its nearness to the natural world, nearly all my life." Indeed, for Wilkinson, the human body was an organism infused with the spiritual "truths of Revelation."[95]

Wilkinson's belief in the spiritual world was, of course, a fundamental article of his Swedenborgian faith. It was buttressed, for a while, by the spiritualist beliefs of his brother William, their friends the Howitts, and the circle of middle-class intellectuals and professionals in London who avidly attended séances from the early 1850s. Like his brother, who edited the *Spiritual Magazine* and published a volume of *Spirit Drawings*, Garth Wilkinson edited an issue of the *Spiritual Herald* in 1856 and the following year published a volume of poems, *Improvisations from the Spirit*, written in a Blakean mode, allegedly without effort on his part.[96] Nonetheless, Swedenborgianism remained the paramount influence in his life, and by 1893 he could write to a correspondent that the only spiritualism he still accepted was that which was consistent with "the New Religion commissioned through Swedenborg."[97]

There was much in Wilkinson's cluster of convictions that appealed strongly to working-class, non-Christian spiritualists. His Swedenborgianism alone placed him among the ranks of the unorthodox and put him in sympathy with those radical workers who found in Swedenborg an inspired voice against the establishment of church and society. Self-taught artisans could, furthermore, find particular sympathy with Wilkinson's sustained attack upon a professional section of that establishment, the doctors. The tyranny of the medical profession over the bodies of its patients provided a reiterated complaint in the literature of working-class spiritualism; in the curriculum of the Progressive Lyceums, the emphasis on "self-help in the area of health" and "democratic self-care" underscored the non-Christian spiritualists' determination to maintain control over their own bodies. They favored natural forms of medicine over the dangerous practices of allopathy and expressed rage at medical treatment imposed by the state.[98] Despite a different social background, Wilkinson shared their outrage and figured as one of those outspoken middle-class reform agitators who recur throughout the history of Victorian social crusades. It was the social reformer in Wilkinson that led him to espouse Fourierism in the late 1840s and prompted him, during the revolution of 1848, to visit Paris, where he met Robert Owen.[99] As with Wyld, diverse forms of protest – professional, social, and spiritual – coalesced in Wilkinson to give particular potency to his views on public welfare and individual health.

In their endorsement of medical treatments that challenged orthodox views and assumptions, doctors like Wyld and Wilkinson were making implicit statements about the human condition that found confirmation in their spiritualist beliefs. They fundamentally opposed any materialist interpretation of the body's functions and embraced instead a view that emphasized the intangible vital essence of each person, the distinct mind existing apart from bone, muscle, and flesh. The medicine that they practiced upheld their antimaterialism; the spiritualism in which they believed celebrated it. This combination of spiritualist medicine and religion occurred in other unorthodox doctors who brought similar preconceptions to the problem of body and mind. John Ashburner, for example, phrenologist, mesmerist, and spiritualist, derided that "system of Positive Philosophy, the great object of which is to prove that the Brain, or nervous matter, is the *Basis* of the thinking faculty. In other words that *matter thinks!*" He had nothing but scorn for those "who prate about the eternity of motion and of matter," and he insisted that matter, by itself inert, could only be activated by some "impelling force."[100] His medical theory was utterly dominated by a vitalism that rigorously separated the life principle from the organic matter to which it adhered. Jacob Dixon, a London physician, practiced both homeopathy and mesmerism, and contributed to the British spiritualist press from the middle of the century. He was a friend of Thomas Shorter, who told the London Dialectical Society that, at one séance, a levitating table kept up rhythmical motions in time with a tune that the doctor played on his concertina. Dixon's spiritualism impinged on his medical, as well as his musical, interests, and something of his bizarre approach to matters of sanitation may be gleaned from the title of his book, *Hygienic Clairvoyance*.[101] James Manby Gully played a leading role in popularizing hydropathy in England during the 1840s. He presided over a hydropathic spa at Malvern that attracted such renowned patients as Darwin, Tennyson, Carlyle, and D. D. Home. The latter produced "some spiritualistic manifestations" for Gully's customers,[102] and, in due course, the doctor became a convert. Nor were hydropathy and spiritualism his only enthusiasms. As Darwin wrote to his cousin, William Darwin Fox, in September 1850:

> It is a sad flaw, I cannot but think, in my beloved Dr. Gully, that he believes in everything. When Miss ——— was very ill, he had a clairvoyant girl to report on internal changes, a mesmerist to put her to sleep – an homoeopathist . . . and himself as hydropathist![103]

Gully found in all these approaches to medical treatment – clairvoyance, mesmerism, homeopathy, and hydropathy alike – a means of making manifest his deep conviction that physical distress had a psychological side, that the body could not be soothed unless the mind, too, rested at ease.

PSYCHOLOGY AND PSYCHICAL RESEARCH

While some doctors sought to address the riddles of the human organism through theories and methods of medical treatment that challenged the established practices of their profession, others followed well-worn paths in their efforts to base the study of man's mind on a scientific foundation. Throughout the nineteenth century, numerous disciplines contributed to the development of psychology, from the epistemology that had been the traditional focus of English mental philosophy to the neurology that imposed ever more insistent physiological questions on students of psychology. By the 1880s, if not earlier, the heyday of philosophical psychology had passed, and the predominant goal of psychological inquirers in Britain was to expose the study of the mind to the methods, vocabulary, and concepts of the physical sciences. The older associationist school of British psychology accepted the change in direction away from metaphysics and toward laboratory experimentation, while the newer physiological approach to psychology was, of course, predicated on it. "The causes of this change," declared the *Fortnightly* in 1879, in an essay on "The New Psychology,"

> may be summed up in one word – the study of biology. It is biology which has brought about the recognition of the "organism" as one of the elements of psychological research. It is biology which has introduced into the text-books . . . such terms as "nerve" and "tissue," "organ" and "cell," "neural tremor" and "muscular reaction." It is biology, again, which has suggested, if not initiated, the application of the law of development to the phenomena of the human mind.[104]

Despite increasing interest in physical problems of the brain and nervous system, of reflex action, and comparative anatomy, British psychologists before World War I never entirely abandoned ontological and ethical concerns. The potent impact of evolutionary thought and advances in the field of genetics might place the role of environment and heredity in the foreground of psychological inquiry; statistical analysis might gradually emerge as a major tool of the new "psychometrician"; Freudian psychoanalysis might even figure in at least one university course of lectures in psychology;[105] nonetheless some of the leading figures in British psychology – Alexander Bain, Henry Maudsley, W. B. Carpenter, James Ward, G. F. Stout, and James Sully, for example – still grappled with such concepts as will, volition, and purpose, and considered moral problems as legitimate areas of inquiry, even for the so-called "scientific psychology." British zeal for a morally neutral experimental psychology was blunted by the theological implications of a physiological, laboratory-based approach to the mind.[106]

Although it might be argued that these concerns gave British psychology a more humane, and humanistic, coloring than the German or American varieties, they helped to retard the expansion of experimental psychology in Britain right down to World War II. British universities were slow to make room for psychology in their course of studies and to provide adequate laboratory facilities. It is true that both Cambridge and University College, London, had something resembling psychology laboratories by the turn of the century, but Oxford resisted until the 1930s and did not establish a chair of psychology until just after World War II.[107] Thus, before 1914, it was virtually mandatory for young British scholars who intended a career in psychology to study in Germany, either in Leipzig where Wilhelm Wundt had founded the first psychological laboratory in 1879 or in one of the newer excellent German facilities for psychological research established by the turn of the century. Britain could not boast a professionally conscious psychological association until 1901, when the British Psychological Society was launched in London, nor a learned journal that sought to address an audience of trained specialists until the founding of the *British Journal of Psychology* in 1904. Within the British Association for the Advancement of Science (BAAS), established in 1831, a separate section for psychology was only provided in 1921. Until that time, papers that dealt with psychological subjects had to be presented under the aegis of anthropology, education, or physiology.[108] Although enthusiasts could take heart at the decision, made by the editors and staff of the *Encyclopaedia Britannica*, to devote a separate article to "Psychology" in the encyclopedia's ninth edition, rather than relegate the subject to "Metaphysics," as in the past, there was little occasion for British psychologists to celebrate at the turn of the century. It remained sadly true that in the field of psychology, "the best work was being done everywhere but in England."[109]

Perhaps it is anachronistic to speak of professional consciousness in a field that still lacked precise contours before 1914. Even within the limited circle of those who could be legitimately labeled psychologists – people whose full-time work was focused on the study of the human mind – there was no clear definition of their subject. While Maudsley, for example, believed that psychology formed one aspect of the physiology of the nervous system, James Ward insisted that psychology was nothing less than "the science of individual experience."[110] Their widely differing views reflected profound disagreement over such fundamental issues as the existence of a separate psychological self and the validity for psychology of categories adapted from the physical sciences. If disputes within the corps of trained psychologists took on such striking dimensions, it is no wonder that the intervention of outsiders only muddled the issue further. It was psychology's misfortune that, just as its practitioners were seeking scientific credentials, their reputations for serious scholarship were jeopardized by less reputable claimants to the title of "psychologist." The widespread use of the terms "psychology" and "psychological" by spiritualist groups in the second half of the nineteenth century left scars on the psychological profession for many years to come. In November 1914, at a council meeting of the British Association for the Advancement of Science, Oliver Lodge listened in amazement while a fellow scientist, lambasting the inclusion of psychology in the Association's proceedings, made explicit his ill-informed assumption that "Psychics" formed an integral part of psychology.[111]

Even before spiritualists adapted psychology to their own needs, mesmerists had laid their claims to the newly emerging field of study. J. C. Colquhoun, a Scottish barrister and writer on animal magnetism, announced in the 1830s that mesmeric experiments could help to enlarge contemporary "psychological knowledge," and his contention found many subsequent echoes. Ashburner was repeating a well-established claim of mesmeric literature when he insisted in 1867 that

> all psychology is based on Animal Magnetism. He who writes on psychology makes a sorry hash of his book in these days of facts if he be ignorant of Animal Magnetism. This is indeed the essential science to the Psychologist. All the old bases of metaphysics have been blown to the winds, and he who would make himself master of the science of mind, must first become familiar with the science of magnetism.[112]

In the second half of the century, spiritualists evidently assumed that psychological expertise was one of the legacies bestowed on them by their mesmeric predecessors.

The varied records of spiritualism during these decades amply illustrate the use of the term "psychological" as synonymous with "spiritualist" and the application of the word "psychology" to the pursuit of evidence supporting human immortality. Groups engaged in spiritualist investigations, such as E. W. Cox's, frequently dubbed themselves psychological associations, while spiritualist periodicals often employed the label "psychological" in their titles or subtitles. One such journal, the *Psychological Review*, launched in 1878, ran an article by the Newcastle spiritualist T. P. Barkas, which was entitled "Recent Investigations in Psychology" and featured an account of several séances for automatic writing.[113] Other spiritualists likewise confused séance attendance with psychological studies. William Howitt's daughter, for example, called him a pioneer "in the direction of Psychology," while an article in the fourth volume of *Human Nature* categorized as "psychological" the phenomena produced at Home's séances.[114] The BNAS declared that its founding purpose was "to promote the study of Pneumatology and Psychology."[115] The gentlemen who formed the Ghost Club in 1882 apparently equated ghost stories with "psychological experiences," or so the conditions of membership in the club would suggest, and Wyld, when asked by *Borderland* in 1893 to comment on the whole movement of modern spiritualism, responded: "Unless this experimental psychology is conducted with prudence, reverence, truthfulness, and unselfishness, . . . it may lead to widespread and multiform phases of devilry."[116] As late as 1905, Thomas Colley entitled his lecture on spiritualism, delivered at Weymouth during the week of the Church Congress, "Phenomena, Bewildering, Psychological." When spiritualists called séance manifestations "psychological" phenomena, they used the adjective to indicate the presence of psyche, not to suggest that a medium's accomplishments existed only in the minds of the audience. The label was not for a moment designed to imply that real spirit agents were absent from the proceedings. The various meanings of psyche in Greek – breath, life, soul, spirit, or mind – allowed the spiritualists considerable freedom in their application of the term, but always they intended by it the nonmaterial essence of each individual human being. For the Victorian spiritualists, "psychology" did not mean, as it does today, the scientific study of the mind and its functions, because they did not believe that mind could be analyzed from the perspective of scientific naturalism. Nonetheless, from their own perspective, they maintained that they had important contributions to make to the exploration of mind and tenaciously clung to the conviction that they were clearing new ground in the virgin forests of psychology, as they defined that word.[117]

To many an observer, their efforts appeared as simply a travesty of psychological endeavor. While physicians and specialists in several scientific fields were trying to map localization of function within the brain and to probe the role of the nervous system in mental activity, spiritualists seemed to be pursuing an occult fantasy that had no evident connection with the realities of contemporary psychology. Those researchers who worked on questions of cerebral pathology and the treatment of insanity had no time for the supernatural; even where psychology still accepted contributions from philosophy, as in the journal *Mind*, which first appeared in 1876, other-worldly forces had no legitimate role to play. To put the same point in the succinct words of W. K. Clifford, when commenting on the brain for the *Fortnightly* in 1874: "It is made of atoms and ether, and there is no room in it for ghosts."[118]

Clifford's determination to keep psychology within the confines of scientific naturalism won wide endorsement. Five years later in the *Fortnightly*, William L. Courtney, philosopher and subsequent editor of that journal, argued that the "Spiritualistic hypothesis" writ large – the belief that spirit as well as matter composed the cosmos and that fundamental reality lay with the former – had nothing further to offer

British psychology. Henceforth, that discipline would have to find its strength in "Science and Experience."[119] In 1906, the president of the British Association for the Advancement of Science likewise dismissed the relevance of the spiritualist hypothesis writ small, or the belief in the possibility of communion with disembodied spirits. E. Ray Lankester, director of the natural history departments at South Kensington and determined opponent of Henry Slade three decades earlier, explained in his presidential address that psychology had emerged "as a definite line of experimental research" in the past twenty-five years. It had done so, he continued, because the

> physiological methods of measurement (which are the physical ones) have been more and more widely, and with guiding intelligence and ingenuity, applied . . . to the study of the activities of the complex organs of the nervous system which are concerned with "mind" or psychic phenomena. Whilst some enthusiasts have been eagerly collecting ghost stories and records of human illusion and fancy, the serious experimental investigation of the human mind, and its forerunner the animal mind, has been quietly but steadily proceeding in truly scientific channels.[120]

Lankester's scarcely veiled contempt for the work of the SPR carried on the tradition of abuse promoted for over thirty years by William Benjamin Carpenter. While pursuing a distinguished career in physiology from the late 1830s until his death in 1885, he worked zealously to further popular scientific education and to assist the development of the University of London, which he served as registrar from 1856 until 1879. He also waged an unremitting campaign against the "Epidemic Delusions" of his day, by which phrase he particularly meant to describe mesmerism and spiritualism, although he had also helped significantly to discredit phrenology during the 1840s.[121] His own work embraced such fields as marine zoology and geology, but it was as an expert on the brain and nervous system that he declared war against occult forces in any guise whatsoever. "My first Paper on the Nervous System, which contained the germ of all I have done since, was written 40 years ago," he pointed out to Barrett in 1876, in a letter clearly designed to make Barrett feel the novice in a field that Carpenter had already thoroughly plowed. "My Thesis," he continued, "applying the doctrine of Reflex Action to invertebrate animals (now universally accepted) gained one of the University Gold Medals 38 years ago."[122] His was, indeed, an authoritative voice on the subject of the mind, and his *Principles of Mental Physiology*, first published in 1874, was a standard text for many years. In his capacity as psychologist, Carpenter represented that school of writers who sought to have it both ways: He wanted to infuse psychology with a heavy dose of physiology, but his sincere religious convictions kept him from endorsing any position that might be deemed materialist or determinist. In the preface to *Principles of Mental Physiology*, he insisted that "a Physiological Psychology" did not destroy "the root of Morals and Religion," because man's will remained free. Carpenter was certain that "Intellectual and Moral capacities" were as much a part of the Creator's endowment to man as was his "Bodily Organism."[123]

If Carpenter tried to preserve some room for the exercise of free will in his psychological theory, he certainly yielded no space to the supernatural. From the early 1850s, in one polemical article after another, he reiterated his unalterable conviction that the phenomena of mesmerism and spiritualism could be explained by a combination of mundane causes. Some, like the human tendency toward self-deception, especially when under the influence of a dominant idea, scarcely required much elaboration from Carpenter, although he belabored the point with characteristic heavy-handedness.[124] Others, such as his theories of ideomotor activity and unconscious cerebration, needed further explanation, which he provided at every possible opportunity. The former he defined as "the involuntary response made by the muscles to ideas with which the mind may be possessed when the directing power of the will is in abeyance."[125] He summoned the concept to explain the table-turning mania of the early 1850s and expounded the theory at the Royal Institution in March 1852, in a lecture entitled "On the Influence of Suggestion in Modifying and Directing Muscular Movement, Independently of Volition." Such phenomena as rotating tables, he argued, or the movements attributed

> to the action of an "od-force," . . . have been clearly proved to depend on the state of *expectant attention* on the part of the performer, his will being temporarily withdrawn from control over his muscles by the state of abstraction to which his mind is given up, and the *anticipation* of a given result being the stimulus which directly and involuntarily prompts the muscular movements that produce it.

He aired the same arguments again the following year, when he published an extensive article on "Electro-Biology and Mesmerism" in the *Quarterly Review*.[126]

"Unconscious cerebration," a phrase that became a cliché with both Victorian spiritualists and their critics, was Carpenter's term for a mental process that could also be described as "latent thought" or "the reflex action of the brain." When challenged by spiritualists to explain how mediums, through raps on furniture, planchette, automatic writing, or other means of communication, could give the correct answers to questions, when neither they, in their normal waking state, nor any of the séance participants knew those answers, Carpenter did not have to turn to spirits for a reply. He pointed out that "much of our *mental work* is done *without consciousness*"; waxing more technical, he informed his readers that studying "the anatomical relation of the Cerebrum to the Sensorium or centre of consciousness" had led him to

> the conclusion that *ideational* changes may take place in the cerebrum of which we may be at the time *unconscious* through a want of receptivity on the part of the sensorium, . . . but that the results of such changes may afterwards present themselves to the consciousness as *ideas*, elaborated by an automatic process of which we have no cognizance.[127]

Thus could the medium, at the critically important moment in a séance, produce information that he or she had no conscious knowledge of possessing.

During his long battle against irrational explanations of physiological phenomena, Carpenter developed a handful of stock responses to spiritualist claims. In addition to the ideomotor principle and unconscious cerebration, he contended that alleged thought reading was nothing more than careful muscle reading, such as police detectives practiced when interrogating suspects and surmising the truth or falsity of their testimony from the most infinitesimal changes of expression, movement of the eyes, or other signs visible only to the trained expert.[128] Supporters of spiritualism claimed that his so-called explanations only covered a fraction of the phenomena that appeared regularly at séances, but which he was too biased to witness with an open mind. To this charge, Carpenter replied that he took "every opportunity of seeing any *new* phenomena that could be shown," only to be "told over and over again, 'Oh, your atmosphere of incredulity prevents these manifestations and therefore nothing comes.'"[129] Either, he complained, he attended séances where manifestations occurred under highly suspicious circumstances, but which he was barred from investigating, or he was permitted to impose effective controls on a medium, who proved incapable of inducing phenomena under such conditions.

There were elements of Carpenter's antispiritualist critique that were tedious and irrelevant. His repeated insistence that a certain, specialized, and rigorously scientific education was requisite to guard against the tricks of mediums seems naive in light of the facility with which skilled conjurors can baffle virtually everyone, except their professional colleagues. Furthermore, Carpenter's definition of a rigorous scientific education varied with his intended victim, so that, in effect, he came close to saying that everyone who rejected spiritualism was sufficiently educated, whereas everyone who did not lacked adequate schooling, no matter what their professional position or title. A. R. Wallace expressed a general frustration shared by Carpenter's spiritualist adversaries when he objected wryly that, as far as Carpenter was concerned, "nobody's evidence on this particular subject is of the least value unless they have had a certain *special early training* (of which, it is pretty generally understood, Dr. Carpenter is one of the few living representatives)."[130] Carpenter's level of argument frequently sank to a nasty *ad hominem* approach that did little to focus attention on the scientific principles he claimed to invoke. Yet, as the paramount critic of spiritualism in Britain, his anathemas could not be ignored. Other men of science might denigrate mesmeric and spiritualist follies, but Carpenter took the time to denounce them at length in print, year after year. Although he packaged them unpleasantly, his arguments were the ones that spiritualists had to refute and psychical researchers to address. The manifold references to Carpenter in the writings of both groups suggest how dangerous a foe he was.

Carpenter's sustained onslaught against spiritualism was not merely motivated by a confirmed distaste for tricksters and willing dupes. He was also impelled by his belief that many of the phenomena associated with spiritualism were manifestations of psychopathological conditions. Indeed a considerable medical literature developed in the second half of the nineteenth century that related hallucinations, clairvoyance, automatism, and entranced states in general – including somnambulism and hypnotism – to abnormalities of the nervous system. Physicians were tracing these connections before the spiritualist movement took hold in the early 1850s,[131] but the growth of spiritualist audiences and the emergence of spiritualist mediums provided the medical profession with a rich assortment of nervous disorders to consider, or so numerous doctors believed. Their writings dealt with a wide range of morbid states, such as hysteria, epilepsy, and insanity, and they likewise placed the inspired utterances of the trance medium, the brilliant visions, the spectral illusions all within the category of disease. Maudsley was a vigorous contributor to this school of medical thought, ascribing sensory delusions to disturbed "relations of the nervous centres" and disordered cerebral functions. When the prominent London physician Sir Henry Holland published his memoirs in 1872, he confirmed an already familiar view in asserting that "mesmeric visions and prophecies, clairvoyance, spirit-rappings, table-turnings and liftings" could best be explained as "morbid or anomalous states of the brain and nervous system."[132]

Not all medical opinion, however, endorsed the psychopathological interpretation of conditions associated with spiritualist experiences. Dr. Elizabeth Blackwell, for example, the first woman listed on the

British medical register, was an associate member of the SPR who appreciated the Society's efforts. She wrote warmly to Sidgwick in 1889:

> With a firm conviction of the double nature of Man – and my inside view of the needs and dangers of the medical profession, I realise the imperative need of your line of Research; and I know of no organi- zation which is endeavouring to meet a deeper human need than this Psychical Research Society.[133]

Blackwell's view, needless to say, was the one which the SPR ac- knowledged, and it ably countered the denigration of its work by Car- penter and like-minded assailants. In the face of criticism and abuse, the Society firmly maintained that its inquiries unquestionably placed it in the vanguard of psychological studies.[134] Nor was its claim utterly unfounded. Not only did Sidgwick play a major role in introducing psychology into the Cambridge curriculum in the 1870s and 1880s, but he also presided over the Second International Congress of Experi- mental Psychology, held in London in August 1892. On that occasion, he and Myers reported, "the majority of the English members attending were either members of the SPR or at least in avowed sympathy with its aims." At the Congress, Sidgwick discussed the Census of Hallu- cinations, Myers likewise read a paper on hallucinations, and Mrs. Sidgwick gave a talk on experiments in thought transference. There was no question in their minds but that the subjects that they pursued so intently were pertinent to the interests of an international gathering of psychologists, among whose sections were both "Neurology and Psychophysics" and "Hypnotism and cognate questions."[135]

The very membership of the SPR confirmed that the Society's work commanded international attention among psychologists. Liébeault, Bernheim, Janet, and Richet of France, Lombroso of Italy, Schrenck- Notzing of Germany, Flournoy of Switzerland, G. Stanley Hall and William James of the United States, among others, all participated as corresponding members. James even became a vice-president of the Society a few years after its establishment and served as its president in 1894–5. Shortly before World War I, the name of Dr. Freud of Vienna joined the list of eminent foreigners affiliated with the SPR, and in 1913, at the Society's request, he contributed "A Note on the Uncon- scious in Psycho-Analysis" to its *Proceedings*. Freud had good reason to appreciate the Society's work, for it was Myers who had first pub- licized in England the Breuer–Freud studies of hysteria.[136] The SPR similarly boasted indigenous British psychologists and medical doctors who specialized in treating the mind. Charles Lockhart Robertson, an editor of the *Journal of Mental Science* during the 1860s, medical su-

perintendent of the Sussex County lunatic asylum, and chancery visitor
in lunacy, joined the SPR at its foundation and became a member of
the governing council, where he remained until his death in 1897.[137]
Dr. John Milne Bramwell and Dr. C. Lloyd Tuckey, both influential
in promoting renewed interest in hypnotism as a form of medical treatment in England, were also council members during the 1890s and the
years preceding World War I. William McDougall, one of the leading
names in British psychology in the early twentieth century, likewise
held office in the Society. Clearly there was no lack of informed and
reputable psychological opinion to suggest that the SPR had a contribution to make to the study of the mind, and not only with regard to
pathological conditions.

While specific individuals within the SPR had their own interests to
pursue in the field of psychology, the totality of their efforts pointed
generally in a single direction. The Society's inquiries supported the
movement within late nineteenth-century psychology that sought to
challenge the reductionism of both the earlier associationist and the
later physiological psychologists. When members of the SPR studied
the workings of the mind, they were not prompted by the goal of reducing mental functions to physicochemical laws or canons of biology.
Like the medical doctors who embraced spiritualism, they harbored,
in the vast majority of instances, a profound distaste for the mechanistic
approach to psychology; they hoped instead to prove how insufficient
were all attempts to explain mental phenomena as part of a strictly
deterministic process. In their experiments, subjectivity was a substantive factor, and when they spoke of "self," or "ego," they tended
to endow it with a spiritual nature – to use it as virtually synonymous
with "soul."

"The semi-theological conception of the soul as an extraneous principle put into the body to govern it, the idea of Self as a spiritual
substance," may have struck advanced thinkers as "crudities of
thought" worthy of scorn,[138] but other Victorians clung to these notions, for theological, philosophical, and even scientific reasons. Spiritualists, of course, used the concept of the ego for obvious antimaterialist purposes, as when Stainton Moses described "the Spiritual
Principle in man" as "the Self, the *Ego*, the Inner Being – which,
acting through the material frame, is . . . independent in its existence,
and will survive the death of the body."[139] Nonspiritualists might
equally invoke an immaterial ego, both to contend on behalf of the
immortal soul and to uphold the ethical foundations of the cosmos. In
1877, Lord Selborne, contributing to "A Modern 'Symposium'" in the
Nineteenth Century, explained that

> To a man who believes in a moral government of the Universe, in
> the distinctness of the *Ego*, the real man, from his bodily organisation,
> and in the doctrines of moral responsibility and moral adjustment in
> a future state, nothing can be more real, nothing more intelligible,
> than this moral instinct or sense, with its suggestions of right and
> wrong, of duty, guilt, and sin, and its judicial conscience.[140]

Within the ranks of British psychologists themselves, the ego could be
invoked in an effort to explain the inexplicable interaction of body and
mind, and to straddle the proverbial fence between excessive emphasis
on matter and undue reliance on spirit. When W. B. Carpenter asked
rhetorically, "Does the body of man constitute *his whole self*, or is
there an *Ego* to which that body is in any degree subservient?" an
affirmative response was soon forthcoming. "If we are led," he reasoned,

> by physiological evidence to recognize in the cerebrum a power of
> directing and controlling the automatism of the axial cord, I do not
> see on what ground we are to reject the testimony of direct con-
> sciousness, that the automatism of the cerebrum is itself directed and
> controlled by some higher power.[141]

The very concept of the ego defied reduction to ultimate physical
categories. To invoke it implied an unwillingness to strip man of "some
higher power" and, in consequence, a fundamentally unsympathetic
attitude toward the neurophysiological school of psychology. In the
late nineteenth and early twentieth centuries, this distinctive attitude
within British psychology, articulated most emphatically in the writings
of James Ward, provided Britain's major contribution to the contemporary international reaction against positivism and scientific naturalism as the guiding assumptions with which to study man and the universe. The psychological investigators of the SPR played their part in
that reaction, arguing that the categories and terminology of the physical sciences were not basically applicable to the mind.[142]

There was, nevertheless, no official Society position to be maintained in the discipline of psychology. Members who undertook experiments along psychological lines did so with complete freedom, in
the absence of any particular orthodoxy to uphold. At the same time,
two of the Society's earliest committees were established to inquire
into thought reading and mesmerism, and the latter played a significant
role in resurrecting the serious study of hypnotism in England during
the 1880s and 1890s. While in France, the work of Liébeault, Richet,
Bernheim, Janet, and Charcot had heralded an extensive examination
of hypnosis, medical scientists on the other side of the Channel avoided

the suspect topic after midcentury. When Charles Lloyd Tuckey went to Nancy in 1888 to observe Dr. Liébeault at work in his dispensary, he could lament that, although "treatment by psycho-therapeutics" was well-known and widely discussed on the continent, it was "unknown commonly or misunderstood" in his own country.[143] Such hypnotic experiments as were proceeding in England were conducted by the SPR Committee on Mesmerism, or by SPR members, like Milne Bramwell, Gurney, and Myers. In these years, Podmore subsequently reminisced, "medical men would come to our rooms to see a boy's body stiffened and laid across two chairs, or to test the 'alleged insensibility' in the trance by driving pins into him." Gradually, the message hit home, and doctors in Britain at last gave a respectful hearing to the massive evidence, accumulated from around the world, that hypnosis was a valuable, if still enigmatic, remedial tool. By 1903, Bramwell gladly acknowledged that, at present, "the treatment of hypnotism by the [medical] profession is not only fair but generous."[144]

Experiments with hypnosis provided an opportunity to examine the elusive realm of secondary, and even tertiary, consciousness, a subject of compelling fascination to SPR investigators. Not only did it seem to contain clues to the mysteries of mediumship, but, more important, it promised to shed some light on the organization and functioning of the human psyche in general. Podmore devoted a chapter to secondary consciousness in his *Studies in Psychical Research*, where he found the existing physiological theories inadequate to explain the phenomena in full. Gerald Balfour likewise grappled with the curious manifestations of what appeared to be more than one consciousness in a single physical frame; he used the term "polypsychism" to designate his belief that the human personality represented "a very complex association of selves with the conscious self at the head."[145] Bramwell, in yet another view, contended that experiments with hypnosis proved conclusively the existence of a primary, secondary, and sometimes tertiary consciousness, but that no existing evidence justified the assumption of more than one personality "in the same human being." As he bluntly pointed out, all levels of consciousness in a single individual depended upon "the life and activity" of the same brain.[146]

If Bramwell unhesitatingly acknowledged the physiological support necessary for all mental phenomena, he was nonetheless very far from suggesting that the mind worked automatically in response to external stimuli. He held instead that the mind moved freely over a range of choices, selecting the object of its attention according to its own dictates. External suggestion no doubt played a role in influencing the selection, but the ultimate choice came from within "the hidden workings of the mind." In developing his theory of the self, Bramwell set "aside all the mechanical theories which [had] hitherto held the field in fashionable science." It was his antimechanism that placed him in close sympathy with the body of psychological theory developed by the SPR, although Bramwell could not accept the belief, widely held in the Society by the early twentieth century, that telepathy was a proven mental phenomenon.[147] He could, however, appreciate the persistent theme that ran through the writings of leading members of the SPR. When Sidgwick defended "the independent existence of mind"; when Leaf paid tribute to "that extraordinary complex which we call Self"; when Barrett described "our Ego" as "a composite structure embracing a self that extends far beyond the limit of our conscious waking life"[148] – they were all refusing to be limited in their exploration of human mental faculties to the formulae and concepts of physical science. If the tools of physics, chemistry, and biology could not measure and analyze consciousness, whose existence both common sense and empiricism upheld, then physical scientists must expand the scope of their inquiries. Psychologists should certainly not curtail their own.

THE WORK OF GURNEY AND MYERS

Of all the diverse members of the SPR who turned their attention to psychological experiments, two stand out in particular. In a series of reports, articles, and books that explored virtually every aspect of hypnosis, and much else besides, Gurney and Myers doggedly attempted to bring psychical research into the recognized domain of experimental and theoretical psychology. "Those who engage in this as in other branches of psychical research," they wrote in an article about mesmerism in 1885, "must be prepared to face much wearisome failure, much deceptive ambiguity. Yet thus, perhaps, may they with most reason hope to lay the corner-stone of a valid experimental psychology."[149] Few would credit either Gurney or Myers with laying such a corner-stone, but their efforts commanded attention and received warm recognition from numerous fellow laborers in psychological fields.

When Gurney died prematurely in 1888, he had not yet articulated an overarching theory of mind, brain, or personality to bind together the many provisional hypotheses to which his experiments in the preceding six years had directed him. His goals, nonetheless, were clear. In the first place, he was determined that his work should bear the impress of his scientific expertise. He had, after all, studied medicine for several years and was as fully competent to discuss the physiological structure of the human organism as to speculate upon its psychological life. He could write persuasively about apparitions and hallucinations in part because he understood the workings of the sense organs and knew the current theories to explain how these interacted with the brain. Indeed, he placed his own work in the context of "the new psychology," with its strong physiological emphasis, and where "the line between the normal and abnormal has become so shadowy that not the smallest or rarest abnormal phenomenon can be safely neglected, by those who aim at the fullest possible realisation of human nature and development."[150]

Gurney's aim to realize human nature as fully as possible was not actually predicated on the elucidation of its physical functions. On the contrary, his constant endeavor in all the psychological subjects he explored was to reveal the limited vision of the "new" psychologists and to expose the inability of neurological reductionism to probe the human mind in all its intricacies. He displayed his familiarity with man's material aspect, not only because he acknowledged the inseparability of body and mind, but because he sought to demonstrate his scientific credibility. His critique of physiological psychology was all the more effective for its accompanying display of physiological knowledge.

Gurney's repudiation of simplistic mechanical modes of thought pervaded all levels of his work in psychology. Just as he deplored the imposition of "mechanical rules" in the analysis of music,[151] so he rejected purely mechanical interpretations of human behavior. A persistent argument in his extensive writings on hypnotism was that actions performed under hypnosis were not merely involuntary reflex responses to the hypnotist's commands. With a variety of subjects, he carefully tested recollection, when awakened, of conversations and actions carried out under the different stages of hypnotism. He found enough evidence of memory, particularly from the "alert" or lighter stage of hypnotic trance, to challenge those "theories which assume hypnotism to be a state of mere unconscious automatism, on the ground that no true memory ever exists of what happens in it."[152] The great riddle of hypnosis, according to Gurney, its outstanding peculiarity, was precisely that it prompted behavior at once conscious *and* reflex. "The central problem of Hypnotism lies in the combination of those two adjectives," he wrote in 1884,[153] and he remained convinced that both were essential to understanding the way in which hypnosis affected people.

The emphasis in his own work, however, fell far more heavily on the conscious aspects of hypnotic behavior than on the reflex. He designed a number of experiments to suggest that intelligence and even volition, in some cases, were at work while the subject appeared blindly to obey the hypnotist's commands. Such demonstrations of "intelligent automatism," as he called it, often involved the following scenario: A subject, while hypnotized, would be given a simple arithmetical problem to perform, typically one of multiplication. Immediately thereafter, he was awakened, and in the majority of cases had no recollection of the task set him.

> He was then made to place his right hand on the planchette, his attention being occupied by reading aloud, or occasionally by counting backwards, leaving out alternate numbers, or some similar device. The planchette meanwhile was writing. In most of the cases where the writing did not prove to be the correct answer to the sum, the figures were sufficiently near the mark to make it apparent that an intelligent attempt had been made to work out the given problem. The paper and instrument were always kept concealed from the "subject's" eyes, and he was never told what the movements of the planchette produced. As a rule, he was afterwards offered a sovereign to say what the writing was, but the reward was never earned. On rehypnotisation, he recalled the whole process – a clear indication that we have had to do with "secondary intelligence," not with unconscious cerebration.[154]

Although the notion of secondary intelligence was fraught with difficulties, Gurney preferred it to a mechanistic concept of the brain. Indeed, his experiments with stages of hypnotic memory confirmed his preference; he found that, when subjects alternated between the alert and deep stages of hypnotic trance without normal wakefulness intervening, they usually could remember what transpired in each stage only when the hypnotist returned them to that stage. Thus, for example, if told a set of cricket scores in the alert stage, the subject had no recollection of the information in the deeper, lethargic stage. Once returned to the lighter hypnotic condition, however, the subject recalled the scores at once. Similarly information conveyed to the subject in the deep stage, Gurney found, could only be retrieved in that stage. "Perhaps," he suggested in 1884, "the clearest interest of the hypnotic alternations of memory is rather as illustrating the spontaneous alternations in cases of '*double consciousness*,' where a single individual lives in turn two (or more) separate existences."[155]

In subsequent writings on this issue, Gurney's language was more restrained. He avoided such vivid and startling ideas as several "separate existences," at least for healthy hypnotic subjects, and spoke instead of "a secondary memory and secondary play of mind," "the latent or secondary consciousness," "the secondary 'self'," and "secondary intelligence." In 1887, he claimed a new significance for hypnotism "as the ready means for establishing a secondary train of consciousness." Psychological research seemed to be producing more and more evidence "of the instability and divisibility of consciousness,"[156] and, in doing so, Gurney believed, it was underscoring with increasing cogency the divergence between brain and consciousness. If a single brain could house two or more disparate strands of consciousness, clearly the proposed identity of brain and mind was fallacious. This was the theme of a long essay, "Monism," which Gurney published in *Tertium Quid* in 1887. There he argued that the material development of the brain proceeded simultaneously with, but separately from, the development of consciousness and that all attempts to reduce them to the same order of experience were doomed to ultimate frustration. As far as the ancient puzzle of body and mind was concerned, Gurney made no pretense of solving it. "All materials for approximate solutions are wanting," he maintained. "The riddle is not one in respect of which facts can be discovered and evidence amassed, but remains perfectly isolated and ultimate, without a single avenue leading to it or from it."[157]

Gurney, while quick to point out "the hopelessness of dogmatic Monism," was not a convinced dualist himself. He asserted that modern science had failed to bridge "the radical distinction between brain and mind," but did not rule out such an achievement in the future. Yet, if success were someday to result, he felt certain that the triumph would not belong to monist modes of thought. Rather, he surmised, it would come with the discovery of some third form of existence, neither mental nor physical – "an at present unguessed *Tertium Quid*." Gurney, a triadic theorist of sorts, believed in something "essentially vital and *sui generis*," touching on the realms of both physical and psychical research.[158]

Gurney's experiments with finger anesthetization, by means of hypnosis, convinced him that the process of hypnotism involved something more than the power of suggestion. If the subject, with his hands thrust through a screen that blocked his vision, could not see which particular finger was being hypnotized, how could the subject's heightened state of expectation, or fixity of attention, be summoned to explain the fact that, in the majority of cases, the hypnotized finger alone became stiff and insensitive to pain? The hypnotist, of course, did not actually stroke the finger in question, but simply held his own hand just above one of the subject's fingers that were spread apart on a table in front of the screen; a purely physical explanation was, accordingly, out of the question. However, experiments showed that, if the hypnotist merely concentrated his fixed gaze on the selected finger, without placing his hand near to it, rigidity and anesthetization failed to occur. Yet, at the same time, the hypnotist's attention also *had* to be focused on the finger. If Gurney guided the hypnotist's hand over the subject's finger while the hypnotist looked away, "the selected finger remained perfectly sensitive and flexible." Thus whatever force was at work required both a physical and mental foundation in order to function properly, Gurney was convinced.[159] Throughout his years of affiliation with the SPR, he pondered the problem thus created and struggled to identify the elusive agency that was neither wholly material nor wholly psychical.

With his recognition of multiple levels of consciousness, and his willingness to posit a *tertium quid* between mind and matter, Gurney might appear to have welcomed the fragmentation of the human personality into diverse selves. But such was not Gurney's ultimate vision for humanity. If he challenged the strictly physical explanation of the human organism at work, he did not do so to leave man strewn in psychic bits and pieces over the psychologist's laboratory. In an article that he published in 1887, Gurney contemplated the meaning, for the healthy individual, of all the recent studies on dissociated personalities and double consciousness. He recognized how deeply the sense of "I" colored all perceptions of life's experiences, despite changes that inevitably occurred as one grew older. His experiments in the stages of hypnotic memory had not been wholly reassuring on this score, with their suggestion that distinct channels of memory functioned during the different stages of hypnosis, as if a single "I" could not preside over the disparate bundles of experience. Yet, he pointed out,

> experiments show that a separation of impressions which appears as distinct and complete as if they belonged to different individuals may in time, and by a spontaneous process, be dissolved away, and the two pieces of experience may merge into the general store over which the mind has unrestricted control.

Gurney could not prove a final fusion of memory in all cases, particularly those of shattered identity that were beginning to fill the annals of abnormal psychology, but he could recognize how critically important it was to hope for such a fusion. "If not," he observed with customary candor, "then I confess that I see no manner in which our faith in the continued identity of the persons concerned, or, ultimately, in our own, can be sustained." He derived what comfort he could from this possibility, that separate strands of memory might ultimately unite under the aegis of a single consciousness. "If the superiority of man to the brutes depends on personality," he reasoned,

> and if personality depends essentially on memory, then those who desire that man's dignity should be maintained, and that personality should be continuous, can hardly afford to despise the smallest fact of memory which exhibits the possibilities of union and comprehension as triumphing over those of disruption and dispersion.[160]

Although he did not achieve enduring fame in psychology, Gurney's psychological inquiries have had their admirers, both during his lifetime and after. Myers claimed that Gurney was the first Englishman adequately to investigate the psychological ramifications of hypnotism and that, although Pierre Janet was first to publish his findings on hypnotic memory, Gurney's work proceeded independently of his French colleague's. Myers claimed, too, that Gurney's work on hallucinations was unique in the English language. If Myers erred on the side of hyperbole, he did so in the name of grieving friendship. The same cannot be said about the author of an article published in 1939, who, presumably without personal attachment to Gurney, described him as "laying the foundations on which the psychology of abnormal mental states during the next twenty years was to be based."[161] Gurney, of course, had his critics, even within the SPR, but it is difficult to deny his significance among English students of hypnosis.[162] His approach to psychology, furthermore, was one shared by such leading figures in turn-of-the-century psychology as Janet, William James, Alfred Binet, and Carl Jung. They recognized the mind as

> a dynamic entity whose principles of operation are not likely to be discovered simply through an analysis (introspective or psychophysical) of short-term sensations. Secondly, all of them shared in the conviction . . . that genuine psychological phenomena had to be dealt with at the level of psychology rather than at the (largely unknown) level of neurophysiology . . . Finally, all of them argued in behalf of a *useful* scientific psychology; one able to explain mental phenomena as these occur in the real world and to actual persons.[163]

Myers shared Gurney's conviction that hypnotism should play a central role in the development of experimental psychology. "Hypnotism is in its infancy," he proclaimed in 1885, to a general meeting of the SPR, "but any psychology which neglects it is superannuated already." Like Gurney, he believed that the hypnotic trance and the multiple personalities that hypnotism often revealed were not necessarily morbid symptoms.[164] On the contrary, Myers used hypnotism to help him unravel the intertwined components of the human personality and to trace the contours of what he labeled the "subliminal self."

The theory of the subliminal self that Myers developed during the 1880s and 1890s was the boldest and best known of the contributions that psychical research made to psychology before World War I. William James, speaking as a professional psychologist after Myers's death, claimed that the formulation of that theory would "figure always as a rather momentous event in the history of our Science." Certainly, where Myers lacked ability to devise original experiments on his own, he compensated with the gift of propounding theories that seemed to unify diverse phenomena into coherent patterns. James was not the only student of the mind for whom Myers's labors reduced to order the chaos of "unconscious cerebration, dreams, hypnotism, hysteria, inspirations of genius, the willing-game, planchette, crystal-gazing, hallucinatory voices, apparitions of the dying, medium-trances, demoniacal possession, clairvoyance, thought-transference, even ghosts and other facts more doubtful." Professor Theodore Flournoy warmly praised the way in which Myers drew together, under his vast theory of the subliminal self, "une foule de théories analogues mais moins élaborées," and he predicted that, if future discoveries confirmed Myers's hypothesis, the Englishman would find his name inscribed in the golden book of great initiators; "et joint à ceux de Copernic et de Darwin, il y complétera la triade des génies ayant le plus profondément révolutionné la pensée scientifique dans l'ordre cosmologique, biologique, psychologique."[165]

Although Flournoy's prediction has a ludicrous ring today, there was, as James pointed out, something akin to Darwin's genius in Myers's passion for collecting and classifying, for finding the phenomenon that linked previously disparate categories, and for ranking his evidence into continuous series.[166] Although his effort to find general categories for psychical phenomena never eclipsed his fascination with individual cases, his concern to introduce scheme and system into the confused findings of psychical research brought him nearer to the methods of contemporary science than his attempts to pass for an experimental psychologist in his own right. He was indeed "one of the great systematizers of the notion of the unconscious mind."[167] Apart from that achievement, however, the means by which he constructed his theory of the subliminal self would have hardly won him acclaim from the scientific confraternity of his day.

Although Myers was closely involved in a number of SPR tests with mediums and hypnotized subjects, particularly during his work with Gurney in the 1880s, the overwhelming impression conveyed by Myers's articles in the SPR *Proceedings* is one of extensive scholarship rather than experimentation. He was an avid reader of the international literature of psychology, and his own writings offer a virtual bibliography of contemporary inquiries into such topics as hypnosis and hysteria. As his appreciation of the Breuer–Freud work on hysteria suggests, Myers had a keen eye for identifying what was novel and significant.[168] He followed developments in French psychology with particular attention, visiting the Salpêtrière in Paris and witnessing experiments in Nancy, both with Bernheim at the Hôpital Civil and with Liébeault in private practice. Many of the building blocks of Myers's theory were drawn from the work of these men and of their professional colleagues, such as Beaunis, Richet, Binet, and especially Janet.[169] Myers knew the lesser names in psychology as well as the renowned, the ephemeral journals as well as the classic texts.

This quantity of source materials all provided grist for the mill of Myers's inventive talents, and he eagerly used the evidence that came to hand. He used it, however, not to refine his own research methods, nor to devise more judicious experiments, but to suggest the analogies that were crucial to the development of his argument. "He was a genius at perceiving analogies," James observed; "he was fertile in hypotheses." Analogies and hypotheses could always serve to bridge the gaps in the explanations that Myers proposed for psychical phenomena.[170] Often the work of European psychologists merely confirmed theories that Myers had constructed without significant data at all, and where he did not rely on mere analogical reasoning, as in his studies of the mediumship of Mrs. Piper and Mrs. Thompson, he tended to allow his enthusiasm for individuals to prompt vast speculations that soared beyond the reach of corroboration. Podmore, who had followed Myers's work carefully, could not help concluding in 1897 that the theory of the subliminal self was "largely founded on assumptions and conjectural interpretations."[171]

The core of Myers's interpretation of the human personality lay in his perception of its complexity. By the mid-1880s, he had concluded that experiments with hypnosis invalidated the traditional criteria for designating an independent, distinct, and unified personality. A finite, self-contained center of consciousness and will, with "continuous memory" and "homogeneous character" dissolved quickly, Myers contended, under the impact of hypnosis, to reveal instead "the multiplex and mutable character of that which we know as the Personality of man."[172] When hypnotized, a subject typically lost the power to choose his own actions freely, Myers pointed out; his central, controlling will became subject to the commands of his hypnotist. Furthermore, his memory proved to be fragmented, not a single stream of recollections binding together the diverse phases of his life. The work of Gurney and Janet on hypnotic memory convinced Myers that the phenomena of alternating memory characterized not just severely disturbed mental patients, but presumably normal men and women as well. Finally, Myers argued that character traits, for long identified as indelible marks of personality, could readily change under hypnotic suggestion. He was certain, in short, that the waking self of daily life represented a small portion of the streams of consciousness associated with a single physical frame.

This was the theme at which Myers hammered away for the remaining years of his life, working out the details in his series on "The

Subliminal Consciousness,'' which appeared in the SPR *Proceedings* during the 1890s, and attempting a grand synthesis of his life's inquiries in the two-volume *Human Personality and its Survival of Bodily Death*, published posthumously in 1903. In Myers's hands, the human individual became a highly elusive entity, no more visible, tangible, or accessible in its entirety than an iceberg, whose protruding tip above the water scarcely indicates the size and shape of the mass below. In a much-quoted passage, Myers suggested that

> the stream of consciousness in which we habitually live is not the only consciousness which exists in connection with our organism. Our habitual or empirical consciousness may consist of a mere selection from a multitude of thoughts and sensations, of which some at least are equally conscious with those that we empirically know. I accord no primacy to my ordinary waking self, except that among my potential selves this one has shown itself the fittest to meet the needs of common life. I hold that it has established no further claim, and that it is perfectly possible that other thoughts, feelings, and memories, either isolated or in continuous connection, may now be actively conscious, as we say, "within me," – in some kind of coordination with my organism, and forming some part of my total individuality.

To label these other thoughts, feelings, and memories "unconscious" or "subconscious," Myers argued, would be to deny them consciousness; to group them under the heading of "secondary self" would accord implicit superiority to the "empirical self," which we know as our common, waking self. Myers therefore adopted the more neutral terms "subliminal" and "supraliminal" to designate, respectively, those parts of the mind that lie below the threshold of normal, waking consciousness and those which remain above it.[173]

With Myers, however, words like "normal," "ordinary," and "common" ceased to have their customarily recognized meaning. For him, one of the great triumphs of his theory of the subliminal self was that, in vastly expanding the potential powers of the human personality, it made room for the so-called abnormal and pathological. Although most contemporary psychologists regarded the state of dissociation, in which a patient revealed two or more apparently distinct personalities, as a pathological condition, Myers contended that multiple levels of consciousness were as much a part of the individual's equipment for life as multiple fingers and toes, albeit a great deal more difficult to investigate. Myers himself never managed to reduce the subliminal strata to anything like precision or clarity. Beneath the layer of waking consciousness, he apparently located both a hypnotic stratum and a "stratum of dream and confusion," while beneath these, he suspected "a still subjacent stratum of coherent mentation as well." But there were no clear lines of demarcation, and presumably considerable overlap, among the regions.[174]

Although Myers could offer no maps to chart the terra incognita of the subliminal, he was confident that he knew what went on in its shadowy depths. "I hold," he announced in the opening essay of "The Subliminal Consciousness,"

> that this subliminal consciousness and subliminal memory may embrace a far wider range both of physiological and of psychical activity than is open to our supraliminal consciousness, to our supraliminal memory. The spectrum of consciousness, if I may so call it, is in the subliminal self indefinitely extended at both ends.[175]

In other words, the subliminal consciousness could receive information through channels, such as telepathy and clairvoyance, that were inaccessible to the supraliminal, restricted as it was to sensory modes of perception.

Yet the supraliminal was not barred from sharing the insights that the subliminal self garnered, for the latter could send messages to the former by a variety of methods. Here, perhaps, lay the most significant part of Myers's theory, insofar as psychical research was concerned, for it was the aspect that helped him explain a wide array of psychical phenomena. All forms of automatic writing and drawing, all kinds of automatic speech, table raps and tilts, hallucinations, crystal vision, and even dreams, he believed, were manifestations of the subliminal self conveying messages to the supraliminal. "They present themselves to us," he explained, "as messages communicated from one stratum to another stratum of the same personality."[176]

"Nunciative" or message-bearing automatisms between an individual's subliminal and supraliminal selves could not, however, account for everything that Myers witnessed over many years of attending séances. With Mrs. Piper and Mrs. Thompson, he confirmed the working hypothesis on which he had pinned his hopes and based his research since the start of his psychical inquiries – the hypothesis that spirit survives the death of the physical body. In the case of both women, he satisfied himself that a spirit presence controlled their automatisms. In 1893, after visiting Mrs. Piper in Cambridge, Massachusetts, Myers announced to Lodge that information, conveyed to him through the medium, was of such a nature as to persuade him that he stood "in the presence of an authentic utterance from a soul beyond the tomb." He had no doubt, he assured Lodge, "that spirits are talking & writing to us thro her." They similarly conveyed their messages through Mrs. Thompson, as Myers argued in an article published posthumously in 1902.[177]

The way in which discarnate spirits were able to employ mediums to communicate their messages involved Myers in an elaborate theory of alien invasion. In the first place, he evidently believed that an individual's subliminal consciousness could depart from its physical frame and, under unusual circumstances, travel considerable distances. Such travels, Myers thought, might explain the apparition of that individual at times and places when its material body was occupied elsewhere. Secondly, and as a direct consequence of the first assumption, Myers became convinced that the subliminal consciousness of someone else might directly dominate another person's body, while the latter's own subliminal was temporarily absent. In the case of mediums, Myers hypothesized that during the period of domination the "invading spirit" might control both the entranced medium's mind and motor faculties, leading the medium to convey those messages that the spirit was concerned to communicate. When a medium's information was garbled and incoherent, Myers reasoned that it probably came from the subliminal strata of the medium's own personality. When, however, the medium delivered messages containing information both truthful and beyond the scope of the medium's own knowledge, Myers proposed several possible explanations. On the one hand, the medium's subliminal consciousness might have telepathically received information from a participant at the séance and conveyed that information to the medium's supraliminal self; on the other, the medium's subliminal consciousness may have received information from the traveling subliminal of another living person, either through an encounter in space or through the actual invasion of the medium's personality by the alien subliminal. Thirdly, Myers was convinced that there were occasions when the medium's subliminal self received messages from deceased persons, by meeting their spirits "in the unseen world" or because such a spirit controlled the medium's thought processes and nervous system for a short span of time.

Call it "spirit-control," "substitution of personality," or "possession," Myers believed that the phenomenon could also be accommodated within the category of the "normal" in human behavior. "I claim that a spirit exists in man," he wrote in connection with Mrs. Thompson,

> and that it is healthy and desirable that this spirit should be thus capable of partial and temporary dissociation from the organism; – itself then enjoying an increased freedom and vision, and also thereby allowing some departed spirit to make use of the partially vacated organism for the sake of communication with other spirits still incarnate on earth.[178]

It was certainly an ingenious theory. In one masterful stroke, it seemed to offer proof that some portion of the human personality survived physical death, and apparently it did so in terms that the modern science of psychology could accept. The trouble was, however, that Myers never could decide exactly which portion of the human personality survived. He used varying terms, such as "invading intelligence," "surviving spirit," or "surviving personality," in his attempt to describe the control of a medium by a mind, memory, or subliminal consciousness once connected to a body that had died. The confusion of terminology was merely part of a larger confusion in which Myers hopelessly entangled the word "spirit" in the two distinct concepts of "not matter" and "soul."[179]

Even when trying to proceed with resolute empiricism, Myers could not avoid the concept of soul because he had landed himself in an ironic, and insoluble, dilemma. Aiming above all else to prove that the human personality survived bodily death, he had virtually destroyed the human personality. In Myers's theory of the subliminal self, man emerged as a not particularly well integrated bundle of many parts; strata and streams of consciousness did not form one seamless web, but remained distinct entities. Myers vastly confused the question of what distinguished one single personality from another. Was personality composed of all the layers of subliminal consciousness taken together, or of one in particular? Was it, perhaps, the sum total of subliminal and supraliminal selves combined? Whatever its constitution, it was liable to abandon its own home, leaving that vulnerable to invasion and possession by an alien personality. Leaf was expressing an understandable opinion when he remarked that Myers's work weakened his own sense of personality. Myers had definitely not, Leaf explained, proved "the survival of what we call the living spirit, the personality – a unit of consciousness, limited and self-contained, a centre of will and vital force, carrying on into another world the aspirations and the affections of this."[180]

Inevitably, there was a tension in Myers's work, as he struggled against the tendency, inherent in his hypothesis, to shatter the unity of the human personality. Like Gurney, Myers realized the implications of his views and sought to counter them by affirming the existence of a deeper self or higher principle of unity. His aim was to affirm the existence of "soul," but he could only use that term at the risk of considerable confusion, since it was clear that the soul did not correspond to the subliminal self, or to any other specific portion of the human personality that figured in Myers's psychological theory.[181] He took refuge, therefore, in metaphor and evaded the problems of definition by resorting to images, such as that of flame, invoked to suggest a unity underlying the component parts into which personality was divided, according to his assumptions. "It must suffice to say," he informed his readers, "that I believe that there *is* an incandescent solid, but that that solid is beneath our line of sight."[182]

Myers's metaphors could be both effective and moving, but they remained in the realm of poetry, not proof. He might flatly insist that "each man is at once profoundly unitary and almost infinitely composite,"[183] but such assertions could neither be verified nor denied. There is a strange ambivalence in Myers's last writings, as he hesitated between a mystical vision of universal oneness that powerfully attracted him and a certainty of individual immortality that remained tremendously important to him. Sittings with Mrs. Thompson at the end of his life had persuaded him once again, as he had been persuaded in the early years of his spiritualist inquiries, that the distinct personality survives death.[184] Since Myers passionately believed that the distinct personality whom the medium contacted was Annie Marshall, he was determined not to allow that precious individuality to slip away, either amidst the confusion of its many selves or in some cosmic process of absorption. In trying to reconcile his cosmological and psychological theories, Myers again produced an awkward problem for himself. As a way of balancing the sense of division and disunity created by his writings on the subliminal self, he played with the idea of an *anima mundi*, or world soul, a kind of total, universal consciousness that provided a basic principle of organization and harmony in an apparently fragmented world. If perceived as a kind of cosmic memory, as Myers suggested, it became the agent whereby individual consciousness, once free of its physical abode, might be subsumed in that "transcendental Self," which Myers tantalizingly glimpsed, but could not satisfactorily explain.[185] The relationship of transcendental self and *anima mundi* never emerged from the intricacies of Myers's prose; it seems clear, nevertheless, that this strand of his thought, while designed to underscore a fundamental psychic unity at large and to suggest the means by which the individual psyche might achieve world consciousness, in fact threatened to swallow up the separate, unique immortal spirit altogether. Myers died before he could emend his writings to rescue that spirit, but even in his 1899 paper to the Synthetic Society, among all the ecstatic expressions of Eastern occultism, a voice speaks out for the "organic unity of Soul."[186]

Myers's mind always worked its way back to the soul. He wanted to establish its existence beyond all doubt, as the crucial first step toward proving the immortality that he craved. Psychology, he hoped, could help him in his difficult task because it seemed to promise an access route to the mind that enjoyed the approval of modern science.

If he could ascertain, through its systematic procedures, that mind was not identical with the brain, nor coterminous with any other organ of the physical and perishable body, then the scientific community would have to acknowledge the limits of its positivism. He longed for his psychological theories to receive the blessings of the reigning scientific authorities, not so much for the personal honor involved, but because he wanted to enlarge the scope of scientific naturalism and to bring the phenomena of psychical research into its authoritative domain. In 1893, he evidently believed he was on the brink of new insights. He had been studying the notebooks of Stainton Moses and was writing a two-part article on the medium for the SPR *Proceedings*. Referring to the intended date of publication, he wrote exuberantly to Lodge: "I am analysing 31 books of MSS of the late Stainton Moses . . . *Most* important! Don't lay down any definite & final limits to the Universe before that date, if you can avoid it!"[187]

It is hardly surprising that contemporary psychologists were not, for the most part, impressed with Myers's fully elaborated theory of the subliminal self. They dismissed the role Myers wanted psychology to play in the modern world and disliked the assurance with which he claimed the endorsement of science for his personal beliefs. There were, of course, a few psychologists, like James and Flournoy, who shared Myers's close involvement in psychical research and were deeply sympathetic to his aims, but not even all Myers's colleagues at the SPR accepted his hypotheses. Andrew Lang, for one, rejected outright Myers's theory of possession, while Gerald Balfour, in his 1906 presidential address to the SPR, had to confess "in all humility" that he had "never yet succeeded in forming a clear idea" of what Myers intended by the subliminal self.[188] The most crushing review of *Human Personality*, because the best informed, the most clearly argued, and the least ill-tempered, came from G. F. Stout, professor of logic and metaphysics at St. Andrews, and not a member of the SPR. In the *Hibbert Journal* for October 1903, Stout explained how Myers's theory of the subliminal self diverged sharply from earlier doctrines of subconscious or unconscious mental states, and how the common psychological understanding of the "subliminal factor" in mental life did not necessitate a secondary personality or alternative self. Stout could only regard "the hypothetical agency called the 'Subliminal Self' as an addition to his ignorance rather than to his knowledge." As a hypothesis, it was "baseless, futile, and incoherent."[189] William McDougall, about to assume the post of reader in mental philosophy at Oxford, made many of the same points as Stout in his own review of *Human Personality*. Writing in the October 1903 issue of *Mind*, McDougall likewise found that the subliminal self unnecessarily complicated proposed explanations of the phenomena that Myers studied. Furthermore, he justly and significantly observed that Myers had "excogitated" the concept of the subliminal self expressly "in order to harmonise the evidence of survival with the general body of accepted scientific belief."

McDougall's review, however, did not strike an exclusively negative note. He appreciated Myers's "literary grace, the brilliant use of analogy, the subtlety of exposition, the lofty and eloquent speculation that adorn every chapter." He was certain that posterity would "accord to Myers a place in the history of the intellectual development of mankind."[190] These were not hollow words of praise from McDougall, a polite formula of respect for the dead psychical researcher. Despite the physiological emphasis of his early work in psychology, McDougall was deeply interested in humanity's psychical aspect. By the time he wrote his review of *Human Personality*, he had joined the SPR. He became a council member the following year and served as the Society's president in 1920–1. His contributions to psychology, from his early years at University College, London, at the turn of the century, through his controversial career at Oxford, Harvard, and Duke Universities, in fact underscore how much his own approach to the study of mind shared the controlling convictions and perspective of the psychical researchers.[191]

McDougall's message to the psychological profession of the early twentieth century was reducible to a single point: Mechanistic explanations could not elucidate the workings of the human organism. The inadequacy of mechanistic models to explain human behavior became an essential tenet in his psychological theory, a theory that defined psychology as the "positive science of the behaviour of living things." Behavior, he asserted, was "the characteristic of living things" and was itself characterized by "manifestation of purpose or the striving to achieve an end." Purposive activity, then, was for McDougall the central fact about the human organism that psychology had to address, and he found no "plausibility" in any contemporary attempts "to explain such facts mechanically."[192] Although McDougall developed his reaction against mechanistic psychology more fully after World War I, when he went to battle with Behaviorism, his lines of argument were clearly arrayed before the war, not only in *Psychology, the Study of Behaviour*, but also in his highly controversial *Body and Mind*, published in 1911.

Combined with its sustained critique of mechanistic modes of thought in the life sciences, *Body and Mind* contained an extended argument in favor of animism, the theory that the human body is vitalized by something nonmaterial, something traditionally dubbed a soul. After tracing in detail the history of animistic beliefs from the ancient world through the nineteenth century and reporting the developments of modern science that were hostile to animism, McDougall marshaled the best evidence he could to confound the opposition. He was convinced that "the differences between the living human organism and the corpse are due to the presence or operation within the former of some factor or principle which is different from the body and capable of existing independently of it." Although he realized that this distinction could not yet be demonstrated to the satisfaction of his scientific colleagues, he could at least hope to unsettle them with the evidence provided by the SPR. In the phenomena of telepathy and the mysteries of the cross correspondences, McDougall believed that he had found indubitable evidence utterly incapable of reconciliation "with the mechanistic scheme of things." Hypnotic phenomena, too, he contended, largely remained "refractory to explanation by mechanical hypotheses," and they went far to uphold the belief that ordinary organic processes "are in some sense controlled by mind, or by a teleological principle of which our conscious intelligence is but one mode of manifestation among others."[193] Had Gurney and Myers only lived to see it, they would have delighted in the knowledge that one of the creators of social psychology was building upon foundations furnished by the SPR.

The inquiries of psychical researchers similarly buttressed the psychological theory, or philosophy of life, constructed by William James, the man whom McDougall recognized as "the largest influence affecting my intellectual life." William James's influence cut deep into psychology and philosophy in the late nineteenth and early twentieth centuries, in Europe and America, and there is scarcely room in this study to do justice to that impact. What is significant, for the purposes of the argument developed here, is the combination of certain tenets of psychology with James's longstanding commitment to psychical research. Like the SPR, his "deepest lying aim" was "the reconciliation of science and religion on empirical grounds," and the psychology that he proposed was fundamentally aligned with the view of man implicit in SPR investigations. He steadfastly refused, McDougall recalled,

> to reduce all consciousness to the level of an epiphenomenon or silent spectator; he insisted always on the real efficiency of our consciousness, our feelings, our efforts, our thoughts as teleological co-determinants of our bodily movements. He discovered an unsuspected wealth of detail in the stream of consciousness; and he restored to that stream its unity, rejecting root and branch the mechanical descriptions and explanations of associationism. He restored effort, activity, desire, in short, the will, to its rightful place.

James would give credence to no theory of mind–body interaction that denied this "real efficiency" of consciousness, thought, feeling, and will in the physical world.[194]

James's relationship with the SPR was very much a reciprocal one. In general, his sweeping rejection of determinism, as applied to humanity, strengthened the Society's theoretical position when it insisted that mechanistic doctrines could not fathom all the secrets of the universe. More specifically, particular aspects of James's theory of the mind enriched the hypotheses of the Society's own psychological theorists, especially Myers.[195] Equally, however, James's involvement in psychical research aided him in developing his psychological principles. His acquaintance with Mrs. Piper convinced him of the reality of supernormal powers. That minds could communicate telepathically became for him a demonstrated fact, and it was critically significant for his entire system of thought that he believed the insufficiency of "mechanical categories" had thus been revealed by empirical proof. Yet he was never persuaded that the individual personality survived the death of the body, and he articulated instead a theory of cosmic consciousness, in which the single consciousness figured as a beam "filtered out from the universal consciousness by our brains." It was, McDougall noted, a way to avoid materialism, but it was far from a reassuring promise of immortality.[196]

Orthodox and traditional views of immortality were certainly not a constant element among the psychologists who found value in the work of the SPR near the turn of the century. If anything, there was a tendency toward pan-psychism in their writings on life after death; they suspected that survival of the complete, unique personality was not a justifiable belief, in light of the evidence, but held varying concepts of universal mind to which the individual consciousness somehow adhered, both during the life of the physical body and after its demise. What united these psychologists even more closely was their refusal to be trapped into materialist views of the human organism. They rejected the old determinism of associationist thought, just as they sought to ward off the new neurological determinism preached by the prophets of instinct and heredity.[197] They upheld the primacy of consciousness as a psychological fact and emphasized in varying degrees the workings of will, of purposive activity, in human behavior.

Psychologists with affiliations to psychical research were not, needless to say, the only ones to favor this approach to the study of psychology. Men like Ward, Sully, and Stout, for example, shared many of the same assumptions about the mind, without involvement in psychical inquiries. Yet what cannot be overlooked is the serious effort by numerous members of the SPR to bring their insights and investigations into a mutually fruitful dialogue with the pioneers of psychology in the late Victorian and Edwardian years. Many psychical researchers were deeply committed to probing the operations of the human mind and hoped that their inquiries into hypnosis, automatic writing, trance speech, hallucinations, and the like would further the endeavor. Sadly for them, the majority of professional psychologists in the years before World War I held them suspect, believing that psychical research still reeked of the discredited practices of mediums, the pseudomagic of the occult, and the naive enthusiasm of the amateur. Certainly the physiological tendency within modern psychology was not deterred one whit by the opposition of the psychical researchers. After the war, furthermore, the vogue for Freudian psychoanalysis rendered the SPR line of inquiry largely irrelevant.

Decades of further research into the workings of the brain have suggested alternatives to the stark choice between determinism and free will, which psychologists thought they faced at the turn of the century. In the 1980s, it appears that even the most elaborate mechanistic model may not be able to impose rigidly predictable patterns and order on the random activities of the brain's individual neurons. In the still fierce debate among philosophers, physicists, neuroscientists, and computer specialists, it is just possible that indeterministic physics may yet leave some slight area of choice, intent, and purpose among the complex machinery and chemical codes of the brain. The possibility may prove illusory, but it is more than Gurney and Myers had to work with. Given the limits of their knowledge and the intellectual conventions in which they worked, the members of the SPR who tried to contribute to psychology before 1914 deserve credit for the way they conducted their inquiries. Their goals were enormous, for they sought to resolve the mind–body puzzle by finding a *via media* between Cartesian dualism and a monism that threatened to eliminate mind entirely. It comes as no surprise that they failed, nor that only a small group of professional psychologists valued the attempt. But to those particular psychologists, the work of the SPR mattered tremendously.

7

The problem of evolution

THE EVOLVING BRAIN

Anyone who seriously studied the human mind in the second half of
the nineteenth century had to struggle with the intricate problem of its
development over time. The problem was thorny, for it clearly could
not be divorced from the deeper significance attending any discussion
of humanity's place in the natural world. Man's brain, of all his organs,
was widely accepted as the crucial difference that separated the human
species from the rest of the animal kingdom, and it was an invidious
task at midcentury for any naturalist to suggest a connection between
the human and simian brains in the remote past. Yet that possibility
assumed ever greater likelihood as the principles of evolution, of uniformity and continuity in nature, took hold of the public imagination
in subsequent decades. In concluding *On the Origin of Species*, Darwin
predicted that the psychology of the future would illuminate the growth
of human intellectual capacities, and most of the important psychological research undertaken thereafter was indeed informed with evolutionary assumptions that brought psychologists into close contact
with the work of biologists, zoologists, and anthropologists. British
spiritualists and psychical researchers responded with uncertainty to
the study of the human race within an evolutionary framework. Broadly
speaking, such a context supported the generally progressive view of
human development that characterized spiritualist thought in this period. At the same time, however, too convincing a revelation of man's
animal origins threatened the very foundations of faith in his immortal
spirit, for if he had not been specially created, it was scarcely plausible
that he was specially endowed.[1]

An evolutionary framework molded the theories of leading British
psychologists, from the profoundly influential Herbert Spencer through
Francis Galton and George John Romanes to C. Lloyd Morgan at the
turn of the century, to mention only a handful of the men who studied
the human and animal brain in this period. Taking many of their cues
from Darwin's rich data, and from the psychological speculations in
which he occasionally indulged,[2] they established the field of animal

psychology, pioneered the psychology of individual human differences, and probed the role of heredity in mental development. They were working, of course, at a time when scarcely any discipline escaped the influence of evolutionary metaphors and motifs. Whether applied to psychology, to the even newer subject of sociology, or to the ancient study of history, evolutionary concepts were irresistible. As has been pointed out repeatedly, Darwin's theory of evolution was infinitely pliable; it could be twisted to justify militarism and pacifism alike, imperialism and cooperation, unbridled laissez-faire capitalism and socialism. Perhaps the chief reason for the ubiquity of evolutionary modes of thought at this time lay expressly in the capacity to appeal to all ideologies.

At a popular level, however, the message of evolutionary theory was not ambivalence, but optimism. The story of natural development over the aeons appeared to be one of ongoing improvement, just as the history of western civilization in the nineteenth century seemed a chronicle of boundless advance – in technology, medicine, and morality, in political and religious freedom, and in the abstract realms of theoretical science. It made no difference that there was, in fact, nothing necessarily progressive about evolution through the transmutation of species; nonetheless, public opinion generally came to associate evolution with the progressive development of the natural world, particularly the human species. The man who had triumphed in the struggle to survive was presumably a better man.[3]

Psychical researchers were drawn to the idea of a "better" man. They were not, by any means, uncritically complacent in their views of contemporary society; like other astute and anxious observers who could not endorse the facile popular optimism of their day, they knew all too well how questionable was the record of human progress. Yet they hoped that the evolutionary process held out some reassurance of a promising future for humanity and looked to their study of the mind for supporting evidence. On no less an occasion than his presidential address to the British Association for the Advancement of Science, in September 1898, William Crookes cited the therapeutic powers of hypnotism as just such evidence. "The human race has reached no fixed or changeless ideal," he told his audience at Bristol;

> in every direction there is evolution as well as disintegration. It would be hard to find instances of more rapid progress, moral and physical, than in certain important cases of cure by suggestion. This is not the place for details, but the *vis medicatrix* thus evoked, as it were, from the depths of the organism, is of good omen for the upward evolution of mankind.[4]

In contrast to Crookes's view, Arthur Balfour assumed that the extraordinary powers that psychical research discovered in a few specially gifted people had no particular evolutionary value. Podmore went even further and suggested that evolutionary change had actually eliminated certain faculties, like telepathy, from conscious human control, thereby depriving modern man of some prowess enjoyed by his remote ancestors. "It is likely enough," he commented, "that the hap-hazard process of selection which the contents of our consciousness have undergone causes the loss of much that we would not willingly let pass away."[5] Yet, while it was always possible to surmise that natural selection had wrought a diminution in human abilities, there was no need to adopt that point of view. The ambiguous evidence favored Crookes as much as Podmore, and if one looked at the question from a Lamarckian perspective, there seemed reason to believe that repeated efforts to use such skills as telepathy and clairvoyance might eventually spread these talents more widely within the human species.

Myers's psychology relied heavily on evolutionary concepts. Along Podmore's line of argument, he hypothesized that the knowledge and capabilities that he attributed to the subliminal self might be remnants of former conscious powers. The demands of utility had relegated these powers to the subliminal, so that the supraliminal consciousness could concentrate on functions and memories essential for survival. As the human race evolved, the supraliminal self acquired the intellectual equipment appropriate for an advanced creature, while the subliminal self retained certain "psychical acquisitions" of an earlier state, such as the capacity to affect one's own physiological condition. In short,

> the recollection of processes now performed automatically, and needing no supervision, passed out of the supraliminal memory, but might be retained by the subliminal. The subliminal, or hypnotic, self could exercise over the nervous, vaso-motor, and circulatory systems a degree of control unparalleled in waking life.[6]

The lack of compelling evidence to support this elaborate theory did not keep Myers from propounding it; nor was he deterred from suggesting other, more positive, roles for evolution, so that eventually the average human being might be brought to that higher stage of development which, in Myers's day, only the man of genius had attained.

When Myers affirmed his belief "in a progressive moral evolution" as part of his final faith, he was not merely thinking of earthly change. As always, his thoughts were more than half directed toward the future life, when the progress of the human spirit would be "no longer truncated by physical catastrophes."[7] His writings on postmortem evolution reflected his readings in Eastern theology and, even more, his contacts with contemporary spiritualists and Theosophists. He shared their conviction that the disembodied spirit could ascend to ever higher levels of knowledge and virtue, ever greater understanding of the bonds that unified the cosmos. The ultimate goal of this "cosmic evolution" was, as Myers succinctly stated, "Light at once and Love."[8]

The great majority of British spiritualists harbored no unsettling doubts about the implications of evolution. They eagerly embraced the idea of organic change, finding that Darwin served to confirm their own scenario of progressive development beyond the veil. "I for one accept the truth of Mr. Darwin's theory of man's origin," wrote Gerald Massey in 1871,

> and believe that we have ascended physically from those lower forms of creation . . . But the theory contains only one-half the explanation of man's origins, and needs Spiritualism to carry it through and complete it . . . Mr. Darwin's theory does not in the least militate against ours – we think it necessitates it; he simply does not deal with our side of the subject.

The spiritualists' side of the subject did not emphasize whence humanity had come, but looked instead whither it was going – to "progressive spiritual states in new spheres of existence" after physical death. There the perfection that fell to no individual in this world was the attainable reward of all upward-striving souls, for evolution in the spirit world proceeded "at a rate more rapid and under conditions more favourable to growth" than those encountered on earth.[9] It gave spiritualists no small pleasure to illustrate how evolutionary theory gave the lie to repressive beliefs concerning man's fall from grace. In 1888, for example, in a talk to the London Spiritualist Alliance, Reverend J. Page Hopps defined evolution as

> a persistence of some tendency which has, gently but irresistibly, forced all things upward into higher forms of life. Man did not begin perfect, and end in a 'fall'; he began imperfect, and is steadily going on in the onward and upward path, out of the animal's darkness into the angel's marvellous light. He is not a fallen but a rising creature.

In an earlier work, Hopps had already announced that "the law of evolution" foreshadowed the spiritualists' afterlife. "That life, properly understood," he explained, "is only another step in the wonderful development of man's being: it is evolution still, but evolution into and in the sphere of mind."[10]

Yet there were severe obstacles for spiritualists and psychical researchers who closely investigated the workings of evolution. It was a simple matter for someone like Hopps, or Gerald Massey, to proclaim the utter compatibility of Darwinian and spiritualist theories concerning human development before and after death. Their knowledge of biology was extremely limited; they were in no position, nor had they any inclination, to evaluate critically what they read about evolution in popular treatments of the subject. It was far more difficult for Alfred Russel Wallace, who worked out the theory of natural selection at the same time as Darwin and who well understood its implications for human history. Wallace realized that, if natural selection alone could account for man's place in nature, then the human species, like all others, had evolved from some earlier organism. If man had evolved from some earlier, prehuman organism, he was not qualitatively different from other, higher denizens of the animal kingdom. The human brain was merely more developed than theirs, having undergone further changes that rendered it capable of intellectual endeavor. If there was nothing basically distinctive about the human brain, apart from this further development, then there existed nothing characteristic of humanity alone among all other creatures – neither mind, nor consciousness, nor spirit. Wallace indeed knew that, far from supporting the spiritualist vision of eternal progress in the hereafter, "Mr. Darwin's theory" could be seen to undermine the most basic tenets of spiritualism. Although it is true that some spiritualists embraced the Eastern belief in transmigration of souls and assumed that other animals, besides man, possessed an eternal spirit, they were only a small portion of the British spiritualist population. The great majority continued to be influenced by the Christian conviction that the human race occupied a unique place in nature.

Whether one greeted with enthusiasm or trepidation the application of evolution to the human brain, by the end of the nineteenth century it was impossible to doubt that it had evolved during the course of man's emergence on earth. The Great Chain of Being, central to eighteenth-century biological thought, had acknowledged that the ape, of all God's creatures, stood closest to humanity and was indeed similar to the human species in many respects. Yet there was no idea of change in the theory, no asumption that one species might gradually give rise to another. A few intrepid naturalists were beginning to think along evolutionary lines, but change was not generally recognized as the preeminent fact of terrestrial life until the second half of the nineteenth century. Only then was the idea successfully challenged that man's place in the universal order was fixed permanently from time immemorial, at the head of the natural world. In 1879, Henry Maudsley welcomed the knowledge that "the slow and gradual operation of processes of natural evolution going on through countless ages" had produced the brain of civilized man; shortly thereafter, the growth of studies in animal psychology appeared to underscore ever more incisively human links with a bestial past. In his presidential address to the British Association for the Advancement of Science in 1906, Lankester hailed "the valuable interaction of the study of physical psychology and the theories of the origin of structural character by natural selection." "The relation of the human mind to the mind of animals, and the gradual development of both," he announced, "is a subject full of rich stores of new material, yielding conclusions of the highest importance."[11] Nor were such statements characteristic only of men who leaned toward a materialist frame of mind. McDougall, for one, asserted bluntly in 1912, "that the mental functions of man present no such radical difference in kind as would forbid us to believe in the continuity of mental evolution; for evidence of some rudiment of every type of mental function may be discovered in animal behaviour." Even Myers, after all, assumed that man's mental evolution stretched far back to "the amoeboid movements of the primitive cell."[12]

The task, therefore, for those scientific thinkers who subscribed to evolutionary theory, but worried about the consequent position of humanity on earth and in heaven, was somehow to modify that theory. Perhaps they could prove that gaps interrupted the continuity of nature, that genuinely novel features did, under the rarest of circumstances, intrude into the evolutionary process, as when life emerged from inorganic matter. If they could find some weakness in "the whole Argument from Continuity against the independent existence of mind,"[13] they could challenge the inexorable movement of the evolutionary process itself. Not denying the evolution of the human mind, they could nonetheless propose that it had developed under conditions different from the rest of the natural world and was, therefore, different in kind from the rest of animal intelligence.

ROBERT CHAMBERS

In the decade before Darwin and Wallace proposed the theory of natural selection to explain how evolution worked, a Scottish popularizer of scientific thought was groping toward a synthesis of evolutionary and spiritualist beliefs. Robert Chambers, who established the famous Edinburgh publishing firm of W. & R. Chambers with his brother William, was a man of widely varied interests. His extensive writings covered diverse aspects of Scottish history, literature, and folklore, while his publishing company produced not only the highly successful *Chambers's Edinburgh Journal*, but such compendia of knowledge as *Chambers's Information for the People*, *Chambers's Educational Course*, and *Chambers's Encyclopaedia*. In 1844, anonymously and under conditions of the strictest secrecy, Robert Chambers published *Vestiges*

of the Natural History of Creation, his account of evolutionary change
in the natural world. The abuse that the book received in the periodical
press did not hinder its brisk sales.[14]

Although the volume paid tribute to the uniformity and regularity of
creation, the orderly design exhibited throughout all portions of the
universe, Chambers's book merited the hostility with which clergymen
greeted its publication. It was, in fact, an argument for the place of
humanity within the greater evolutionary context of natural history,
and Chambers did not hesitate to underscore the close kinship between
man and beast. It did not matter that *Vestiges* offered no persuasive
explanation of how the transmutation of species occurred, nor that
much of the supposedly scientific evidence amounted to nothing more
than ridiculous tall tales. What mattered was that Chambers dared to
articulate a theory of evolution that not only repudiated the fixity of
species since the creation, but also denied humanity's special relationship with God, while suggesting that man would be more comfortable with a tail than a halo. In conjunction with these offences,
Chambers appeared to propound a view of the brain and nervous system that all but eliminated mind as an independent entity.[15]

Wallace, reading *Vestiges* before setting off on his long expedition
to the Amazon in 1848, could not have known that he and Chambers
would one day share more than a common interest in evolution. It was
still a few years before spiritualism attracted public attention in Britain,
and almost two decades before both men became known as supporters
of the movement. Wallace did not even begin his serious investigation
of the subject until the mid-1860s. Chambers, however, began to attend
séances early in the 1850s. He visited several mediums during the
spring of 1853, including the Americans, Mrs. Hayden and Mrs. Roberts. Although cautiously impressed by some of the phenomena that
he witnessed, he did not abandon his critical faculties. Commenting
on the messages transmitted through Mrs. Hayden by alphabet raps,
he observed in *Chambers's Edinburgh Journal*:

> I was not conscious at the time of acting in such a way, in my pointings
> at the letters, as to give any hint of which were the true ones; but I
> became fully convinced next day, on reflection, that a clever person
> in the capacity of Medium might in most cases detect a significant
> pause at the letter which the experimenter knew to be the right one,
> and would thus be able easily to spell out the expected words and
> sentences.

He suggested that the very slightest pressure from fingertips would
suffice to make tables tilt, reminded his readers of the marvelous feats
performed by conjurors, and concluded that the majority of mediums
were "credulous people, visionaries and enthusiasts, who first impose

upon themselves, and then upon others." "Those who have studied the profound deceitfulness of the human heart," he added, "and seen how shadowy are the divisions between self-delusion and active deluding, will find less difficulty in the case."[16]

These, clearly, were not the thoughts of a convert to spiritualism. Chambers's biographer has persuasively surmised that he attended his first séances with the particular purpose of collecting information for a *History of Superstition* which he was writing in the early 1850s. The internal evidence of the *Chambers's Journal* article suggests some such intention, as when he compared "Spirit Manifestations" to "one of the manias of the middle ages."[17] Yet the *History of Superstition* was never published; indeed, Chambers destroyed the manuscript. Instead, ten years later he wrote an anonymous introduction and appendix to the first volume of D. D. Home's *Incidents in My Life*.

Both the affiliation with Home and the anonymity of Chambers's contribution to the 1863 volume were significant. Mme. Home implied that her husband was chiefly responsible for converting the Scottish skeptic to spiritualism after they met in 1859, and there is some truth in her claim.[18] Home's talents as a medium were certainly many cuts above those that Chambers had hitherto observed, and he seems to have provided the final impetus that drove Chambers into the believers' ranks. Chambers, however, was already headed in that direction before 1859. As early as June 1853, when he witnessed table tilts and turns in his own home, he revised his critical stance of the previous month and assumed that the phenomena were not produced by trickery.[19] He came to believe that the natural order could somehow accommodate such events, if only one expanded the definition of "natural." He continued to attend séances in the following years, meticulously keeping notes on his experiences and increasingly unable to explain away the manifestations of spirit power. Gradually, his confidence that séance phenomena belonged within the framework of the physical world yielded to the growing certainty that they represented pure spirit at work, untouched by material forces.[20]

Thus Chambers was ready for Home and for the marvels of Home's séances. On one occasion, in 1860, Home placed him in touch with a spirit who claimed to be his father. To test the spirit's identity, Chambers asked it

> to play upon the accordion, which was lying on the floor, the favourite air his father was accustomed to play on the flute. The accordion immediately – *no one touching the keys* – played "Ye banks and braes o' Bonny Doon." "That," said Mr. Chambers, "was my father's favourite Scotch tune, and that the manner in which he used to play it." He then said, "Play my father's favourite English air," and the

accordion immediately played "The Last Rose of Summer." Mr. C. acknowledged that it was his father's favourite English tune.[21]

A few years later, after Chambers had lost his first wife and a beloved daughter in 1863, he was even more vulnerable to the messages that Home claimed to convey from beyond the grave. One such communication came in 1866, during a séance at S. C. Hall's London home. Chambers was in Scotland, and Hall was instructed to convey information to the absent man, from the spirit of a deceased daughter. When Hall requested "some token of identity that he might furnish to Dr. Chambers, the words, 'Tell him, *Pa, love*,' were spelt out." Those words, Chambers subsequently informed Hall, were the last his daughter "pronounced in life."[22] It is perhaps unfair to accuse Home of shamelessly profiting from an old man's grief, for the messages from a mourned daughter doubtless provided Chambers with much badly needed solace. Nonetheless, the fact remains that Chambers was willing to make a public declaration of his spiritualist faith only after he received this communication.

Until Home's legal dispute with Mrs. Lyon in 1867–8, Chambers had tried hard to keep his spiritualism out of public discussion. Not only had he refused to sign his name to the parts that he contributed to Home's autobiography, but he had also refrained from publishing any of the several manuscript essays on spiritualism that remained among his papers after his death. He was, of course, known for his support among spiritualist circles, but he did not inform the general public of his new beliefs until Home asked him for an affidavit in the impending lawsuit of *Lyon v. Home*.[23] Then, apparently, Chambers's fears of bringing ridicule and discredit upon himself and his business yielded to his sense of obligation to the medium. In September 1867, he gave a sworn deposition attesting to Home's "irreproachable character," and his testimony was used at the trial the following spring.[24] Having revealed his position, Chambers naturally became one of the celebrated figures in art and science whom British and American spiritualists loved to boast about. The London Dialectical Society solicited his opinions when undertaking its inquiry into spiritualism in 1869, while the *Year-Book of Spiritualism for 1871* cited him twice, in two separate listings of eminent people who embraced the faith. Chambers's reply to the London Dialectical Society, in a letter written from St. Andrews on 8 March 1869, suggested that he was uncomfortable in the role of proselytizer for a new faith. After naming a couple of books to read on the subject, he concluded rather brusquely: "Every man must examine and attain conviction for himself. It is well, however, that new students should be warned against trusting in the *dicta*, for these are as often false as true."[25] The note of caution in this public pronouncement contrasts sharply with a letter Chambers wrote to Wallace on 10 February 1867. In response to a letter from the naturalist, Chambers wrote:

> It gratifies me much to receive a friendly communication from the Mr. Wallace of my friend Darwin's "Origin of Species," and my gratification is greatly heightened on finding that he is one of the few men of science who admit the verity of the phenomena of spiritualism. I have for many years *known* that these phenomena are real, as distinguished from impostures; and it is not of yesterday that I concluded they were calculated to explain much that has been doubtful in the past, and when fully accepted, revolutionize the whole frame of human opinion on many important matters.[26]

It was, presumably, concern for his standing in the literary world and for the reputation of his publishing firm that kept Chambers from participating more actively, and vocally, in a cause that clearly commanded his full allegiance. Yet it is also possible that he hesitated for other reasons as well. Despite the confidence with which he wrote to Wallace in 1867, it may well be that he never experienced the utter certainty about séance phenomena that came so readily to S. C. Hall or William Howitt, another of Chambers's spiritualist friends. Although it is true that Chambers's interest in scientific subjects declined sharply from the mid-1850s, it would be inaccurate to claim that he abandoned the battles fought so vigorously in *Vestiges*.[27] Huxley, in fact, would not have faced Wilberforce at the Oxford meeting of the British Association in 1860 had not Chambers urged him to champion evolution against the bishop's denunciation.[28] Nor would Chambers have continued to revise *Vestiges* if the point of view and the argument contained in that volume had become repugnant to him.

The reconciliation of *Vestiges* and spiritualism was Chambers's task from 1853 until his death in 1871. It was not as arduous an undertaking as might appear at first glance, because Chambers never occupied the extreme position in either set of beliefs. He was not "one of the hardest and most dogmatic of materialists," as Mme. Home tried to maintain,[29] nor was he utterly gullible where spiritualist phenomena were concerned. In the *Memoir of Robert Chambers*, which appeared in 1872, his brother William tried to minimize Robert's embarrassing attraction to spiritualism in the following terms.

> One of the more bulky papers which he left is a species of inquiry into the so-called manifestations of spiritualism. Without pronouncing an opinion dogmatically, he considered the subject worthy of patient investigation. "The phenomena of spiritualism," he says, "may be the confused elements of a new chapter of human nature, which will only require some careful investigation to form a respectable addition

to our stock of knowledge. Such, I must confess, is the light in which it has presented itself to me, or rather the aspect which it promises to assume." Acknowledging so much, perhaps he thought of a saying he had heard used by Sir Walter Scott, that "If there be a vulgar credulity, there is also a vulgar incredulity." In his anxiety for fairplay, he perhaps leant to the side of credulity.[30]

This is scarcely an adequate explanation of Robert Chambers's attraction to spiritualism. William, far more the careful and successful businessman than his brother, was obviously attempting to belittle Robert's commitment to that faith; a sense of fair play might seem a harmless, and peculiarly British, excuse for endorsing so unreasonable a belief, from a reasonable man's perspective. Nonetheless, despite William's misrepresentation, he captured something of the balance that Robert consistently, and with great effort, sought to strike between the naturalism of his early years and the spiritualism of his later ones.

Robert Chambers's biographer, Milton Millhauser, has eloquently charted that struggle, portraying him as a man under whose "discreet deism" lurked both a strong predisposition to religious faith and a lifelong hope for the soul's immortality.[31] Even in *Vestiges*, Chambers had spoken of "things above this world" and suggested that each human being enjoyed connections, not only with the larger world, but "with something beyond it." He talked of "supra-mundane things"[32] in a way that implied definite chinks in his naturalistic armor. Although Chambers lost all patience with the Christianity of sects and dogma, he never lost a belief in some sort of deity, even if that belief took the form of a highly impersonal ultimate law. From the 1850s on, he moved away from this cold, rationalist faith to a warmer personal credo that allowed him, in his old age, to write a catechism and life of Jesus for children. He had never actually ruled out personal immortality, but its vivid likelihood had eluded him until spiritualist experiences assured him on that score. Perhaps, as Millhauser contends, Chambers would have found it impossible to return to Christianity without spiritualism; perhaps grief and loneliness would have driven him back to the Christian faith in any case. One can only assume that spiritualism substantially eased the way.

Chambers's approach to spiritualism was not solely predicated on religious need. He believed that a scientific lesson might also be learned around the séance table and vigorously objected to people, especially scientists, who would not accept evidence confirmed by their own senses. Chambers's entire concept of the scientific enterprise rested on the validity of the visible, tangible record left by actual events. He believed in the evolutionary process because the geological and biological record persuaded him to do so. Similarly, he came to believe in the record of spirit existence produced at séances. He knew about the opportunities for fraud in the medium's profession, and he tried to stay on guard against those possibilities. Nonetheless he adamantly refused to deny the actuality of what he witnessed simply because spiritualist phenomena could not be reproduced in a laboratory.

When he first began his spiritualist inquiries and became convinced of the reality of the manifestations, Chambers sought to fit them into his view of the natural world by enlarging his notions of matter and physical force to include the mysterious séance occurrences. After he accepted their spiritual character, he toyed with ideas of electricity or magnetism as links between the worlds of spirit and matter. Always he clung steadfastly to his belief in law; he was sure that laws were at work in the spirit realm, as in the physical, but they were not readily accessible to the probings of the scientist. There was nothing dogmatic in Chambers's thoughts on these questions – hence, perhaps, his reluctance to serve as publicist for the spiritualist faith – but only a determination openly to approach all subjects that might offer clues to the enigmas of nature. By the end of his life, the "supra-mundane things" he mentioned vaguely in *Vestiges* had become real presences to him. The evolutionary process, he came to suspect, included more than the scientific naturalism of his younger years had revealed. Perhaps Chambers's scientific naturalism was never as narrow as his critics had assumed.

ROMANES AND COLLEAGUES

Chambers, like Wallace later, found in spiritualism some resolution of the troublesome questions raised by the evolutionary connection between man and animal, but most inquirers into the development of the human organism were not attracted to the spiritualist refuge. A few, however, particularly interested in mental development, seriously considered the likelihood that psychical research might prove relevant to their investigations. George John Romanes, FRS, Fullerian Professor of Physiology at the Royal Institution, innovative student of animal psychology, and founder of the Romanes lectureship at Oxford, was one such inquirer who, for a time, looked to psychical research to liberate himself from the confines imposed by scientific naturalism.

Yet in the public mind, Romanes appeared to embrace the latter point of view. His own religious aspirations remained private, while his studies in physiology at Cambridge, under Michael Foster in the early 1870s, and his subsequent work in the same field at University College, London, under Sir John Burdon Sanderson, prepared him for a career devoted to the examination of nervous systems. Early inquiries into

the nerves of jellyfish, starfish, and sea urchins led to broader considerations of animal intelligence, and thence to the mental evolution of
animals and man. The work of this physiologist, who has been labeled
"the first systematic comparative psychologist,"[33] unfolded against
what he perceived as a clearly defined background of evolutionary
progress. As far as the development of mental faculties was concerned,
Romanes wrote, "the method of nervous evolution has everywhere
been uniform; it has everywhere consisted in a progressive development of the power of discriminating between stimuli, combined with
the complementary power of adaptive response."[34]

Certain that mental evolution had followed the same pattern throughout the animal kingdom, Romanes tried to study the behavior of man
and beast within the framework of a single theory of character and
motivation. When he was allowed to bring home a monkey from the
London zoo, he confessed to Darwin in 1880 that he "wanted to keep
it in the nursery for purposes of comparison, but the proposal met with
so much opposition that [he] had to give way." Later in the 1880s, he
trained another zoo monkey, this time a chimpanzee named Sally, to
count to five, and he wrote to the *Times* that "so far as 'counting' by
merely sensuous computation is concerned, the savage cannot be said
to show much advance upon the brute."[35] Many of Romanes's comments about his intelligent monkeys indeed appear to confirm the observation by one historian of psychology who found that, in his eagerness to span the psychological gap between gorilla and gentleman,
Romanes "scaled the gorilla up rather than the gentleman down."[36]

Romanes's persistent weakness in his studies of animal behavior lay
precisely in his inept handling of their comparative aspects. Not only
did he rely too heavily on unsubstantiated anecdotes concerning clever
animals in order to support his arguments about brute intelligence, but
he also invested his bestial subjects too freely with human emotions
and designs.[37] In this respect, Romanes may have misapplied or distorted some of the arguments in Darwin's own *Expression of the Emotions in Man and Animals*, for Darwin's influence was profound in
many areas of Romanes's work. Although nearly forty years apart in
age, the two men became warm friends in the decade before Darwin's
death. They corresponded regularly on the many subjects that intrigued
them both, and Darwin encouraged the much younger man to turn his
physiological talents to the study of the mind.[38]

When he did so, however, Romanes decided not to rely solely on
natural selection to explain the transformation of species and the process by which mental evolution in animals led to mental evolution in
the human race. Although he considered himself the staunchest of Dar-
winians, he maintained that Darwin's evolutionary theory needed to

be expanded. Not only did Romanes consider Lamarck's arguments for the inheritance of acquired characteristics partly valid – as he believed that Darwin himself came to agree after 1859 – but Romanes also put forward his own theory of evolution. His concept of "physiological selection," first proposed in a paper to the Linnean Society in 1886, contributed to the debate over the processes of evolution that intensified after Darwin's death in 1882. The theory suggested that mutual sterility between animal groups played a larger role than hitherto recognized in determining whether or not a beneficial variation would be swamped by the effects of free intercrossing among individual members of a species. In contrast to Darwin, Romanes argued that mutual sterility was "the cause, not the result, of specific differentiation." He was confident that his theory of physiological selection in no way opposed that of natural selection, but rather worked alongside it and explained away certain problems left unresolved by the Darwin–Wallace hypothesis. For example, he pointed out how many variations among species had no apparent connection with usefulness in the struggle to survive; "in fact, a large proportional number of specific characters, such as minute details of structure, form, and colour, are wholly without meaning from a utilitarian point of view." Survival of the fittest could not elucidate the existence of these features, and therefore the theory of natural selection could not be summoned into service. If, however, a bar of sterility had intervened to protect the possessors of chance variations from being reabsorbed into the parental species, even a seemingly trivial distinctive trait could be preserved, eventually to become characteristic of a new species. As to the cause of the initial sterility among members of the same parental form, Romanes was content simply to point out the "highly variable" nature of reproductive systems throughout the natural world.[39]

Physiological selection begged as many questions as the theory it was supposed to complement, and the scientific community remained largely unimpressed by it. Nonetheless, Romanes's contribution to the debate over the actual causes of evolutionary change showed him eagerly seeking answers in terms that contemporary scientists could readily accept. When they objected to physiological selection, it was on grounds of insufficient evidence, not because the theory violated any precepts of scientific naturalism. In fact, in 1878 Romanes had pseudonymously published a work that appeared to proclaim his utter allegiance to those precepts. Using the name "Physicus," he had offered *A Candid Examination of Theism* in which he attempted thoroughly to demolish all arguments for the existence of God. There was, he explained, no valid, scientific reason to support a belief in First Cause, universal mind, or any such cosmic concept. Mind was fully explicable in terms of force and matter; nor could any other metaphysical entity be considered outside the workings of the scientific method. Whatever could not be probed by that method could not be shown to possess objective reality.[40]

Despite the assertive stance adopted in *A Candid Examination of Theism*, Romanes derived no satisfaction from his conclusions – quite the contrary. He found the universe without God a barren and frightening place, but was incapable of reconciling his scientific beliefs, particularly the evolution of the human species, with the former Christian faith that had once prompted him to consider taking holy orders.[41] That his loss of faith pained him deeply is suggested, not only by his own statements in *A Candid Examination*, but also by the help he sought from spiritualists. In the year that saw the publication of Romanes's renunciation of religious comfort, C. C. Massey wrote to Stainton Moses:

> Romanes was with me to-day in great distress about his favourite sister, who is dying. It was easy to see he had come in the hope of being infected by my robust belief in a future life. Poor fellow; I was deeply moved by the simple pathos of his yearning for this conviction, the materialistic habit of his scientific mind in conflict with the intense desire of his affectionate nature, and his unverified presentiment of the truth. He looked terribly ill and cut up, telling me his hopes and doubts with an emotion scarcely suppressed.[42]

By the spring of 1878, Romanes had been investigating spiritualism for well over a year. In 1876, he had written several letters to Darwin on the subject. In one, which was included in Romanes's *Life and Letters*, he discussed Lankester's exposure of Slade, observed that the saying "Once a thief always a thief" probably applied to that medium, and concluded, in a flippant tone: "I do hope next winter to settle for myself the simple issue between Ghost *versus* Goose."[43] But in two other letters to Darwin which were not published by Romanes's wife, he expressed much greater credulity. Wallace reminded him of these letters in July 1890, when the two men were involved in an acrimonious dispute over the "absurdity" of believing in spiritualism. Wallace had been allowed to see Romanes's letters, and he had taken full notes of their contents. He referred to those notes when he informed Romanes, in private correspondence:

> You told him [Darwin] that you had had *mental questions* answered with no paid medium present . . . And you declared your *belief* that some non-human intelligence was then communicating with you. You also described many physical phenomena occurring in your own house with the medium [Charles] Williams. You saw "hands," apparently human, yet not those of any one present. You saw hand-

bells, etc., carried about; you saw a human head and face above the table, the face with mobile features and eyes. Williams was held all the time, and your brother walked round the table to prove that there was no wire or other machinery (in your own room!), yet a bell, placed on a piano some distance away, was taken up by a luminous hand and rung, and carried about the room!

Can you have forgotten all this?

In your second letter to Darwin you expressed your conviction of the *truth* of these facts, and of the existence of spiritual intelligences, of *mind without brain*. You said that these phenomena had altered your whole conceptions.

Romanes replied that the opinions expressed in those two letters were "*provisional.*" Further experiments with Williams had revealed him to be an impostor, and Romanes, accordingly, had revised his views.[44]

In summarizing her late husband's inquiries into spiritualism, Mrs. Romanes merely stated that "he worked a good deal at spiritualism for a year or two, and he never could assure himself that there was absolutely nothing in spiritualism, no unknown phenomena underlying the mass of fraud, and trickery, and vulgarity which have surrounded the so-called manifestations."[45] In fact, however, Romanes "worked at" spiritualism for at least four years. In February 1880, in an anonymous letter to *Nature*, he expressed interest in further investigating séance phenomena and particularly requested assistance from scientists in his endeavors. Wallace responded, and Romanes wrote to him in reply that "one or two facts, which you might consider almost commonplace, have profoundly staggered me, and led me to feel it a moral duty no less than a matter of unequalled interest, to prove the subject further." He assured Wallace that he was not "blind to the importance of the testimony already accumulated." In a subsequent letter, before their first meeting in April, Romanes explained to Wallace what he principally sought: "If I could obtain any definite evidence of mind unassociated with any observable organization, the fact would be to me nothing less than a revelation – 'life and immortality brought to light.'" Williams's trickery, and his public exposure in Amsterdam in 1878, had evidently not shaken Romanes's hopefulness as much as he later implied.[46]

When Romanes and Wallace met, Romanes told the spiritualist champion a little about his own experiences. He described how a member of his family had discovered "considerable mediumistic power" in herself, how he had witnessed, with her help, the communication of messages by raps, how sometimes even his mental questions were answered. Although Romanes withheld from Wallace much of the information that he conveyed to Darwin in the two letters of 1876, Wallace still received the clear impression that Romanes was "absolutely convinced that the sounds and motions – the physical part of the phenomena – were not caused in any normal way by any of the persons present, and almost equally convinced that the intelligence manifested was not that of any of the circle."[47] Romanes's account of his meeting with Wallace, however, struck a very different note. He wrote to Darwin on 22 April 1880 that, while he had enjoyed his recent first encounter with Wallace, he had not learned "anything new about Spiritualism." Wallace, Romanes joked,

> seemed to me to have the faculty of deglutition too well developed. Thus, for instance, he seemed rather queer on the subject of astrology! and when I asked whether he thought it worthy of common sense to imagine that, spirit or no spirits, the conjunctions of *planets* could exercise any causative influence on the destinies of children born under them, he answered that having already "swallowed so much," he did not know where to stop!![48]

Romanes, it appears, vacillated between an intense sympathy with spiritualism and an equally strong inclination toward skepticism. The trained scientist in him kept calling for more, and more irrefutable, evidence before he could wholeheartedly proclaim himself a spiritualist. Professional ambition also held him back; he feared the reaction of his scientific colleagues, and the effect on possible career advancement, if he became publicly associated with spiritualism. In a letter to Barrett late in 1881, Massey explained Romanes's reticence concerning spiritualism explicitly in terms of professional caution. "He is, or recently was, greatly interested," Massey wrote,

> and had a series of séances at his own home, at which I was present, about 3 or 4 years ago. He was at that time on his scientific promotion, and very unwilling to connect his name with the subject of Spiritualism. But he always told me that with satisfactory results he should come forward when his position in the scientific world was more established.[49]

If the spiritualism practiced at séances could not provide him with the rigorous evidence he craved, Romanes hoped that the infant SPR, through systematic investigations and fastidious record keeping, could perform that essential service. Romanes evidently attended the founding meeting of the SPR in January 1882 and, along with Stainton Moses, Massey, Myers, and others, voted in favor of a resolution calling for its establishment.[50] He did not, however, join the Society at once, but with characteristic hesitation waited to see what sort of work it would accomplish. When the first issue of its *Proceedings* appeared, Barrett sent him a copy, which Romanes read with care. It largely concerned a series of experiments in thought reading, carried out with the family of the Reverend Andrew M. Creery, of Buxton, Derbyshire. The Reverend's daughters and a female servant in his household displayed unusual prowess in the "willing game," a popular form of parlor entertainment at the time. They could, Creery claimed, divine without benefit of hypnosis, and with remarkable success, whatever object, person, name, or playing card had been selected by the group of players while the percipient was out of the room. Several members of the SPR had been investigating the Creery sisters since the spring of 1881 and had come to the conclusion – erroneous as it later turned out – that they were witnessing the operation of genuine thought reading, without help from any of the normal means of sense perception.

Romanes's response to the evidence presented in the first SPR *Proceedings* was at once enthusiastic and guarded. "If," he told Barrett, "you can maintain such a quality in the material and such a tone in the style of your subsequent issues, I think that your new Society is bound to make a serious impression upon the science and the philosophy of our generation." Still, there were problems in the report on the Creery sisters. "I do not see it stated," Romanes observed, forecasting the later revelations about the girls' *modus operandi*,

> that the conditions under which the experiments were made were such as *absolutely* to exclude the possibility of deception. For in all the cases of marked success it appears that Mr. Creery was present, if not also some of his daughters, besides the one selected to guess. Although it is stated that no words were spoken, signs might still have passed unobserved by the experimenters.

"I make these remarks," he explained,

> not in a hostile spirit, but rather in a friendly one, because I feel what kind of criticism is likely to arise in the minds of those who have not witnessed the experiments, and, on the other hand, if the facts really are genuine, I should like to see their evidence . . . the best that can be brought into court.

And, finally, he concluded:

> I should like very much to join your Society, if by so doing I could get access to any of your committees of enquiry. I should work well and without prejudice, for I have become very much staggered about the whole subject, and so feel that for me it has assumed more importance than any other.[51]

Romanes joined the SPR shortly thereafter, or at least that is the impression conveyed by another letter to Barrett, written in December 1882. After reiterating his belief that psychical phenomena, if proved genuine, would unquestionably "be of more importance than any other to the science and philosophy of our time," Romanes turned his attention to the SPR membership. The Society was formed, he asserted, expressly "to include persons entertaining all shades of opinion concerning the genuineness of these alleged phenomena, and it is on this account that I have joined it." He regretted being unable to attend the forthcoming meeting of the Society, of which Barrett had written to remind him, but in every other respect his letter implied that Romanes intended to be an active, involved member of the SPR.[52] Yet when the Society published its initial membership list, dated December 1883, at the end of the first volume of its collected *Proceedings*, Romanes's name appeared neither as a member nor as an associate. Something must have happened to bring an abrupt end to Romanes's affiliation with the SPR.

What evidence exists to shed light on this problem suggests that Romanes became disillusioned with the work of the SPR when he participated in the Society's thought-transference tests with Smith and Blackburn early in 1883. On that occasion, the two men had been brought to London from Brighton, and the experiments were held in the SPR rooms in Dean's Yard, Westminster. Many years later, the *Westminster Gazette* published an account of the proceedings by one of the participants, James Crichton-Browne, Lord Chancellor's visitor in lunacy at the time. Having forgotten some of the details, he limited his account to those aspects of the experiments that he could definitely recall:

> I was invited to join the conference by George Romanes, who told me that he had been asked by some members of the Psychical Research Society to bring together a few friends interested in such matters to witness a remarkable manifestation of thought-reading by a youth who could without contact receive and reproduce, not words but figures, diagrams, or pictures, present to the mind of an operator with whom he was in some way *en rapport*.

Other spectators that day, Crichton-Browne recollected, included Sidgwick, Myers, Gurney, Wyld, and Francis Galton. After the percipient, "Mr. S.," had been blindfolded, the participants selected a committee, including Romanes, Galton, Crichton-Browne, Sidgwick, and Myers, "to prepare and conduct the experiments." The committee retired into a back room, and Romanes first sketched an owl on a sheet of paper. "Mr. B.," the agent, was then called into the back room; he studied the sketch intently for a few minutes. When he returned to the main room with the committee members, the sketch remained behind.

> Mr. B. stood behind Mr. S. at the distance of about a couple of yards, and gazed at the back of his head. I remember distinctly that he had

his hands in his trousers pockets and that he contracted his brows from time to time and made faces. This went on for, I suppose, about five minutes, and then Mr. S. drew on the paper before him a crude and clumsy outline of an owl. It was very different from Romanes' sketch, but it was undoubtedly suggestive of an owl.

After a few such simple tests, Romanes and Crichton-Browne suspected "that some kind of code might be in use" and purposely drew "a figure without a name, a sort of nondescript arabesque, simple enough, but not easily describable in words." As they anticipated, Smith was utterly unable to reproduce the design. A further test convinced the two men that Smith was inadequately blindfolded,

> and that it was practicable for Mr. B. to communicate with him both by sight and hearing; so Romanes and I asked permission, which was granted, to blindfold him anew. We proceeded to do so *secundum artem*. Cottonwool was procured, the sockets were packed, the ears were plugged, and a large handkerchief made all secure. After that several experiments were tried as before, but there never was the smallest response on the part of Mr. S. to Mr. B.'s volitional endeavours. There was no more flashing of images into his mind. His pencil was idle. Thought-transference was somehow interrupted.

Crichton-Browne went on to explain how he, Romanes, and Galton all had the impression that the Morse code was somehow being employed between Blackburn and Smith, possibly by the clinking of coins in Blackburn's pocket. It was certainly not coincidental that "the moment that Mr. S.'s senses were thoroughly occluded all transference stopped." Perhaps what alarmed Romanes most, however, was not that he had discovered two charlatans practicing "patent imposture," but that the SPR officials present that day did not seem particularly grateful for his, and Crichton-Browne's, acuity. "The last scene of all, or passage-at-arms," remained vividly in the latter's memory. "Mr. Myers, standing in front of the fireplace said: 'It must be allowed that this demonstration has been a total failure, and I attribute that to the offensive incredulity of Dr. Crichton-Browne.' "[53]

Romanes was never listed among the members of the SPR during the remaining decade of his life, and he seems to have given up active psychical research after the distressing experience in Dean's Yard. He did not, however, abandon all interest in the subject. In 1894, for instance, as he was dying of arterial disease in his midforties, he enjoyed sharing ghost stories with a friend, "and he and she used to 'cap' each other's narratives."[54] In 1888 or 1889, just before the onset of his fatal illness, Romanes had explored another alternative to the materialist philosophy with which he was associated professionally: He had visited

Madame Blavatsky "to discuss the evolutionary theory set forth in her *Secret Doctrine*."[55]

Romanes had indeed strayed far from scientific naturalism if he was prepared to consult such a source on such a subject as evolution. That consultation, however, was very much in keeping with the direction of his thought for some years past. Even during the period when he most sweepingly rejected the religious interpretation of the natural world, Romanes had not abandoned the outward forms of religious practice. When in London, he usually attended Christ Church, Albany Street, on Sunday; while at Geanies, his home in northern Scotland, "he had a short Evening Service for guests and servants who could not drive ten miles to church."

> This service, unless a clergyman happened to be staying at Geanies, he conducted himself, and ended it by reading a sermon. He had all his Presbyterian ancestors' love for a good discourse, and serious efforts had to be made to prevent him from reading too long a sermon.[56]

Although it may be true that these religious performances by Romanes were the motions of "a social and political conservative" who considered religion useful for curbing "mankind's lower yearnings,"[57] it seems that Romanes threw more energy into them than was absolutely necessary to set a good example. Perhaps they may also be interpreted as the actions of a man who did not want to lose hold of the lifeline leading back to faith. Although the line never led him back to orthodox Christianity, it led to a metaphysical compromise that accommodated God in the universe. The conversation with Madame Blavatsky in the late 1880s was a fitting stop to make along the way.[58]

Psychical research, although it failed to furnish incontestable proof that spirit shared the cosmos with matter, nonetheless appears to have played an important role in persuading Romanes to abandon the "Physicus" point of view. Surely it was not coincidental that he published his essay on "The Fallacy of Materialism" in 1882. The article, which explicitly rejected materialism as an adequate explanation of ultimate facts, appeared in the *Nineteenth Century*[59] at the very time when Romanes was eagerly studying the early work of the SPR. Even after the disappointments of the following year, he was not driven back to the desolate position of *A Candid Examination*. In 1883, he offered to send a copy of that volume to Professor Asa Gray, the eminent American botanist, but carefully explained: "I do not now hold to all the arguments, nor should I express myself so strongly on the argumentative force of the remainder." In a subsequent letter to Gray, Romanes wrote that he fully endorsed the view that the doctrine of the human mind having been proximately evolved from lower minds is not incompatible with the doctrine of its having been due to a higher and supreme mind. Indeed, I do not think the theory of evolution, even if fully proved, would seriously affect the previous standing of this more important question.

Furthermore, Romanes continued in this resounding departure from his earlier position,

> I cannot help feeling that it is reasonable (although it may not be orthodox) to cherish this much faith, that if there is a God, whom, when we see, we can truly worship as well as dread, He cannot *ex hypothesi* be a God who will thwart the strong desire which He has implanted in us to worship Him, merely because we cannot find evidence enough to believe this or that doctrine of dogmatic Theology.[60]

The ideas that Romanes sketched in the second letter to Gray matured in the final decade of his life. His persistent refusal to embrace dogmatism expressed itself against restrictive religious beliefs in the letter, but it was equally capable of turning against the narrow orthodoxies of science. As the 1880s progressed, Romanes increasingly perceived many of his professional colleagues as dogged apologists for a highly limited view of reality – that associated with the scientific naturalism that he had outgrown. In the second letter to Gray, Romanes also touched upon the idea of "a higher and supreme mind," which, within a few years, blossomed into an argument for the existence of God. The argument took many twists and turns, and Romanes died before he could reduce them to order.[61] They not only included the vision of a monistic universe whose fundamental substance synthesized matter and spirit, but they also suggested Romanes's final belief that a First Cause was, after all, a necessary postulate in nature.[62]

Romanes justified all his departures from scientific naturalism in the name of "pure agnosticism," a concept that he first developed in his Rede Lecture at Cambridge in 1885.[63] For Romanes, agnosticism meant a total absence of dogmatic judgment, a refusal to rule out questions, and possible answers, on a priori grounds. The pure agnostic took his orders neither from the scientific laboratory nor from the cathedral pulpit. Problems such as the existence of God might not be provable in empirical terms, but they were not thereby irrational and worthy of dismissal. On the contrary, they deserved dispassionate inquiry as fully as the problems of physics and biology. The pure agnostic recognized that the human mind was not equal to the task of unraveling many cosmic mysteries and believed man should avoid the kind of arrogance that implied otherwise. Romanes advocated his pure agnosticism on numerous occasions after 1885,[64] and it was as a pure agnostic that he made his peace with the Anglican church before he died; it would have been dogmatic to refuse to try the path of faith. As Charles Gore, Anglican theologian and editor of Romanes's posthumous *Thoughts on Religion*, explained at the close of that volume: "George Romanes came to recognize . . . that it was 'reasonable to be a Christian believer' before the activity or habit of faith had been recovered."[65]

Although he had never managed to capture spirit at the séance table nor to isolate it in tests devised by the SPR, Romanes ended by positing the existence of a universe imbued with spirit, over which presided an immaterial cosmic and unifying mind. Yet Romanes trusted that his revised belief in the reality of mind, or spirit, did not violate the scientific laws he revered. In a note written during the summer of 1893, he sketched the following brief resolution of his long intellectual and spiritual conflict:

> The result (of philosophical inquiry) has been that in his millennial contemplation and experience man has attained certainty with regard to certain aspects of the world problem, no less secure than that which he has gained in the domain of physical science, e.g.
> Logical priority of mind over matter.
> Consequent untenability of materialism.
> Relativity of knowledge.
> The order of nature, conservation of energy and indestructibility of matter within human experience, the principle of evolution and survival of the fittest.[66]

Romanes found these "aspects of the world problem" incompatible in the 1870s, yet he held them all as equally certain articles of faith by the early 1890s. They may still seem incompatible to us today, but he had somehow learned to reconcile them, to his own satisfaction, before his death.

At the end of his life, Romanes would have felt little sympathy for the views expressed by Karl Pearson, student and biographer of Francis Galton, when he lamented the foolishness of the scientist who "forgets that his senses have been developed to grasp physical phenomena, that his concepts are deductions from his sensuous perceptions, and that neither his sensuous nor mental outfits are adapted for sensating, perceiving and conceptualising the hyper-phenomenal." "Some men," Pearson continued, warming toward a system of classification,

> grasp this truth by the logic of reasoning, others by the logic of experience, others by a healthy instinctive appreciation, and some never grasp it at all. To the first group we may, perhaps, say Huxley belonged, to the second Galton, to the third Darwin, and to the fourth Crookes and Alfred Russel Wallace.[67]

With the exception of Crookes, Pearson took his examples from among men who studied patterns of evolutionary change, biological growth, and inheritance. Like Romanes, they were drawn by their interests – and not always willingly – to consider the place that the "hyper-phenomenal" might occupy within or beyond nature.[68]

T. H. Huxley's opinions about spiritualism and psychical research were not quite as simple as Pearson suggested. It was not the logic of reasoning alone that persuaded Huxley to regard séances as a waste of his time. He also had some firsthand experience to help mold his point of view. His son and biographer, Leonard Huxley, noted that he had regularly attended séances with Mrs. Hayden at his brother George's home in the early 1850s.[69] Huxley must have been even less impressed with the medium than Chambers was, for there is no evidence to suggest that he continued his experiments in spiritualism at that time. That the woman was a fraud, who relied on clues from her sitters in spelling out the rapped answers to their questions, seemed well established by the spring of 1853,[70] and no doubt Huxley decided the subject was unworthy of further effort on his part.

In 1866, when Wallace wrote to invite Huxley to some private séances for investigative purposes, Huxley replied with a certain weariness of tone:

> I am neither shocked nor disposed to issue a Commission of Lunacy against you. It may be all true, for anything I know to the contrary, but really I cannot get up any interest in the subject. I never cared for gossip in my life, and disembodied gossip, such as these worthy ghosts supply their friends with, is not more interesting to me than any other. As for investigating the matter, I have half-a-dozen investigations of infinitely greater interest to me to which any spare time I may have will be devoted. I give it up for the same reason I abstain from chess – it's too amusing to be fair work, and too hard work to be amusing.[71]

He gave a similar, although even blunter, reply to the London Dialectical Society when it invited him to join its inquiry into spiritualism in 1869. He could expect nothing but trouble and annoyance from such an inquiry, he snapped. Furthermore, his one previous inquiry into the subject had revealed "as gross an imposture as ever came under my notice." Even if the phenomena were genuine, they did not interest him, for the spirits' conversation was as boring as "the chatter of old women and curates in the nearest cathedral town." "The only good that I can see in a demonstration of the truth of 'Spiritualism'," he concluded, "is to furnish an additional argument against suicide. Better live a crossing-sweeper than die and be made to talk twaddle by a 'medium' hired at a guinea a *séance*."[72] Nothing, it seemed, could have induced Huxley to pay further attention to spiritualism.

He did, however, attend at least one more séance: In late January 1874, he and Darwin's son George sat with Charles Williams. Huxley wrote a lengthy report on the session for Charles Darwin in which he explained how he had surmised the movements of the medium's body, despite total darkness, by fixing his eye on "three points of light, coming from the lighted passage outside the door." When the light rays were suddenly blocked, Huxley knew that the medium had changed his position, as when he leaned over the table to twang the guitar strings with his mouth (or so Huxley guessed). All the while, Huxley and George Darwin, sitting on either side of the medium, kept a firm grip on Williams's hands, with their feet resting against his. Very little of note occurred, apart from the twanging guitar strings, until Darwin was asked to change places with one of the other sitters. Then Huxley was hardly surprised to hear furniture moving – indeed a chair climbed on to the séance table. Huxley, who reported to Charles Darwin that his "attention was on the stretch for those mortal two hours and a half," concluded that Williams was "a cheat and an impostor."[73] Twenty years after his first séances, Huxley's initial poor opinion of mediums and spiritualism was confirmed, but it is curious that he subjected himself to two and a half hours of tedium over a subject in which he claimed absolutely no interest. Probably he participated at the séance as a favor to Charles Darwin, who had become uneasy over spiritualism by 1874.

Darwin's own direct exposure to mediums was slight. In the 1840s and early 1850s, he had been skeptical about mesmerism and clairvoyance, and for the next twenty years he apparently saw nothing in spiritualism to make him abandon that frame of mind. In the early 1870s, however, Crookes's endorsement of the movement gave Darwin pause. Francis Galton, his gifted cousin, began attending séances with Crookes and sent Darwin enthusiastic and admiring reports in 1872. Darwin must have written at least a mildly encouraging reply, for Galton informed him at the end of March 1872: "Your letter will be a great encouragement to Crookes and I have forwarded it to him to read, telling him what I had written." In May, Galton thanked Darwin for his "very kind" letter about D. D. Home. Darwin trusted his cousin's judgment and did not know what to make of the new developments that seemed to challenge all his assumptions about natural processes.[74]

At last, in January 1874, Darwin himself was persuaded to attend a séance, held in London, at 6 Queen Anne Street, the home of his brother Erasmus. Two of Darwin's cousins, Galton and Hensleigh Wedgwood, were also present, as were George Darwin, G. H. Lewes, and George Eliot, with Williams as the medium. Darwin left before any phenomena occurred, but he was clearly bothered by reports of what transpired and by Galton's verdict that it was "a good séance."[75] Troubled by what he could neither understand nor accept, lacking the stamina to investigate the medium as closely as circumstances demanded, he turned to his champion-in-arms, Huxley, and asked him to test Williams's mediumship. Huxley's report of 27 January, describing his own experiences with Williams, reassured Darwin immensely. He thanked Huxley warmly and announced with palpable relief: "Now to my mind an enormous weight of evidence would be requisite to make one believe in anything beyond mere trickery."[76]

When Romanes, still in the throes of amazement over Williams's powers, wrote to Darwin two years later, Darwin's opinions on spiritualism had become fixed. He had never warmed to the subject and had always resisted the possibility that his understanding of the natural world would have to make room for the strange phenomena of the séance chamber. Yet, as a student of life's myriad forms, Darwin could not entirely ignore those phenomena, at least not until Huxley and George Darwin – men who knew valid evidence from invalid – confirmed his strong desire to do so. Such was the basis of that "healthy instinctive appreciation," which Pearson claimed kept Darwin from bothering with the "hyper-phenomenal."

In Galton's case, according to the same source, it was the logic of experience that turned him away from spiritualist pursuits. Sir Francis Galton, knighted in 1909 when he was in his late eighties, had followed the logic of experience in diverse directions. Enjoying an independent income, he had made his own place in British science without regard to the categories of academic appointments. A gentleman scientist in the venerable British tradition, he was known for his African travels and explorations before he undertook a systematic study of meteorology in the early 1860s. Darwin's work was stirring his interest at the same time, and by the middle of the decade Galton was already involved in the studies of inheritance with which he was subsequently associated.[77]

While psychology in that period was struggling to chart the workings of the mind in general, Galton turned his keen attention to the differences that characterized individual minds. He became convinced that mental traits were as much affected by heredity as were physical features, but he was also fully aware of the role played by environment in forming the distinctive aspects of personality. His study of *English Men of Science*, published in 1874, marked the first appearance of the terms "nature" and "nurture" to summarize the debate.[78] Very early in his studies of individual differences, Galton also realized the "pressing necessity of obtaining a multitude of exact measurements relating to every measurable faculty of body or mind, for two generations at least, on which to theorise." He accordingly became an avid anthropometrist, seeking to collect data, for example, on "Keenness of Sight and of Hearing; Colour Sense, Judgment of Eye; Breathing Power; Reaction Time; Strength of Pull and of Squeeze; Force of Blow; Span of Arms; Height, both standing and sitting; and Weight."[79] He was convinced "that quantitative measurement is the mark of a full-grown science" and, building on the work of the Belgian mathematician Quetelet, sought to apply statistical laws to psychology.[80] The same impetus drove Galton to lead the way in developing mental tests, the ancestors of all the psychological testing imposed upon twentieth-century children and adults alike.

Galton's was not an abstract intellectual interest in the puzzles of inheritance. A highly practical concern underlay the researches that produced *Hereditary Genius* (1869), *Inquiries into Human Faculty and its Development* (1883), and *Natural Inheritance* (1889), among his other books and scores of papers on the subject. He cared passionately about "Race Improvement" and is known as the father of eugenics, a term he coined in 1883, in *Inquiries into Human Faculty*.[81] In eugenics, or the practice of replacing natural selection with the purposeful selection of the fitter members of society in order to produce an improved race, the Darwinian influence behind Galton's work blended with the statistical. Clearly many aspects of Britain's human resources had to be measured and quantified before the intelligent selection of optimal breeding partners could occur.[82] Even before that keenly desired point had been reached, however, Galton hoped that his fellow countrymen would realize the follies of their philanthropic ways. "It is known," he complained at the end of his autobiography, published in 1908, the year that Parliament passed the Old Age Pensions Act,

> that a considerable part of the huge stream of British charity furthers by indirect and unsuspected ways the production of the Unfit; it is most desirable that money and other attention bestowed on harmful forms of charity should be diverted to the production and well-being of the Fit.

So seriously did he take the cause of eugenics that he believed "its principles ought to become one of the dominant motives in a civilised nation, much as if they were one of its religious tenets."[83]

In the midst of his myriad inquiries, Galton found time, intermittently for over a decade, to examine the phenomena of spiritualism and psychical research. It was certainly not a religious impulse that prompted him to attend séances. Things of the spirit figured only slightly in the world of this zealous quantifier. He had, as one historian of science has commented, a "predisposition toward deterministic science and away from current theology."[84] Why, then, did he find it worth his while to sit for hours in dim rooms, awaiting the chance rapping, luminous hand, or spirit figure?

Galton's autobiography is silent on the subject, and such evidence as exists points to his total repudiation of spiritualism long before he wrote his memoirs. Indeed, apart from his hopeful letters to Darwin in 1872, Galton's behavior at séances reflected little but contempt for all the proceedings. Initially, however, he approached the experience with what Pearson called an attitude of agnosticism. In March 1872, he wrote to Darwin about three séances that he had recently attended, two at Crookes's house and one at Serjeant Cox's. They "utterly confounded" him, he reported, and he was "very disinclined to discredit them." He praised Crookes's method of conducting the séances, and although he recognized much "rubbish" and "absurdity" in the proceedings, he confessed to be staggered by "the extraordinary character of the thing." By the time Galton wrote again to Darwin in April, he had met the great Home and sang his praises. He reiterated his conviction that Crookes was "thoroughly scientific in his procedure"; Galton was certain that no mere "vulgar legerdemain" was involved. Thereafter Galton reported having difficulties in obtaining invitations to séances, but seems nonetheless to have kept in touch with Crookes, of whom he continued to entertain a good opinion.[85]

If spiritualists were growing wary of Galton, his behavior at a séance in January 1873 helps to explain why. On this occasion, the sitting took place at the home of one Miss Douglas, a spiritualist who resided in fashionable South Audley Street in Mayfair. Galton's colleagues that evening were numerous and included Stainton Moses, Crookes, E. W. Cox, Mrs. De Morgan, Mrs. Makdougall Gregory, C. Maurice Davies, and Lord Arthur Russell. The mediums were Mr. and Mrs. Nelson Holmes, who suffered a humiliating exposure in Philadelphia two years later. At the very beginning of the séance, Galton antagonized confirmed spiritualists among the participants when he protested against the terms that the mediums imposed on the conduct of the sitting. (Mr. Holmes, for example, had strong objections to being tied up in the back room designed to serve as materialization cabinet.) Galton announced firmly that "he wished to get rid of conditions which were manifestly fraudulent or favorable to fraud." When some phenomena occurred, Galton asserted "that it was all a trick." In such an unsettled atmosphere, no spirit faces appeared, and Moses, who took notes on the séance, attributed much of the failure to Galton's disruptive conduct.[86]

What happened to sour Galton's attitude between the spring of 1872 and early winter of 1873 is anybody's guess. It may have had something to do with a growing suspicion on his part that he was being denied the opportunities for a full investigation. Yet when he attended Williams's séance with Darwin in January 1874, he pronounced it a good one and did not, apparently, offer any revealing criticisms to ease Darwin's uncertainty. The following year, Galton took part in Crookes's tests with Annie Eva Fay, but Pearson makes no reference to that inquiry in Galton's biography. According to Pearson, very shortly after Huxley's report to Darwin exposing Williams's chicanery, Galton "became a despiser of 'spiritualistic' séances."[87]

Yet Galton was not entirely finished with his investigations. In the early 1880s, he was known to retain an interest in psychical research, and his name figures in the early history of the SPR, although he never joined the Society. Romanes wrote to Barrett in July 1881 proposing Galton, as a man with "the reputation of being critical," for a committee to examine the Creery family.[88] Romanes and Galton had already joined forces the previous May and June to test the prowess of the popular performer W. I. Bishop, who claimed great ability in thought reading. They found no evidence of such power and said so in the pages of *Nature*.[89] In 1883, Galton again played the role of critical observer, this time in the Smith–Blackburn experiments in Dean's Yard, and when Crichton-Browne published his recollections of that drama in 1908, he was confident that Galton would "be able to confirm or amend" the report.[90]

The year after the Smith–Blackburn fiasco, Galton published a brief essay on free will in the journal *Mind*. In it, he described an introspective analysis of his own decisions and the motives for his actions during a six-week period. For so baffling a subject, his conclusion was surprisingly definite. "The general results of my introspective inquiry," he explained,

> support the views of those who hold that man is little more than a conscious machine, the larger part of whose actions are predictable. As regards such residuum as there may be, which is not automatic and which a man however wise and well informed could not possibly foresee, I have nothing to say, but I have found that the more carefully I inquired, whether it was into the facts of hereditary similarities of conduct, into the life-histories of very like or of very unlike twins, or now introspectively into the processes of what I should have called my own Free-will, the smaller seems the room left for the possible residuum.[91]

It is tempting to speculate that Galton's inquiries into spiritualism and psychical research also contributed, in a negative sense, to his conclusions about man as machine.

The reasons why Galton became interested in spiritualism early in the 1870s remain obscure. In the 1840s, he had concluded that the effects of animal magnetism were "purely subjective on the part of the patient," and presumably he could have explained away much of spiritualism in a similar manner.[92] Crookes's work on the subject, however, seems to have persuaded him that serious results might be anticipated from séances, and he embarked on his own researches without prejudice. Having decided that mind was subject to the workings of evolution and heredity, he may have investigated spiritualism in the belief that it was his obligation to do so. Before he could reduce the chaos of contemporary thinking about mental attributes to statistical order, perhaps he felt compelled to make sure that he understood the functioning of mind as thoroughly as possible. If séance phenomena could be traced to genuine spirit origins, then the assumptions of scientific naturalism could explain neither the workings of mind nor its development over time. Much of the work of the SPR, furthermore, was germane to Galton's own, on such matters, for example, as mental associations and imagery.[93] After his experiences among spiritualists and psychical researchers, however, Galton had no reason to deviate from the road he had chosen to follow, for he had found no indubitable evidence of spirit or mind acting independently of body. He did not embrace utter determinism or materialism,[94] but he was never tempted to doubt that the tools of mathematics were suited to tasks that involved the unique and highly variable qualitities of individual lives. His route did not again swerve from the clarity and precision of numbers into the mists and fogs of spirit.

WALLACE AND THE SPIRITUALIST FAITH

By contrast, Alfred Russel Wallace appeared particularly at home among those vapors. He stands as the archetypal example of a man who managed to reconcile the Darwinian view of nature, the inexorable workings of evolution, with a belief in the efficacy of spirit agents at large. Co-discoverer of the theory of natural selection in the 1850s, he was its most ardent champion, defending it against all challengers who threatened to introduce modifications in the theory, or to diminish the pervasive influence of natural selection on the formation of new species. Yet Wallace did endorse one stunning exception to the impact of evolutionary change: He came to exempt the human mind and to argue for a distinct process of mental growth in man, entirely separate from the rest of the natural world.

Nor is Wallace like the other men in this chapter who investigated spiritualist phenomena out of a sense of scientific responsibility. Once he began to attend séances in 1865, he seems, unlike them, quickly to have attained unquestioning belief. Even Chambers denied irrefutable knowledge of the spirit world, but with Wallace there is no evidence of subsequent doubt concerning the phenomena that he observed. He was convinced that the testimony of his own senses, combined with the records of countless other investigators over the centuries, provided an adequate empirical base on which to establish the validity of spiritualism. Unlike Chambers, too, Wallace's commitment to spiritualism was widely publicized over a period of decades. He became the leading apologist for the movement in England, testifying on behalf of every beleaguered medium and writing endless letters to editors in defense of spiritualist manifestations. He shirked no battle against his colleagues in science who, he claimed, arrogantly refused to investigate spiritualism because of a priori assumptions about the natural world.

Where mediums were concerned, Wallace always extended them the benefit of the doubt. At the *Colley v. Maskelyne* trial, in April 1907, Wallace insisted that he "had never met a medium who was a scoundrel" and that, in most cases, he placed no credence in their so-called exposure as cheats. He testified that Maskelyne's alleged imitation of Monck's séance phenomena was "ludicrous"; for one thing, the materialized young woman walked away at the end of Maskelyne's act, while, in Monck's case, the materialized figure was reabsorbed into the medium's side. Wallace told the court that he himself had seen such materializations emerging from the entranced medium, just as in 1877 Wallace told the editor of the *Spectator* that the messages that Monck obtained on closed slates were demonstrably free of imposture. In 1876, Wallace appeared in court to speak out for Slade, and his testimony revealed the entirely trusting frame of mind in which he had attended séances with that medium.[95] In Wallace's own writings on spiritualism, not even a shadow of suspicion darkened his descriptions of the mediums with whom he sat. Florence Cook, for example, emerged as a paragon of probity, Kate Fox an exemplar of lifelong sacrifice to her calling.[96] Both in his personal correspondence and his public statements, he ardently praised all members of the profession, whether famous or obscure.

Because he placed his faith so confidently in mediums, it is scarcely surprising that Wallace similarly accepted all the physical and mental manifestations of mediumship, together with all the phenomena that were supposed to arise spontaneously, without a medium's intervention. He drew the line at nothing: The so-called Reichenbach phenomena, spirit photography, second sight, crystal seeing, apparitions of all sorts, ghosts and haunted houses, levitation, clairvoyance, and clairaudience – all received his enthusiastic endorsement. The facts of spiritualism, he maintained as early as 1874, had been proved in their entirety. They required no further confirmation, and the burden of proof had shifted to their opponents.[97] He remained undaunted by the persistent barbs of W. B. Carpenter and other critics, contending in reply that a negative result now and again could not invalidate a series of positives. Every inquirer realized, he explained with some disdain, that spiritualist phenomena were too delicate and uncertain to be relied upon for systematic tests. Indeed, even if failures outnumbered successes, the reality of those successes could not thereby be negated. "As well deny that any rifleman ever hit the bull's-eye at one thousand yards, because none can be sure of hitting it always, and at a moment's notice."[98]

With such an outlook, Wallace was virtually unshakable in his spiritualist convictions, to an extent that even sympathetic psychical researchers could not condone. Myers of all people, scarcely a model of critical judgment in such matters, tarred him with the brush of "resolute credulity." "I regard Mr. Wallace's testimony with regard to the character of public mediums," he wrote in 1895,

> as precisely on a par with the testimony of Marcus Aurelius Antoninus with regard to the character of Faustina. There are natures . . . which stand so far removed from the meaner temptations of humanity that those thus gifted at birth can no more enter into the true mind of a cheat than I can enter into the true mind of a chimpanzee.[99]

The previous year Myers had written to Lodge, after attending a gathering at which Wallace was present:

> Wallace was vigorous, sensible, cheery, manly – full of anecdotes and ideas. The general feeling was that – just as Lord Chancellor Brougham wd have known a little of everything if only he had known a little *law*; – so also Wallace, if he had only had just a trace of scientific instinct or training, wd have been one of the most robust minds of our time.[100]

Lodge agreed about Wallace's intellectual inadequacies. In 1914, when he read Wallace's autobiography, he found the naturalist "a crude, simple soul, easily influenced, open to every novelty and argument." Lodge regretted that Wallace had trusted "his own judgment continually when it cannot really have been founded upon adequate knowledge."[101]

For Wallace, however, formal scientific training or adequate knowledge was irrelevant to the pursuit of spiritualist truth. Like Chambers, he trusted his eyes and ears, his sense of touch and smell, and staunchly maintained the sole sufficiency of sense perception to decide whether or not genuine spirit forces were responsible for séance phenomena.

"We persist in accepting the uniform and consistent testimony of our senses," he announced in *Fraser's* in 1877, at the start of an article that he wrote in reply to one by Carpenter the month before. Carpenter's essay included a sustained and snide assault on Wallace's grasp of the concept of demonstration, and the criticism was valid. Wallace's capacities for logical deduction seemed to atrophy when spiritualist phenomena were under consideration. It was not that his powers of observation failed him. He could describe in great detail, for example, a bundle of flowers and ferns that allegedly dropped onto the séance table from spirit sources.

> Counting the separate sprigs we found them to be forty-eight in number, consisting of four yellow and red tulips, eight large anemones of various colours, six large flowers of *Primula japonica*, eighteen chrysanthemums mostly yellow and white, six fronds of Lomaria a foot long, and two of a Nephrodium about a foot long and six inches wide.

The problem was that the naturalist who identified flora so carefully refused to accept the idea that his senses could be deceived. He might grudgingly admit that there were impostors who managed to imitate mediums, but he could never acknowledge treachery in any medium whom he knew personally. He preferred to assume that conjurors were really mediums, rather than vice versa, because he rejected the idea that the hand of the performer might move faster than the eye of the observer. When Carpenter pointed out, on information provided by Home, that a medium might conceal objects in the lining of a cloak until they were needed for sudden, mysterious séance appearances, Wallace dismissed the remark as irrelevant: Home was not talking about the medium who had precipitated flowers for Wallace.[102] No matter how precise the evidence marshaled to disprove, no matter how definitive the revelation of trickery, Wallace found a way to nullify its sting and to maintain the validity of spiritualist manifestations. In evaluating séance phenomena, he would neither acknowledge the fallibility of human powers of perception, nor recognize the sitter's role in affirming that what did not happen did.

When Wallace served as a member of the London Dialectical Society's committee to investigate spiritualism in 1869, he aired a number of his objections to the lines of reasoning habitually employed by his critics. He began by challenging their favorite belief in collective delusion; never, he insisted, had "any large amount of cumulative testimony of disinterested and sensible men" been assembled "for an absolute and entire delusion." Furthermore, he faulted the antispiritualist commentators for arguing in circles. They began, he explained, by saying that the uncertainty of spiritualist phenomena, the fact that people could not study them in a systematic way, placed them outside the scope of natural law and hence made them unbelievable to the scientific community. If they could be shown to follow definite laws, then skeptics might consider them more seriously as belonging to the vast family of natural phenomena. This was nonsense, Wallace expostulated:

> The essence of the alleged phenomena . . . is, that they seem to be the result of the action of independent intelligences, and are therefore deemed to be spiritual or superhuman. If they had been found to follow strict law and not independent will, no one would have ever supposed them to be spiritual. The argument, therefore, is merely the statement of a foregone conclusion, namely, "As long as your facts go to prove the existence of unknown intelligences, we will not believe them; demonstrate that they follow fixed law, and not intelligence, and then we will believe them." This argument appears to me to be childish, and yet it is used by some persons who claim to be philosophical.[103]

Wallace's dislike, or at least distrust, of persons who practiced philosophy at the séance table kept him from an active role in the SPR. From the start in January 1882, when Barrett urged him to join, Wallace expressed serious doubts about an association headed by Sidgwick. "He has been enquiring into the subject *years before* you began," Wallace pointed out to Barrett.

> He has had ample means of experiment, and he has, *I know*, from records kept by Mr. Myers and which he showed me, witnessed a number of *most extraordinary* phenomena. Yet during all this time not one word has he said or printed to support those who bear the obloquy of a belief in the phenomena, and when friends have enquired what he has seen he has thrown cold water on the whole thing and has given them to believe it was all a *fiasco*! . . . I very much doubt if he will give his name to any *facts* without such *qualifications* and *doubts* as to render them worse than useless, – in fact damaging rather than otherwise.
> . . . Mr. Myers assured me that Sidgwick was always making suggestions, of *how* the phenomena might *possibly* be produced without ever attempting to show that they could *possibly* be so produced. If this is allowed there is an end to rational enquiry.

He declined to join the SPR right away, preferring to wait and see whether it publicized the results of its research "in a way to carry weight."[104] Shortly, however, Wallace became one of the Society's honorary members, a position he retained despite increasing distaste for the tenor of its investigations.[105] On at least two occasions after the turn of the century, Barrett tried to interest Wallace in the Society's presidency, but to no avail. An elderly man in his late seventies, Wallace was not eager to make the trip from his Dorset home to London, and he assured Barrett that he would never attend meetings. He was certain, furthermore, that the Society would not rise in the public esteem under the leadership of such a widely known "crank" and "faddist" as himself.[106]

It was not just concern for the Society's good name that prompted Wallace to turn down the honor of its presidency, nor the nuisance of the necessary travel to London. By the 1890s, he found that he had less than ever in common with the Society and particularly objected to the concept of the subliminal self that he considered altogether too prevalent in the writings of its members. He had no patience with elaborate theories that demanded the postulation of a secondary personality. Such notions, he thought, muddied the waters far more seriously than the simple spiritualist hypothesis. Wallace's basic criticism on this score appeared in an essay that he contributed to the Boston *Arena* in February 1891. He argued there that the hypothesis of a double personality in every one,

> a second-self, which in most cases remains unknown to us all our lives, which is said to live an independent mental life, to have means of acquiring knowledge our normal self does not possess, to exhibit all the characteristics of a distinct individuality with a different character from our own, is surely a conception more ponderously difficult, more truly supernatural than that of a spirit-world, composed of beings who have lived, and learned, and suffered on earth, and whose mental nature still subsists after its separation from the earthly body.[107]

To dismiss the active agency of disembodied intelligence in human affairs as an unjustified and unproven assumption, and then to embrace the utterly speculative explanation of a second self was simply absurd, Wallace asserted. Even to his close friend Barrett, he could not refrain from a gentle reprimand when the "erroneous theory of the 'subconscious self'" appeared in the physicist's work. "I totally dissent from it," Wallace explained in February 1911. "To me it is pure assumption, and, besides, proves nothing." Five days later he was writing again to emphasize how vehemently he objected to the concept, which he labeled "the most gross travesty of science."[108]

It is difficult, at first, to explain the tenacity with which Wallace clung to his belief in spirit agency and the vigor with which he defended it against all comers, whether in the form of Carpenter's skepticism or Myers's subversive alternate hypothesis. Wallace stubbornly refused to accept even the outside chance that man's mind might work in ways his senses could not fathom – might, in fact, alter the power of his senses to record veridical impressions – and this refusal led him to naive and credulous assumptions that were not characteristic of his work as a naturalist. In a letter of 1908, written in answer to a question concerning the possibility of high forms of life existing in an atmosphere of carbon dioxide, Wallace firmly set forth his approach to such hypothetical inquiries: "I . . . consider that any postulate of 'possibility' merely because we can not assert *'impossibility'*, should never be introduced into any problem of science. In fact, to do so renders all rational and truly scientific _reasoning_ impossible."[109] Why, then, where spiritualism was concerned, did he vehemently defend the postulate of "possibility," without giving the claims of "impossibility" prolonged and balanced consideration?

The answer to the question lies more in Wallace's social conscience than in any unsatisfied religious hunger. He abandoned Christianity in his youth, and, unlike most of the men and women who figure in this study, was not tormented by religious doubts thereafter. Apparently he did not embark on his spiritualist explorations seeking reassurance of his own immortality.[110] But, despite a period in his life when he endorsed the individualism of Herbert Spencer, he never ceased to worry about the social imbalance that he saw all around him. He could certainly find no evidence of justice and equity prevailing in Victorian England, with its extremes of opulence and squalor; nor were the Christian churches altogether encouraging about equality of opportunity in the life to come. Even nature, which he examined so minutely, seemed to operate arbitrarily, crushing out life as extravagantly as it produced new forms. Into this vision of suffering, spiritualism burst on Wallace like the proverbial shaft of sunlight. Its impact, quite simply, devastated his former assumptions. Wallace embraced spiritualism with all his heart, because it perfectly suited his deepest desire to design a new world view for himself – one where the influence of Robert Owen, Henry George, and Edward Bellamy figured prominently, and where the theory of natural selection underwent substantial modification.[111]

WALLACE AND THE EVOLUTION OF MAN

It is sometimes hard to follow Wallace's beliefs concerning the impact of natural selection on human development because, several times, they changed dramatically during the second half of the nineteenth century. In the 1850s, scientific naturalism was the philosophy that molded his initial articulation of the theory of natural selection, and it remained the assumption underlying an important paper that he delivered to the Anthropological Society of London in 1864. By the end of that decade, however, apparently thanks to the influence of spiritualism, he had a strikingly different explanation to offer of human evolution. An essay published in 1870 suggested that both the physical and mental traits of man bore the unmistakable evidence of design; in place of natural selection at work, he posited the activity of a higher intelligence which, from the very start, had supervised the emergence of the human organism. In the late 1880s, in *Darwinism*, Wallace enunciated yet à third position. No longer exempting man's physical development from the impact of natural selection, he argued only that human mental growth remained utterly inexplicable in naturalistic terms. This, his mature and final theory, represented a middle ground between his materialist views of the 1840s and 1850s, and the all-important role in human development that he assigned to nonmaterial forces in the late 1860s and 1870s.

According to Wallace's own account, his loss of Christian faith caused him no anguish and occurred quite simply as he subjected Christian doctrine to rational consideration. It is true that, as a boy, he had experienced some tremors of religious passion when he attended chapel with Nonconformist friends. These glimpses of Dissenting enthusiasm provided "a welcome change" from the Anglican services that he attended with his parents, usually twice on Sunday. In chapel, "the extempore prayers, the frequent singing, and the usually more vigorous and exciting style of preaching," Wallace recollected, were "far preferable to the monotony of the Church service; and it was there only that, at one period of my life, I felt something of religious fervour." There was, however, nothing to sustain that fervor, "no sufficient basis of intelligible fact or connected reasoning" to satisfy Wallace's inquiring intellect. His training in the Anglican faith provided no satisfaction. "The only regular teaching" Wallace received from his parents "was to say or hear a formal prayer before going to bed, hearing grace before and after dinner, and learning a collect every Sunday morning, the latter certainly one of the most stupid ways of inculcating religion ever conceived."[112] This shaky Christian foundation could not compete with the opposed system of thought that he first encountered as an adolescent.

Wallace was just fourteen when he was introduced to the advanced thought characteristic of working-class secularists in the first half of the nineteenth century. Early in 1837, he joined a brother who was living in London, apprenticed to a master builder. The two brothers frequently attended the Hall of Science close to Tottenham Court Road, and Alfred Wallace there received his first exposure to the principles of socialism and secularism, the ideas of Robert Owen and Tom Paine. Owen himself once gave an address at this Hall of Science, and Wallace was profoundly impressed. Although he did not embrace socialism for many years to come, he cherished a special attachment for Owen. "I have always looked upon Owen," Wallace acknowledged in his autobiography, "as my first teacher in the philosophy of human nature and my first guide through the labyrinth of social science. He influenced my character more than I then knew."[113] Perhaps what was most important in Owenism at this stage of Wallace's life was Owen's conviction that the individual human being was no worse or better than the environment in which he was reared. That belief gave a practical and powerful focus to Wallace's developing dismay over the injustices of early Victorian society. Subsequent opportunities as a land surveyor to observe a range of living conditions only increased his dismay. Furthermore, the utopian strain in Owenism, the optimistic forecast of human progress, both individual and social, may have seized Wallace's imagination at this impressionable time in his life.

While significant stimuli were working on Wallace's social conscience, other ideas were busily demolishing his Christian beliefs. Among the literature available at the Tottenham Court Road Hall of Science, Wallace encountered the old dilemma concerning the origin of evil. He pondered long over the problem: was God not benevolent? was He not omnipotent? Wallace could find no answers to resolve his perplexity; least helpful of all was his father's reassurance that the human mind could not fathom such mysteries. Wallace became convinced, as a result of these questions, "that the orthodox ideas as to His nature and powers cannot be accepted." At the same time, he read a tract by Owen's eldest son, Robert Dale Owen, which vehemently condemned the doctrine of eternal damnation, still a frequent feature of sermons and exhortations from both Anglican and Nonconformist pulpits. Having heard such ghastly warnings himself, Wallace could sympathize warmly with the unorthodox pamphlet. "I therefore thoroughly agreed with Mr. Dale Owen's conclusion," Wallace wrote many years later,

> that the orthodox religion of the day was degrading and hideous, and that the only true and wholly beneficial religion was that which inculcated the service of humanity, and whose only dogma was the brotherhood of man. Thus was laid the foundation of my religious scepticism.[114]

The foundation was strengthened when, a few months later, Wallace joined another brother in Bedfordshire, to begin learning the skills of land surveying. He spent seven years with that brother, working on a variety of surveying jobs and further imbibing the "spirit of scepticism, or free-thought." He read a number of provocative books, including an old edition of Rabelais, but nothing influenced his departure from

Christianity more profoundly than a series of lectures on Strauss's *Life of Jesus*. His brother fully accepted the argument, which the lectures developed, by which the miracles recorded in the Gospels were reduced to myths for the ignorant, and the Son of God became merely a great moral guide whose disciples had exaggerated when they composed the chronicle of his life. Wallace was likewise convinced, and the lectures "helped to complete the destruction of whatever religious beliefs still lingered" in his mind. He continued to attend church for a while after moving to Bedfordshire, but his skepticism only became more firmly entrenched, as he "found no attempt in any of the clergymen to reason on any of the fundamental questions at the root of the Christian and every other religion." By the time Wallace reached his twenty-first birthday, he was "absolutely non-religious."[115]

In charting his growth toward religious skepticism, Wallace also ascribed a causative role to science. "My growing taste for various branches of physical science," he explained, "and my increasing love of nature disinclined me more and more for either the observances or the doctrines of orthodox religion."[116] In the early 1860s, when he was known as the coauthor of the theory of evolution by natural selection, he regarded nature from a materialist and largely determinist perspective. He had painted a picture of the natural world that seemed to deny the efficacy of free will in the inexorable functioning of both environmental and hereditary influences. The very concept of will scarcely seemed to belong in Wallace's categories of thought. Down to the mid-1860s, he "was so thorough and confirmed a materialist," he confessed, "that I could not at that time find a place in my mind for the conception of spiritual existence, or for any other agencies in the universe than matter and force."[117]

Wallace's materialism found support, not only in the writings of free thinkers and the records of nature, but also in mesmerism and phrenology. He endorsed both movements for the first time in the 1840s and remained their advocate for the rest of his life. Although they later helped to ease Wallace smoothly into spiritualism, they initially complemented his materialist cast of mind. Spencer Hall introduced Wallace to mesmerism in a series of lectures that the magnetist presented in 1844 at Leicester. At the time, Wallace was employed there as a schoolteacher, and he attended Hall's demonstrations with a few of his older students. Both he and the boys were fascinated by the mesmeric phenomena and tried some experiments of their own. With certain students, Wallace could produce the usual range of effects: "partial or complete catalepsy, paralysis of the motor nerves in certain directions, or of any special sense, every kind of delusion produced by suggestion, insensibility to pain, and community of sensation" with himself.[118] He was sure that this was a genuine physical reaction, "a real action upon the muscles," not to be accounted for by "any preconceived ideas."[119] He gave credence to what he saw with his own eyes: boys whose limbs could not move even to retrieve a promised coin, who felt pain when Wallace secretly pinched himself, or who tasted whatever Wallace put in his own mouth. Here, surely, by all the canons of empirical reasoning, a physical force was working to bind mesmerist and subject together in mysterious union.

Mesmerism soon led Wallace to phrenomagnetism. He had already read some of George Combe's work[120] and was eager to test the phrenologist's claims. He purchased "a small phrenological bust" to serve as guide and set to work with a mesmerized schoolboy. Here, too, the results convinced Wallace that something other than the power of suggestion was involved. Even when he looked away, placing his finger at random on the boy's head, the boy's behavior correctly reflected the cerebral organ that had been aroused, according to the phrenological map of the brain. "I thus established," Wallace reported,

> to my own satisfaction, the fact that a real effect was produced on the actions and speech of a mesmeric patient by the operator touching various parts of the head; that the effect corresponded with the natural expression of the emotion due to the phrenological organ situated at that part . . . and that it was in no way caused by the will or suggestion of the operator.[121]

Even after Wallace left his job at Leicester and returned to land surveying, he continued to investigate phrenology. Sometime in 1846 or 1847, he attended two lectures on the subject and on each occasion paid the speaker to draw up a written delineation of his character. Both times he was astonished at the accuracy with which a total stranger could estimate his faculties. He was more than ever assured that human character was inextricably linked to "the positions of all the mental organs."[122]

During the years that Wallace spent surveying land, he found frequent opportunities to indulge his growing interest in botany. He began to collect plant specimens and to study the system of their classification. Through his own reading, supplemented by lectures and discussions at Mechanics' Institutes, he picked up what knowledge he could of zoology, and geology as well. He read Chambers's *Vestiges* and was strongly impressed. "I do not consider it a hasty generalization," he wrote to his friend Henry Walter Bates, probably in 1847, "but rather as an ingenious hypothesis strongly supported by some striking facts and analogies, but which remains to be proved by more facts and the additional light which more research may throw upon the problem."[123] It seems certain that Wallace was already convinced of species' mutability before setting forth with Bates on their voyage to Brazil in 1848.[124] For four years Wallace collected specimens along the Amazon and Rio Negro, returning to England in 1852, but only briefly. Early in 1854, he was off again, this time to the Malay Archipelago, where he stayed for eight years and where he hypothesized the theory of natural selection.

From the earliest of Wallace's writings on the subject of evolution, the sustaining principle of all his theorizing was the concept of law. Like Chambers, he was profoundly committed to the idea of natural law creating order out of the profusion of life. The essay that he published in 1855, "On the Law Which Has Regulated the Introduction of New Species," proposed a comprehensive law to elucidate all the facts of change in the organic world. Wallace stated it simply: *"Every species has come into existence coincident both in space and time with a preexisting closely allied species."* The law was designed to explain the geographical distribution of animal and vegetable life over the face of the earth, but it made no attempt to explain how species changed over time. The essay attracted little attention when it appeared in the *Annals and Magazine of Natural History*; nonetheless it already revealed the direction of Wallace's evolutionary thought. Not only was he convinced of continuous uniform and gradual changes in the earth's surface, but he was also certain that the organic changes that occurred to create new species were similarly gradual and slight. The law decreed, he explained, "that no new creature shall be formed widely differing from anything before existing; that in this, as in everything else in Nature, there shall be gradation and harmony."[125]

The story of how Wallace surprised and alarmed Darwin with his sketch of the theory of natural selection in 1858 is too well known to need repeating. By that time, Wallace had put together all the pieces of the puzzle to his own satisfaction. He had added to the foundations of his 1855 essay a recognition of the role played by the struggle for existence and by the requisite adaptation of each species to its environmental conditions. He had reasoned that useful variations would tend to increase within a given species, while useless or harmful changes would decrease in number, as their bearers lost out in the competition to survive. Furthermore, he assumed that the variety, if "more perfectly developed" and of a "more highly organized form," would eventually prevail over the parent species from which it diverged.

> It would be in all respects better adapted to secure its safety, and to prolong its individual existence and that of the race. Such a variety *could not* return to the original form; for that form is an inferior one,

and could never compete with it for existence . . . But this new, improved, and populous race might itself, in course of time, give rise to new varieties, exhibiting several diverging modifications of form, any of which, tending to increase the facilities for preserving existence, must, by the same general law, in their turn become predominant. Here, then, we have *progression and continued divergence* deduced from the general laws which regulate the existence of animals in a state of nature, and from the undisputed fact that varieties do frequently occur.

He could not explain how those varieties occurred, but he was confident that will and effort on the part of the species had nothing to do with the process. He rejected Lamarck's hypothesis that "progressive changes in species have been produced by the attempts of animals to increase the development of their own organs, and thus modify their structure and habits." Using the obvious example of the giraffe, Wallace insisted:

> Neither did the giraffe acquire its long neck by desiring to reach the foliage of the more lofty shrubs, and constantly stretching its neck for the purpose, but because any varieties which occurred among its antitypes with a longer neck than usual *at once secured a fresh range of pasture over the same ground as their shorter-necked companions, and on the first scarcity of food were thereby enabled to outlive them.*[126]

These arguments about species and varieties, and about the evolutionary function of a limited food supply, formed the substance of the fateful communication that Wallace sent to Darwin in 1858. They likewise formed the starting point of the paper that Wallace delivered to the Anthropological Society of London in 1864 – "The Origin of Human Races and the Antiquity of Man Deduced from the Theory of 'Natural Selection.'" This paper must be understood against the background of a bitter dispute between the Anthropological Society, founded in 1863, and its parent body, the Ethnological Society of London. The latter organization included a group of members who, like Huxley, relished the Darwinian label and were willing to see the theory of evolution extended to human development. The Anthropological Society, by contrast, under the presidency of its founder, Dr. James Hunt, vigorously rejected Darwinian theory, particularly its relevance to man. Hunt held that humanity had multiple origins and was organized, from the start, into separate and unchanging races. The Anthropological Society would appear a strange forum indeed for Wallace, although he never did actually join the ranks of its fellows.[127] Perhaps he was stimulated by the prospect of trying to convert so hostile an audience.

Wallace was clearly arguing several points when he addressed the Society in 1864. In the first place, he challenged the static view of the natural world shared by some of its members and opened his paper with a concise summary of the gradual workings of natural selection. Secondly, he sought to argue the case for human monogenesis against the "polygenists" of the Anthropological Society. He regretted that the science of ethnology should be split into two contradictory schools of thought concerning man's origins, with the monogenists arguing for the initial homogeneity of the human race and "the other party maintaining with equal confidence that man is a genus of *many species*, each of which is practically unchangeable, and has ever been as distinct, or even more distinct, than we now behold them." Wallace well understood the significant implications of both positions. From the polygenist viewpoint, the white European race had no bonds of original kinship with the presumed "lesser" races of the world, nor could the multiple species of humanity be traced back to any single animal source. The monogenist viewpoint argued, in opposition, that evolutionary change had produced the present diversity of the human race from an original common form, a form that, in its turn, had very gradually diverged from a bestial forebear. Wallace recognized that millions of years were needed "to bridge over the difference between the crania of the lower animals and of man," but he believed that such a bridge would someday be traced. In less time, but by an equally gradual process of change, Wallace explained that natural selection could effect the physical differences that characterized the diverse races of humanity.[128]

One of the main difficulties that Wallace encountered in trying to prove the original unity of mankind was the strong evidence that the various human races have scarcely changed within the time span of recorded history. "The Egyptian sculptures and paintings show us that, for at least 4,000 or 5,000 years," Wallace conceded, "the strongly contrasted features of the Negro and the Semitic races have remained altogether unchanged."[129] It became his task, therefore, to show how the process of gradual physical change, according to the laws of natural selection, might be arrested in man's case, leaving the different races virtually unaltered over thousands of years. This line of reasoning in Wallace's 1864 paper marked the first time he exempted any aspect of animal or vegetable life from the inexorable workings of natural selection.

His thesis was simple and clear. Mankind, unlike most other animal species, was a "social and sympathetic" creature. Among the overwhelming majority of animals, each member of a species had to struggle alone throughout life; once grown, no other member would care for it when feeble and ill; it had only itself to depend upon and was totally

vulnerable to the process of selection that nature constantly made between weak and strong. Within the human race, however, social organization and the division of labor, even in the most primitive tribes,
soon obviated the need for each member to fight for food. The stronger
members of the tribe protected the weaker, and natural selection ceased
to affect their physical characteristics. At a very early stage in human
development, therefore, physical traits became less important, and in
their place "mental and moral qualities" exercised "increasing influence on the well-being of the race." The ability to place public over
personal good, the capacity to sympathize, to share, to practice self-
restraint, to plan for the future – all these qualities became critical in
the struggle of tribe against tribe. "For it is evident that such qualities
would be for the well-being of man; would guard him against external
enemies, against internal dissensions, and against the effects of incle-
ment seasons and impending famine, more surely than could any
merely physical modification." The tribes possessing such qualities
would, in time, outnumber those which lacked them; the latter unfortunate peoples "would decrease and finally succumb."[130]

In short, man's mental faculties took over the burden of adaptation
to a shifting environment before human history was very old. Physical
changes among the various human tribes were no longer necessary;
the ongoing alterations that had increasingly distinguished a people
living in tropical Africa, say, from another living in the Sahara Desert
came to an end. In these terms, Wallace could explain the seeming
constancy of racial types without denying the possibility of earlier development away from a common stem. Such divergence must, indeed,
have occurred during the very infancy of the human race, "when man
was gregarious, but scarcely social, with a mind perceptive but not
reflective, ere any sense of *right* or feelings of *sympathy* had been
developed in him." At such a time, Wallace maintained, man "had not
mind enough to preserve his body from change, and would, therefore,
have been subject to the same comparatively rapid modifications of
form as the other mammals."[131] With the development of intellect,
however, with the production of clothing and weapons, with the ability
to grow food for a relatively constant supply of nutrition, and with the
expanding forms of social cooperation as well, the several varieties of
humanity learned to escape the influence of natural selection and thus
managed to retain their respective appearances unchanged.

Wallace's defense of monogenesis in "The Origin of Human Races"
unfolded alongside his beliefs concerning mental evolution. In that
essay, he argued that the superior "mental and moral qualities" of the
human race actually interfered with natural selection as that process
affected man's physical appearance. Yet, in 1864, Wallace still believed

that the human mind itself, and its organ the brain, were subject to evolutionary change. The initial appearance of a sophisticated organ for thinking and reasoning was, Wallace maintained, the product of natural selection, and thereafter "every slight variation in his mental and moral nature" that helped man to overcome hostile circumstances was "preserved and accumulated." The human body evolved, thanks to natural selection, up to the point where the human mind took over and deflected the inexorable struggle for physical survival. But Wallace was sure that the human mind continued to evolve. Natural selection, he reiterated, "acting on his mental organisation, must ever lead to the more perfect adaptation of man's higher faculties to the conditions of surrounding nature, and to the exigencies of the social state." To this extent, then, the human mind manifested the sway of that "one great law of physical change" that Wallace saw all around him.[132]

Wallace's essay of 1864 offered a biologically untenable hypothesis that the human mind might alter while the human body stood still, in evolutionary terms. But it was for Wallace an emotionally satisfying hypothesis, for it allowed him to argue for human progress. Despite the fact that adaptation to a changing environment did not inevitably entail qualitative improvement, Wallace's application of natural selection to man explicitly involved an ongoing amelioration of the human condition. Closing his paper to the Anthropological Society on a note of visionary optimism, Wallace forecast the brilliant future awaiting man:

> While his external form will probably ever remain unchanged, . . . his mental constitution may continue to advance and improve till the world is again inhabited by a single homogeneous race, no individual of which will be inferior to the noblest specimens of existing humanity. Each one will then work out his own happiness in relation to that of his fellows; perfect freedom of action will be maintained, since the well balanced moral faculties will never permit any one to transgress on the equal freedom of others; . . . the passions and animal propensities will be restrained within those limits which most conduce to happiness; and mankind will have at length discovered that it was only required of them to develope the capacities of their higher nature, in order to convert this earth, which had so long been the theatre of their unbridled passions, and the scene of unimaginable misery, into as bright a paradise as ever haunted the dreams of seer or poet.[133]

There was very little of the naturalist speaking in that closing paragraph, and a good deal of the phrenologist. Loudest and clearest of all, however, was the voice of the social utopian sharing Owen's vision.

The year after Wallace presented his paper to the Anthropological Society, he attended his first séance in July 1865. He had been home from the Far East since 1862; had he been anxious to meet the spirits before then, he surely could have found an occasion during those three years. The evidence, in fact, points in the other direction – that Wallace had no pressing interest in spiritualist phenomena at the time. While in the Malay Archipelago, he had read in newspapers of the "strange doings of the spiritualists in America and England," but they struck him as "wild and outré," the mere "ravings of madmen." Some phenomena, he admitted, seemed too well authenticated to be readily dismissed. Yet he persisted in concluding "that such things *must* be either imposture or delusion."[134] When he began to attend séances, solely in private circles among friends and family, however, he could not account for the tapping sounds and the vibratory motions of the table in those terms.[135] Soon he sought out professional mediums, and at a séance with Mrs. Mary Marshall, information about one of his dead brothers was communicated to an impressed Wallace. "I could have told how I had seen Wallace and his friends humbugged by Mrs. Marshall," W. B. Carpenter wrote to Barrett in 1876,[136] but Wallace thought that he had taken every precaution against humbugging. He carefully examined furniture in the séance room, for example, and marked pieces of paper before spirit writing tests to preclude stealthy substitution when the lights went out. Despite his precautions, he found no evidence of trickery and by 1866 was moving closer to accepting the reality of spirit agents.

Wallace at first supposed that rappings and table vibrations might be attributed to some nameless force emanating from the sitters. It was impossible, however, to extend that argument to the more advanced phenomena that he witnessed with Miss Nichol, the medium who later became famous as Mrs. Samuel Guppy, and later still as Mrs. Guppy Volckman. Miss Nichol was a family friend, living with Wallace's sister, Fanny Sims, when her mediumistic powers were discovered in November 1866. Thereafter Wallace and Mrs. Sims "had constant sittings" with her, and her talents developed rapidly. A small table rose completely off the floor in her presence, although Wallace had devised means to guarantee that no embodied human foot could raise it. Even more startling was the levitation of a table with the very stout medium sitting upon it. Most striking of all the phenomena said to occur through her mediumship "was the production of flowers and fruits in closed rooms." Suddenly bare tables would be strewn with flowers and ferns whose freshness was underscored by the drops of dew on petal and leaf. Their fragrance filled the séance room to such an extent that, Wallace argued, Miss Nichol could not possibly have concealed the flowers on her person for the preceding hours of the séance.[137] These were the facts which Wallace could not explain away, facts which, he pointed out in his autobiography,

> did not fit into my then existing fabric of thought. All my preconceptions, all my knowledge, all my belief in the supremacy of science and of natural law were against the possibility of such phenomena. And even when, one by one, the facts were forced upon me without possibility of escape from them, still . . . "spirit was the last thing I could give in to." Every other possible solution was tried and rejected. Unknown laws of nature were found to be of no avail when there was always an unknown intelligence behind the phenomena – an intelligence that showed a human character and individuality, . . . Thus, little by little, a place was made in my fabric of thought, first for all such well-attested facts, and then, but more slowly, for the spiritualistic interpretation of them.[138]

By the end of 1866, therefore, it seems that Wallace had decided that no theory of physical force could elucidate what he had witnessed. Nor was some unidentified, impersonal psychic force – whatever that might be – adequate to explain the information that mediums transmitted, the passage of matter through matter, or the defiance of gravity. Wallace knew that he had crossed the borderland between respectable scientific endeavor and the ridicule of his professional colleagues; he knew that he had left the canons of contemporary physics behind. He could not, however, repudiate his own observations. All his adult life, his own perceptions, and the inferences he drew from them, had molded his work as a naturalist. They had raised him to a position of respect within the British scientific community, and he had no reason to question what his senses told him transpired at séances. His spiritualist experiences had begun with the simplest phenomena – "rapping and tapping sounds and slight movements of a table" – in a friend's home, where to suspect trickery was unthinkable. Wallace tested the phenomena as carefully as he knew how, both at his friend's house and at his own, and "similar phenomena were obtained scores of times."[139] He believed he had definitively ruled out the possibility that any physical action by any person present caused the sounds and movements.

Once having accepted the veracity of the simple phenomena, Wallace had a foundation on which to base his acceptance of the more complex. He was confident that his observations had been correct at the earliest séances; why should they not be reliable at the later ones? When Wallace attended his first séance with Home in 1871, when he witnessed full-form materializations in the mid-1870s, and saw a spirit form grow out of Monck's body, he was equally assured "that no form of legerdemain [would] explain what occurred."[140] As he grew older, his participation at séances declined sharply. "I know nothing of London mediums now," he wrote to a correspondent in 1899.[141] Whether he knew them or not, he continued to defend spiritualism with his customary vigor. He had seen enough back in the 1860s to convince himself that, while there might be impostors abroad who performed poor imitations of the real thing, spiritualist phenomena were as real as the orangutans of Borneo.

He was further prepared to accept the reality of spirit manifestations, despite their challenge to contemporary science, because his experiences with mesmerism had furnished a lesson he never forgot. "The importance of these experiments," he recalled about his mesmeric investigations in Leicester,

> was that they convinced me, once for all, that the antecedently incredible may nevertheless be true; and, further, that the accusations of imposture by scientific men should have no weight whatever against the detailed observations and statements of other men, presumably as sane and sensible as their opponents, who had witnessed and tested the phenomena, as I had done myself in the case of some of them.[142]

Wallace never tired of reminding his fellow scientists that their profession had been wrong to condemn mesmerism out of hand, and he looked forward to a similar vindication of spiritualism. His acceptance of mesmeric phenomena, some of which closely resembled the behavior of entranced mediums, made it easier to endorse the phenomena of spiritualism. His knowledge of mesmerism made him curious to investigate spiritualism; his success with the former may have given him confidence to tackle the latter. Like so many investigators before him, Wallace found that mesmerism, and phrenology too, were peculiar systems of thought, marvelously pliant and capable of suiting altered philosophies of man and cosmos.

In the second half of his life, under the influence of spiritualism, Wallace utterly rejected the design of the universe as sketched by the architects of scientific naturalism. He envisioned instead a world in which humanity could achieve its fullest moral and spiritual potential because a superior mind designed all of nature to further human development at every stage of growth, throughout life and afterlife.[143] Wallace no longer saw natural selection as the agent of human progress, as he had in 1864, and the great change in his interpretation of evolution occurred in the years immediately following his paper to the Anthropological Society. By 1869, in an article that he contributed to the *Quarterly Review*, Wallace was ready to suggest that natural selection could neither explain the emergence of human intelligence nor account for man's distinguishing moral qualities. These, he indicated, might better be explicated by the directive action of a guiding mind.[144] Wallace did not elaborate upon this startling theme in 1869, but he went far enough to cause Darwin deep distress. Darwin could scarcely believe that Wallace himself had composed the article. "I *defy* you to upset your own doctrine," he wrote to Wallace in March 1870. For Wallace, however, it was not a question of defiance. He maintained that he was simply following the course of action required of any scientist in the face of surprising new evidence. "I can quite comprehend your feelings with regard to my 'unscientific' opinions as to Man," he told Darwin in April 1869,

> because a few years back I should myself have looked at them as equally wild and uncalled for . . . My opinions on the subject have been modified solely by the consideration of a series of remarkable phenomena, physical and mental, which I have now had every opportunity of fully testing, and which demonstrate the existence of forces and influences not yet recognised by science.[145]

The truths of spiritualism, Wallace believed, mandated a dramatic alteration in the theory of natural selection as he had first articulated it. That alteration was fully evident when he rewrote the ending of the Anthropological Society paper, for inclusion among the essays published in 1870 as *Contributions to the Theory of Natural Selection*. Instead of the Owenite utopian vision, Wallace now looked round and saw mediocrity flourishing, "for it is indisputably the mediocre, if not the low, both as regards morality and intelligence, who succeed best in life and multiply fastest." How, then, could Wallace continue to anticipate progress in humanity's moral and intellectual growth? Since he could no longer look to "survival of the fittest" for the inexorable improvement of the human species, he was compelled to conclude that such advance arose from

> the inherent progressive power of those glorious qualities which raise us so immeasurably above our fellow animals, and at the same time afford us the surest proof that there are other and higher existences than ourselves, from whom these qualities may have been derived, and towards whom we may be ever tending.[146]

In 1864, Owen's influence, mingled with phrenology, had prompted Wallace's optimism about the impact of natural selection on man's future. In 1870, spiritualism allowed him to repose unbounded hope in the nonmaterial aspect of existence. Wallace's expertise as a naturalist counted for little in both cases. An astute observer of nature, he knew that natural selection did not automatically imply improvement. He had stretched the facts a little in 1864 to satisfy a yearning for human justice and happiness. In 1870, he no longer had to stretch the facts, for he believed he had found new ones to justify his conviction that a spiritual agent worked to further man's intellectual and moral growth. Wallace no longer needed natural selection for that all-important purpose.[147]

From 1869, Wallace accordingly sought to elucidate the emergence of the human mind, not in terms of natural selection, but in terms of the universal activity of spirit. In "The Limits of Natural Selection as Applied to Man," which first appeared among his collected essays published in 1870, Wallace offered the earliest sustained explanation of the "unknown higher law" working to develop humanity to its highest capacity. He began by discussing several physical traits that suggested the inadequacy of natural selection in human evolution. If Darwin were correct in arguing that natural selection had "no power to advance any being much beyond his fellow beings, but only just so much beyond them as to enable it to survive them in the struggle for existence," how could one explain the fact that the brain of primitive man was so much larger than that of anthropoid apes? And how, by contrast, could one explain the close similarity in size between the brain of the savage and that of the cultured European in the nineteenth century? Wallace could not avoid the conclusion that the savage possessed a brain "quite disproportionate to his actual requirements." Man's comparative hairlessness, furthermore, pointed to an agency in human development distinct from natural selection; thick hair across the back, Wallace reasoned, would have been useful to early man, who had to fashion alternative forms of cover in its absence. Similarly the "specialization and perfection of the hands and feet," and the human voice as well, defied explanation in terms of natural selection. Wallace did not see how the earliest man benefited from his "purely erect locomotion," nor made use of the infinite "latent capacities and powers" of his hands, nor needed for survival a larynx that produced a range of sounds, including speech and song. In all these cases, Wallace clearly conveyed the implication that some higher intelligence had provided savages with physical traits to serve the future requirements of civilized man.[148]

Even more central to Wallace's argument, however, were his observations about man's mental faculties. It was not possible to explain certain intellectual capacities in terms of their utility in the natural competition to survive. The ability "to form ideal conceptions of space and time, of eternity and infinity," the gift of artistic sensibilities, or the capacity to reason mathematically – these surely were not the result of chance variation in nature and survival of the fittest. Nor were the evolutionary laws that aptly applied to the rest of the animal world adequate to explain the profound moral sense evident even in savages. At this point in the argument, one can again distinguish the voice of the phrenologist, as Wallace cited the cerebral faculties unique to the human race and demanded to know how any could have emerged through natural selection. Before Wallace's introduction to spiritualism, phrenology constituted no threat to his confidence in the efficacy of natural selection. Yet after he had accepted the reality of spirit throughout the universe, phrenology offered powerful corroboration that the human mind, with all its rich complexities, could not have evolved in a purely naturalistic way, and by accident, as it were, from the matter of the animal brain. On the contrary, the only inference which Wallace could draw from the evidence he produced was "that a superior intelligence has guided the development of man in a definite direction, and for a special purpose."[149]

In the last section of the 1870 essay, Wallace attempted to explain the evolution of consciousness, a task which, he observed, had hitherto baffled philosopher and physiologist alike. The problem was, of course, to elucidate the means whereby sensation, perception, and intelligence could arise from the physical organization of the brain. Specifically, Wallace's self-assigned job was to show how Huxley's assertion, that thoughts are occasioned by molecular changes in the brain's protoplasm, was utterly "inconsistent with accurate conceptions of molecular physics." Wallace was not a physicist; his attempt to redefine the basis of matter had little to recommend it to the attention of contemporary science.[150] His argument rested on the conclusion that "matter is essentially force," and he distinguished between two kinds of force: the primary forces of nature – gravitation, repulsion, cohesion, and so on – and man's own "will-force." Following this division further, he suggested that some exertion of will-force was necessary to set the other forces in motion and that, fundamentally, "all force may be will-force." If all force were will-force, he continued, "the whole universe is not merely dependent on, but actually *is*, the WILL of higher intelligences or of one Supreme Intelligence." Since single molecules lacked consciousness, no combination of them, no matter how complex, "could in any way tend to produce a self-conscious existence." That consciousness had to come from without. "We have no difficulty," he stated, "in believing that for so noble a purpose as the progressive development of higher and higher intelligences, those primal and general will-forces, which have sufficed for the production of the lower animals, should have been guided into new channels and made to converge in definite directions."

Excepting the human species, Wallace continued to maintain the general applicability of natural selection for the plant and animal kingdoms. He refused to acknowledge that it invalidated the entire theory to argue that "the laws of organic development have been occasionally used for a special end," such as the introduction of consciousness into the physical organization of. man.[151] Yet the very notion that an overarching mind had created the universe in the best interests of humanity was, clearly, incompatible with any theory of evolution based on chance. It was not possible to assert foresight and plan where human needs were concerned, and random variation in the rest of the natural world. Theologians might manage to synthesize the claims of design and chance in order to preserve religious faith in modern times, but evasive compromises were not expected of scientists in the late nineteenth century.

Nonetheless, like Romanes defending physiological selection, Wallace insisted that his new considerations on human development supplemented, but by no means undermined, natural selection. Fellow naturalists thought differently, and Wallace's deviation from the theory in 1869–70 provoked outraged protests.[152] Wallace responded to some of this criticism in subsequent years by abandoning the claim that certain physical features of man could not be explained by natural selection. In *Darwinism*, a lengthy and extremely lucid exposition, published in 1889, he limited to the brain the area of human development outside the scope of natural selection, exempting only man's mental, moral, and intellectual capacities from its impact. Between 1870 and 1889, arguments concerning the latent abilities of physical organs – evolved for a particular function, yet capable of performing others as well – had effectively challenged Wallace's views that some human organs, too advanced for savage needs, must have been designed by a superior mind for civilized human society. Wallace reverted instead to his earlier insistence that evolution by natural selection could explain all changes in the physical structure of every species. He remained absolutely convinced, however, that the outside intervention of a higher intelligence was required to produce man as a thinking, reasoning, and ethical creature. Even if Wallace acknowledged consciousness and sensation to be characteristic of the animal kingdom in general, he resolutely asserted the unique attributes of the human mind. "The real problem," he wrote to Lodge in 1898,

> is to account – first, for the infinitely complex constitution of the material world and its forces which rendered living organisms possible; then, the introduction of consciousness or sensation, which alone rendered the animal world possible, – lastly, the presence in man of capacities and moral ideas & aspirations which could not conceivably be produced by variation & natural selection.[153]

Wallace was never fundamentally influenced by criticism. Lacking any professional post or academic attachment that gave him a position to maintain in scientific circles, he was not unduly concerned about his reputation. He followed certain lines of investigation and developed particular hypotheses because they seemed to reveal important truths about the human condition and the natural world, not because they promised to gain him professional advancement. He believed that spiritualist phenomena were essential to any satisfactory understanding of the mind–body relationship,[154] and he wrote about them accordingly. The views that he set forth in *Darwinism*, and reiterated in his letter to Lodge in 1898, illuminate how far he had traveled from the scientific naturalism of the 1850s and early 1860s. Wallace had grown certain that designing mind had deliberately interrupted the ordinary course of natural evolutionary law to endow humanity with the intellectual and moral qualities needed for ongoing spiritual development. Such development was nothing less than the goal toward which the whole world of nature was striving. He had begun his public career as a naturalist by propounding a persuasive theory of material evolution. He continued it with a far less persuasive theory of spiritual evolution. Although the latter embarrassed many friends and supporters, Wallace always upheld the inseparability of the two theories. He came to realize, his biographer explained,

> that, indeed, there were two lines of development – one affecting the visible world of form and colour and the other the invisible world of life and spirit – two worlds springing from two opposite poles of being and developing *pari passu*, or, rather, the spiritual dominating the material, life originating and controlling organisation. It was, in short, his peculiar task to reveal something of the Why as well as the How of the evolutionary process, and in doing so verily to bring immortality to light.[155]

It is difficult to ascertain exactly how much Wallace's work as a student of the spiritual universe transformed his investigations of the physical world. Many of Wallace's scientific labors did not even touch on spiritualism – his distinguished contributions to zoogeography, for example, in which he theorized about the distribution of flora and fauna in geographic locations around the globe. Equally, it is important to remember that, despite what they thought of his spiritualism, the scientific and academic communities of Victorian and Edwardian England recognized Wallace's achievements and rewarded them accordingly. Among much else, he received numerous medals from scientific societies, an honorary degree from Oxford, Fellowship in the Royal Society, and from the Crown, the Order of Merit.[156] He was president of the Biology Section of the British Association for the Advancement of Science in 1876. He wrote as prolifically for the scientific press as for the spiritualist, but did not explicitly mention spiritualism in his scientific writing until the publication of *Darwinism*. It can scarcely be doubted, however, that Wallace's whole interpretation of the natural world was radically altered by his acceptance of spiritualism.

Wallace's approach to science became imbued with spiritualist assumptions because he rejected the validity of any dividing line between science and spiritualism. His was the standard spiritualist claim that beliefs concerning the existence of the spirit world could be verified empirically and constituted a body of scientific knowledge as demonstrable as the laws of physics. Wallace never abandoned his faith in the prevalence of law throughout the universe, and he maintained that the truths of spiritualism, far from violating natural law, provided a more accurate definition and a fuller conception of it. Not only was "the Great Mind of the universe," when it intervened to accelerate the growth of human intellect, "acting through natural and universal laws," but the "conventional 'white-sheeted ghost'" of horror tales similarly "had a foundation in fact, . . . dependent on the laws of a yet unknown chemistry."[157] The "great law of 'continuity'" that pervaded the whole universe, furthermore, mandated intermediate levels of intelligence between man and the supreme mind, so that Wallace could posit the existence of numerous grades of "preterhuman" discarnate beings acting as spiritual guides to assist men and women still in bodily form. Like numerous other spiritualists, Wallace was trying to eliminate the aura of the supernatural that clung to spiritualist phenomena. "The whole history of the progress of human knowledge shows us," he pointed out, "that the disputed prodigy of one age becomes the accepted natural phenomenon of the next, and that many apparent miracles have been due to laws of nature subsequently discovered."[158] The laws of nature that Wallace wanted recognized were those that governed nothing less than the progressive destiny of the human race.

So certain was Wallace that the phenomena of spiritualism rightfully belonged within the realm of science that he exerted considerable effort to persuade fellow scientists to join him in examining them. To Huxley he extended an invitation to study "a *new branch* of Anthropology," and Carpenter and Tyndall were also pressed to probe the phenomena with an open mind.[159] Wallace's failure to interest any of these men in sustained spiritualist inquiry simply confirmed him in his views on scientific bigotry; he was by no means convinced that his pursuits were in any way inappropriate or unworthy of a scientist's attention. For a correspondent in 1899, he set forth a course of spiritualist study, explaining: "The only way to gain any real knowledge of spiritualistic phenomena is to follow the course pursued in all science – study the elements before going to the higher branches."[160] Always he stressed that spiritualism appealed to his reason, distinguishing it in this respect from the merely fanciful theories of occult religions. "I have tried several Reincarnation and Theosophical books," he told a friend in 1897, "but *cannot* read them or take any interest in them. They are so purely imaginative, and do not seem to me rational."[161]

Wallace insisted that spiritualism was "a science of human nature" because it was based on facts and experiments, took no belief on trust, and emphasized the importance of investigation. Such an experimental science, he continued,

> affords the only sure foundation for a true philosophy and a pure religion. It abolishes the terms "supernatural" and "miracle" by an extension of the sphere of law and the realm of nature; . . . It, and it alone, is able to harmonise conflicting creeds; it must ultimately lead to concord among mankind in the matter of religion, which has for so many ages been the source of unceasing discord and incalculable evil; – and it will be able to do this because it appeals to evidence instead of faith, and substitutes facts for opinions.[162]

Inevitably, Wallace's science of human nature became his religion. Like Myers, whose theories of the subliminal self he condemned, Wallace was increasingly drawn to the idea of founding a "pure religion," especially as he grew older. It was not that he became uncomfortable without religion, in the doctrinal meaning of that word. He no more needed dogmatic faith at the end of his life than he did in the 1840s. But he had always, in a sense, worshipped the mysterious workings of the natural world and marveled at the infinite forms of life that abounded on earth. The idea of a natural, rational religion that could reveal the order and pattern under these myriad varieties appealed powerfully to Wallace, particularly a natural religion based on sentiments of universal brotherhood and social justice, and verified by his newly discovered laws of the spirit realm.

Wallace's attraction to a rational religion, like his concern for social justice, underscored his deep sympathies with the non-Christian branch of British spiritualism. He came from the same environment as many working-class spiritualists, that intense atmosphere that Owenite secularists and self-taught naturalists imbibed in halls of science and Mechanics' Institutes. It is understandable that he embraced their brand of the faith, rather than the more genteel variety professed in London drawing rooms. With characteristic decency, Wallace never indulged in the vitriolic abuse of Christianity that delighted some of his progressive spiritualist comrades, but it was for him just one among many world religions, with no claim to special status.[163] He considered its moral teachings vastly inferior to those of spiritualism, particularly its doctrine of eternal punishment or salvation, which he had abhorred since adolescence. By contrast, he praised the progressive spiritualist doctrines concerning life after death as a "pure system of morality," which placed the responsibility for future joy or sorrow firmly on the individual.[164] This vision of the afterlife helped him to solve the problem that had troubled him in his youth: why or whether God created evil. Misery, suffering, and cruelty, Wallace came to reason, were the means by which the designing intelligence forged in man the moral qualities requisite for his complete spiritual development.[165] Instead of arbitrary judgment, the non-Christian spiritualist account of eternity offered regularity and lawfulness, with the promise of infinite progress after death. Best of all, Wallace believed that ample empirical proof supported every aspect of the account, and for all these reasons he found the vision irresistible.

Although individualism and the concept of just deserts remained the hallmarks of Wallace's thought concerning life after death, his commitment to social reform on earth took an increasingly socialist turn as his life advanced. In his social radicalism, too, Wallace remained close to the values of plebeian spiritualism. He had been deflected from his early Owenite sympathies by reading Spencer whose "individualist views" Wallace adopted for several decades.[166] But Spencer's influence on Wallace is not easy to follow in a single line of reasoning about man and his social environment; indeed Wallace applied Spencer's thought in ways that scarcely reflected individualist inspiration. For example, he traced his passionate involvement in the land nationalization campaign back to a chapter on the use of the earth in Spencer's *Social Statics*. That campaign was part of a much broader criticism that Wallace leveled against the institutions of capitalist society. The tyranny of the capitalist system had distressed him all his life, but in the 1880s and 1890s he waged a veritable "frontal attack on property" as he converted to full-scale socialist beliefs.[167]

Late in the 1880s, when Wallace read Bellamy's *Looking Backward*, the principles of individualist economics were finally eliminated from his thinking. "Every sneer, every objection, every argument I had ever read against socialism was here met and shown to be absolutely trivial or altogether baseless," he recalled about the impression that Bellamy's book made upon him, "while the inevitable results of such a social state in giving to every human being the necessaries, the comforts, the harmless luxuries, and the highest refinements and social enjoyments of life were made equally clear."[168] Wallace began to probe further into the literature of socialism, finding the theories expounded both intellectually compelling and emotionally satisfying. He was drawn to the idea of a "Co-operative Commonwealth," which he considered "the shortest and most accurate definition of what socialism really is."

His definition of socialism revealed that, for him, economic theory took a subordinate position to anger at the exploitation of human labor and pity for the degradation of human life.[169]

It comforted him both as a spiritualist and a social reformer, therefore, to view the natural world as the work of an infinite intelligence whose one determining goal was man's full evolution, both in his physical life and in his eternal spirit existence. Yet this belief, as we have seen, destroyed the theory of natural selection that Wallace first sketched in 1858. How could he widely publicize his conviction that the universe represented the plan of a supreme mind and still defend natural selection at the same time? Far more than Darwin, Wallace fought for natural selection as the sole necessary means of evolution in nature, except for the special case of man. Where Darwin came to acknowledge the importance of sexual selection in the evolutionary process and always conceded that the inheritance of acquired traits might play some part as well, Wallace would have no rivals to the supremacy of natural selection in the nonhuman world. It seemed a bizarre inconsistency, entirely out of touch with late nineteenth-century biology, to uphold one law of development for all the rest of nature and a different law for humanity. Nevertheless Wallace's work as a naturalist and his inquiries as a spiritualist persuaded him, by the late 1860s, to believe in two very distinct evolutionary scenarios.

With Wallace, it is dangerously easy to find oneself arguing in circles. He embraced spiritualism, one can say, because he believed what his senses told him and thought he had empirical proof of its validity. But why did he so readily and uncritically accept evidence which others warned him was highly suspicious? Because, one ventures, he wanted to believe in spiritualism. The line of argument is frustrating; yet it is not altogether useless. Wallace did want to believe in spiritualism once he was persuaded of its "facts," but not because he sought any traditional religious comfort. He ran to the spiritualist camp because it fitted the fabric of his thought far better than he initially realized. Materialist though he may have been, he had a lofty, indeed an idealized, vision of man, derived in large part from phrenology. He wished to see humanity liberated from economic want and social antagonism. He longed to know that man was realizing his quintessentially human nature, with intellectual and moral traits far above the rest of the animal kingdom. Spiritualism, with its message of man's spirit essence and its doctrine of progressive spirit evolution after death, promised to fulfill Wallace's dream for mankind. At the same time, he profoundly sympathized with the radical social convictions of many non-Christian spiritualists who tried to challenge the established institutions of Victorian society.

It remains unclear whether spiritualism alone can explain Wallace's rejection of natural selection as the sole agent of evolutionary change where the human race was concerned. Because he did not specifically combine his spiritualist convictions with biological arguments until 1889, he seemed to base his revised evolutionary views on considerations of utility, and on such issues as length of time in which man's particular organs could, or could not, develop. Yet whether these physiological considerations really influenced him is doubtful. Most significant is the fact that Wallace never raised the same questions of time and utility with regard to animals. At the very time when the study of animal intelligence was breaking new ground for psychological inquiries, Wallace persisted in treating man as utterly unique in all creation.[170] Perhaps spiritualism prompted in Wallace the inclination to consider human development as distinct from the animal kingdom, in order to explain what he accepted as disembodied intelligence at work among the living. Thereafter, analysis of certain aspects of the hypothesis that he and Darwin had articulated confirmed Wallace in his belief that natural selection could not adequately explain human development. It does appear that Wallace's doubts about natural selection first arose from evidence acquired at the séance table, not from biological or geological discoveries that forced him to reconsider his initial theory of evolution in respect of humanity.

There was, however, no ambiguity in Wallace's consistent claim that a strong foundation of evidence supported his spiritualist beliefs. It was not yet impossible in the late nineteenth century to call oneself an empiricist and still maintain the independent role of mind, or spirit, in the physical world. Indeed, all the evolutionary thinkers and writers who investigated spiritualism or psychical research in that period would have claimed the empirical label. They would not all have endorsed Wallace's final interpretation of evolution, but they nonetheless shared a common hesitation about the placement of the human brain on the evolutionary scale that they were sedulously constructing. Even C. Lloyd Morgan, a leader in the fields of animal and comparative psychology, found natural selection insufficient to explain human development. Although in the 1890s Morgan looked to neurophysiology for fundamental information about the workings of the mind, he could not accept the sole adequacy of naturalistic agencies in the case of man's mental growth. "Is mental development in all its phases entirely, or even mainly, dependent on natural selection?" he rhetorically asked in his text, *An Introduction to Comparative Psychology*. The reply came with "an emphatic *no* . . . I see no evidence to show that commanding intellect, mathematical or scientific ability, artistic genius or lofty moral ideas, are attributable solely to natural selection."[171] Per-

haps the recurring appeal of Lamarck's evolutionary theories throughout the nineteenth century lay particularly in their incompatibility with
strict scientific naturalism, for they emphasized the role of the organism's own effort, will or intelligence – something presumably nonmaterial – in the process of adapting its physical body to a changing
environment.

8

Physics and psychic phenomena

AN END TO ABSOLUTES

In 1908, the physicist Edmund Edward Fournier D'Albe explained, in
the preface to *New Light on Immortality*, why a man of his profession
should publish a book designed to illuminate "the 'question of questions,' the possibility of human immortality." "Lest it appear pre-
sumptuous," the author apologized,

> for a physicist to venture an opinion on such a subject, which is usu-
> ally associated with psychology and theology rather than experimen-
> tal science, I may plead that the relations between mind and matter
> require for their elucidation an extensive acquaintance with what is
> actually known about matter and what is *not* known about it. And
> every one, I think, will acknowledge that the relations between mind
> and matter are at the very root of all possible theories concerning
> immortality. Now the physicist is permanently confronted with prob-
> lems concerning the ultimate nature of matter.[1]

During the course of the nineteenth century, scientists had added
vast amounts of information to what was actually known about matter.
Energy, heat, magnetism, electricity, and light were all subjected to
the probing scrutiny of physicists and reduced to regularity within laws
and models that celebrated the order of the physical world. For most
of the century, the models were based on the mechanical analogy that
gave so attractive an air of predictability to the findings of modern
physics. To many workers in that field, it seemed by the 1850s and
1860s that the major theoretical formulas, the methodological principles, and the fundamental concepts had all been set forth. The remaining labors would, of course, involve elaboration and amplification,
but always along the lines already established. The mid-Victorian physicists did, indeed, feel that "they had left their successors with little
more to do than to clean up a few minor problems." They confidently
placed their faith in the much vaunted criteria of positivistic science
"operating on objective and ascertained facts, connected by rigid links
of cause and effect, and producing uniform, invariant general 'laws'
beyond query or wilful modification."[2]

During the last quarter of the nineteenth century and the opening years of the twentieth, however, certainty and confidence yielded to probability and doubt. In that period, physics underwent a revolution so profound as to shatter the unified explanatory model of the natural world toward which physicists had been optimistically working. The Newtonian vision of the universe controlled by temporal and spatial absolutes gave way to a cosmos of relativity. The very composition of matter demanded reinterpretation as atoms proved to be divisible, and not, after all, the unalterable building blocks of nature. Indeed, disintegration of atomic nuclei released forces that helped to burst asunder the confines of classical physics. Radioactivity, X rays, and the electron loomed large in the scientific consciousness at the turn of the century, hinting of new forces to be harnessed and new mysteries to be fathomed. In the twentieth century, the quantum theory and relativity completed the destruction of all the old assumptions on which physics had been built. Just as biologists had had to reconcile themselves to "a changing and indeterminate world" of ongoing evolution, physicists, too, had to accept tentative hypotheses in place of eternal natural laws. They had to acknowledge that the answers that they proposed to nature's riddles would be revised with every new generation of inquirers.[3]

These riddles took on entirely novel dimensions as physics descended to the subatomic level of investigation. The behavior of particles within the atom defied ready explication in mechanistic terms, as did other developments in twentieth-century physics. The deterministic quality of mechanical models was inappropriate in a field often characterized by motion that apparently followed no permanent, repeating pattern. Although the full extent of the radical shift in perspective did not emerge in physics until after World War I, physicists in the late nineteenth century were already experiencing the first powerful tremors of the upheaval to come. Those who sought to contest what they perceived as the detrimental stranglehold of positivism on scientific thought derived much comfort and strength from the emerging reinterpretation of the physical world. Not least articulate among these scientists were the men whose interest in psychical research, or commitment to spiritualism, benefited significantly from the realization that contemporary physics had not yet pinned labels on all the phenomena of nature.

Arrayed against them were some of the most distinguished names in Victorian physics. Tyndall, Faraday, and Kelvin summoned their great ammunition of logic and derision against spiritualism. With mid-Victorian assurance that the physical laws, which they helped to reveal, were uniform and universal, they vehemently defended such important principles of scientific inquiry as repeatability of experiments and economy of explanation. Spiritualism violated these principles, they contended. John Tyndall, superintendent of the Royal Institution, voiced the frustrations that they all experienced when trying to reason scientifically with spiritualists. In the conclusion to a brief essay on "Science and Spirits," written in 1864, he explained:

> When science appeals to uniform experience, the spiritualist will retort, "How do you know that a uniform experience will continue uniform? You tell me that the sun has risen for 6,000 years: that is no proof that it will rise tomorrow; within the next twelve hours it may be puffed out by the Almighty." Taking this ground, a man may maintain the story of "Jack and the Bean-stalk" in the face of all the science in the world.[4]

Tyndall's mentor at the Royal Institution, the renowned Michael Faraday, followed Carpenter's lead in 1853 and attributed table turning to nothing more esoteric than the unconscious muscular effort of the sitter, "the mere mechanical pressure exerted inadvertently by the turner." He was equally hard on mesmerists and phrenologists who liked to toss about such terms as "electro-biology" or "electro-psychology" to describe the forces at work in their own fields of inquiry. Faraday made it clear that he had no patience with ignorant fools who eagerly attributed unusual phenomena "to electricity and magnetism, yet know nothing of the laws of these forces, . . . or to some unrecognized physical force, without inquiring whether the known forces are not sufficient."[5] The great Kelvin bundled animal magnetism, table turning, spiritualism, and clairvoyance all together and dismissed the entire package as "wretched superstition."[6]

In their defense of the sufficiency of known forces against the unknowns of spiritualism, these giants of Victorian physics found ready support from lesser known colleagues. At the 1871 meeting of the British Association for the Advancement of Science, held at Edinburgh, for example, Professor Peter Guthrie Tait scoffed at "Spiritualists, Circle-squarers, Perpetual-motionists, Believers that the earth is flat" in his address to the Mathematics and Physics Section, while Professor Allen Thomson, addressing the Biology Section, sharply concluded that spiritualist beliefs were "scientific impossibilities."[7] All these curt, denigrating comments reveal the determination of an increasingly professional group of scientists to protect their domain from quacks and false practitioners. The scientific claims of spiritualists, as of mesmerists and phrenologists, threatened to make a mockery of scientific method; they had to be rendered absurd, and therefore harmless, as effectively as possible.

Only three years after P. G. Tait and Allen Thomson dismissed spiritualism in their BAAS papers, Tyndall very nearly dismissed theology
in his notorious Belfast Address as president of the British Association
in 1874. In this most assertive declaration of the sole sufficiency of
scientific naturalism as a means to comprehend the entire universe,
Tyndall made it implicitly clear why spiritualism was repugnant to him.
"Is there not," he asked,

> a temptation to close to some extent with Lucretius, when he affirms
> that "Nature is seen to do all things spontaneously of herself without
> the meddling of the gods?" . . . Believing, as I do, in the continuity
> of nature, I cannot stop abruptly where our microscopes cease to be
> of use. Here the vision of the mind authoritatively supplements the
> vision of the eye. By a necessity engendered and justified by science
> I cross the boundary of the experimental evidence, and discern in
> that Matter which we, in our ignorance of its latent powers, and not-
> withstanding our professed reverence for its Creator, have hitherto
> covered with opprobrium, the promise and potency of all terrestrial
> Life.

If any of Tyndall's audience on that occasion were relieved by his
reference to the Creator, they must have shuddered a few minutes later
when, warming to the conclusion, he asserted:

> The impregnable position of science may be described in a few words.
> We claim, and we shall wrest from theology, the entire domain of
> cosmological theory. All schemes and systems which thus infringe
> upon the domain of science must, in so far as they do this, submit to
> its control, and relinquish all thought of controlling it.[8]

In his declaration of scientific hegemony over religion, Tyndall also
defended the scientist's right – indeed duty – to supplement empirical
findings with fertile conjectures from his imagination. It was only in
this way, Tyndall would have agreed, that the scientist could render
meaningful the bewildering profusion of information that his equipment
delivered to him. But Tyndall's imagination, when it directed him toward materialism, was not as authoritative as his visual perception
when it reported evidence from the laboratory, and there were members of the scientific community who were by no means prepared to
endorse what amounted only to Tyndall's considered opinions. It was
widely believed that Tyndall had abused the opportunity of the British
Association presidential address to expound personal views on a subject about which official scientific pronouncements were neither desirable, nor possible. Far from aiming to banish all effective notion of
deity from the cosmos of science, many a scientist in the late nineteenth
and early twentieth century hoped to preserve some small corner safe

for divinity. Professor Balfour Stewart of Owens College spoke for a number of colleagues when he claimed "strong grounds for supposing that our environment is something very different from that to which the atomic materialists would wish to confine us."[9]

Stewart joined the SPR shortly after expressing that opinion. He and the several other physicists who also became members of the new Society were, like Tyndall, applying their imaginative vision to the data of physics and chemistry, but their conclusions differed dramatically from Tyndall's. Although they reached no single conclusion, all concurred that psychic phenomena deserved earnest scientific inquiry on the outside chance that they might provoke new insights into the workings of the universe. Marie Curie, whose brilliant discoveries helped reveal the workings of radioactivity and won her two Nobel Prizes, became an honorary member of the SPR before World War I. Other Nobel Prize winners – Rayleigh, J. J. Thomson, William Ramsay – brought further scientific distinction to the Society's membership lists in its first thirty years. So did John Couch Adams, discoverer of the planet Neptune, professor of astronomy at Cambridge, and president of the Royal Astronomical Society; Arthur Schuster, successively professor of applied mathematics and physics at Owens College, and significant contributor to the study of cathode rays; and more of equal eminence.[10] There were skeptics among the SPR physicists, just as there were men who became confirmed spiritualists. In all cases, however, each scientist refused to accept uncritically the fiat that spirit must necessarily be banished from the world revealed by modern science.

THE OPEN-MINDED

The skeptical party among the Society's scientists boasted two of Britain's most renowned physicists, Lord Rayleigh and Joseph John Thomson. Thomson succeeded Rayleigh as Cavendish Professor of Experimental Physics at Cambridge in 1884 and took over the directorship of the Cavendish Laboratory where Rayleigh, a few years earlier, had redetermined the electrical units of absolute measurement, the ohm, ampere, and volt. "Lord Rayleigh," Thomson reported, "by the experiments he made when he was Cavendish Professor, raised the standard of electrical measurement to such a high level that it may be claimed that here he has changed chaos into order." Rayleigh, never one to rest on his laurels, went from those labors to another set of exacting measurements that led, in the mid-nineties, to his most famous scientific discovery. While ascertaining the density of the principal atmospheric gases, his investigation of nitrogen revealed discrepancies between its weight when prepared chemically and when derived from the air. Further experiments, which also involved William Ramsay, professor of chemistry at University College, London, resulted in 1895 in their joint announcement of the existence in the atmosphere of a hitherto unnoticed element, the gas argon.[11] During the course of his scientific career, Rayleigh examined an extensive array of problems in physics, with his customary attention to detail, and his reputation was such that King Edward VII is said to have accosted him at a social gathering with "Well, Lord Rayleigh, discovering something I suppose?"[12] After the turn of the century, Rayleigh received distinction after distinction. He was one of the twelve initial members of the Order of Merit, established in 1902. He received the Nobel Prize for physics in 1904, at the same time that Ramsay was awarded the chemistry prize. From 1905 to 1908, he served as president of the Royal Society. In 1908, he was chosen Chancellor of Cambridge University, to which he had donated his Nobel Prize money, largely for the extension of the Cavendish Laboratory. Throughout this decade of public honors and international recognition, he remained a vice-president and council member of the SPR, as he had been during the 1880s and 1890s. Indeed, one of his last public appearances before his death in 1919 was the occasion of his presidential address to the Society.[13]

In that address, he confessed to his audience that he had "no definite conclusions to announce." Initially stimulated by Crookes's investigations, Rayleigh had been interested in psychic phenomena since the 1870s; yet nothing impressive enough had occurred to shake him from "forty-five years of hesitation" – neither séances with Kate Fox Jencken, Annie Eva Fay, Mary Showers, and Henry Slade in the mid-1870s, nor encounters with Palladino at Cambridge in 1895.[14] He had approached the subject enthusiastically at first, willing and even eager to believe the phenomena genuine. After a conversation with Crookes and a sitting with Mrs. Jencken, he wrote to his mother in May 1874, "My mind is still in suspense, but I rather expect to be converted." The following month, after more sessions with Mrs. Jencken and a few with Mrs. Guppy, Rayleigh admitted to Sidgwick that his impression was "in favour of the genuineness of the phenomena." In August, his wife explained to his mother that ten days of Mrs. Jencken as a guest at Terling, the Rayleigh home in Essex, had helped them make "a good deal of progress towards believing."[15]

But Rayleigh never advanced much further down the road to belief. For him, as for his friend and brother-in-law Sidgwick, it was never possible to eliminate doubt about psychic manifestations. Although he candidly admitted that some phenomena occurred at séances that he could not easily explain, he was not thereby convinced that spirits were necessarily at work. He had little more to tell his listeners in 1919 than his recognition that "we are ill equipped for the investigation of phenomena which cannot be reproduced at pleasure under good conditions."[16] Why, then, did this "last of the great polymaths of physical science"[17] retain an abiding interest in the work of the SPR right down to his death? He was not, admittedly, an active officer of the Society; apart from his presidential address, its *Proceedings* benefited from no reports or papers from his pen. Yet the fact that a scientist of such eminence did not hesitate to have his name associated with psychical research was in itself noteworthy. For Rayleigh, the significance of that research was closely tied to his Christian beliefs.

The place that religion occupied in Rayleigh's life is not easily summarized, although it may be suggested by his behavior during the last frenetic weekend before he and Ramsay announced the discovery of argon late in January 1895. "Though the work had to be hurried through in a short time," his son reported, "Rayleigh was not to be diverted from his usual attendance at morning church. However, he compromised by coming out before the sermon." The practice of his faith clearly mattered to Rayleigh, but his was a faith untinged by fanaticism or rigidity of any sort. His "deep sense of the mysteries of existence" kept him fully alert to the absurdity of pressing any definitive claims of faith over science, or vice versa.[18] Instead he took refuge in the conviction, shared with another brother-in-law, A. J. Balfour, that "true science and true religion neither are nor could be opposed."[19] As a Cambridge undergraduate in 1862, Rayleigh had studied Paley, Butler, Mansel, and Whewell, among his mathematical and classical readings, as he sought "a basis for definite religious convictions." A letter to Sidgwick twelve years later suggests that he turned to psychical research for similar reasons. "I am quite amazed," he wrote on 7 June 1874, "at the little interest most people take in the question. A decision of the existence of mind independent of ordinary matter must be far more important than any scientific discovery could be, or rather would be the most important possible scientific discovery."[20] Such a discovery would, of course, challenge the materialist position more profoundly than all the sermons in Christendom could ever manage, and Rayleigh had no use for materialism. "I have never thought the materialist view possible," he wrote to a correspondent in 1910, "and I look to a power beyond what we see, and to a life in which we may at least hope to take part."[21]

As a physicist, Rayleigh displayed enormous capacities for suspending judgment. "It was easy to arouse his interest in dubious phenomena, but very difficult to get his final verdict either for or against them."[22] Exactly the same words could describe his position as a psychical researcher. Although keenly alive to the implications of proving that telepathy functioned as an extrasensory means of communication, for example, he could not concede that the reality of telepathy between the living had been adequately demonstrated. Yet he did not abandon hope of a future convincing demonstration.[23] Similarly, as a Christian, he acted in the belief that Christ's message was true, that there was a life beyond the visible present, and that prayer was not a meaningless ritual, although he could not be certain of its efficacy. In the same letter of 1910 in which he dismissed materialism, he wrote:

> The great thing is to pray, even if it be in a vague and inarticulate fashion. Surely we can ask a blessing on those we love and (in the words of the Collect) that the Holy Spirit may in all things direct and rule our hearts. The vaguest attitude of aspiration and resignation seems better than no attitude at all.[24]

Was it because his religious beliefs *were* vague – nurtured as alternatives to no belief at all – that Lord Rayleigh persisted in finding psychical research important and looked forward to the day when telepathy would be established beyond doubt?

J. J. Thomson agreed with Rayleigh about both the significance and the uncertainty of telepathy. He described the subject as "one of transcendent importance," but one for which nothing more could be claimed than "the Scottish verdict – not proven."[25] Thomson had been affiliated with the SPR for decades by the time he expressed his views in the volume of memoirs published in 1936. He joined the Society in 1883 and for more than thirty years served on its council, a body always glad to welcome Fellows of the Royal Society. Although, like Rayleigh, Thomson was not an active psychical researcher, his experiences included exposure to some of the more artful mediums of his day. During the 1890s, on Myers's urging, he attended séances with Eglinton and shared with Rayleigh the frustrations of Palladino's Cambridge sittings.[26]

Like Rayleigh, too, Thomson attained the greatest heights of scientific eminence. He had first studied physics in Balfour Stewart's classroom at Owens College, and his professional career saw him radically influence the direction of modern physics. During the course of experiments with cathode rays in 1897, Thomson became convinced that they consisted of electrically charged particles smaller than an atom. He had, in fact, discovered the electron, thereby shattering the long-held assumption that the indivisible atom was the fundamental unit of matter. He earned the Nobel Prize for physics in 1906, was knighted two years later, and in 1912 received the Order of Merit. His colleagues in the scientific profession made him president of the British Association in 1909, and of the Royal Society from 1915 to 1920. At Cambridge, working on the foundations established by Rayleigh as Cavendish Professor before him, Thomson transformed the Cavendish Laboratory into a pioneering experimental physics lab for the actual instruction of undergraduate and graduate students. He became Master of Trinity College in 1918 and inhabited the Master's Lodge until his death in 1940. Yet, with all the other demands on his time, he never lost his enthusiasm for psychical research, if more as observer than participant. It is noteworthy that the man who probed the secrets of the atom devoted five pages of his memoirs to the subject of water dowsing, in the efficacy of which he fully believed.[27]

Thomson, undaunted by the mysteries of the universe, candidly admitted that the mysteries of man seemed to him far more awesome. "The most complicated physical apparatus," he confessed, "is simplicity itself compared with a human being." In contrast to many fellow scientists at the séance table, he did not insist that mediums submit to scientific testing – or, at least, he did not expect them to respond favorably under such conditions. Mediums, he explained,

> are very psychic and impressionable, and it may be as unreasonable to expect them to produce their effects when surrounded by men of science armed with delicate instruments, as it would for a poet to be expected to produce a poem while in the presence of a Committee of the British Academy.

With his appreciation of the complexity and sensitivity of the human instrument, Thomson was hardly the man to make definitive statements about the mental phenomena of mediumship, including telepathy. He distinguished sharply between telepathy among the living (which he termed "short-range thought transference") and telepathy that linked the living and the dead. He certainly favored ongoing investigation of the former, but it was the latter, not surprisingly, that fascinated him most, for it raised "much deeper and more important questions." If only a few reported psychical occurrences of that sort could be proved actually to have happened, Thomson asserted, they would "revolutionise our ideas about physical as well as spiritual things."[28]

Although Thomson did not often choose to discuss spiritual things, they were apparently as important to him as they were to Rayleigh. Rayleigh's son, in his biography of Thomson, describes his subject as "a regular communicant," a man who knelt daily in private prayer, and who, on retiring to bed at night, "always paused at the bookcase near the door wherein his small Bible stood, and spent two or three minutes in reading it." He was intrigued by ultimate enigmas and is said to have expressed "great sympathy with Pontius Pilate, and his question, What is truth?''[29] It was a natural question for the man who invalidated one of the cardinal truths of nineteenth-century science. It may be that he looked to psychical research for solutions to his cosmic puzzles because of his association with Trinity College, which he attended as an undergraduate in the 1870s and which elected him a Fellow in 1880. But Trinity had no monopoly on psychical researchers, and the fact that both his first mentor in physics, Balfour Stewart, and the Cavendish Professor, Lord Rayleigh, were interested in the subject may have helped arouse Thomson's curiosity more than his Trinity affiliation. Thomson did not, as it turned out, find in psychical research any solutions to any riddles. Yet it is nonetheless clear that, for him, the phenomena that might revolutionize ideas about physical and spiritual things were never cause for light speculation or idle concern.

Neither Rayleigh nor Thomson ever allowed their sympathy with the goals of psychical researchers to compromise their judgment as scientists. Both knew that psychical research, despite its implications for their Christian faith, had to prove its value in terms that the scientific profession could respect. When it failed to meet the standards of evidence established by contemporary science, they were left no choice but to hand in the verdict "not proven."

Other scientists, however, interpreted their professional duty less strictly. Without wishing to challenge the authority of the experimental method, these men believed that the methodology of the physical sciences need not rigorously exclude psychical phenomena. They were not disposed to accept uncritically the spiritualist belief in communication with the dead, but they could find no valid reason to deny the possible existence of natural forces about which physicists as yet had no knowledge. Augustus De Morgan, logician and professor of mathematics at University College, London, proved an articulate defender of this viewpoint. Although his wife Sophia was a confirmed spiritualist, De Morgan himself was not. Instead, as he wrote in the preface to her book, *From Matter to Spirit*,

> Thinking it very likely that the universe may contain a few agencies – say half a million – about which no man knows anything, I cannot but suspect that a small proportion of these agencies – say five thousand – may be severally competent to the production of all the [spiritualist] phenomena, or may be quite up to the task among them. The physical explanations which I have seen are easy, but miserably insufficient: the spiritual hypothesis is sufficient, but ponderously difficult. Time and thought will decide, the second asking the first for more results of trial.[30]

De Morgan was expressing the wait-and-see position of a thoughtful man who had attended séances, who had found no existing natural explanation satisfactory, but who was not therefore driven to assume that supernatural agents were responsible for what he observed. He was willing to suppose that all the physical mysteries of the cosmos had not yet been resolved and was particularly annoyed by scientists who investigated novel occurrences with fixed antecedent notions of what was possible or impossible in nature. A lecture on "Mental Training," delivered by Faraday in 1854, prompted De Morgan to comment: "We thought that mature minds were rather inclined to believe that a knowledge of the limits of possibility and impossibility was only the mirage which constantly recedes as we approach it."[31] In debating Faraday's pronouncements concerning the nature of evidence in scientific demonstrations, De Morgan implied that any investigator embarking on so new and controversial a subject as séance phenomena was not well served by incredulity. On the contrary, a degree of openness was requisite, a willingness to suspend disbelief for a while, lest one reject out of hand, without impartial investigation, blinded by a predisposition to ridicule.

The quality of mind that most endeared Balfour Stewart to his students and fellow scientists was precisely such openness, an "absolute freedom from preconceived ideas."[32] No narrow specialist, he pursued investigations and published work in the disciplines of physics, mathematics, chemistry, meteorology, and astronomy – indeed, he was director of Kew Observatory before assuming the physics professorship at Owens College. In all his inquiries, he considered preconceived ideas to be the greatest obstacle threatening the advancement of science. "Let us suppose," he wrote in July 1871,

> that a man comes before us as a witness of some strange and unprecedented occurrence. Here it is evident that we are not entitled to reject his testimony on the ground that we cannot explain what he has seen in accordance with our preconceived views of the universe, even although these views are the result of a long experience; for by this means we should never arrive at anything new.[33]

While Tyndall denounced this "Jack and the Bean-stalk" frame of mind, Stewart, like De Morgan, insisted that it alone could promote an atmosphere conducive to the full expansion of scientific knowledge. Both of Tyndall's opponents on this issue well understood how often current prejudices within the scientific community molded the course of their colleagues' research and dictated what subjects received professional benediction or anathema. Stewart's own prejudices were resolutely antimaterialist. They led him to the SPR, which he served as council member, vice-president, and president, and whose studies of thought transference he avidly followed. He never expressed absolute conviction on the subject of telepathy, but he found the evidence which the SPR adduced in support of extrasensory communication increasingly persuasive.[34] He was entirely certain that unknown forces were at work in the universe, which it was his duty as a scientist to try to identify.

His attempts at identification led to several hypotheses, but never to the spiritualist contention of communication with the dead. He vigorously reiterated his disbelief in messages from deceased people and groped for alternative explanations of séance manifestations. One result was *The Unseen Universe or Physical Speculations on a Future State*, first published anonymously in 1875, and in numerous editions thereafter. The volume's coauthor was P. G. Tait, professor of natural philosophy at Edinburgh, the same man who four years earlier had spoken contemptuously of spiritualism in his comments to the BAAS. His partnership with Stewart, however, did not represent a *volte-face*, for *The Unseen Universe* made no concessions to spiritualism. Rather it was an expression of the authors' desire to conjure up a cosmos that both science and theology could accept. They believed in the existence of an invisible universe, which science was not competent to investigate, but they linked it to the visible world by bonds of energy, a concept with which modern physics was becoming increasingly familiar. Their book was, fundamentally, an argument for design and continuity in nature, an argument for theism even, but also an attempt to place theism in an intellectual landscape whose contours were recognizable to their professional colleagues.[35] On other occasions, Stewart suggested the possible existence of "intelligent beings besides man," or hinted at the workings of electro-biology.[36]

In using a term like "electro-biology," Stewart was guilty of that loose reliance on scientific language that Faraday condemned in 1853. The "electro-biological state" was merely a grandiose way of describing the mesmeric trance. Stage performers employed such jargon to give their patter an erudite polish; Balfour Stewart employed it out of genuine perplexity. Loathing materialism and seeking to justify his faith in God and immortality, he could find no language as persuasive as that of the scientific naturalism he denounced. He was driven to the concepts of energy and electricity because they had the physical basis that the spiritualist hypothesis lacked. Both were physical phenomena of such uncharted extent and impact that they might plausibly manage to elucidate psychical occurrences.[37]

The scientific profession could not take issue with the notion that unknown forces – or known forces working in still mysterious ways – might explain psychical phenomena without recourse to spirit agents. Force was, of course, an entirely legitimate scientific term, one used to designate a range of physical processes, both those that science had illuminated and those that remained to be explored. The difficulty arose with regard to psychical research because not everyone was as restrained as De Morgan in refusing to place limits on the natural and the possible. Even Stewart, for all his care to shun spiritualism, appeared excessively gullible to Arthur Schuster, while Tait dismissed some of his ideas as "weird and grotesque."[38] At a time when discoveries in electromagnetism commanded the attention of scientists and educated public alike, it was all too tempting to borrow bits and pieces of electrical theory and hitch them to fantastic, inappropriate partners. Fournier D'Albe, for example, worked up the theory of "psychomeres," or "soul-particles" – the "really vital part of the cell" that constitutes the individual's immortal spirit – and decked it out with ionization and condensation, electric polarity, molecular dimensions, and electrons for good measure.[39] It is easy to understand why few scientists were persuaded that the answers to the questions of the séance chamber lay in confused speculations concerning unidentified forces acting according to hitherto unsuspected laws of nature.

WILLIAM CROOKES

In William Crookes, the nameless psychic force gained an unrivaled advocate, for he courted controversy with zest and aggression. Fundamentally no more certain than Stewart about the identity of the force behind psychic phenomena, Crookes did not hesitate to insist upon its existence, nor to uphold the reality of the phenomena themselves. For a few years in the 1870s, he landed himself in the midst of a row that not only threatened to tarnish his scientific reputation, but also cast doubts upon his personal conduct and code of honor.

Crookes's career as a scientist is one of the most variegated and fascinating of the Victorian and Edwardian eras. He followed no prescribed paths to success, and blazed his own highly individual trail to knighthood in 1897, the Order of Merit in 1910, and the presidency of the Royal Society from 1913 to 1915. The son of a prosperous London tailor, Crookes attended no university, but in 1848, when scarcely sixteen, enrolled in the newly opened Royal College of Chemistry, where he studied under A. W. Hofmann. His attempts to establish himself at an academic post were not successful, and after his marriage in 1856, he became a "freelance chemical consultant" working out of a laboratory in his own home.[40] He also devoted much of his time to scientific journalism; in 1859, soon after editing some photographic journals, Crookes launched a new weekly, *Chemical News*, which he edited until 1906 and which gave him an influential voice in English chemical circles. He likewise edited the *Quarterly Journal of Science* in the 1860s and 1870s, and thus never lacked opportunities to publish his findings, whether orthodox or otherwise.

All the while that Crookes was busy editing, he was also actively pursuing the role of public advisor on scientific matters, playing his part in the transformation of the scientist from "solitary philosopher" to the technical expert with whom the twentieth century is so familiar.[41] During the cattle plague of 1865–7, he was appointed by the government to report on the use of disinfectants in curbing the disease. A director of the Water Inspection Laboratory at the end of the century, he studied urban sewage disposal systems and had to scrutinize the quality of London's water supply. He served on the Explosives Committee of the Ordnance Board, and, as president of the British Association in 1898, he publicized his concerns about the world food supply. He was very much a public-minded scientist, a man for whom the theoretical splendors of laboratory experimentation had direct and urgent application in the world at large. The fact that Crookes gleefully sought remuneration for such services whenever possible throws little discredit on the sincerity of his belief in the public responsibility of science. He was a self-employed businessman, with a large family to support, and he liked to get paid for his time and effort, for without the advantage of prestigious academic affiliation, Crookes fretted considerably about problems of money and status. When he was knighted in 1897, he ordered ornate new stationery that proclaimed "From Sir William Crookes, FRS," in a design that highlighted the word "Sir."[42] He was disappointed that his knighthood merely carried the distinction of Knight Bachelor and felt considerable relief when the Order of Merit added luster to his title in 1910. Financial anxieties and a fondness for honors are, however, venial weaknesses that scarcely diminish the value of Crookes's scientific labors.[43]

During nearly seventy years of laboratory investigations, Crookes established himself as both a pioneering chemist and physicist. His discovery of the element thallium in 1861 earned him election to the Royal Society two years later; his painstaking examination of the properties of the new element – particularly his determination of its atomic weight – set high standards of precision and won him public recognition as one of Britain's foremost chemical analysts. Indeed, when Rayleigh and Ramsay discovered argon, and when Ramsay isolated helium shortly thereafter, both gases were sent to Crookes for analysis. Yet what is most extraordinary about Crookes's career is not his eminence as a chemist, a profession for which he had specifically trained, but his stature as a physicist. J. J. Thomson recorded that his own researches were inspired by Crookes's "beautiful experiments on cathode rays," although, in physics, Crookes was a rank amateur. "His knowledge of mathematics was of the most rudimentary description," Thomson noted, "and he had never been through any course of instruction in physics. He picked up his physics as he wanted it for the research in which he was engaged." When necessary, he picked the brains of Sir George Gabriel Stokes, professor of mathematics at Cambridge and for three decades secretary of the Royal Society, while from the late 1890s, Crookes turned to Oliver Lodge to keep him informed of advances in physics that he no longer felt capable of following in the fullest detail.[44]

As with Wallace, handicaps of education never daunted Crookes's curiosity, nor limited the abundance of his scientific hypotheses. With his acute powers of observation, he was quick to notice slight discrepancies in measurement and devised experiments, often of great elegance, to explore further their physical causes. In this manner, he discovered that small differences in temperature could affect the weight of substances in a vacuum, and in 1875 he invented an instrument, the radiometer, that demonstrated the influence of radiation on objects in vacua. These investigations prompted him to undertake a close study of electrical discharge through highly attenuated gases – that is, through gases in high vacua – and the subsequent experiments on cathode rays that won Thomson's admiration. Although Crookes missed the momentous conclusion that Thomson himself reached in 1897, with the discovery of the electron, Crookes's work clearly proved a stepping-stone to that breakthrough. His interest was also drawn to the chemical puzzle of the rare-earth elements, and, around the turn of the century, he devoted his attention to radioactivity. The Royal Society rewarded him over the years with three of its prestigious medals; the French Académie des Sciences in 1880 bestowed upon him a gold medal and special prize, and he held the presidency of numerous scientific organizations. For an independent chemical consultant, he had not done badly.

In a long scientific career, Crookes's researches into psychic phenomena occupied only a small span of time, some five years of active investigation between 1870 and 1875. Yet the work of that half-decade became nearly as well publicized as all his chemical and physical studies combined. When a famous Hollywood director contemplated making a film about Crookes in the 1960s, it was not the discovery of thallium that was to be featured, but the scientist's relations with Florence Cook.[45] The relationship did indeed represent an interlude in Crookes's life that was fraught with dramatic possibilities, but the time that he devoted to psychical research also bore directly on his ongoing attempts to explore "the shadowy realm between Known and Unknown." That region, Crookes confessed, always held for him "peculiar temptations."[46]

The controversy over Crookes's inquiries into psychic manifestations has simmered at several levels of debate, among spiritualists, antispiritualists, scientists, and historians of science alike. All the discussions of necessity consider the role played by Cook, with whom the scientist undertook a long series of séances in 1874 and 1875.[47] When an eminent scientist, famed for his powers of observation and analysis, reports that he has, on a number of occasions, seen the fully materialized form of "Katie King" together with her entranced medium, Florence Cook; when he claims to have clasped "Katie" in his arms and felt a substantial body, yet still asserts that she was pure spirit; when he announces that he has repeatedly photographed this spirit – more than a few eyebrows will be raised.[48] Either, one assumes, the man was so utterly gullible as to believe the crassest of deceptions, or he was part and parcel of the deception, playing a huge joke on the British public. If the latter case, his most probable motive would have been romantic involvement with the medium, as Trevor Hall contends. There is certainly no doubt that Crookes considered Florence a highly attractive young woman and may have been unable to resist the chance to flirt with her, or "Katie," during the séances. He waxed positively lyrical about "Katie" in a letter to the *Spiritualist Newspaper*, on 5 June 1874, where her "perfect beauty" prompted him to quote lines drawn from Byron's *Don Juan* – perhaps not the wisest source, under the circumstances.[49]

The evidence, however, does not fall preponderantly on the side of Mr. Hall, who himself occasionally recalls that his hypothesis remains only "a matter of surmise," and far from proven.[50] We cannot assert that Cook was Crooke's mistress. In fact, we cannot assert much of anything about events in Crookes's library and laboratory at 20 Mornington Road, Regents Park, London, where some of the most successful séances with Florence were held. It is impossible to doubt that "Katie" was either the medium herself or, when circumstances permitted, a confederate. Circumstances would certainly have favored the use of a partner at the many séances held in the Cook family's Hackney home, and Podmore observed, "It was admitted even at the time, and by believers, that 'Katie's' appearance differed widely on different occasions." In Crookes's own home, however, no such conspiracy could have been perpetrated without his consent, and it is far likelier that the spirit figure was Florence in disguise. Yet surely a man who discovered a new element could detect when a fleshly maiden masqueraded as a ghost. If Crookes was not involved in Florence's chicanery, how could he declare: "I have the most absolute certainty that Miss Cook and Katie are two separate individuals so far as their bodies are concerned"?[51]

Although there is no compelling reason to accuse Crookes of duplicity when he wrote those words, there is every reason to assume that he was duped. He would scarcely be the first scientist whose self-confidence led to self-deception and whose technical expertise gave him a false sense of security. Proud of his own astute powers of observation, he may well have failed to grasp how easily he could be misled. Apparently he did not hold himself accountable to his readers in the way that a less distinguished psychical researcher would, if only to prove his credibility, for Crookes left no records of his séances with Florence that would satisfy a skeptic's thirst for precise data and minute details. He was, as he candidly acknowledged, "accustomed . . . to have [his] word believed without witnesses";[52] he grew frankly annoyed when critics, even fair-minded ones, asked whether Crookes might not possibly be mistaken about the events that he alleged to have witnessed. If he claimed to have seen them, they had indubitably occurred. This frame of mind, when combined with his genuine fondness for Florence Cook and his complete faith in her honesty, must have helped to place him less on his guard than he ought to have been during the séances that he supervised.[53] Crookes's description of séances in the spring of 1874 raises the suspicion that Florence was more in charge of the show than the investigator realized. "During the time I have taken an active part in these séances," Crookes wrote to the *Spiritualist Newspaper* in his letter of 5 June,

> Katie's confidence in me gradually grew, until she refused to give a séance unless I took charge of the arrangements. She said she always wanted me to keep close to her, and . . . I found that after this confidence was established, and she was satisfied I would not break any promise I might make to her, the phenomena increased greatly in power, and tests were freely given that would have been unobtainable had I approached the subject in another manner. She often consulted me about persons present at the séances, and where they should be placed, for of late she had become very nervous, in consequence of certain ill-advised suggestions that force should be employed as an adjunct to more scientific modes of research.[54]

One would like to know what promises Crookes gave "Katie" and to what extent these, in fact, limited "more scientific modes of research."

Crookes, then, may well have been the ideal scientist for Florence Cook – a man of substantial reputation whose vanity enabled him to endorse her mediumship without equivocation. Although a scientist is no better equipped than any other equally intelligent person to perceive conjuring tricks or similar deceptive practices, Crookes's support of Cook in 1874 nevertheless proved a tremendous boost to her career. It was generally assumed at the time that men of science *did* have special investigative talents that enabled them to identify fraud in the séance chamber as readily as they might isolate rare gases. Spiritualists were, therefore, particularly jubilant when they could embrace a scientist as convert to their faith; if a scientist believed in the validity of séance phenomena, they argued, he provided living proof that no fraud tarnished the proceedings. Great was the celebration among spiritualists when it appeared that Crookes had joined their ranks. They hailed him throughout their publications and broadcast his "conversion" as widely as possible. Even today, a plaque on the wall in the library of the Belgrave Square headquarters of the Spiritualist Association of Great Britain bears the inscription: "This Library is named after Sir William Crookes, FRS, Eminent Scientist and Spiritualist."

Whether Crookes can, in fact, be labeled a spiritualist is, however, far from certain. What he believed with regard to psychic phenomena defies summation, as his views varied over time and with the emotional circumstances of his personal life. Drawn to the subject when his youngest brother, Philip, died of yellow fever, Crookes seems initially to have endorsed the spiritualist contention that the dead communicate with the living. But after sustained investigation, his opinions became less certain, more interrogatory than declarative, and it was not until the end of his life, following the death of Lady Crookes, that he again reverted to pure spiritualism. Bereavement must have loosened for Crookes whatever restraints he ordinarily imposed upon himself as a psychical researcher.

Philip Crookes died in the autumn of 1867, and, although Crookes did not begin any kind of systematic study of spiritualist phenomena until 1869, his biographer dubs him "a convinced spiritualist" by December 1870.[55] In that month, Crookes sailed to Spain and North Africa as part of the government-sponsored scientific expedition that studied a total solar eclipse, and, since Tyndall was a fellow passenger, Crookes took advantage of the opportunity to proselytize. He boasted in his diary, on 6 December, that he had persuaded Tyndall to take seriously the subject of spiritualism. It is an unlikely story, but there is no doubt that Crookes himself approached the topic in deep earnestness. On 31 December, still away from home, Crookes recorded in his diary:

> It is getting on for midnight, . . . I cannot help reverting in thought to this time last year. Nelly [his wife] and I were then sitting together in communion with dear departed friends, and as 12 o'clock struck they wished us many happy New Years. I feel that they are looking on now, and as space is no obstacle to them, they are, I believe, looking over my dear Nelly at the same time. Over us both I know there is one whom we all – spirits as well as mortals – bow down to

as Father and Master, and it is my humble prayer to Him . . . that
He will continue His merciful protection to Nelly and me and our
dear little family, . . . May He also allow us to continue to receive
spiritual communications from my brother who passed over the
boundary when in a ship at sea more than three years ago.
    . . . Nelly darling and my dear children, Alice, Henry, Joe, Jack,
Bernard, Walter, and little Nelly baby, I wish you all many, many
happy New Years, and when the earthly years have ended may we
continue to spend still happier ones in the spirit land, glimpses of
which I am occasionally getting.[56]

What Crookes confided to the privacy of his diary, he did not articulate to a public audience. In July 1870, when he had officially announced that he was undertaking an investigation of spiritualism, he
vouched for the reality of the physical occurrences – "the movement
of material substances, and the production of sounds resembling electric discharges" – but he stoutly maintained: "I have seen nothing to
convince me of the truth of the 'spiritual' theory." His position was
that of the unbiased scientist who wanted to introduce "the modes of
reasoning of scientific men" into the field of spiritualist inquiry in order
to "promote exact observation and greater love of truth among enquirers," and to "drive the worthless residuum of spiritualism hence
into the unknown limbo of magic and necromancy."[57] This was the
tone that he maintained for the next five years, as he informed the
British public of his ongoing psychical research. He was the disinterested expert, the man with the skills to probe the mystery, the authority
who did not like to have his pronouncements questioned.

Crookes probed the mystery with a variety of mediums, both professional and private, particularly focusing on those who produced physical phenomena. His record for attending séances of this kind is, it
appears, unmatched by "any technically qualified investigator before
or since," and the list of mediums with whom he sat, in addition to
Cook, includes Home, Kate Fox, Annie Eva Fay, J. J. Morse, Stainton
Moses, Mary Showers, Herne, Williams, and numerous others.[58]
Home, of all these men and women, predictably made the most lasting
impression on Crookes. His talents led Crookes boldly to proclaim "the
existence of a new force, in some unknown manner connected with
the human organisation, which for convenience may be called the
Psychic Force."[59]

Crookes's series of séances with Home between 1871 and 1873 have
achieved a special status in the annals of psychical research and have
been described as nothing less than "the foundation stone of the evidence in support of the reality of the so-called 'Physical Phenomena'."[60] Despite their fame, however, and despite the superior skill of

the medium, these sittings with Home were in many ways characteristic of Crookes's general approach to experimental séances. First of all, the appearance of scientific control figured as an important feature of the proceedings. As Crookes recounted his investigations with Home in the *Quarterly Journal of Science*, July and October 1871, he emphasized the mechanical contrivances devised to test the medium's powers, as though the apparatus itself eliminated the possibility of human error or deceit.[61] Furthermore, Crookes was always delighted when he could report that his painstakingly arranged experiments had won the endorsement of other scientists. Deeply disappointed that more of his professional colleagues could not be persuaded to participate in his psychic investigations, he capitalized on the presence of allegedly trained observers whenever possible. William Huggins, for example, the distinguished amateur astronomer and future president of the Royal Society, witnessed at least one of Home's séances at Crookes's home in 1871, and obliged Crookes by writing a brief letter confirming the account that appeared in July of that year.[62]

Yet Home was not likely to allow Crookes and his scientific friends real authority over conditions prevailing at his séances. In 1889, when Crookes published some of his actual séance notes in the SPR *Proceedings*, it became clear that his sittings with Home had not been as rigorously supervised as the essay of 1871 implied. If spirit commands ordered "Hands off the table," Crookes remembered, all hands, including his own, were removed at once, just as in the most ordinary séances that did not benefit from expert observers.[63] Like Cook a few years later, Home may have easily imposed his own conditions, while seeming to offer the utmost cooperation to the scientist. After all, if Home did not approve of Crookes's arrangements, he could always see to it that no manifestations whatsoever occurred, blaming the failure on uncongenial circumstances. When phenomena did occur, including Home's customary repertoire of levitation, disembodied hands, and the handling of hot coals, there is no guarantee that Crookes was really directing the séance.

The same pattern recurred in Crookes's experiments with other physical mediums: the attempt to introduce foolproof scientific tests, the inadequate reports about those tests, the suggestion that the medium was utterly subject to Crookes's supervision, but the possibility that Crookes was more tractable than he knew. With Cook, the mechanical devices included not only an elaborate barrage of cameras for the purpose of photographing "Katie," but also an electrical apparatus, called a galvanometer, by means of which the young woman became part of a mild electric circuit. When Florence was attached to this contraption, Crookes assumed that she could not impersonate a materialized spirit without producing telltale fluctuations in the galvanometer readings. Nonetheless "Katie" emerged undaunted. The following year, in 1875, Crookes applied similar tests to Annie Eva Fay, in the presence of Huggins, E. W. Cox, Galton, and Rayleigh. Fay sat behind a curtain, with her hands attached to the apparatus in such a way that she could scarcely move them without altering the galvanometer so noticeably as to alert the sitters in front of the curtain. Under these circumstances, a hand bell rang on Fay's side of the barrier, a hand emerged to offer Cox a book, and later to throw a box of cigarettes at the sitters. Another time, several musical instruments played at once, while the galvanometer "index was very steady, which proved that the medium was still while these things were occurring." Crookes was delighted with the results, for, as he reported to the *Spiritualist Newspaper*, the galvanometer test had

> the advantage of *absolute certainty*, since, if the medium has her hands or body removed from the wires, in a state of trance or otherwise, the galvanometer outside [i.e., on the spectators' side of the curtain] lets the spectators know the moment that the circuit is broken. On the other hand, if the wires should be joined together so that the current can still pass, the effect is quite as surely made evident by the galvanometer.[64]

Crookes believed that he had identified all possible forms of attempted fraud and had successfully forestalled them with the electrical tests. Subsequent psychical researchers, however, have suggested ways in which both Cook and Fay could have used other parts of their bodies, or even a resistance coil, to maintain the electric circuit intact, while freeing their hands for other purposes.[65] These alternative arrangements do not explain the appearance of "Katie King" while Cook was supposedly attached to the sensitive galvanometer, but they do suggest that Crookes had less control over the experiment than he confidently assumed.

Crookes hated to admit that he might be hoodwinked by a clever medium, but he was not incapable of realizing the extent to which deceit generally abounded in spiritualist circles. Even in July 1870, when he was just beginning his investigations, he described spiritualism as "a subject which, perhaps, more than any other lends itself to trickery and deception,"[66] and he never had cause to revise that estimate. His experiments with Mary Showers in 1874 and 1875, and his discovery that she consistently produced fraudulent manifestations, only confirmed what he already knew about mediums. The discovery, nevertheless, seems to have been a hard blow for Crookes, as his letters to Home in November 1875 suggest,[67] and it may have hastened his decision to abandon further active investigations into psychic phenomena.

He certainly had had enough of allegations concerning his conduct with winsome young female mediums. "For myself," he confided to Oliver Lodge many years later, in a letter marked *Private*, "I have been so troubled by hints and rumours in connection with Miss Cook, that I shrink from laying stress on what I tried with her mediumship and rely on phenomena connected with Dan. Home's mediumship when saying anything in public."[68] Perhaps he had ceased to stress Cook's phenomena because her mediumship was implicated in Showers's treachery: "Katie King" had, after all, on one memorable occasion in 1874 promenaded arm in arm with Mary's materialized spirit, "Florence Maple," in Crookes's own laboratory. Crookes never specifically repudiated Cook – that was not his style – but he did manage to suggest that his absolute faith in the genuineness of her talents had been shaken. At a meeting of the SPR in November 1895, during a discussion of Palladino's feats, Crookes ventured the opinion "that among 100 or more 'mediums' he had had to do with – with hardly any exception – all more or less at times resorted to 'trickery'."[69] The exception was Home, whose manifestations Crookes always maintained were utterly free of suspicion; where other mediums were concerned, Crookes apparently learned not to trust a pretty face.

Crookes's state of mind in the 1870s is exceedingly difficult to gauge. His biographer, Fournier D'Albe, insists that in the early 1870s "Crookes *was* a spiritualist at heart, and was known to be such by a number of his friends, but he had not published the fact, and evidently did not intend to do so until spiritualism was officially recognised by the scientific authorities in power."[70] Certainly, Crookes's diary from the eclipse expedition supports Fournier D'Albe's contention, as does a letter that Crookes wrote to a fellow psychical researcher and chemist, Professor Boutlerow of St. Petersburg, in April 1871. "Since I wrote my article in the *Quarterly Journal of Science*," explained Crookes in reference to his initial essay of July 1870, "I have not had many opportunities of pursuing any experiments. All that I have since seen have carried my opinions further towards those held by advanced spiritualists, but before I publish anything I want to confirm what I say by an appeal to experimental evidence."[71] Nor is there any doubt that Crookes was fascinated by the phenomena associated with spiritualism. In addition to his involvement with the SPR – he served on its council, as an honorary member, and as its president in the late 1890s – he also joined the Theosophical Society and the Ghost Club.[72] But fascination with spiritualist phenomena does not necessarily amount to an ongoing belief in the reality of communication with the dead. Other evidence suggests that, although Crookes briefly subscribed to "advanced" spiritualist opinions, his attitude had changed by the mid-1870s and he suspended judgment on the subject until 1916.

There is good reason to believe that, from virtually the start of his psychical inquiries, Crookes was seeking a satisfactory naturalistic explanation for the allegedly supernatural phenomena that he observed. He was unalterably convinced of their objective reality and was looking for a way to incorporate them within the known confines of the physical world. "I must have some conversation with you respecting these obscure phenomena," he wrote to William Barrett in May 1871. "If you could help me to form anything like a physical theory I should be delighted. At present all I am quite certain about is that they are *objectively* true." Later that year, Crookes ended a letter to Huggins with the flat assertion: "There is a new force, or a new form of a known force."[73] Admittedly, in writing to professional colleagues, Crookes may well have felt the imprudence of the outright spiritualist hypothesis, even though both Barrett and Huggins were sympathetic to Crookes's research. The far less guarded letter to the foreigner Boutlerow does seem to indicate that during a few years Crookes genuinely accepted the full range of spiritualist beliefs, including communication with the dead, and relished the comfort that they offered. At the same time, however, he never definitively eliminated explanations within the purviews of scientific naturalism. With increasing experience of mediums and their methods, the spiritualist faith evidently faded for Crookes, while the immense range of alternative explanations remained deeply impressed on his inventive mind. That Crookes had abandoned the spiritualist hypothesis by August 1874 is strongly confirmed by the letter that he wrote that month to a Mme. B. in St. Petersburg:

> During this whole time I have most earnestly desired to get the one proof you seek – the proof that the dead can return and communicate. I have never once had satisfactory proof that this is the case. I have had hundreds of communications *professing* to come from deceased friends, but whenever I try to get proof that they are really the individuals they profess to be, they break down . . .
> I am extremely sorry I can give you no more comforting assurances. I have passed through the same frame of mind myself and I know how earnestly the soul craves for one little sign of life beyond the grave.[74]

His repudiation of spiritualism did not, however, bring Crookes any closer to understanding the psychic force whose existence he had confidently announced in 1871. At the end of the following decade, he could still write, with evident frustration, that "the widening of the circle of our definite knowledge does but reveal the proportionately widening circle of our blank, absolute, indubitable ignorance." In 1897, when he was starting to draft his upcoming address as president of the British Association, he confided to Lodge: "I should like to bring in 'occult' matters, but can't for the life of me find anything definite enough for a peg to hang remarks on."[75] He never ceased believing that psychical research was revealing the operations of a new force, but he was no longer certain that scientists could assign it a place on the map of modern physics. He had attempted to locate psychic force on that map during his tests with Home: He had placed the medium "in a helix of insulated wire through which electric currents of different intensities were passed"; he had brought strong magnets near to Home and to the objects that moved in Home's presence; he had illuminated the experiments with different colored lights – all to see what effect the known agencies of physics might have on the medium's manifestations. Crookes had noticed no effect whatsoever.[76]

Having failed to establish clear connections between psychic phenomena and the laws of physical science, Crookes turned with ever greater hope to the unfolding study of the mind. By the 1890s, his interest had veered sharply from the physical to the mental aspects of mediumship. In both his presidential address to the SPR in 1897 and to the BAAS the following year, he stressed the role of the brain as the linchpin holding together the material and psychical realms, and in particular emphasized telepathy as the agency that facilitated the union. He confessed to the British Association that, were he seeking to acquaint science with psychical research for the first time in 1898, he would proceed quite differently than he had back in 1870–1.

> It would be well to begin with *telepathy*; with the fundamental law, as I believe it to be, that thoughts and images may be transferred from one mind to another without the agency of the recognised organs of sense – that knowledge may enter the human mind without being communicated in any hitherto known or recognised ways.[77]

Clearly Crookes had become convinced that telepathy was an ascertained phenomenon, as assuredly operating according to orderly natural laws as the phenomena of electromagnetism, say, or gravitation. As with his earlier pronouncements concerning psychic force, he was confident that telepathy existed, although he could not yet explain how it worked. The recently discovered Röntgen, or X, rays were one possible influence on the transmission of telepathic messages, and he explored that possibility in his 1897 presidential address to the SPR. "It seems to me," he argued,

> that in these rays we may have a possible mode of transmitting intelligence, which with a few reasonable postulates, may supply a key

to much that is obscure in psychical research. Let it be assumed that these rays, or rays even of higher frequency, can pass into the brain and act on some nervous centre there. Let it be conceived that the brain contains a centre which uses these rays as the vocal chords use sound vibrations (both being under the command of intelligence), and sends them out, with the velocity of light, to impinge on the receiving ganglion of another brain. In this way some, at least, of the phenomena of telepathy, and the transmission of intelligence from one sensitive to another through long distances, seem to come into the domain of law, and can be grasped.

While he recognized that these speculations were "strictly provisional," he nonetheless repeated them in his address to the BAAS. Developing an hypothesis that involved Röntgen ray vibrations and molecular movements in the brain, he announced firmly that "it is unscientific to call in the aid of mysterious agencies."[78]

It would seem from these speeches that Crookes, in shifting the focus of his attention from matter to mind, was still eager to bring psychic phenomena – particularly telepathy – within the domain of law acknowledged by the physical sciences. He shunned the invocation of the supernatural and appeared to be arguing that the mysterious workings of the mind might some day prove to be controlled by familiar physical agencies. Yet, as always with Crookes, it is dangerous to attempt a generalized summary of his views in the late 1890s. Like De Morgan, he believed that no scientist could predict what secrets the universe might still shelter from human knowledge, and at times he inclined toward the view that *natural* law might not be the key needed to unlock those enigmas. In April 1899, at a meeting of the Ghost Club that he attended as a guest,

> Crookes said that a natural law is a sequence of events – certain causes are followed by certain effects – But apparently spiritualistic phenomena do not come under these conditions. He hoped that some law might be discovered governing them, but he was not sanguine that the discovery would be made just yet.[79]

Even as he explored the potential links between Röntgen rays and cerebral ganglia, he toyed with the idea of spiritual beings that were "centres of intellect, will, energy, and power, each mutually penetrable, whilst at the same time permeating what we call space." These centers bore no resemblance to their former physical counterparts, Crookes surmised, but each nonetheless preserved "its own individuality, persistence of self, and memory."[80]

During the years when Crookes rejected, as beyond the range of proof, the spiritualist claim of communion with the dead, he did not therefore repudiate all belief in some sort of individual survival after death. On the contrary, the possible existence of spirit entities continued to enlist his support, although he no longer sought to get in touch with them. In 1874, when he wrote to Mme. B. in St. Petersburg that he had never obtained satisfactory evidence of contact with a deceased person who could be positively identified, he nonetheless added: "All I am satisfied of is that there exist invisible intelligent beings, who *profess* to be spirits of deceased people, but the proofs which I require I have never yet had."[81] Crookes also felt certain that invisible intelligent beings of nonhuman origin inhabited the universe, and he articulated opinions to that effect as early as 1871. C. C. Massey, writing to Colonel Olcott in December 1875, emphasized Crookes's belief in nonhuman spirits, describing him as a full-fledged occultist. On one memorable occasion, Crookes claimed to have exorcised a "fiend" who was distressing Florence Cook. Many years later, Crookes still enjoyed speculating about spiritual beings and wondered whether they might not reside somewhere in "4-dimensional space."[82]

Crookes ceased his variegated speculations about spiritual beings after his wife's death in May 1916. The event, following sixty years of marriage, left him "prostrated with grief"[83] and proved the stimulus that propelled him back to an earnest, totally uncritical spiritualist faith. In his eighties, Crookes no longer had the tough-mindedness to resist the tempting promises of human immortality and communication across the abyss of death. He became convinced that the "spirit photographer" William Hope had captured his wife's spirit on film and in December 1916 wrote enthusiastically to Lodge:

> At a Seance I had with Mr. Hope, at Crewe on the 10th inst. I obtained a Spirit photograph of my dear departed . . . I recognise it as very like what she was ten years ago, by comparing it with photos. taken by me about that date. There could not possibly be any trickery as the plate never left my possession and I did all the manipulating and developing myself. I am glad to say the possession of this definite proof of survival has done my heart much good.

Lodge, who had had some experience with Hope's chicanery in the past and knew that he was not above tampering with photographic plates, was extremely surprised to receive Crookes's endorsement of the "spirit photographer." He wrote a carefully phrased response to Crookes, which provoked further assurances from Crookes that Hope could not possibly have indulged in fun and games at his expense. "I went into the question of photographic trickery many years ago," Crookes informed Lodge, "and from confessions and admissions I had from tricksters I am acquainted with all the dodges possible." Ever the expert, Crookes would not believe that anyone could fool him, especially now, when the stakes were so high.[84] His assistant at the time, J. H. Gardiner, told Crookes's biographer that the negative from which Hope's photo of the spiritual Lady Crookes was produced, "showed clear signs of double exposure," but Crookes ignored the signs. He considered the picture "a Sacred Trust," as he told his close friend Alice Bird, and found in it "a slight consolation" for his loss, "as it proves the continuity of the Self after passing through the change called Death." He evidently resumed attending séances for physical phenomena and, from the results, was further persuaded that his wife was close by.[85]

It may be that, even before Lady Crookes's death, her husband was beginning the return to his earlier belief in contact between the dead and the living.[86] Such a change of heart would be understandable in an elderly man approaching the end of his life. If so, Lady Crookes's death only accentuated his need for reassurance about the continuing bonds of human communion. Whether the appeal of spiritualism attracted the octogenarian before May 1916, however, or whether his renewed spiritualist commitment must be seen as a direct response to profound grief, his was not the reconversion of an erstwhile skeptic. Since 1869, Crookes had balanced two approaches to the subject of séance phenomena: He had continued to believe in the activities of spirit beings while he simultaneously explored a range of naturalistic solutions to the spiritualist puzzle. At different times in his life, and under varying circumstances, he was more inclined toward one stance than the other, but neither point of view ever disappeared from the Crookesian perspective. After 1916, he simply lost the desire, and perhaps also the energy, to maintain the two approaches in some kind of equilibrium.

He was certainly not the schizoid personality that at least one contemporary critic suggested. W. B. Carpenter, who never missed a chance to berate Crookes for his spiritualist researches, deplored the fact that so gifted a scientist should waste his time, and stifle his critical faculties, in an effort to prove the reality of a psychic force. The "curious 'duality' of Mr. Crookes's mental constitution," Carpenter lamented, had led the physicist into the quicksands of quackery and deception. Crookes had great fun spoofing the theory of his two personalities and suggested that Carpenter ought to use proper scientific vocabulary, referring to them as the "Ortho-Crookes and Pseudo-Crookes."[87] In truth, Crookes's work was all of a piece. There was no tidy division between his physical and psychic experiments. Although the former were far more successful than the latter, they were motivated by the same curiosity and informed by the same aspirations. For Crookes, they were integral, intertwined parts of his lifelong fascination "with other conditions of existence than the familiar."[88] They shared a place in his efforts to elucidate the still mysterious forces that shaped the universe.

Crookes's unorthodoxy as a scientist was by no means reserved solely for mediums and séances. In the realm of physical matter, he likewise proposed provocative hypotheses that surprised his professional colleagues. Always he hankered after new discoveries; having launched his career with the discovery of an element, he was ever on the lookout for similar achievements and sometimes jumped too readily to novel conclusions. During his first studies of the radiometer – in which four vanes in a vacuum glass bulb rotated on a vertical axis in the presence of light – he believed that he had revealed a hitherto unknown motive force in light.[89] When it became apparent, thanks to experiments by Arthur Schuster among others, that residual gas molecules in the vacuum were responsible for the movement of the vanes, rather than the direct action of radiation on the vanes themselves, Crookes had to abandon his startling proposal. But his scrutiny of the radiometer was already prompting him to frame another hypothesis, one that deeply marked his approach to physics in the last decades of the nineteenth century. After protracted study of the behavior of gases in vacua, he became convinced that gases within a highly exhausted container were qualitatively different from the same gases before attenuation. Under normal atmospheric conditions, Crookes reminded the audience at the British Association meeting of 1879, gas molecules cannot

> move far in any direction without coming into contact with some other molecule. But if we exhaust the air or gas contained in a closed vessel, the number of molecules becomes diminished, and the distance through which any one of them can move without coming in contact with another is increased, the length of the mean free path being inversely proportional to the number of molecules present. The further this process is carried the longer becomes the average distance a molecule can travel before entering into collision; or, in other words, the longer its mean free path, the more the physical properties of the gas or air are modified. Thus, at a certain point, the phenomena of the radiometer become possible.

With further rarefaction, Crookes continued, phenomena occurred that were "so distinct . . . from anything which occurs in air or gas at the ordinary tension, that we are led to assume that we are here brought face to face with Matter in a Fourth state or condition, a condition as far removed from the state of gas as a gas is from a liquid."[90]

Crookes particularly applied his theory of matter in a fourth state – neither solid, nor liquid, nor gaseous – to the study of cathode rays, the rays of particles that stream from the cathode, or negative pole, when an electric discharge passes through a highly evacuated vessel. He thought that the particles were molecular, called the stream propelled from the cathode "radiant matter,"[91] and exposed himself to sustained criticism from physicists who found that his theory of the fourth state did nothing to clarify, and plenty to confuse, the phenomena under consideration. He was, of course, wrong about the molecules (cathode rays are composed of electrons), and his commitment to the concept of matter in a fourth state may well have prevented Crookes from discovering either X rays or electrons himself.[92] What matters most in terms of his own work, however, is the manner in which he used the concept, and how much he thought it helped to bring him closer to the heart of nature's mysteries. "I admit," he told a Friday evening audience at the Royal Institution in April 1879,

> that between the gaseous and the ultra-gaseous state there can be traced no sharp boundary; the one merges imperceptibly into the other. It is true also that we cannot see or handle matter in this novel phase. Nor can human or any other kind of organic life conceivable to us penetrate into regions where such ultra-gaseous matter may be supposed to exist. Nevertheless, we are able to observe it and experiment on it, legitimately arguing from the seen to the unseen.[93]

The passage from the seen to the unseen was the particular territory that Crookes sought to chart, and he faced the task with all the excitement of a pioneer embarking on great adventures into hidden realms. Whether his route led him into the bewildering regions of the mind or the unfathomed recesses of matter was of no significance to Crookes, for the adventure was a unified endeavor. In the 1870s, for example, if he wanted to examine action at a distance, which he considered the "most mysterious of all natural forces,"[94] a study of mind reading or tests with the radiometer seemed to him equally fruitful ways to approach the problem. His gaze was concentrated on "the border land where Matter and Force seem to merge into one another," and where, he assumed, "the greatest scientific problems of the future will find their solutions." There, it seemed to Crookes, lay "Ultimate Realities, subtle, far-reaching, wonderful."[95] He hoped that the future would witness "a profounder science both of Man, of Nature, and of 'Worlds not realized,'" a science that would recognize the importance of psychical research.[96] At times, he seemed to fancy himself the guardian of a new truth – a new dispensation, one is tempted to say – but it was never a purely spiritualist truth, confined to the possibilities of communication from the dead. It was the vision of a universe in which the bonds between matter and spirit were forged by forces whose secret intricacies Crookes yearned to master.

WILLIAM BARRETT

After the complexity and ambivalence of Crookes's opinions about the realm of spirit, the views of William Fletcher Barrett and Oliver Lodge appear comparatively simple and straightforward. Both men, in the course of their long affiliation with the SPR, made the transition from cautious psychical researcher to outspoken spiritualist. Theirs was a direct progression from less to more credence in spiritualist teachings, with none of the backtrackings and apparent detours that characterized the development of Crookes's attitudes. Unlike Crookes, too, religion figured overtly in the ultimate interpretation which both men imposed on their psychic inquiries. Neither Barrett nor Lodge could study the phenomena of mediumship in a morally neutral context: Each brought to their work for the SPR a deeply rooted Christian faith.

Among the scientific luminaries of the SPR, W. F. Barrett occupied a minor niche. He won no prestigious awards and held no distinguished academic appointments. Although he was a Fellow of the Royal Society and received a knighthood in 1912, he did not even earn a listing in the *Dictionary of National Biography*. His research on sensitive flames and on the electromagnetic properties of iron alloys, while meritorious, scarcely placed him among the pioneers of contemporary physics. He was, by contrast, a pioneer in psychical research and took the initiative in launching the SPR in 1881–2. Yet he embarked on psychical research without much enthusiasm – indeed, if his reminiscences can be credited, with decidedly negative expectations, thanks to the professional company he kept. "Between the years 1862 and 1867," he recalled,

> I was assistant to Professor Tyndall at the Royal Institution. The atmosphere surrounding my early years there was entirely opposed to any belief in psychical phenomena . . . Faraday I saw almost daily, before he left the Royal Institution . . . Faraday had published about 1855 his famous experiment on table-turning, showing how unconscious muscular effort accounted for what *he* saw . . . Tyndall also had denounced spiritualism as an imposture. Both Huxley and Herbert Spencer were frequent visitors to the Royal Institution laboratory, and both of these eminent men treated all psychical phenomena with contemptuous indifference.[97]

Against this background of scorn at worst and indifference at best, Barrett was astonished to witness experiments in mesmerism at the county Westmeath home of an Irish acquaintance. Astonishment, and initial incredulity, changed to lively interest, and Barrett was soon sending sensitive subjects into mesmeric trances, while he studied how the power of suggestion left its extraordinary impress upon them. On one occasion, in London, he persuaded a hypnotized lad to believe that the scientist had floated around the room, and this experience led Barrett to espouse the hallucination theory to explain allegedly spiritualist phenomena. Barrett was not, therefore, first drawn to these phenomena by the spectacular physical effects produced at séances in the 1870s, but by the possibilities of exploring the impact of mind on mind. It was the "divining power" that Barrett wanted to fathom, and his efforts to do so provide a common theme throughout the decades of his psychical research. Nonetheless, in touch with and stimulated by Crookes, Barrett did broaden his psychical experiences to include physical phenomena. He studied rappings; he sat with Slade; his conviction grew that, despite occasional fraud associated with physical mediumship, objectively real physical phenomena also sometimes manifested themselves.[98]

So convinced, he had the audacity to propose a paper on thought transference and spiritualist phenomena for presentation at the Glasgow meeting of the BAAS in September 1876. It was a courageous step for the thirty-two-year old physicist to take. He had recently been appointed professor of physics at the Royal College of Science in Dublin and had a reputation to make. The psychical investigations of Crookes, who was far better known and better established than Barrett, had provoked derision and even abuse just a few years before. Although the Biology Section of the British Association wanted nothing to do with Barrett's proposal, the Anthropology Department, one of its subsections, agreed to give his paper a hearing. Within the committee of the Anthropology Department, the decision to allow Barrett's paper was secured in large part through the efforts of Colonel Augustus Lane Fox, anthropologist, antiquary, and archaeologist, and by the deciding vote of the Department's chairman that year, who happened to be Wallace.[99]

The paper itself was a model of caution. Entitled "On Some Phenomena Associated with Abnormal Conditions of Mind," it dealt largely with Barrett's mesmeric experiments and introduced the subject of spiritualist manifestations only toward the end. Of these latter, Barrett suggested that many could be explained by the observer's self-deception, but added that he had heard rappings whose existence he could not doubt. "Is it not possible," he asked his audience, "that there may be *some* foundation for the stories of occasional supernatural irruptions into the present visible universe?" Barrett was by no means prepared to accept the spiritualist explanation of the phenomena he had seen, but, in concluding his paper, he warned his fellow scientists to

> be careful lest in a too hasty rejection of phenomena that seem incredible and inexplicable, according to received opinions, we are not

laying ourselves open to that same spirit of bigotry that persecuted Galileo. Surely the motto of every man of science ought to be found in Sir John Herschel's words, "The natural philosopher should believe all things not improbable; hope all things not impossible."

The significance of Barrett's paper, however, did not lie in his brief discussion of spirit rappings and the appropriate attitude for scientists to adopt in the face of such curiosities. It lay in his observations concerning the mesmeric trance and the powers of clairvoyance, or thought reading, which seemed to emerge in certain sensitives under trance conditions. "I convinced myself," Barrett explained,

> that the existence of a distinct idea in my own mind gave rise to an image of the idea in the subject's mind; not always a clear image, but one that could not fail to be recognised as a more or less distorted reflection of my own thought. The important point is that every care was taken to prevent any unconscious muscular action of the face, or otherwise giving any indication to the subject.[100]

In short, by 1876 Barrett was working toward a theory of thought transference, "the action of one mind upon another, across space, without the intervention of the senses."[101] He was certain that much more was involved than the normal processes of suggestion between two people – more even than heightened sensitivity and abnormally acute perceptive powers on the part of his subjects. What was involved he did not pretend to know, but he proposed a tentative hypothesis based on the model of electrical induction, or influence, across space. Defining thought as a form of nervous action, Barrett inquired at Glasgow: "May not nerve energy, whatever be its nature, also act by influence as well as conduction?" If the nerve force were "a radiant energy of some kind," might it not be "capable of throwing the nerve tissue of passive, receptive individuals into states of activity corresponding to the states existing in an active adjoining mind"?

The hypothesis was no doubt unusual in the context of a British Association meeting, but Barrett's approach was unexceptionable. Eschewing the supernatural and the vocabulary of the occult, he sought fruitful lines of inquiry by drawing analogies from other fields of physics. Since animal magnetism first drew Barrett's attention to clairvoyance and thought reading, it was not surprising that he initially looked for clues in the vast new regions of electromagnetism. In the discussion following his paper at Glasgow, Rayleigh praised Barrett for his "careful experiments," and shortly thereafter the *Spectator* expressed relief that "his paper does not make such immense drafts on the belief of men of science as Mr. Crookes's various papers on kindred subjects have made."[102] Nevertheless, Barrett's call for the British Association to appoint a committee of scientists to investigate mesmerism and spiritualism fell on deaf ears. For the next six years, until the founding of the SPR, Barrett continued his experiments in thought transference as his leisure time permitted, alone or in informal association with the Trinity group of Cambridge psychical researchers. Indeed, Barrett's work so impressed Myers that he wrote to the physicist in 1877, "Remember that if in any way your experiments can be helped by money, I shall be positively glad to know of what I consider so good a way of spending some."[103]

In later years, Barrett claimed virtually to have discovered thought transference as a subject of serious inquiry.[104] If that claim is too sweeping for the historical record to support, he clearly does deserve a prominent place among the early innovators in the field. "I remember his coming to Cambridge," recollected Mrs. Sidgwick after his death, "and reading to us a large budget of cases, experimental and spontaneous, collected by himself (largely by means of appeals through the press) and pointing to what we now call telepathy."[105] Nor did Barrett limit his audience to the sympathetic group at Cambridge. In July 1881, he published in *Nature* a provisional announcement of his findings, an announcement that emphasized the inadequacies of Carpenter's theories on thought transference. After years of persistent inquiry, Barrett asserted with confidence that thought reading could not be dismissed with Carpenter's favorite explanation – "namely, unconscious muscular action on the one side, and unconscious muscular discernment on the other." Again, Barrett spoke of "nervous induction," of the "somewhat analogous phenomena of electric and magnetic induction," but did not attempt any more definitive causal explanation. All he asked the readers of *Nature* was that they "suspend their judgment on this question" until he could bring before the public the startling evidence that justified his challenge to Carpenter.[106]

The strongest part of that evidence, as far as Barrett was concerned, focused on the Creery household. Since the physicist's interest in thought reading had been well advertised by 1881, Reverend Creery had notified him of his daughters' alleged talents, and at Easter of that year Barrett made his first investigative journey to the Creery home in Buxton. He was immensely excited by what he witnessed, so much so that he made his brave announcement in *Nature*. In November 1881, and again in February 1882, the Creery family received another scientific visitor, Balfour Stewart, who came accompanied by Alfred Hopkinson, professor of law at Owens College. Both men were convinced that, with Mary, Alice, Emily, Maud, and Kathleen Creery, as well as with Jane, the servant, they had "obtained results which neither . . . was able to account for by any received hypothesis."[107] Thereafter, the visits to Buxton continued, and the spring of 1882 brought Myers and Gurney who added their voices to the chorus of wonderment emanating from the Creery house. In the early months after the founding of the SPR, its initial Committee on Thought-Reading made the Creery sisters the focus of its first investigations. In the summer, the sisters arrived in Cambridge to be tested outside the domestic setting, and in November Barrett performed more experiments with them in Dublin. The results of the girls' "thought-reading" far exceeded what pure chance would have obtained, whether cards, names, or numbers were used as targets. The members of the SPR Thought-Reading Committee, of which Barrett was the honorary secretary, believed that they stood on firm ground when they concluded that

> there does exist a group of phenomena to which the word "thought-reading," or, as we prefer to call it, *thought-transference*, may be fairly applied, and which consist in the mental perception, by certain individuals at certain times, of a word or other object kept vividly before the mind of another person or persons, without any transmission of impression through the recognised channels of sense.[108]

Barrett and his SPR colleagues were, of course, fully aware that many alleged examples of thought transference resulted from nothing more extraordinary than "the interpretation by the so-called 'Reader' of signs, consciously or unconsciously imparted by the touches, looks, or gestures of those present."[109] They ruled out this humdrum explanation in the Creery case, however, because individual tests had been made with the sisters, when none of the other siblings were apprised of the object in mind, when none could therefore offer help in guessing it, and when the sister being tested was unable to see the person whose thoughts she was to read. Barrett accordingly hailed the Creery experiments as the first incontrovertible proof of the reality of thought transference, and for a while his conviction was contagious within SPR ranks. But in 1887, the career of the Creery sisters came to an inglorious end. Their talents, in any case, had been less and less apparent since 1882, and in October 1887, the Sidgwicks and Gurney caught two of them cheating during experiments at Cambridge. "I am very sorry to have to tell you," Sidgwick wrote sympathetically in breaking the news to Barrett,

> that we have undoubtedly detected the two Creery girls – Mary and Alice – in the use of a code of signals to produce spurious 'thought-transference' phenomena: or rather *two* codes – one of *visual* signals used when both the girls were together in the room without any barrier between them, and the other of signals to the *ears*, used when they were separated by a screen . . . We agreed that Gurney should go off

> to Manchester – before the girls returned from Cambridge – and endeavour to find out whether a similar code had ever been used before. He did so: and the result was a confession from Maud Creery, of which I enclose a copy. The signals she describes are exactly the same as those we discovered. So there can be no doubt that the code is of long standing, I fear. Meanwhile we kept Mary and Alice here, without saying anything of our discovery till the evening before their departure, when my wife taxed them with the code, in two private interviews. They at first denied it indignantly, but when she convinced them that she knew the signals they admitted it: but still said that it had never been used before this visit. This, I fear, cannot be believed after Maud's confession.[110]

If Sidgwick was disheartened by the exposure, Barrett could have been devastated. The sisters were his discovery. The foundations for his belief in the actual occurrence of thought transference among people in a normal mental state rested in large part on the genuineness of their powers. Barrett had had no doubt that the phenomena connected with the Creery sisters were valid and over the preceding few years had begun to formulate a system of beliefs predicated on that certainty. He did not intend to dismantle the system, and his response to the discrediting of the Creery sisters represented his first deviation from the attitude of impartial scientific inquiry that characterized his earlier ventures into psychical research. It was only reasonable that the unmasking of trickery in 1887 should cast a highly unfavorable light on the investigations of 1881–2, but Barrett would not accept that verdict. While deeming the revelations of 1887 a regrettable blemish on the Creery girls' characters, he cautioned Sidgwick not to overreact. Late in October 1887, Barrett wrote to him:

> I expect the natural alarm which the Cambridge Expts. has [sic] caused in our minds is probably apt to make us unjust in our judgment of the Earlier Experiments. For my own part I am convinced that enough entirely trustworthy expts. exist with the Creery family to make it unwise to expunge the whole of their evidence.[111]

Nor did Barrett ever expunge the Creery sisters from his records. When he contributed a volume on *Psychical Research* to the Home University Library of Modern Knowledge in 1911, the Creery family figured prominently in his chapter on "Thought-Transference in the Normal State of the Percipient." He described the Creery experiments as "unimpeachable," a view he tacitly reiterated in a letter to the *Times*, on 20 December 1924, where he traced his belief in the genuine nature of thought transference back especially to the experiments with the Creery family.[112] Clearly, by the late 1880s, Barrett's critical stance of the neutral scientist had vanished, giving way to a vehemently partisan apologist. He had grown emotionally committed to the reality of psychic phenomena and was, in fact, becoming a spiritualist. The standards of exactitude that he practiced as a physicist no longer controlled his performance as a psychical researcher.

Barrett's disagreement with Sidgwick over the decision to disregard all of the Creery experiments exemplified a growing strain in his relations with the Cambridge group. He never enjoyed the close friendship of the Sidgwick circle, and he exasperated them periodically with what they perceived to be the poor quality of his work. From 1882 on, Myers's letters to him were filled with exhortations, sometimes heavily underlined, for Barrett to write up his cases with fuller details, to meet printing deadlines with completed reports, and not to leave loose ends dangling for the harassed SPR editorial staff.[113] Barrett, for his part, after the turn of the century grew increasingly disgruntled with the Society's management, which he believed to have fallen into the hands of Alice Johnson. His correspondence with Lodge in 1912 and 1913 bristled with criticism of Johnson – with the narrow range of investigations that she encouraged as SPR research officer and with her alleged indifference to all psychic inquiry that did not focus on the cross correspondences, her consuming interest. He threatened to leave the SPR or at least to resign from its council, on which he had served for three decades, and he blamed the "foolish policy" of the Society's officers for provoking his recent heart attack.[114] There can be no doubt that Barrett was quick to take offense and enjoyed assuming an aggrieved air – not the most desirable qualities in a colleague.

Nonetheless, the breach between Barrett and the Cambridge group in the SPR should not be overemphasized. Sidgwick and his friends never failed to give Barrett credit for his initiating role in the Society's history, and early in 1899, with their endorsement, the SPR council elected him an honorary member, to succeed none other than Gladstone, who had died the year before.[115] Sidgwick himself urged Barrett to undertake a study of dowsing for the SPR and to write the report on the divining rod that appeared in the Society's *Proceedings* in 1897 and 1900. Furthermore, he offered to help finance Barrett's research on this elusive topic.[116] Sidgwick was not inclined to select incompetents for important tasks, nor to offer subsidies to an experimenter in whom he had no confidence. Nor was the SPR council likely to have chosen as president for 1904 a man whom they regarded with little respect.[117] Barrett, in turn, admired Mrs. Sidgwick's judgment and occasionally sent her manuscripts in proof so that he could benefit from her comments and criticism before publication.

Ironically, Barrett's long inquiry into dowsing served forcefully to confirm his departure from the role of cautious investigator admired by Sidgwick in the 1870s. From the early 1890s, Barrett devoted much of his time to the divining rod. He prepared two authoritative papers on the subject for the SPR and subsequently continued his research for many years as he composed an exhaustive analysis, published posthumously in 1926.[118] What attracted Barrett to this concrete, physical form of the divining power was the simplicity and ease of its investigation. It seemed to him the most direct and conclusive way of ascertaining whether in fact there existed a supernormal power of perception. With dowsing, none of the problems associated with telepathy tests could develop to complicate and discredit the experiments: no possibilities of codes between agent and percipient, no unconscious facial expressions by the agent to help the percipient along. All that was necessary was the dowser and the divining rod – the forked twig that for centuries had been used by certain sensitive people to locate underground water, mineral lodes, buried treasure, lost archaeological sites, and even missing persons. What, Barrett wanted to know, was the stimulus that triggered the seemingly involuntary movement of the rod over the place where the object of search lay hidden? In the language of psychical research, what stimulated the dowser's "motor automatism" at precisely the crucial moment?

To elucidate the phenomenon, Barrett began by eliminating some obvious explanations. In the first place, his many years of investigation convinced him that purposeful deception by the dowser could be ruled out, except in cases of obvious fraud; so, too, could "chance coincidence or an 'eye for the ground'," acquired through experience in surveying geological configurations.[119] Like Pierre Janet, Barrett maintained that the dowser's conscious effort played no part in the "sudden muscular spasm, which twists the forked rod with a vigour he cannot control." Yet it was only begging the question to assert that unconscious muscular action was at work, allowing the dowser involuntarily to react to "hyperaesthetic discernment of surface signs too faint or complex to be perceived by the ordinary observer." Barrett did not doubt that some extraordinary subconscious acuity of perception might be involved from time to time, but he felt sure that the puzzle could not be fully explained by the working of the sensory organs. He looked, by contrast, to an "avenue of knowledge other than the recognised channels of sense," and sought "the key to the mystery . . . in the *psychical* and not in the physical world." Tentatively at first, in the reports for the SPR *Proceedings*, but with ever greater conviction in his own independent writings, Barrett argued for "some kind of transcendental discernment" that operated in the most successful dowsers, even when they were blindfolded.[120]

Barrett's view, quite simply, was that the dowser possessed a supernormal perceptive power. "I have," he wrote to Wallace in 1905, "a vast amount of material unpublished on 'dowsing' and am convinced the explanation is subconscious clairvoyance."[121] In the last two decades of his life, Barrett's conviction only became more firmly rooted; by the time he published his "Reminiscences" in the SPR *Proceedings*, in 1924, he claimed "conclusive evidence" for such a clairvoyant faculty. This was altogether distinct, he believed, from a telepathic faculty that might explain the perception of distant objects, or the reception of thoughts, reflected through another human mind. Telepathy required the cooperation of at least two minds. The clairvoyance of the dowser worked independently of other human influence and seemed to bear relation to migratory and homing instincts in birds and animals.[122] In the 1920s, Barrett no longer drew his analogies from the inorganic cosmos of physics, but from the organic universe of biology, and he did so in order to hint that the dowser's talent somehow placed him in touch with the vast world of nature. There was something about dowsing that Barrett could merely suggest and never precisely define – a gift, granted only to a small number of people, that gave them profound sensitivity to the most hidden secrets of life.

His conclusions about dowsing led Barrett down paths of inquiry that deviated widely from those of the men and women who guarded the credibility of the SPR. Because those paths seem to reflect unfavorably on Barrett's critical acumen and reflective judgment, it is only fair to recall that he was a respected member of the scientific community throughout his life. His election to the Royal Society came in 1899, the year he announced the discovery of a silicon–iron alloy and long after his avowed spiritualist sympathies could have allowed that body to dismiss him from consideration had his scientific contributions been negligible.[123] As far as his opinions on dowsing were concerned, even today the action of the divining rod continues to baffle investigators, while in Barrett's own day, after all, J. J. Thomson considered dowsing an effective way to locate water. For the purpose of trying to understand Barrett's mature attitude toward the role of modern science, the question to ask is not how his professional colleagues regarded him in the later phases of his career, but how much he valued their good opinion. "About the narrowest minded, most intolerant & least sympathetic minds at present," Barrett wrote to Lodge in 1893, "are those whose eyes are for ever glued to the microscope of their own special branch of science."[124]

Barrett's eyes were hardly glued to the microscope. On the contrary, he wanted to cultivate a totality of vision that embraced both the physical and the spiritual worlds. When he argued for a form of clairvoyance that transcended all previously acknowledged human capabilities, he was involved in constructing an argument that ultimately led him to the immortality of the soul – the culminating conclusion of Barrett's psychical research. He had begun his studies with thought transference and had endorsed the possibility of extrasensory communication between two or more individuals, in both abnormal and normal states of mind. With extensive study of dowsing, however, and wider experience of spiritualist phenomena, he concluded that telepathy could not account for everything that came under the scrutiny of the psychical researcher. He accordingly hypothesized that a supernormal clairvoyant prowess existed, possibly connecting its possessor with the fundamental physical matter of universal life. Yet even that hypothesis could not suffice alone. Barrett came to insist further that there were phenomena observed and messages received through mediums that could only be understood as the communicative efforts of unseen intelligences or discarnate beings. His years of inquiry into the functioning of mind persuaded Barrett, as he wrote to the *Times* a few months before his death, that there dwelt in the human personality "an immaterial entity, a soul or transcendental self, which is not limited to the confines of our body, or conditioned by matter, time, or space."[125]

If two or more minds connected with living bodies could communicate without using sensory equipment, it followed, Barrett reasoned, that the mind could function independently of the body. It further followed that the mind need not cease to exist when its physical frame perished. When he first began to explore this line of reasoning, Barrett realized how imponderable it was. By 1890, he provisionally accepted the spiritualist hypothesis, but admitted the difficulties associated with it. "The simplest solution to me, at present," he wrote to Lodge concerning certain problems connected with telepathy,

> is the survival in some form, without the clothing of gross matter, of the mind & individuality of deceased friends. How we came across their path, or they ours, or why communication is so difficult & so unsatisfactory are points that time will I hope clear up.[126]

As time passed, Barrett was less troubled by unanswered questions and more serenely confident that deceased friends were getting in touch. In April 1908, Mrs. Sidgwick returned to him the proofs of his book *On the Threshold of a New World of Thought*, which he had asked her to read. In doing so, she expressed surprise that Barrett was so ready to proclaim the operation of an unseen intelligence on one occasion under discussion and so firmly convinced that the intelligence belonged to Myers.[127] But Barrett had already slipped beyond the influence of Mrs. Sidgwick's caution, for he was ready to embrace spiritualism without restraint or condition. In 1924, in his "Reminiscences," he unequivocally proclaimed his certainty that evidence published by the SPR "decidedly demonstrates (1) the existence of a spiritual world, (2) survival after death, and (3) of occasional communications from those who have passed over."[128] Indeed, his posthumously published *Death-Bed Visions* was little more than a long series of cases marshaled to illustrate one kind of such communication: the presence of spirit forms at the deathbeds of beloved friends and relatives.

Barrett's ultimate assurance that the human spirit survives death was not the result of scientific inquiry. "It seems to me highly probable," he mused, "that the experimental discovery of the survival of human personality after death will always elude *conclusive* scientific demonstration."[129] He appears, instead, to have found the decisive evidence in theology, specifically in his own Christian faith. In an extraordinary passage toward the end of the obituary of Crookes, which he wrote for the SPR *Proceedings* in 1920, Barrett compared the disappearance and reemergence of "Katie King" to "the appearance of our Lord after His resurrection. If we accept that cardinal fact of the Christian religion," he observed, "we have less difficulty in accepting Crookes' statements." A decade earlier he had agreed with C. C. Massey that religion alone "can demonstrate immortality, and only through that can it be rightly conceived."[130] For Barrett, psychical research was evidently not the support desperately needed to shore up a sagging religious faith, as it was for some of his colleagues in the SPR. Christianity, by contrast, was the prop that gave authority and significance to his psychical research.

Like Sidgwick, Myers, and Gurney, Barrett was the son of a clergyman. Unlike them, however, he seems never to have suffered the agony that came with questioning the fundamental Christian doctrines in which he had been raised. He could accept with equanimity the idea that modern biblical scholarship had cast into doubt "the Divine authorship and verbal inspiration of the Pentateuch, Psalms, and Apocalypse."[131] Such knowledge failed to undermine the foundations of Barrett's Christianity, for he was able to endorse, in a way that Sidgwick could not, the liberal Christian contention that the essence of religious truth remained untouched by purely historical or literary arguments. There is no evidence that Barrett ever tormented himself with theological controversy. In short, he never had to turn to spiritualism as a surrogate religion, because he never outgrew the faith of his fathers. "During his whole mature life he was a devout and earnest Christian and he died a communicant of the Church of England."[132] His writings on psychical research, especially after the turn of the century, frequently reveal how closely his Christian faith underlay his deepening spiritualist convictions. It was no coincidence that, in an address to the Spiritualists' National Union in 1910, Barrett paraphrased sacramental language to describe the motion of the divining rod in the dowser's hands as "the outward and visible sign of [an] inward and singular gift."[133]

By the 1890s, Barrett belittled the expertise of the scientific specialist, not because it failed to provide answers to certain, limited problems, but because those problems no longer satisfied him. Like Crookes, he was attracted to larger questions and wanted to probe nothing less than the riddles of life and death. Adamantly insisting that the assumptions of scientific naturalism could not begin to address those mysteries, he sought to expand the physicist's horizons with inspirations from theology. Untroubled by the disapprobation of fellow scientists, Barrett was fully satisfied with the conclusions he drew from his synthesis.

Barrett's very contentment, however, presents difficulties for anyone seeking to understand the motives behind his scientific unorthodoxy. If as a Christian, Barrett already accepted the promise of immortality, what other reason impelled him to devote so many years to the obscure and frustrating phenomena of psychical research? And if he already believed in the soul's survival of physical death, why did he explore the shadowy realms of psychic phenomena for more than a decade before the lines of his reasoning began to converge in the direction of spiritualism? A belief in human immortality does not, of course, necessitate a belief in the possibility of communion between living and dead, and it seems reasonable to accept Barrett's claim that he began psychical research without spiritualist convictions, in a mood of scientific curiosity and eager to learn what mesmerism and thought reading tests could tell him about the human mind. He was, additionally, intrigued by a topic that challenged all the "givens" of science in the second half of the nineteenth century. One can only speculate here, but the evidence suggests that an explanation in these terms is tenable. Apparently as Barrett proceeded with the Creery investigations, he was not fully conscious of the extent to which his religious predisposition colored his reaction to the sisters. When the presumably scientific evidence in their favor collapsed, the evidence furnished by Christian doctrine – always latent in Barrett's thoughts – replaced the earlier justification for his belief in the reality of mental telepathy. Perhaps Barrett did not realize at the time what a vast alteration had occurred in his thinking. He certainly realized it by the turn of the century, however, and no longer claimed that scientific demonstration proved the independence of mind from body and the survival of the human spirit.

Furthermore, Barrett realized that psychical research could be employed on behalf of a cause that he valued immensely: the survival of religious institutions in the ever more materialist civilization of the West. "It is evident," he affirmed several decades after countless other spiritualists had already made the same point, "that psychical research will ere long be regarded, by all thoughtful men, as the most valuable handmaid to religion."[134] If Barrett's own faith did not need support, he knew many people who miserably sought some meaning in the whirling particles of matter that composed the cosmos. He hoped that the records of psychical research might offer enough reassurance of the spirit's immortality to preserve for them the religious frame of mind. The antimaterialist refrain had been sounded for so long, with such little success, by the time that Barrett contributed his belated comments that they appear banal, or worse ridiculous, in 1924. Nonetheless the way in which Barrett wove together his psychic inquiries and his Christian impulses to protest materialism forms the principal element of his spiritualist thought.

For all Barrett's sterling record of church attendance and his steady belief in the chief tenets of Christianity, it is fair to say that he was not quite an orthodox Christian. As he grew older, his religious opinions came to be heavily tinged with pantheism. Distressed by "the barren formalism" of official religion, Barrett hoped for a revival of the true spirit of religion, "an awakening to the consciousness of the indwelling spirit of God."[135] He had earlier sketched his thoughts on the union of the many and the One in letters to such correspondents as Lodge and Mark Twain, to whom he confided his belief that "whatever is most gracious and ennobling in art, literature and science is doubtless but the faint responsive echo within the human mind of the ineffable music of the Divine Idea."[136] By the last decade of Barrett's life, scarcely a piece of writing flowed from his pen that did not explore the subject of cosmic unity. He was not content merely to reiterate the mystical doctrine of an intimate link between the individual soul and God; he also sought to shed light on "the existence of a *soul in Nature*," "a purposive and a pervasive factor, running throughout the whole realm of nature," binding the separate parts to the totality of the universe.[137] Perhaps this factor had something to do with the dowser's faculty of sightless perception. In any case, there might well exist a universal bond similar to telepathic communication between individual minds. "We may conceive the human race," Barrett wrote in 1918,

> as constituent cells, the many members, of one Body, to which all are related, and yet all transcended in one Immanent and Supreme

> Being. And may there not be some telepathic inter-communion be-
> tween the Creator and all responsive human hearts, to some being
> given the inner ear, the open vision, and the inspired utterance?[138]

Just as Barrett had grown sure that ideas could be communicated telepathically across space from one mind to another, without the usual help of sensory perception, so he became certain that religious truth, knowledge of the ultimate meaning of God, could be grasped intuitively, by an internal experience of illumination totally distinct from rational thought processes. The theories that Barrett developed as a psychical researcher and the pantheistic beliefs that he came to espouse as a deeply religious man were mutually supportive in every respect. It would be difficult to say which set of opinions preceded the other; they seem to have unfolded together, over a number of decades. Underneath both lay Barrett's fascination with the divining power, with the ability to probe the profoundest mysteries of nature, to understand the most fundamental relations among human beings on the one hand, and between mankind and the nonhuman world on the other. He wanted to learn what connected the infinite parts of the universe to each other and to the whole, and once having done so, to proclaim the unity beneath the seemingly endless multiplicity of life.

In a sense, his study of thought transference set the pattern for his pursuit of these boundless questions, for while he first suggested electromagnetic induction as a model with which to explain the puzzling exchange between two minds, he ended by rejecting any theory of "mechanical transmission."[139] Processes and functions derived from the physical world, he emphasized in his later writings, were inapplicable to the riddles of the psychical world. He first delineated the unalterable incompatibility of the two realms in his presidential address to the SPR in January 1904, when he explained why the methods of laboratory science could not truly contribute to psychical research, despite SPR claims to the contrary. "Psychical experiments," he insisted,

> depend on the mental state of the subject; you may tell a person to
> do something, but whether he does it or not depends on the person
> addressed. Physical experiments are independent of our volition; a
> magnet attracts iron, or sets itself in the magnetic meridian, irre-
> spective of our mental condition. This obvious difference between the
> two sets of phenomena is constantly overlooked. Physical science
> excludes from its survey the element of personality, with which we
> have to deal and over which we have little or no control.[140]

After 1904, Barrett hammered out the same theme incessantly.

It is easy to lose sight of the physicist in the writings of Barrett's later years. He seemed at times almost to loathe physical science and to be anxious above all else to preserve the domain of psychical research from its encroachments. He no longer urged other scientists to join him in exploring the subject, as he did in the 1870s, for he had concluded that their skills were antipathetic to an investigation of spirit. The conviction "that the physical plane is the whole of Nature, or at any rate the only aspect of the universe which really concerns us," was totally unacceptable to Barrett. It was "paralysing," a "false and deadly assumption," under which "all wider views and spiritual conceptions wither and die as soon as they are born."[141]

At heart, however, Barrett was no opponent of modern science; nor did he ever cease to consider himself a professional physicist. He revered the accomplishments of science, above all its establishment of "the great world order," "a universal reign of law." What he abhorred in contemporary science was simply its role as censor, its assumption of the authority to dictate what was, or was not, within the possible realm of natural law, and its exclusion of what he insisted were not supernatural, but merely supernormal, phenomena. After all, he contended, "all phenomena – however novel and inexplicable they may appear to be – are really *natural*; only God is above and beyond Nature."[142] It was the standard distinction by which spiritualists claimed the authority of science for what they alleged to happen at their séances. If the phenomena of psychical research were supernormal, then they belonged in the natural world order and could assert their rightful place in the universal reign of law. Just as Sidgwick, Myers, and their friends hastened to point out the philosophic and ethical inadequacies of materialism, yet could not wholly dispense with the interpretative mold of scientific naturalism, so Barrett likewise tried to have it both ways. At one and the same time, he could acknowledge the paramount importance of naturalism as the guiding philosophy behind modern science and yet could extend the boundaries of that philosophy to infinity. He could claim to respect scientific naturalism, while utterly ignoring its limitations and radically undermining its very meaning. Although he wanted to keep psychical research free of meddlesome intrusion by unsympathetic scientists, he was nonetheless arguing that psychical research legitimately belonged within the scope of an enlarged natural science.

Such ambivalence suggests that Barrett did not want to relinquish his membership card in the confraternity of scientists who practiced their profession in the early twentieth century. Yet it also suggests that he would have belonged much more comfortably in a far older scientific tradition than the one emerging in his lifetime. Although he was a career physicist who supported himself by his professional skills, his fundamental sympathies lay with the seventeenth-century virtuosi. Those natural philosophers had viewed science and religion in harness, pulling together toward the same transcendent goal: veneration of the Deity and elucidation of His glorious cosmic design. For Barrett, science and theology were still endeavors so intimately related as to become, at times, inseparable, and he recognized that, for intellectual kinship, he should look back to the likes of Robert Boyle and Joseph Glanvill in the early years of the Royal Society, rather than to his own contemporary Fellows.[143] Indeed Glanvill was a particular favorite of Barrett's, a scarcely surprising choice given Glanvill's resistance to Hobbesian materialism and his deep interest in spirits, apparitions, poltergeists, and even thought transference. It was Glanvill who, in *The Vanity of Dogmatizing* (1661), introduced the story of the Scholar Gypsy made famous two hundred years later by Matthew Arnold's poem. To explain this early tale of telepathic powers, Glanvill suggested a mechanical transmission of ideas between the scholar and his friends – a transmission that made use of that elusive substance ether, which might, in some mysterious way, transfer brain signals across great distances. For an alternate explanation, Glanvill drew on the seventeenth-century Cambridge Platonist notion of a world soul, which might likewise serve, by equally unfathomed means, as the medium of transfer for thoughts across space.[144]

The intellectual context had vastly altered by the time Barrett set forth his own views on thought transference, eventually rejecting the idea of mechanical transmission and embracing the concept of a "soul in Nature" as the central pillar of his own personal philosophy. Yet one wonders if Barrett fully recognized the gulf that separated the relativistic world of modern physics from the milieu of Restoration science. By the end of his life, he seemed more at home in the latter than in the former. When making some point about the significance and validity of psychical research, it was Glanvill whom Barrett would quote, as if Glanvill's was a voice contemporary critics would listen to respectfully.[145] To summon Glanvill in the war against materialism in Barrett's day was as effective as deploying archers against a bombing raid. While his contemporaries in physics were overhauling the Newtonian universe, Barrett yearned for the natural world of his intellectual ancestors, where every animalcule was pregnant with divine purpose.

In keeping with the received wisdom of his own century, Glanvill saw the exploration of the natural world as a form of religious inquiry, a means of proving the existence of the Creator by revealing the underlying pattern beneath the apparent chaos of nature. In Barrett's day, such arguments had long outworn their utility, and he sought instead to defend his Christian faith by stretching the boundaries of the natural world to include what today is dubbed the paranormal. The hope of leavening contemporary science with the vision of universal oneness that delighted him as a man of religion and a spiritualist also inspired him to attempt a new harmony of science and faith. In the process of working out that synthesis, Barrett slipped from the role of scientist to that of pseudoscientist, allowing belief to precede valid demonstration. The will to believe replaced the empirical method, for Barrett found the former effective and the latter useless in the fight against the dead end of materialism.

OLIVER LODGE

In the July 1912 issue of the scientific quarterly *Bedrock*, Dr. Ivor Tuckett acerbically protested Barrett's proclamation of telepathy as an established fact, and, while he was on the subject, cast his net of condemnation widely enough to catch Sir Oliver Lodge as well. "It is not necessary," he concluded,

> either to regard the phenomena of so-called telepathy as inexplicable or to regard the mental condition of Sir W. F. Barrett and Sir Oliver Lodge as indistinguishable from idiocy. There is a third possibility. *The will to believe* has made them ready to accept evidence obtained under conditions which they would recognise to be unsound if they had been trained in experimental psychology.[146]

Tuckett's was the language of partisan debate, but the close juxtaposition of Barrett and Lodge was justifiable enough. In many respects, their approach to psychical research and the conclusions that they derived from their investigations were strikingly similar. Like Barrett, Lodge refused to believe that modern science had adequately defined what was possible within the order of nature and deplored the construction of "a premature fence or boundary" beyond which scientists could not be allowed to venture in their attempts to understand man and the cosmos. "As to 'impossibility'," Lodge told a correspondent in 1912, "nothing is *a priori* impossible but the self-contradictory," and the following year, in his presidential address to the British Association at Birmingham, he urged his fellow scientists to remember that "the universe is a larger thing than we have any conception of, and no one method of search will exhaust its treasures." Like Barrett, too, Lodge insisted that the findings of psychical research never contradicted established knowledge, but enlarged and expanded it, since they also belonged to the natural world.[147] Above all, Lodge shared with Barrett the Christian background against which he pursued his psychical studies and which gradually helped transform Lodge's initial skeptical curiosity about psychic phenomena into an unalterable spiritualist faith.

As Barrett had been influenced by the great men of the Royal Institution, so Lodge's earliest attitudes as a scientist were molded by Huxley, Tyndall, and W. K. Clifford, whose lectures he attended at their respective institutions – the Royal College of Science, the Royal Institution, and University College, London – during the course of adolescence and young adulthood in the 1860s and 1870s. Under their tutelage, he had learned to concentrate on the material world and to consider that sufficient subject for a scientist's inquiry. He was not particularly interested in, nor receptive to, the concept of thought transference when it was first explained to him by Edmund Gurney, who quite coincidentally attended some lectures on mechanics and physics that Lodge gave at University College. Through Gurney, Lodge met Myers, heard about the launching of the SPR in 1882, but made no effort to join the ranks of its members. In the previous year, at the age of thirty, he had been appointed to the chair of physics at the newly established University College, Liverpool, and he was busy with his "duties of teaching orthodox physics and mathematics."[148]

The unorthodox, however, hovered close at hand. In 1883, Liverpool's citizens were treated to clever "thought transference" performances by W. I. Bishop. Capitalizing on the popularity of the willing game, Bishop occasionally took his show out of the theater and onto the streets of the city where, blindfolded, he could nonetheless locate hidden objects. Since he could only find the object in question when a member of the audience – or someone who knew its hiding place – held Bishop's hand, SPR spokesmen inclined toward a physical explanation of his talents. For once they agreed with Carpenter that unconscious muscular reactions on the part of the hand-holder might give the showman signals that indicated whether or not he was approaching the hidden object.[149] But when Malcolm Guthrie, JP, the head of a leading Liverpool drapery firm, reported to the SPR that two of his young female employees were particularly successful at the willing game, the Society began to pay more attention to the Liverpool excitement. In May 1883, Barrett traveled to Liverpool, and a report on Mr. Guthrie's experiments appeared in the first volume of the Society's *Proceedings*.[150] Like the Creery sisters, the young women, Miss Edwards and Miss Relph, seemed to have an uncanny gift for naming objects, words, letters, numbers, and even entire scenes imagined by the agent or agents in the experiments. They could reproduce target drawings with a high degree of accuracy and, although some of the experiments were undoubted failures, no evidence of chicanery or collusion was ever uncovered. Guthrie was sufficiently impressed by the experiments to seek further scientific confirmation of the women's abilities. Naturally, he turned to Liverpool's new University College, of whose governing body he was an active member, and thus Lodge was drawn into the psychical research that was to fascinate and inspire him for the rest of his life.

Lodge's debut into psychical research was not a particularly challenging one for a scientist of his caliber. He repeated Guthrie's experiments with the young women in a normal waking state, varying the tests only slightly, so that the investigation remained Guthrie's, as tested and corroborated by a trained physicist. The report that Lodge then submitted to the SPR *Proceedings* was pedestrian, except for the fact that it recorded Lodge's belief in the actual transfer of thoughts by extrasensory means. After testing Edwards and Relph both at the drapery firm and at University College, after testing, too, for the transference of tastes, smells, and even pains, Lodge came to the significant conclusion that "one person may, under favourable conditions, receive a faint impression of a thing which is strongly present in the mind, or thought, or sight, or sensorium of another person not in contact." Anything more than that he was not prepared to say in 1884. "How the transfer takes place," he continued,

> . . . or what is the physical reality underlying the terms "mind," "consciousness," "impression," and the like; and whether this thing we call mind is located in the person, or in the space round him, or in both, or neither; whether indeed the term location, as applied to mind, is utter nonsense and simply meaningless, – concerning all these things I am absolutely blank, and have no hypothesis whatsoever.[151]

Lodge had, temporarily, more questions to pose than answers to suggest to the members of the SPR, but he could no longer claim that psychical research held no interest for him. Early in 1884, he had joined the Society and he remained closely involved in its activities until his death in 1940. He was a council member for years, served as the Society's president in 1901–3, and then again assumed that office in 1932, with Mrs. Sidgwick, on the fiftieth anniversary of the SPR.

Throughout most of the 1880s, Lodge remained content to believe in the possibility of telepathy between the living and did not proceed to argue for potential communication with the minds, spirits, or souls of the dead. He only took that momentous step after 1889, following séances with Leonora Piper, a trance medium from Boston, Massachusetts, who had attracted the favorable attention of both William James and Hodgson. Despite the latter's reputation for exposing frauds and blowing the whistle on chicanery, Hodgson had nothing but respect and praise for Mrs. Piper's mediumship. Her ability, when entranced, to provide detailed information about her sitters' lives, habits, friends, and relatives, both living and dead, persuaded him, as it persuaded

James, to credit her with some supernormal power. In order to put that power to an effective test, it was decided to invite her to give sittings in England. There an entirely new set of surroundings would help determine how much of her success in Boston arose from familiarity with the general environment and perhaps from subconscious knowledge concerning the lives of fellow Bostonians.[152]

Piper's first visit to England in 1889–90 added Lodge, recently elected a Fellow of the Royal Society, to the list of her admirers. Together with Myers and Leaf, he formed a committee to study her mediumship, an inquiry that they pursued in Cambridge, London, and Liverpool. Usually they invited other guests to meet Piper and these, introduced under assumed names, frequently marveled at her ability to recount personal information about themselves and their families that, they were convinced, she could not have acquired through normal means. Sometimes it seemed that her knowledge could only have come to her from the deceased. Trickery or purposeful deception on her part appeared out of the question. In Boston, Hodgson had hired detectives to follow her and ascertain whether she had confederates who supplied her with information, or whether she herself did research on potential sitters. Piper successfully passed that test and while in England was most cooperative in allowing SPR investigators to search her luggage and to scrutinize her mail. There was nothing to suggest that she turned to outside sources, human or literary, for the contents of her trance conversations. "I took every precaution that I could think of," Lodge recalled in his autobiography,

> and on the whole the result of the Piper enquiry was conclusive. The report is to be found in volume vi. of the *Proceedings*, and directly after her visit I went with my wife to Alassio in Italy, at length thoroughly convinced not only of human survival, but of the power to communicate, under certain conditions, with those left behind on the earth.[153]

Lodge was actually more cautious in that 1890 report than he recollected forty years later. Coauthored with Myers, Leaf, and James, the report had a long section by Lodge, but it did not attest to the immortality of the soul. On the contrary, while he clearly expressed his confidence in the genuine nature of Piper's trance condition, Lodge argued that her control – a self-proclaimed French doctor named Phinuit – was "probably a mere name for Mrs. Piper's secondary consciousness." Lodge conceded that Phinuit occasionally "fished" for the right information from his sitters, and Lodge held that well-developed thought transference between the living was the best way to explain Mrs. Piper's astonishing command of obscure, intimate infor-

mation about total strangers.[154] It was only in a sort of appendix, tacked
on near the end of the report, that Lodge allowed himself to speculate
on the meaning of a certain class of facts that, on rare occasions,
emerged from the medium's trance discourses. These, he maintained,
could not have been known to any sitter present and thus could not
have been conveyed telepathically to Piper. What possible alternative
explanations existed? Perhaps, Lodge tentatively proposed, telepathy
from distant persons might be involved, although he was not yet convinced that thought transference could operate between minds geographically separated. Only "as a last resort" did he suggest "telepathy
from deceased persons."[155] His attitude at the time appears to have
been marked by a desire not to stray too far from the scientific phenomena and explanations with which he felt most familiar. "You will
see," he wrote to Myers in the autumn of 1890,

> that it is by sticking definitely to Physics that I shall be of most service
> [to the SPR], for only so shall I carry any weight. Without saying a
> word against Barrett it is clear that with all his enthusiasm & ability
> his [comparative?] neglect of Physics has made him less powerful than
> he ought to be.[156]

The communication of ideas from deceased persons, which Lodge
put forward as a last resort in 1890, became an increasingly plausible
hypothesis for him in the following years. He avidly followed the further developments of Piper's mediumship throughout the 1890s and
was deeply impressed when, thanks to her, Hodgson came to embrace
the reality of spirit communication.[157] The conversion of that arch-
skeptic remains a difficult mystery to unravel. Piper's mediumship is
not above criticism, and C. E. M. Hansel delivered a thoroughgoing
condemnation in *ESP: A Scientific Evaluation*. He dismissed "Phinuit"
as ridiculous – the "doctor" could scarcely speak French and knew
precious little about medicine. As for Piper's later, more famous control, "G. P., " Hansel pointed out that the family of the young man
whose spirit G. P. supposedly represented utterly repudiated the medium. Indeed, most of Hansel's case against Piper comes from the
family, who obviously resented their member being dragged into prominence through association with a spiritualist medium. On the other
side of the argument, Sidgwick himself was prompted by Piper's mediumship to write in December 1890: "I think we are on the verge of
something important."[158]

Back in Boston after her trip to England in 1889–90, Piper came
under the virtually exclusive supervision of Hodgson. He was the principal channel through which news of her alleged feats reached the SPR.
Her high standing in SPR records is largely a result of his endorsement.

If she was a fraud, merely making shrewd guesses and, when possible, obtaining advance information about her sitters, he would have had to know about it. Yet he repeatedly vouched for her integrity, and, unlike Crookes, Hodgson knew how to guard against deception at séances. Certainly nothing else in his career as a psychical researcher suggests that Hodgson was capable of betraying his colleagues by purposely foisting a fraudulent medium on them. Rather the evidence suggests that, whatever Piper's psychic gifts, Hodgson underwent a genuine conversion at her hands. After detecting trickery in medium after medium, he apparently found in Piper a case he could not crack. He became literally a changed man. The American psychologist Morton Prince, who knew Hodgson well, commented that the promotion of Piper's mediumship "wrecked Dick Hodgson who had one of the most beautiful minds I ever knew."[159]

Lodge's similar espousal of the possibility of spirit communications doubtless prompted some of his fellow scientists to wonder if his mind, too, had not been wrecked. In the early years of the twentieth century, Lodge became a well-known physicist – recent winner of the Royal Society's Rumford Medal, president of the Physical Society at the turn of the century, principal of the new University of Birmingham since 1900. Yet this very public scientific figure was publishing books with such titles as *The Immortality of the Soul* (1908) and *The Survival of Man* (1909). In 1902, the very year that he was knighted, Lodge told the SPR, in his second presidential address, that he was

> for all personal purposes, convinced of the persistence of human existence beyond bodily death; and though I am unable to justify that belief in a full and complete manner, yet it is a belief which has been produced by scientific evidence; that is, it is based upon facts and experience.[160]

For Lodge, human immortality had passed from the uncertainty of a working hypothesis to the status of scientific fact.

In the first decade of the new century, the unfolding of the cross correspondences helped confirm Lodge's inclinations toward spiritualism. He became certain that the complex system of cross references and allusions represented "the efforts of what was evidently a sort of SPR on the other side," directing their discarnate minds in concert to persuade living colleagues of the reality of human immortality.[161] The threads of proof seemed to be drawing together, forming a meaningful, coherent pattern of evidence. By 1912, Lodge could confidently state "that, without accepting all the tenets of the people who call themselves Spiritualists, I am convinced of human survival and the persistence of personality; and I have (I consider) ascertained that under certain exceptional conditions, and with considerable difficulty, communication is occasionally possible."[162]

In later years, after the death of his son Raymond, killed in Flanders in 1915, Lodge became a leader of the spiritualist revival associated with World War I. His book *Raymond*, published in 1916, was a runaway success, going through six editions in two months. With its assurance of postmortem communications from fallen soldiers, it provided solace to countless numbers of grieving families, not inclined under the circumstances to evaluate critically the flimsy evidence on which that assurance was made. Lodge has, ever since, enjoyed a fame entirely unconnected with his work as a physicist; indeed, many of his admirers have little idea what sort of a scientist he was. The assumption is easily made that grief over Raymond's death dulled Lodge's critical faculties and allowed him to associate himself with avowed spiritualists. But such was not the case. He was a firm believer in human survival before Raymond's death and had concluded, before the war broke out, that the dead could, on infrequent occasions, make contact with the living. His son's death and his decision to publicize the messages that Lodge felt certain he had received from Raymond merely brought the physicist a wider audience than any of his previous writings had gained.[163]

Lodge reached his belief in survival almost exclusively through acquaintance with mental mediumship. Despite his interest in Palladino, the more flamboyant physical phenomena of spiritualism played very little part in the formulation of his convictions, but their absence scarcely helped placate Lodge's critics. A number of his professional colleagues censured him for what they perceived as his credulity and, predictably, focused their criticism on his apparent eagerness to believe in the immortality of the human spirit.[164] They all erred, however, in trying to isolate that one strand of his thought, for it was integrally attached to his entire philosophy as a physicist and as a Christian. It could not be separated from the rest, held up to ridicule, and tossed aside as worthless – unless one was willing to dismiss as worthless the totality of Lodge's work.

Unlike Barrett in this respect, Lodge saw no sharp distinction between physical and psychical research. The methods of the physical sciences were indeed applicable, Lodge maintained, to the psychical realm. They were the very methods that had led him to his own certainty about human immortality. "I must risk annoying my present hearers," he ventured in his presidential address to the British Association,

> not only by leaving on record our conviction that occurrences now
> regarded as occult can be examined and reduced to order by the meth-

ods of science carefully and persistently applied, but by going further and saying, with the utmost brevity, that already the facts so examined have convinced me that memory and affection are not limited to that association with matter by which alone they can manifest themselves here and now, and that personality persists beyond bodily death.[165]

Lodge was certain that, in the future, the survival of the human spirit would be scientifically validated for a wide audience, as it had already been convincingly established for himself personally. He did not consider the soul's immortality a subject that must always remain beyond the reach of empirical inquiry, as Barrett recognized. Even as he embarked on theological meditations and explored the nature of religious truth, Lodge claimed a firm foundation in the rules of experimental investigation.[166]

Yet Lodge's endorsement of the methods employed by modern science was less opposed to Barrett's viewpoint than it would at first appear. Lodge never envisioned a reduction of psychic phenomena – bound up, as they were, with the mysterious, elusive element of life – to the formulas of physics and chemistry. What he, like Barrett, did foresee was a new, expanded physics, one whose concerns and laws had been increased "to embrace the behaviour of living organisms, under the influence of life and mind." If chemistry and physics, as constituted in the early twentieth century, were inadequate to explain the interrelatedness of all the parts of the universe, then their laws needed to be augmented until they could apply to the phenomena of psychical research.[167] In his first presidential address to the SPR, delivered in March 1901, Lodge told his fellow members that the particular nature of their "pioneering work" was "the founding and handing on to posterity of a new science." It is clear from his many other comments on the subject that his science of the future was to blend chemistry and physics with biology, physiology, and psychology, in order to speak as authoritatively on mind as on matter. But science did not follow Lodge's blueprint, and in *My Philosophy*, written in the last decade of his life, Lodge expressed his disappointment. "Science is utterly incompetent to explain the existence of the world as we know it now," he lamented. "Existence itself is a problem beyond its scope."[168]

Lodge's call for an expanded physics at the turn of the century came at a time when that science was, in fact, undergoing such alterations as to cast it in a virtually new form. To all appearances, Lodge belonged in the vanguard of the transformers. During his years in Liverpool, his research was "well in the forefront of the science of his day." He came close to discovering electromagnetic radiation, and after Heinrich

Hertz preempted that discovery, Lodge's continuing experiments with Hertzian waves led him to achieve the wireless transmission of signals before Marconi's far more famous, and more commercially successful, ventures into wireless telegraphy. Lodge's many interests in the realm of physics also included the theory of electrolysis, the dispersion of dust or smoke by electrostatic discharge, and the ignition system of the newly invented motor car.[169] Probably the most famous of his experiments were those that concerned the ether, the hypothetical substance upon which physics had for centuries been predicated. Presumed to occupy all space, it served as the omnipresent connecting entity of the cosmos. Its existence had never been proven, but its nonexistence created severe problems, for it would suppose that matter could act upon matter at a distance, across empty space. In 1893, Lodge set two large parallel steel disks whirling at very high speed, an inch apart on a vertical axis, and sent two light rays around the space between the disks. Since it was practically a universal assumption of late nineteenth-century physics that the ether transmitted waves of light, Lodge's experiment was designed to elucidate the controversial question whether ether in the vicinity of moving matter was or was not dragged along with the movement. Specifically, Lodge's experiment was conceived as a response to the much publicized Michelson–Morley experiment of 1887 whose results suggested that ether was indeed carried along by neighboring matter in motion, and that the velocity of light close to the earth's surface was thus affected by the movement of the earth. Lodge's experiment, repeated with numerous variations, yielded the opposite result – "nothing that [he] could do would alter the velocity of light one iota."[170] He could find no evidence that matter in motion measurably affected the ether in its vicinity. The Michelson–Morley and Lodge experiments, with their seemingly insoluble contradiction, provided significant evidence that led to the formulation of the special theory of relativity. If Einstein was the father of that theory, Lodge was, in a direct sense, one of the grandfathers.

Yet it was a grandchild of whom Lodge was very wary. For all his apparent modernity, he regretted the passing of Victorian physics, and as the twentieth century progressed, Lodge found himself less and less in sympathy with contemporary developments in his field of science. He was uncomfortable with the abstract nature of a physics that had moved so far into the world of mathematical formulas as to obviate the need for mechanical models. "Modern physics is peculiar," he complained in his autobiography,

> and during the present [twentieth] century it has been found possible to deal with phenomena without contemplating their detailed machinery. Much of the treatment of modern physics is of this highly

abstract and non-pictorial character, though the reasoning is obscure and difficult to follow. Most of our progress in the past has been due to men who tried to form clear conceptions of what was happening, and who sought the aid of analogies and what have been called "models."[171]

For Lodge, picturing a particular physical phenomenon at work was a giant step toward explaining it. The authors of his Royal Society obituary were quite right in describing him as "the child of an age which rejoiced in the explanation of physical processes by means of mechanical models."[172]

It was philosophical predisposition, more than education or training, that brought Lodge into opposition with the trends of twentieth-century physics. Other physicists, such as J. J. Thomson, who were about Lodge's age and who had been similarly trained in the heyday of Victorian physics, could accept the new directions of their post-Newtonian discipline. Lodge, however, could not, and by the time he died, he had become a striking anachronism in his profession, not merely clinging to, but still publicizing, beliefs long since discarded by his colleagues. Foremost among these was Lodge's devotion to the concept of ether, a devotion that goes far to explain his dismay over contemporary physics. What he particularly disliked in the theory of relativity was not so much its abstract, mathematical character as its challenge to the theory of ether. "The Principle of Relativity," he informed his British Association audience in 1913, ". . . is a principle of negation, a negative proposition, a statement that observation of certain facts can never be made, a denial of any relation between matter and ether, a virtual denial that ether exists." And more bluntly, two years later, he dubbed the relativity principle "a revolt against the existence of the Ether of Space."[173]

The ether of space was a plausible and highly useful hypothesis, and it still enjoyed intellectual respectability during the first decade of the twentieth century. Indeed, in the last quarter of the nineteenth, it had gained an increased authority, as a new consensus emerged concerning its nature and function. Whereas Faraday, for example, was uncertain whether to envision ether "as a continuum or as constituted of discrete particles," the view that prevailed by the turn of the century was of a nonmaterial, continuous ether into which all particles of matter were integrated. When J. J. Thomson delivered the Adamson Lecture at the University of Manchester in 1907, he called the ether "the invisible universe" that served as "the workshop of the material universe."[174] It was, to put it mildly, extremely disconcerting to abandon the ether for a theory that seemed to leave everything in doubt, that eliminated absolute standards of measurement, and made the observing scientist a factor in his own calculations. "The 'Relativity' people," Lodge grumbled, claimed "that all we could ever tell was the relative motion of Matter to Matter; and that absolute motion, or motion with reference to something not Matter, was meaningless."[175] Nonetheless, gradually during the second and third decades of this century, the ether ceased to be a meaningful concept for modern physics. The necessity for its existence had vanished with the restructuring of the foundations of physical thought.

Lodge, however, refused to concede the point. *My Philosophy*, published in 1933, and intended as an *apologia pro vita sua*, was a sustained argument for the indispensability of the ether. Without it, much in the physical world remained inexplicable to Lodge and, more important, an essential part of his metaphysical speculations was undermined. Although he acknowledged that the ether could "never be demonstrated by any experiment," he found "it necessary to a philosophic contemplation of the sensory universe."[176] He could not conceive of a universe in which discontinuity was a cosmic law, a universe whose parts were not all joined together by some unifying agent. That, Lodge argued, was the supremely important part that the ether played. "The vehicle of transmission of all manner of force, from gravitation down to cohesion and chemical affinity," the ether kept the cosmos from existing only as "a chaotic collection of independent isolated fragments." Because Lodge could not "admit discontinuity in either Space or Time," he clung to the ether despite the latest developments in physics.[177]

Yet he was not merely a stubborn opponent of contemporary physics after World War I. He tried to remain open to what was valuable in the new orthodoxy, and *My Philosophy* includes many attempts at compromise. Lodge accepted quantum theory, calling it "very interesting and most important"; he even made the extreme concession that "all we know at present of locomotion is the locomotion of one piece of matter with references to other pieces"; he admitted that some early nineteenth-century notions of the ether were outmoded. But concerning the existence of the ether itself, he could see no reason for altering his views.

> An ether is still necessary, . . . Ether is the seat of potential energy; it is the recipient of all strain; it alone exerts stress; it is the vehicle of an electric and a magnetic field as well as of all radiation . . . The motions, the accelerations, of inert matter are not intelligible without it. It is essential to all activity.

If Lodge had to accept the theory of relativity as a canon of modern physics, he could at least insist that it "does not abolish the ether. It

382   *A pseudoscience*

only affects some of its imagined properties."[178] Despite his best attempts to endorse what he called "the new knowledge," a wistful note repeatedly sounded through the pages of *My Philosophy*. Lodge identified it himself when, quoting an address on inertia which he had delivered at the Royal Institution in 1919, he wrote:

> As a conservative physicist I may be allowed to lament the extraordinary complexity introduced into physics and into natural philosophy by the principle of relativity, as so remarkably and powerfully developed by the mathematical genius of Einstein, with complication even of our fundamental ideas of space and time. The complications do not commend themselves to all of us, and I for one should be glad to return to the pristine simplicity of Newtonian dynamics, modified of course by the electrical theory of matter.[179]

When Lodge looked back longingly to a simpler period in physics, it was to a time when the continuity of nature was a fundamental assumption of scientific inquiry. Continuity and the ether were the twin pillars of his world view, and he was committed to the latter primarily because it was indispensable to the former. Like Wallace, he abhorred the very notion of gaps in the fabric of the universe, and theirs was a widely shared natural philosophy during the nineteenth century – and earlier, of course. Crookes, for one, was wont to repeat the assertion that "all the phenomena of the universe are presumably in some way continuous,"[180] while Stewart and Tait had treated the "Principle of Continuity" as a central feature in their *Unseen Universe* of 1875. What distinguished Lodge's assumptions from this commonplace Victorian viewpoint was merely the clarity, forcefulness, and persistence with which he expressed them.

His biographer has observed that, for Lodge, the most fascinating and significant part of scientific research "was conducted in the mind rather than the laboratory." When Lodge became principal of the University of Birmingham, and found his time largely occupied by public and administrative duties, he felt less frustrated than other scientists might have been by the lack of opportunity for sustained experimental work. In his odd spare moments, he could still undertake what he enjoyed most – "'brooding'. He liked to take a few observations and then worry away at them in his mind until he could arrive at some conclusion satisfying to himself about their more general significance."[181] It would not be unfair to say that, for forty years after the turn of the century, whenever leisure time allowed, Lodge brooded – about the structure of the universe, the essence of life, the meaning of such words as "mind" and "soul" – and from this long meditation emerged the publications that together spell out his credo as scientist, psychical researcher, and amateur theologian.

When Lodge spoke of continuity as the fundamental condition of nature, he used that concept as a shorthand way to convey two distinct beliefs. He meant, on the one hand, that there is no empty space in the physical fabric of nature and, furthermore, that matter never perishes. It may lose one form, or shape, but from its alterations emerge new forms of matter, thereby providing a continuous link through time.[182] On the other hand, Lodge endowed his concept of continuity with a spiritual significance. "I have made no secret of my conviction," he explained in *Raymond*,

> not merely that personality persists, but that its continued existence is more entwined with the life of every day than has been generally imagined; that there is no real breach of continuity between the dead and the living; and that methods of intercommunication across what has seemed to be a gulf can be set going in response to the urgent demand of affection.[183]

In the face of what he saw as "the modern tendency . . . to emphasise the discontinuous or atomic character of everything," Lodge waved the banner of continuity, between living and dead, between species on the scale of evolutionary development, between particles of matter in space. The whole of the vast universe, he did not hesitate to proclaim, hung together, and it only did so thanks to the ether of space, a sort of cosmic glue without which "there could hardly be a material universe at all."[184]

In keeping with his twofold views of nature's continuity, Lodge assigned the ether two separate tasks in his own thought. There was the physical role already discussed – to enable the universe to function as an integrated whole and to explain the exertion of mechanical force which, Lodge argued, could not occur across empty space. There was also, however, a psychical role, for Lodge found the ether necessary to clarify and validate his ideas about vitality, life, mind, soul, and human immortality. The relationships among these elusive terms intrigued Lodge, and he spent much of his brooding time in an effort to obtain some glimpse of the reality behind each. That his line of argument was thoroughly antimaterialist must, of course, be obvious: It was his constant purpose to demonstrate that life, the enigmatic force that gives each organism its own unique form, was not merely the inevitable result of a certain number of atoms combining in a sufficiently complex way. As he wrote in 1912 to his frequent correspondent J. A. Hill,

> I have tried to use the term "vitality" to signify the interaction between matter and the thing which I suppose to exist and which I call "Life." Certainly, as we know it, vitality is an interaction or a relation

> – a function, one may say, of organized matter – but I think of it
> (whether rightly or wrongly) as an interaction or relation between two
> real things – matter and something else.[185]

Like the ether, that something else could not be pinned down, isolated,
and studied in a test tube. Its analysis was as much the job of a philosopher as of a scientist, and in the many writings after the turn of
the century through which Lodge undertook that task, he made little
or no distinction between the two roles which he had, of necessity, to
perform.[186] He was wearing both hats when he insisted that, although
science could not identify the ingredients of life, no "genuine science"
had ever "presumed to declare that it is purely imaginary." While
immaterial, how could it be otherwise than real, he wondered, when
the physical body undergoes incessant alteration, as cells give way to
other cells. At no one moment in time could a person's material body
be considered "fully representative of the individual" who possesses
perhaps thousands of other physical shells before the last of them finally
dies. Surely then, Lodge reasoned, the essence of the individual, the
locus of personality, must reside elsewhere than in the physical body,
"that flowing and constantly changing episode in material history, having no more identity than has a river."[187]

In working out his own theory of life, individuality, and personality,
Lodge argued for the existence of what he termed the "etheric" or
"etherial" body. Although the term was the same as that which Theosophists used to designate one of the body's several layers, Lodge's
inspiration did not come from Eastern occultism. He first explored the
idea of the etheric body before 1914, but presented its fullest elaboration in *My Philosophy*, after many years of subsequent speculation.
Convinced that "life does not belong to the material world," he sought
to understand, like countless inquirers before him, how vitality or an-
imation is imparted to inert matter, how – in the case of humanity –
emotions, thoughts, and desires find expression in physical speech and
activity. These very feelings and ideas, Lodge contended, indicated
the presence of a spiritual world in conjunction with the material universe. What linked the two, just as it joined matter to matter, was, of
course, the ether, or etheric body. "What we have learnt physically,"
Lodge explained,

> is that the ether can act on matter through electric and magnetic prop-
> erties: we also know that mind can somehow act on matter, though
> probably indirectly. Our assumption is that we possess an ether body
> or animated structure of modified ether here and now, that life or
> mind is closely in touch with the ether body, and that through its
> action on this at present imperceptible body it is able to exert an action
> on the familiar material body. To assume that mind acts on ether and

that ether acts on matter, is I hope an assumption in the direction of
truth: and it appears to be justified by psychical facts, which show
that the action of mind can be independent of matter.[188]

Thus the ether figured as prominently in Lodge's psychical research
as in his physics. As matter and ether were, for him, the fundamental
realities of the physical world, so ether and mind (or soul, or spirit)
were the component parts of the psychic realm. Each human being
might accordingly be perceived, not in terms of mind–body duality,
but rather as a tripartite organization: spirit, flesh, and etheric body
as intermediary between them. Lodge described the etheric body as
extended, like the physical body, and immaterial, like the spirit. It was
thus the perfect go-between, because it shared certain characteristics
of both its partners. "We shall find, I think," he suggested in *My
Philosophy*,

> that we possess, all the time, a body co-existent with this one that
> we know – a body essentially substantial and related to space and
> time, not really transcendental, but yet in no way appealing to our
> present senses. Intangible and insensible, it may yet exist; and if it
> exists it may be detachable and capable of separate existence. It will
> be the etherial aspect or counterpart of our present bodies, but more
> permanent than they. For there is no property in the ether which
> suggests ageing, or wear and tear . . . No imperfection of any kind
> has yet been detected, or even suspected, in the ether of space.

Lodge hypothesized that, after the demise of the physical body, an
individual's etheric body remained with its mind, to enable the individual to retain a strong sense of identity. He reported that messages
"from the other side" seemed to support this hypothesis because physically deceased communicators, "though discarnate," felt "no more
disembodied than we do." They had bodies, Lodge surmised, but instead of fleshly, they were etheric bodies suited to the new postmortem
surroundings in which mind or spirit found itself.[189] It was a strange
theory for a distinguished, much-honored physicist to articulate in the
1930s. It seems even stranger in light of Lodge's sincere belief that,
with the theory of the etheric body, he had made a significant contribution toward illuminating the ancient puzzle of body and mind.

Lodge hoped that scientists and religious thinkers alike would find
acceptable the explanation of human survival that accompanied his
hypothesis. He tried not to use extravagant language in his discussions
of immortality and pointed out the obstacles that barred the way to
complete certainty of continued consciousness after physical death.
He admitted that it was exceedingly difficult to verify the personal
identity of an alleged spirit communicator – although he felt no doubts

about "the existence of a spiritual world" in general. He was even scrupulous in distinguishing between immortality and the survival of personality, conceding that psychical researchers could not speak about eternity, but only of "our individual continuance after separation from this material body."[190] When he expressed confidence that old friends and family members were communicating with him, and that SPR studies in thought transference confirmed the possibility of communication by means unknown to science,[191] he claimed that his findings were "the result of a lifetime of scientific study." As with Barrett, however, Lodge's argument for spirit survival did not rest on laboratory data. It was, rather, the result of a profound emotional commitment to his vision of continuity, a refusal to concede to death the power to terminate finally and absolutely any aspect of the universe. That his perspective was more religious than scientific is clear from the facts that he marshaled to prove the independent reality of spirit. In *Reason and Belief*, for example, Lodge called his readers' attention to the most authoritative example he could name of spirit persisting separately from body: He reminded them that "Christ did not spring into existence as the man Jesus of Nazareth. The Christ spirit existed through all eternity. At birth he became incarnate."[192]

Lodge saw nothing incongruous in citing the Christ spirit to demonstrate a point disputed by science. Nor did he deem it inappropriate to discuss the rules of scientific inquiry in an article entitled "The Christian Idea of God." "You cannot parcel out truth into that which is divine and that which is not divine," he maintained in the article; "the truths of science were as much God's secrets as any other."[193] In characteristic spiritualist fashion, he believed that it was his ongoing mission, his special concern, to reconcile science and religion in the twentieth century, but few spiritualists put such tremendous effort into the endeavor. Like Arthur Balfour, to whom he dedicated *Reason and Belief*, Lodge had no sense that it was a futile task. On the contrary, he claimed it as a duty and privilege, and he even took the time to write what he called "a catechism for parents and teachers," in which he suggested how to instruct children in the fundamental truths of the "universal Christian experience" without swimming "athwart the stream" of science.[194]

In attempting to bring about so momentous a reconciliation, Lodge employed several different approaches. In part, he adopted the methods of nineteenth-century liberal theologians, emphasizing the moral value of the Gospels over their historical accuracy, or treating the historical narratives as parables. Alternately, he sought to strengthen the edifice of Christian faith by freeing its "lost simplicity"[195] from numerous accretions that threatened to mar its splendor and majesty.

Aware that much of the modern repudiation of Christianity arose more from its cruelty than from its irrationality, Lodge rejected certain teachings as nonessential to the fundamental spirit of Christ's message. Original sin, vicarious atonement, and the need to propitiate a wrathful God were all as unacceptable to Lodge as to the non-Christian spiritualists. Either, he asserted, they were vestiges of a barbarous paganism far older than Christianity, that had infiltrated the new faith over the centuries, or they were monkish inventions.[196] The doctrine of hell likewise played no part in Lodge's mature Christian beliefs, although in his youth it had cost him "one or two rather sleepless nights." The God whom the adult Lodge worshiped was not a punitive deity, but a "Loving Father, . . . willing to help us whenever we are willing to be helped."[197] As yet another tactic, Lodge urged a variety of reforms within the Church of England, designed to broaden its scope until the country had once again a truly "National Christian Church." Toward that end, he called for "a great simplification of Church enactments, so as to leave fair freedom of interpretation concerning the meaning of Christian ceremonies." It was "the essence of Christianity" that concerned Lodge, not "oaths and formularies," and he sought to promote a "greater tolerance" among sects in order to effect their union.[198]

Lodge's strategies for achieving the ultimate reconciliation of science and faith tended to concentrate on the necessary revisions of Christianity. When he turned to the scientific side of the reunion, specific suggestions came less readily to him. His thoughts centered on the hope that the future would witness the emergence of "a completer Science," which transcended the world revealed by man's five senses, recognized the reality of both spirit and ether, and, at long last, overcame the dualism of body and soul. The region of Lodge's completer science was, in short, "the region of true Religion," where the ether would blend all elements of the physical and the psychical into one perfect continuum.[199] Never for a moment, however, did Lodge mistake his dream for reality. As we have seen, he came to realize that the new science of matter and spirit combined would not emerge during his own lifetime. He did not expect to see the promised land himself, nor could he claim to have led his fellow physicists out of the wilderness of materialism. Yet Lodge never betrayed the splendor of his own vision, although it had turned him away from the most fruitful fields of scientific inquiry in the twentieth century.

His vision was not only of a universe in which matter, ether, and spirit functioned in harmony, with the law of continuity binding all parts together into an integrated unity.[200] In Lodge's cosmos, progress, too, was a universal law, and design and purpose guided the actions

of every part of the whole. Just as he could not accept a universe
punctuated by discontinuity, so he could not conceive of a meaningless
world. He refused to believe that the Darwinian theory of evolution
banished teleology from the world stage. On the contrary, he saw evolution as ongoing proof that "the universe progresses in time," and
he argued for the presence of a controlling "dominant Power . . . One
Who understands and guides and influences, and Who has brought
everything into existence for some inscrutable end." As a Christian,
Lodge looked on that power as the beneficent Father who sent His
Son to mankind, but regardless of particular religious creeds, Lodge
insisted that "Higher Powers," existing in the realm of spirit, ordered
and gave significance to the movement of matter.[201] Sometimes he
came close to suggesting an angelic sort of guidance for humanity, as
when he wrote in 1910:

> The idea of "angels" is usually treated as fanciful. Imaginative it is,
> but not altogether fanciful; . . . facts known to me indicate that we
> are not really lonely in our struggle, that our destiny is not left to
> haphazard, that there is no such thing as *laissez faire* in a highly
> organised universe. Help . . . is available; a ministry of benevolence
> surrounds us – a cloud of witnesses – not witnesses only but helpers,
> agents like ourselves of the immanent God.[202]

More typically, however, Lodge made no attempt to describe or specify
the agency of protection directing man's path through life and time.
He was content to insist that "nothing is inexorable except the uniform
progress of time," and that there is a "deeper meaning involved in
natural objects" than contemporary science acknowledged.[203] Lodge
never managed to imbibe the pessimism of the twentieth century, but
retained to his death the optimism and faith in progress more characteristic of the nineteenth. Certainly his spiritualist convictions contributed to that hopefulness. He wrote confidently of the soul in "higher
organisms" growing towards "the Divine Being," and he recognized
"in the woven fabric of existence, flowing steadily from the loom in
an infinite progress towards perfection, the ever-growing garment of a
transcendent God."[204]

There was so little of the detached, impartial scientist in the older
Lodge's psychical research that he could criticize William James for
approaching the subject "rather in a cold-blooded manner as a branch
of psychology."[205] For Lodge, it was a branch of philosophy, of theology, of ontology, of all the "ologies" and "osophies" that pertained
to matter, spirit, and nature. Like Barrett, Lodge was incapable of
remaining "cold-blooded" about a pursuit that formed so central a part
of his personal *Weltanschauung*. He could not keep his religious convictions separate from his psychic inquiries, nor from his speculations

concerning the physical world. After his death, the *Times* underscored the degree to which Lodge had deviated from the methods and standards of pure physics – a deviation which Lodge had defended "on the general ground that where discoveries of vast moment to mankind were in question sympathy rather than cold criticism was a more profitable attitude." Lodge was, the *Times* observed, a scientific heretic, for he "was disposed to the pragmatical heresy in argument, believing that if a conclusion were of benefit to mankind statements supporting it were probably true."[206] Convinced, as he profoundly was, that the spiritualist doctrine was of inestimable benefit and comfort to mankind, Lodge accepted probable truth as the best available.

The physicists who figure in this chapter concurred that scientific naturalism was inadequate to account for everything under the sun, but there also existed broad areas of disagreement among them. Those, like Rayleigh and Thomson, who remained psychical researchers only, suspending judgment on ultimate questions, could separate their abhorrence of materialism from their investigations as scientists. They realized that, while a knowledge of physics might help to answer the great problems of philosophy and religion – of time and space, life and death, cause and effect – the relationship was not reciprocal; the metaphysician or theologian could not speak authoritatively on the great problems of modern physics. For those who went further, however, and embraced spiritualism, the repudiation of materialism became an underlying assumption of their physical experiments.

Yet both groups of physicists showed a deep respect for the methodology of their profession. They admired its systematic procedure and endorsed the basic precept that nature was regulated by fundamental laws. Not even the most devoted spiritualist among them ever desired to suggest that the universe bristled with inexplicable irregularities. The dilemma arose, however, when the experimental method, which they all claimed to revere, was perceived to imply atomistic, mechanistic conclusions about the cosmos, which many scientists were not prepared to accept. Indeed, one perceptive commentator on the Edwardian scene has observed that Edwardian science was in part concerned with seeking "new forces and new freedoms, by which the universe might be made teleological again."[207] But if those forces were too free to be analyzed and reproduced in a laboratory, how were the seekers to make good their claim that science must pay heed? This was the quandary that faced those physicists who believed that psychic phenomena might successfully challenge materialism, if only the scientific profession would acknowledge the challenge. In an effort to make their colleagues accept the legitimacy of such phenomena, Barrett and Lodge, among others, contended that psychic occurrences were fully subject to the working of natural laws, albeit laws still hidden from science.

Given the revolutionary changes in physics around the turn of the century, the contention of psychical researchers does not seem outrageous or unreasonable. It was just conceivable that telepathy among the living, for example, might in the future prove to have regular, demonstrable causative agents at work – agents that could be accommodated within the framework of natural laws. The same was not true of the spiritualist hypothesis. Nothing in the expanding horizons of post-Newtonian physics legitimated the assumption that spirits of the dead exist all around us and communicate with us under special circumstances. The scientist who succumbed to the attractions of spiritualism invariably found himself in an unsuccessful struggle to satisfy two demanding mistresses, for science and spiritualism could not be served simultaneously. Fournier D'Albe drew from his own, and Crookes's, experience when he wrote:

> Spiritualism as a religion may legitimately be studied in a section of anthropology, but spiritualism as a science does not exist. To be a spiritualist, the scientist must surrender his wishes, his methods, his views into the hands of his "spirit friends" on the "other side." If he does that he may achieve a certain peace of mind, but his scientific work will be at an end.[208]

Conclusion

The experience of the physicists described in the preceding chapter raises once more the central theme pursued throughout the book. With their fellow spiritualists and psychical researchers, they shared goals that were central to the period in which they lived – a period that perceived the need to bring religion more into line with the teachings of modern science and thereby to reduce the threat that science posed to the fundamental tenets of Christianity. Spiritualists and psychical researchers in the Victorian and Edwardian decades sought to achieve both of these related aims, and many felt that they had obtained considerable success in their endeavors.

In seeking to evaluate those efforts, and in trying to understand the varied response of the British scientific community to them, problems of definition furnish substantial obstacles. One needs to know precisely what science meant to the British public after 1850 and what were perceived as the limits of its jurisdiction. But there are no clear answers to those questions during the period under consideration, for people thought about science from widely diverse perspectives. For some, science was restricted to laboratory research. For others, it included every realm of knowledge concerning man and nature, and every form of technical expertise. There was, furthermore, an infinite number of intermediate positions between the two extremes. Although spiritualism and psychical research, both of which eluded repeatable laboratory tests, were evidently outside the former interpretation of the scientific enterprise, the latter, obviously, was broad enough to embrace them.[1]

The imprecise concept of a scientific "establishment" likewise makes it difficult to grasp accurately the relationship between spiritualism and psychical research, on the one hand, and the scientific profession in Britain, on the other. The notion of a citadel of scientific orthodoxy, and a locus of scientific power, tormented some of the scientists who turned to spiritualism and psychical research, but it is seldom clear exactly what institution or group they identified as their persecutors. Certainly there were organizations, such as the Royal

Society and the British Association, that were presumed to speak for the British scientific profession in general and whose officers were considered, by colleagues and the public alike, to have reached a certain eminence in their fields of inquiry. Similarly, membership in the many British scientific societies and institutes provided a cachet for the work of individual researchers. Furthermore, as scientific endeavor became an increasingly professional pursuit during the course of the nineteenth century, academic programs for training scientists emerged, both at the new universities and at the old, where students learned by participating in experimental research. The professional scientist, like Tyndall, Huxley, W. B. Carpenter, J. J. Thomson, and many others, worked hard at science for a living and made a lifelong career out of it. Yet not one of these conditions of science in late nineteenth-century Britain implies the existence of a single kind of scientific respectability, a sole *cursus honorum* from whose path no deviation could be tolerated.[2]

Nevertheless, at some point in their careers, Barrett, Lodge, and Crookes all felt that they were outsiders in their own profession. They perceived themselves as challenging the prevailing orthodoxy – scientific naturalism – and as therefore incurring the disapproval of the men who occupied the center of authority in Victorian and Edwardian science. In Crookes's case, both the Royal Society and the British Association snubbed his psychical research in the early 1870s, and he deeply resented this treatment at the hands of organized science. More than twenty years later, Crookes wrote to Lodge commending the younger man's boldness in "advocating an unpopular thing," despite possible injury to his career.[3] Barrett, too, believed that "official" and "orthodox" science had discriminated against him.[4] Although he was permitted to read his paper on thought transference to the BAAS in 1876, the Association declined to include it in the published annual report for that year. In 1883, Barrett's application to read a second paper on telepathy at the annual BAAS meeting was flatly vetoed. He finally learned not to expect support from scientific bodies or publications, as far as his psychical research was concerned. Even the newcomer *Nature*, launched as a weekly in 1869, dutifully took its cues from Huxley on that subject, or so Barrett alleged to Lodge in 1894. When, at the end of his life, Barrett lamented the "scientific ostracism" that Crookes suffered for his views, it was obvious that Barrett was licking his own wounds as well.[5]

Even Lodge, the least temperamental scientist imaginable, with a tolerance and breadth of vision that made him admirably slow to take offense, assumed that there existed a bastion of scientific orthodoxy hostile to his psychic investigations. He used the label "orthodox science" to mean "not the comprehensive grasp of a Newton, but science as now interpreted by its recognised official exponents, – by the average Fellow of the Royal Society for instance," and he was resigned to its inevitable "ridicule and dislike." With a somewhat wistful humor, he wrote to Hill in 1914: "That my occasional psychic utterances do harm to my scientific reputation – even so far as causing some of them [fellow scientists] to think me more or less cracked – is manifest, for I have many signs of that."[6] Nor were professed spiritualists the only scientists who assumed that to declare one's belief in the reality of psychic phenomena was to court professional disaster. J. J. Thomson, for one, noted in his memoirs that "by scientific men in general such subjects were regarded as 'untouchables'; anyone touching them would lose caste."[7]

In any historical account, the actors' self-perceptions are, of course, a critically important element in the story, and the historian can ignore them only at great risk of distortion or misrepresentation.[8] Nonetheless, despite their reiterated assertions of discrimination, it is hard to see how scientists who embraced spiritualism or investigated psychic phenomena actually "lost caste." Doubtless all were subjected to some degree of personal abuse in the scientific and periodical press, and each must have felt at one time or another that his psychical research was barred from the channels of dissemination usually available to reputable scientists. But no professional disgrace drove any of them into untimely retirement. None was expelled from a professional society, nor denied the professional recognition that was his due. The identity and location of the scientific authority that sought to punish them for their unorthodoxy remain elusive. Although both the Royal Society and British Association dismissed Crookes's psychical research, each subsequently elected him its president. Indeed between 1900 and 1920, four of the Royal Society's presidents were men who had participated in psychic investigations to a greater or lesser extent: Huggins, Rayleigh, Crookes, and Thomson. Lodge's prestigious awards, honors, and appointments after he had begun to publicize his growing involvement with psychical research, Barrett's knighthood, Wallace's Order of Merit – all the evidence suggests that interest in psychic phenomena affected meritorious scientific careers to no significant degree. Not even outright spiritualist commitment relegated a capable scientist to the ranks of outcast, for his colleagues always thought they could distinguish between the totality of his contributions to science and certain questionable parts of it. E. Ray Lankester, no friend of psychical research, could still include Crookes, with Kelvin, Clerk Maxwell, Rayleigh, and J. J. Thomson, in a list of physicists whose "triumphant handling of nature" filled him with "reverence and admiration."[9] The

very notion of professional outcasts, of insiders and outsiders, ultimately offers little aid in understanding the organization of the British
scientific community in the late nineteenth and early twentieth centuries. With the comparatively meager amounts of state support available for scientific research, there was no central source of patronage
for aspiring scientists.[10] The roads to scientific eminence in Britain
were many.

Yet the self-perception of those scientists who fully accepted the
spiritualist label had some basis in reality. If they were mistaken in
imagining that British science spoke with a single authoritative voice,
they were not misled in assuming that they swam against the current
of mainstream science in their day. In particular, they differed sharply
from most of their colleagues on the fundamental question of evidence.
Over the decades, the same charges were repeated by the critics of
psychical research and spiritualism: The phenomena were incapable
of demonstration by the acknowledged methods of observation and
controlled experimentation, under predictable circumstances, with all
variable factors identified. They were, by contrast, hopelessly erratic
and utterly subjective. Critics who refused to credit stories of nonmaterial agents at work in the universe understood the difference between an hypothesis that could be proven false, if necessary, and one
that could not be. A theory concerning the sensory-motor apparatus
of the nervous system, such as the psychophysiologists of the late
nineteenth century were developing, belonged to the first category.
Speculations concerning the separate existence of mind apart from
body, like arguments for the reality of spirit forces, appertained to the
second.[11] All the fundamental questions that the SPR addressed, and
all the major assumptions of spiritualism, were neither definitely provable nor falsifiable by any conceivable test. For these reasons, the
models of the physical sciences – whether those of physics or of biology
– never really applied to the efforts of the SPR before World War I,
and even less to the pursuits of the spiritualist associations throughout
the country. For these reasons, too, the majority of British scientists
found that psychic phenomena merited neither their investigation nor
their belief. The British Association and the Royal Society never inquired into alleged psychic occurrences, not because they wished to
show disapproval of wayward scientists, but to avoid wasting time on
what seemed fundamentally a futile task that could lead to no reasonably certain knowledge.[12]

Scientists who endorsed the efficacy of spirit, or psychic, agents at
large did not directly deny the arguments of their critics. They agreed
that psychic phenomena were elusive and hardly ideal subjects for the
scientific laboratory. What they did deny was that the characteristics

of psychic phenomena placed these manifestations beyond the pale of legitimate scientific concern. They rejected the assumption that only the predictable, the measurable, the observable, or the tangible should command the attention of contemporary science and serve as valid evidence in the scientist's efforts to understand the world. Such criteria of evidence, Wallace pointed out in 1876, ought to bar science from investigating meteorites – those fragments of the cosmos whose fall could not necessarily be forecast, nor witnessed firsthand, nor reproduced at will by the experienced scientist in the laboratory.[13] Yet there were scientists who eagerly studied meteorites while unhesitatingly excluding psychic phenomena from the range of their interests. Such investigators, Lodge argued twenty-five years later, were trying to impose their own selective notions of order and explicability on the forces of nature. As a result, they severely restricted their ability to comprehend nature.[14] If science in the twentieth century were to continue its triumphant history of discovery, Lodge was certain that it would not be thanks to men of such narrow vision.

> They are like the orthodox mariners of old who limited themselves to the shores of the Mediterranean, cruising round its coasts and gradually becoming familiar with every port. The world as known to the Ancients was their domain, and it was impious to sail out through the Pillars of Hercules into the ocean beyond. Venturesome explorers who transgressed those limits, and from time to time returned with legends of tides and other unusual phenomena, were doubtless received with disapprobation and incredulity; – still more so if they ventured to deduce the possible existence of a new Continent, which as yet confessedly they had not reached, from evidences derived from drifting logs and a Sargasso Sea.[15]

It is, of course, true that scientists in the past have contemptuously dismissed certain theories that in time became the received truths of subsequent generations, and Lodge no doubt had those theories in mind when he spoke of the orthodox ancient mariners. Some of the vocabulary of psychical research that provoked scorn in the nineteenth century seems less laughable in the twentieth. While Faraday, for example, berated spiritualists for their ignorant use of terms like "electro-biology," or Carpenter dismissed Barrett's suggested nerve energy that operated in ways analogous to electricity, scientists today are depicting the brain more and more in terms of infinitely complex electrical circuits activated by "little pulses of electricity."[16] Men like Barrett or Balfour Stewart, when they used analogies from electromagnetism, were not merely trying to mask their ignorance behind impressive jargon, but were genuinely seeking valid scientific answers to the puzzles that fascinated them. Some of their hypotheses may have been less outlandish than most of their contemporaries assumed.

Many of those puzzles remain unsolved. The enigma of action at a distance, the success of the divining rod, the hypnotist's power, the mediumship of D. D. Home and Leonora Piper – questions of both general concern to science and specific interest to psychical research – still await full explanation. Perhaps the explanations if they are ever forthcoming, will suggest that the Victorian and Edwardian psychical researchers glimpsed at least an element of the solution.

What cannot be overlooked is the important point that the British scientific profession as a whole in these decades was not indifferent to the fundamental hopes of its members who embraced spiritualism or pursued psychical research. Most of the scientists who discounted psychic phenomena were certainly not motivated by the desire to preserve a materialist world view against the encroachments of spiritualism. Many were religious men for whom Christianity meant more than a limited number of appearances in church, and they were far from sheltering any a priori commitment to a philosophy of materialism. Nor were they trying to avoid metaphysical queries when they ignored, or ridiculed, investigations into the so-called psychic realm. Understandably, they sought to discriminate between speculation that was conducive to further scientific discovery and speculation that was not, but modern science has by no means barred itself from pondering ultimate questions. As J. J. Thomson explained in 1907:

> There is indeed one part of Physical Science where the problems are very analogous to those dealt with by the metaphysician, . . . To some men this side of Physics is peculiarly attractive, they find in the physical universe with its myriad phenomena and apparent complexity a problem of inexhaustible and irresistible fascination. Their minds chafe under the diversity and complexity they see around them, and they are driven to seek a point of view from which phenomena as diverse as those of light, heat, electricity, and chemical action appear as different manifestations of a few general principles.[17]

The attempt to enunciate those principles, to locate the common denominators of the universe, to find the ever-elusive "basic building block" or "ultimate substance" of nature – these aspirations inspired spiritualists and psychical researchers, just as they inspired scientists who criticized spiritualism and psychical research. The quest for a hidden pattern, a unifying framework, a fundamental theory, to bring together every diverse particle and force in the cosmos, was intrinsically the same, whether one stressed the links between heat, electricity, magnetism, and light, or looked for connections between mind, spirit, and matter. The vision of a "new science," which a number of spiritualists shared, may have been incapable of realization, but the search for a *tertium quid* between spirit and matter, mind and body, still haunted scientific consciousness around the turn of the century. The desire to bridge the gulf between them, which would be tantamount to reconciling science and faith, involved the would-be reconcilers in questions that concerned not simply definitions of matter and mind, but the central inquiries of epistemology and ontology. They became embroiled in problems of proof and demonstration, and they never ceased to ponder, at the back of all their investigations, the meaning of life and of God. Thus spiritualists and psychical researchers addressed, directly and indirectly, the most critical issues of science, philosophy, and religion. Some of their proposed solutions may, in time, seem prophetic; others must always, no doubt, appear absurd. But, fundamentally, their work was neither ridiculous nor even misguided, for through it they hoped to find the means of accepting the changed world around them.

NOTES

So much literature, from such a variety of disciplines, was consulted in the course of writing this book that a satisfactory bibliography becomes impossibly long, and a select bibliography meaningless. To compensate for the absence of a list of sources, the notes have been made as extensive as space would allow. They indicate the primary materials on which the author has based her interpretation and the secondary studies to which she is indebted. Within each chapter, furthermore, the citations are complete; that is to say, in every chapter the first reference to any book or article gives full publication information. The location of manuscript sources and special collections is, however, mentioned only once, and the following list is therefore provided for ease of reference:

Balfour Papers, British Library, London.
W. F. Barrett Papers, Society for Psychical Research, London.
British National Association of Spiritualists, Minute Books, College of Psychic Studies, London.
Thomas Davidson Papers, Yale University Library, New Haven, Connecticut.
Ghost Club, Minute Books, British Library, London.
Harry Price Library, Senate House, University of London.
Houdini Collection, Library of Congress, Washington, D.C.
Oliver Lodge Collection, Society for Psychical Research, London.
London Spiritualist Alliance, Minute Books, College of Psychic Studies, London.
F. W. H. Myers Papers, Wren Library, Trinity College, Cambridge.
Henry Sidgwick Papers, Wren Library, Trinity College, Cambridge.

Abbreviations that appear in the Notes include:

BAAS British Association for the Advancement of Science
BNAS British National Association of Spiritualists
CAS   Central Association of Spiritualists
DNB   *Dictionary of National Biography*
JSPR *Journal of the Society for Psychical Research*
LSA   London Spiritualist Alliance
PSPR *Proceedings of the Society for Psychical Research*
SNU   Spiritualists' National Union
SPR   Society for Psychical Research

Introduction

1. Owen Chadwick, *The Secularization of the European Mind in the Nineteenth Century* (Cambridge: Cambridge University Press, 1975). This par-

agraph is much indebted to the stimulating introduction to Chadwick's volume.

Part I. The setting

1. Mediums

1. Frank Podmore, *Modern Spiritualism: A History and a Criticism*, 2 vols. (London: Methuen, 1902), 2:289. The latest contribution to the general history of modern spiritualism is Ruth Brandon, *The Spiritualists: The Passion for the Occult in the Nineteenth and Twentieth Centuries* (London: Weidenfeld & Nicolson, 1983), a survey that is largely concerned to reveal the degree of fraud and deceit underlying "spiritualist" phenomena.
2. William Henry Harrison, *Spirit People* (London: W. H. Harrison, 1875), pp. 11–13, 30; London Dialectical Society, *Report on Spiritualism, of the Committee of the London Dialectical Society* (London: Longmans, Green, Reader, & Dyer, 1871), pp. 157ff; and the chapter on "Private Mediumship" in Podmore, *Modern Spiritualism*, 2:63–77.
3. Podmore, *Modern Spiritualism*, 2:75
4. For observations about female mediumship in nineteenth-century America, many of which are absolutely germane to British spiritualism in the same period, see R. Laurence Moore, *In Search of White Crows: Spiritualism, Parapsychology, and American Culture* (New York: Oxford University Press, 1977), pp. 102–29.
5. Bulwer-Lytton quoted in Katherine H. Porter, *Through a Glass Darkly: Spiritualism in the Browning Circle* (1958; reprint ed., New York: Octagon Books, 1972), p. 13. Sir Edward was raised to the peerage as Baron Lytton in 1866.
6. "Lyon v. Home," *Times* (London), 2 May 1868, p. 13; "Home, Great Home!" *Punch* 39 (18 August 1860):63; "Fly Away, Home!" *Mask: A Humorous and Fantastic Review of the Month* 1 (June 1868):145.
7. Eric John Dingwall, *Some Human Oddities: Studies in the Queer, the Uncanny and the Fanatical* (1947; reprint ed., New Hyde Park, N.Y.: University Books, 1962). Chapter V is devoted to D. D. Home.
8. John W. Truesdell, *The Bottom Facts Concerning the Science of Spiritualism: Derived from Careful Investigations Covering a Period of Twenty-Five Years* (New York: G. W. Carleton; London: S. Low, 1883), pp. 23–4.
9. Jean Burton, *Heyday of a Wizard: Daniel Home the Medium* (London: George G. Harrap, 1948), pp. 70, 128–9, 161, and passim; Earl of Dunraven, *Experiences in Spiritualism with D. D. Home* (London: Society for Psychical Research, 1924), pp. 46–7; Mme. Dunglas Home, *The Gift of D. D. Home* (London: Kegan Paul, Trench, Trübner, 1890), pp. 124–60; Podmore, *Modern Spiritualism*, 2:225: G. Zorab, "Have We Finally Solved the Enigma of D. D. Home's Descent?" *Journal of the Society for Psychical Research (JSPR)* 49 (June 1978):884–7.
10. John Lewis Bradley, ed., *The Letters of John Ruskin to Lord and Lady Mount-Temple* (Columbus, Ohio: Ohio State University Press, 1964), p. 79.
11. Earl of Dunraven, *Past Times and Pastimes*, 2 vols. (London: Hodder & Stoughton, 1922), 1:10.

12. D. D. Home, *Lights and Shadows of Spiritualism* (New York: G. W. Carleton, 1877), pp. 394–5.
13. Alan Gauld, *The Founders of Psychical Research* (London: Routledge & Kegan Paul, 1968), p. 213; Podmore, *Modern Spiritualism*, 2:258.
14. Account by W. M. Wilkinson, solicitor and friend of Home, quoted in William F. Barrett, *On the Threshold of the Unseen. An Examination of the Phenomena of Spiritualism and of the Evidence for Survival After Death*, 2d ed. rev. (London: Kegan Paul, Trench, Trübner; New York: E. P. Dutton, 1917), pp. 76–7.
15. One recent explanation of the Ashley House performance credits Home with an alpinist's knowledge of abseiling. If Home could have prepared a double rope in advance and secured it from the roof, so the argument goes, he could have literally swung himself in and out of the Ashley House windows. G. W. Lambert, "D. D. Home and the Physical World," *JSPR* 48 (June 1976):311–12.
16. Harrison, *Spirit People*, p. 34.
17. "Spiritualism, as related to Religion and Science," *Fraser's Magazine* 71 (January 1865):25–6. *The Wellesley Index to Victorian Periodicals*, vol. 2, attributes this article to Abraham Hayward, barrister and essayist.
18. Podmore, *Modern Spiritualism*, 2:238, 263; E. B. Tylor, "Ethnology and Spiritualism," *Nature* 5 (29 February 1872):343.
19. Podmore, *Modern Spiritualism*, 2:230. However, rumors persist that Home, in the late 1850s, was caught cheating in Biarritz, in front of the French emperor and empress. See Count Perovsky-Petrovo-Solovovo, "Some Thoughts on D. D. Home," *Proceedings of the Society for Psychical Research (PSPR)* 39 (March 1930):247–65.
20. Eric J. Dingwall, *The Critics' Dilemma: Further Comments on Some Nineteenth Century Investigations* (Crowhurst, Sussex: E. J. Dingwall, 1966), p. 7.
21. Notebooks of William Stainton Moses, quoted in R. G. Medhurst, "Stainton Moses and Contemporary Physical Mediums. 2. Florence Cook, Kate Cook," *Light* 83 (Summer 1963):73–4. "Katie King" claimed to be the daughter of a famous seventeenth-century buccaneer. Materializing for other mediums in addition to Cook, "Katie" was a frequent spirit visitor at this time.
22. "Ghosts and Geese," *Punch* 67 (15 August 1874):65; Dingwall, *Critics' Dilemma*, pp. 28–9; Medhurst, "Moses: Florence Cook, Kate Cook," p. 73.
23. Report from the *Daily Telegraph*, 13 January 1880, quoted in Osbert Sitwell, *Left Hand, Right Hand!* (Boston: Little, Brown, 1944), pp. 294–7. This was not the first time that Cook's masquerade was exposed. Under such circumstances, her supporters could only argue that the entranced medium was entirely unconscious of the fraud that she was perpetrating.
24. Home, *Lights and Shadows*, pp. 387–90.
25. Ibid., pp. 414–15.
26. Dingwall, *Critics' Dilemma*, p. 29.
27. R. G. Medhurst and K. M. Goldney, "William Crookes and the Physical Phenomena of Mediumship," *PSPR* 54 (March 1964):49–50.
28. Charles Maurice Davies, *Mystic London: or, Phases of Occult Life in the Metropolis* (London: Tinsley Brothers, 1875), p. 313.
29. Ibid., p. 330; Dingwall, *Critics' Dilemma*, pp. 30–1, 56, 67–8; Podmore, *Modern Spiritualism*, 2:102.

30. Davies, *Mystic London*, p. 316; Podmore, *Modern Spiritualism*, 2:104; Medhurst and Goldney, "Crookes and the Physical Phenomena of Mediumship," pp. 107–8.
31. Podmore, *Modern Spiritualism*, 2:85, Truesdell, *Bottom Facts*, pp. 238–75.
32. Dingwall, *Critics' Dilemma*, pp. 44, 51; Gauld, *Founders of Psychical Research*, pp. 104–6.
33. Davies, *Mystic London*, p. 253.
34. Earl W. Fornell, *The Unhappy Medium. Spiritualism and the Life of Margaret Fox* (Austin: University of Texas Press, 1964), p. 176; Dingwall, *Critics' Dilemma*, pp. 15n, 48. Moore, *In Search of White Crows*, pp. 115–19, discusses the charges of promiscuity against female mediums in nineteenth-century America and the evidence connecting mediumship, divorce, and marital infidelity. See also pp. 33–5, for Moore's discussion of the group of Harvard scientists who in 1857 warned the public against the "contaminating influence" of spiritualism "which surely tends to lessen the truth of man and the purity of women."
35. Podmore, *Modern Spiritualism*, 2:126, 134–9.
36. Truesdell, *Bottom Facts*, pp. 143–8; Podmore, *Modern Spiritualism*, 2:89; Gauld, *Founders of Psychical Research*, pp. 124–7.
37. "The Prosecution of Dr. Slade," *Spiritualist Newspaper* 9 (6 October 1876):110. For the letters from Lankester and Donkin, see *Times* (London), 16 September 1876, p. 7.
38. "Mr. Munton's Speech in Defence of Dr. Slade," *Spiritualist Newspaper* 9 (3 November 1876):157.
39. Podmore, *Modern Spiritualism*, 2:90.
40. Moore, *In Search of White Crows*, p. 105; Geoffrey K. Nelson, *Spiritualism and Society* (London: Routledge & Kegan Paul, 1969), pp. 119–21.
41. Keith Thomas, *Religion and the Decline of Magic* (New York: Charles Scribner's Sons, 1971), pp. 178, 229–31, 255, 266–7, 277–9.
42. E. H. Palmer, "Robert-Houdin on Conjuring," *Academy: A Weekly Review of Literature, Science, and Art*, n.s., no. 296 (5 January 1878):5.
43. Geoffrey Lamb, *Victorian Magic* (London: Routledge & Kegan Paul, 1976), pp. 34–5, 66.
44. Ibid., p. 88. Mediums were not the only people at this time who "freely awarded themselves professorships and doctorates."
45. Davies, *Mystic London*, pp. 288, 355–61.
46. Coleman's opinion is quoted in John Nevil Maskelyne, *Modern Spiritualism. A Short Account of its Rise and Progress, with Some Exposures of So-Called Spirit Media* (London: Frederick Warne, for the author, 1875), p. 67, and also see pp. 58–69; P. T. Barnum, *The Humbugs of the World* (1865; reprint ed., Detroit: Singing Tree Press, 1970), pp. 51–5, 97–8; Lamb, *Victorian Magic*, pp. 75–6; Podmore, *Modern Spiritualism*, 2:61.
47. Davies, *Mystic London*, pp. 359–60.

2. Membership

1. Marghanita Laski, "Domestic Life," in *Edwardian England 1901–1914*, ed. Simon Nowell-Smith (London: Oxford University Press, 1964), p. 184.
2. Florence Cook's enthusiastic letter about Luxmoore is quoted in Trevor H. Hall, *The Spiritualists: The Story of Florence Cook and William*

Crookes (New York: Helix Press, Garrett Publications, 1963), p. 17n. Also see Eric J. Dingwall, *The Critics' Dilemma: Further Comments on Some Nineteenth Century Investigations* (Crowhurst, Sussex: E. J. Dingwall, 1966), pp. 17, 55.
3. Hudson Tuttle and J. M. Peebles, *The Year-Book of Spiritualism for 1871* (London: James Burns, 1871), p. 156.
4. For biographical information about Cox, see *Dictionary of National Biography (DNB)* s.v. "Cox, Edward William"; Michael Stenton, ed., *Who's Who of British Members of Parliament* (Hassocks, Sussex: Harvester Press, 1976–), 1:93; "The Late Serjeant Cox," *Illustrated London News* 75 (6 December 1879):529–30.
5. E. W. Cox, *Spiritualism Answered by Science; with the Proofs of a Psychic Force*, rev. ed. (London: Longman, 1872).
6. Quoted in R. G. Medhurst, "Stainton Moses and Contemporary Physical Mediums. 2. Florence Cook, Kate Cook," *Light* 83 (Summer 1963):64.
7. Frank Podmore, *Modern Spiritualism: A History and a Criticism*, 2 vols. (London: Methuen, 1902), 2:170. Also see Fraser Nicol, "The Founders of the S.P.R.," *PSPR* 55 (March 1972):349–50.
8. C. C. Massey, *Thoughts of a Modern Mystic. A Selection from the Writings of the Late C. C. Massey*, ed. W. F. Barrett (London: Kegan Paul, Trench, Trübner, 1909), pp. 1–2, 9–11, 13–14, 24, 26, 44, 57, 76, 83, 88–9; Nicol, "Founders of the S.P.R.," pp. 345, 359; Stenton, ed., *Who's Who of British M.P.s*, 1:263.
9. Geoffrey K. Nelson, *Spiritualism and Society* (London: Routledge & Kegan Paul, 1969), p. 109; Podmore, *Modern Spiritualism*, 2:170; Minute Books of the British National Association of Spiritualists, Committees, no. 2, pp. 1, 13 (19 June 1878), College of Psychic Studies, London.
10. Wallace to W. F. Barrett, 18 December 1876, Barrett Papers (box 2, A4, no. 133), Archives of the Society for Psychical Research, London; Massey, *Modern Mystic*, p. 2; "The Slade Prosecution," *Spiritualist Newspaper* 9 (27 October 1876):152–3.
11. Johann Karl Friedrich Zöllner, *Transcendental Physics*, tr. from the German, with a preface and appendices by Charles Carleton Massey, 4th ed. (Boston: Banner of Light Publishing Co., 1901), pp. 9–26 of translator's preface; C. C. Massey, "Zöllner. An Open Letter to Professor George S. Fullerton, of the University of Pennsylvania, Member and Secretary of the Seybert Commission for Investigating Modern Spiritualism" (no. 4 in *Houdini Pamphlets: Spiritualism*, vol. 8, Houdini Collection, Library of Congress, Washington, D.C.). The "Open Letter" was published in *Light*, 13 August 1887.
12. Eveleen Myers to W. F. Barrett, 10 November 1904, Barrett Papers (box 2, A4, no. 58); Massey, *Modern Mystic*, pp. 11, 140, 225–7.
13. See the London Dialectical Society, *Report on Spiritualism, of the Committee of the London Dialectical Society* (London: Longmans, Green, Reader, & Dyer, 1871); Jean Burton, *Heyday of a Wizard: Daniel Home the Medium* (London: George G. Harrap, 1948), p. 177; John Nevil Maskelyne, *Modern Spiritualism. A Short Account of its Rise and Progress, with Some Exposures of So-Called Spirit Media* (London: Frederick Warne, for the author, 1875), pp. 40–1; Nelson, *Spiritualism and Society*, p. 94. The London Dialectical Society was a club recently established for discussion and debate. Its investigative committee reached the cautious

conclusion that spiritualism "is worthy of more serious attention and careful investigation than it has hitherto received." (*Report*, p. 6.)

14. London Dialectical Society, *Report on Spiritualism*, p. 247; Burton, *Heyday of a Wizard*, pp. 151, 181; Alan Gauld, *The Founders of Psychical Research* (London: Routledge & Kegan Paul, 1968), pp. 70, 72–3; Nicol, "Founders of the S.P.R.," pp. 344–5; Podmore, *Modern Spiritualism*, 2:39–40, 44, 163.

15. Estelle W. Stead, *My Father: Personal and Spiritual Reminiscences* (New York: George H. Doran, 1913), pp. 155, 157, 170–3, 177–8, 288–94. Ruth Brandon devotes a long section to Stead in *The Spiritualists: The Passion for the Occult in the Nineteenth and Twentieth Centuries* (London: Weidenfeld & Nicolson, 1983), Chapter 7.

16. E. Stead, *My Father*, p. 304. Also see W. T. Stead's preface to *After Death, A Personal Narrative* (New York: John Lane, 1907) – a "New and Cheaper Edition of *Letters from Julia*" – and his two articles in the *Fortnightly Review*: "How I Know that the Dead Return: A Record of Personal Experience," 85 n.s. (1 January 1909):52–64, and "The Exploration of the Other World," 85 n.s. (1 May 1909):850–61. Stead wrote to Frederic Myers of his personal experience of telepathy as early as 1894. Myers Papers 4[70], Wren Library, Trinity College, Cambridge.

17. Stead to Lodge, 20 February 1902, Lodge Collection (box 10, no. 2524), Archives of the SPR

18. Estelle W. Stead, ed., *Communication with the Next World. The Right and Wrong Methods. A Text-Book Given by William T. Stead from "Beyond the Veil"* (London: Stead's Publishing House, 1921), p. 97.

19. S. C. Hall, *The Use of Spiritualism?* (London: E. W. Allen, 1884), p. 63n (no. 13 in *Houdini Pamphlets: Spiritualism*, vol. 13, Houdini Collection).

20. Martin F. Tupper, "Some Spiritualistic Reminiscences," *Light* 3 (6 January 1883):5. Tupper, a popular versifier, while recording his few encounters with spiritualist phenomena, was not generally impressed with what he saw.

21. John Lewis Bradley, ed., *The Letters of John Ruskin to Lord and Lady Mount-Temple* (Columbus, Ohio: Ohio State University Press, 1964), pp. 41–2, 79.

22. S. C. Hall, *Retrospect of a Long Life: From 1815 to 1883* (New York: D. Appleton, 1883), pp. 191–2, 448–9, 477, 548–74. The *London* and *St. James's Magazines* were among the periodicals which Mrs. Hall edited, during the 1850s and 1860s.

23. Hall, *The Use of Spiritualism?*, pp. 61–77.

24. Hall, *Retrospect*, p. 379.

25. Quoted in Amice Lee, *Laurels and Rosemary. The Life of William and Mary Howitt* (London and New York: Geoffrey Cumberlege, Oxford University Press, 1955), p. 333.

26. Quoted by his daughter, Anna Mary Howitt Watts, in *The Pioneers of the Spiritual Reformation* (London: Psychological Press Association, 1883), p. 314. Also see p. 284 for William Howitt's statement about the Triune God, and Carl Ray Woodring, *Victorian Samplers: William and Mary Howitt* (Lawrence, Kansas: University of Kansas Press, 1952), pp. 212, 216.

27. Podmore, *Modern Spiritualism*, 2:72; Watts, *Pioneers*, pp. 239–40, 246, 261–7; Woodring, *Victorian Samplers*, p. 204, and pp. 190–1, where

Woodring reports that Mary Howitt was converted to spiritualism in 1856, when she received a message from her deceased son Claude.

28. Burton, *Heyday of a Wizard*, pp. 144–5; Watts, *Pioneers*, pp. 205–6, 215.
29. Watts, *Pioneers*, pp. 278–9, 296–7.
30. "Spiritualism, as related to Religion and Science," *Fraser's Magazine* 71 (January 1865):32. *The Wellesley Index to Victorian Periodicals*, vol. 2, attributes this article to Abraham Hayward, barrister and essayist.
31. Frederic Boase, *Modern English Biography*, 6 vols. (Truro: Netherton & Worth, for the author, 1892–1921), 6:159–60; Jule Eisenbud, "The Case of Florence Marryat," *Journal of the American Society for Psychical Research* 69 (July 1975):215–33; Florence Marryat, *The Spirit World* (New York: Charles B. Reed, 1894), p. 175.
32. Marryat, *Spirit World*, p. 7.
33. Florence Marryat, *There is No Death* (London: Kegan Paul, Trench, Trübner, 1891), pp. 111–15, 148. See Trevor Hall, *The Spiritualists*, pp. 60–7, for an estimate of Marryat's reliability as a witness of séance happenings.
34. Emma Hardinge Britten, *Nineteenth Century Miracles; or, Spirits and Their Work in Every Country of the Earth* (New York: William Britten, 1884), pp. 222–3; Gauld, *Founders of Psychical Research*, p. 74.
35. Owen's announcement in his *Rational Quarterly Review* (1853), quoted in J. F. C. Harrison, *Robert Owen and the Owenites in Britain and America: The Quest for the New Moral World* (London: Routledge and Kegan Paul, 1969), p. 250.
36. Ibid., pp. 250–1; Podmore, *Modern Spiritualism*, 2:18–19, 22–3.
37. Boase, *Modern English Biography*, 6:553; Robert Cooper, "Thomas Shorter," *Light* 19 (2 September 1899):410; Harrison, *Owen and the Owenites*, p. 251n; Podmore, *Modern Spiritualism*, 2:34–5.
38. *Report on Spiritualism, of the Committee of the London Dialectical Society*, p. 173.
39. *DNB*, s.v. "Massey, Gerald."
40. E. W. Wallis, *The Story of My Life; and Development and Experiences as a Medium* (London: James Burns, n.d.), pp. 4, 6–10 (no. 3 in volume of collected pamphlets, *Spiritualism and Its Truth*, Houdini Collection).
41. E. D. Rogers, *Life and Experiences of Edmund Dawson Rogers, Spiritualist and Journalist* (London: Office of *Light*, [1911]), pp. 7–10, 12–16, 35–8, 41–7; *DNB*, s.v. "Rogers, Edmund Dawson"; Nicol, "Founders of the S.P.R.," pp. 341–2.
42. Britten, *Nineteenth Century Miracles*, pp. 183, 209; Nelson, *Spiritualism and Society*, p. 99; Podmore, *Modern Spiritualism* 2:166, 63; Tuttle and Peebles, *Year-Book of Spiritualism*, pp. 161–2; Wallis, *Story of My Life*, p. 7. On Burns's background, see Logie Barrow, "Socialism in Eternity: The Ideology of Plebeian Spiritualists, 1853–1913," *History Workshop: A Journal of Socialist Historians* 9 (Spring 1980):40–2.
43. Podmore, *Modern Spiritualism*, 2:165–6; also see Britten, *Nineteenth Century Miracles*, p. 209; Charles Maurice Davies, *Mystic London: or, Phases of Occult Life in the Metropolis* (London: Tinsley Brothers, 1875), p. 277, and *Unorthodox London: or, Phases of Religious Life in the Metropolis*, 3d ed. (1875; reprint ed., New York: Augustus M. Kelley, 1969), pp. 315–21; Gauld, *Founders of Psychical Research*, p. 76.
44. Nelson, *Spiritualism and Society*, pp. 121–3.
45. Cooper, "Thomas Shorter," p. 410; also see Gauld, *Founders of Psychical Research*, pp. 72–3.

46. Information derived from the facts of publication in each volume of the complete set of the *Spiritual Magazine* in the Harry Price Library, Senate House, University of London.
47. Tuttle and Peebles, *Year-Book of Spiritualism*, p. 195; Podmore, *Modern Spiritualism*, 2:165.
48. William Henry Harrison, *Spirit People* (London: W. H. Harrison, 1875), pp. v, 19, 40.
49. Podmore, *Modern Spiritualism*, 2:168, 118–19; Trevor Hall, *The Spiritualists*, p. 8; Dingwall, *Critics' Dilemma* , p. 32.
50. Nelson, *Spiritualism and Society*, pp. 107–10, and the list of major British spiritualist publications on p. 295; Podmore, *Modern Spiritualism*, 2:168–9, 177; Minute Books of the BNAS, Committees, no. 2, pp. 108–10 (3 February 1879), 145–8 (28 April 1879), 153–4 (13 May 1879).
51. Rogers, *Life and Experiences*, p. 45. E. J. Dingwall sets forth the little that is known about J. S. Farmer and argues that the spiritualist and lexicographer were the same man, in *"Light* and the Farmer Mystery," *JSPR* 51 (February 1981):22–5.
52. Podmore, *Modern Spiritualism*, 2:178.
53. "The Annual General Meeting," *Two Worlds* 5 (12 February 1892):73; Nelson, *Spiritualism and Society*, pp. 113–14, 122–4, 273, 281–3.
54. See John L. Campbell and Trevor H. Hall, *Strange Things: The Story of Fr. Allan McDonald, Ada Goodrich Freer, and the Society for Psychical Research's Enquiry into Highland Second Sight* (London: Routledge & Kegan Paul, 1968).
55. Barrow, "Socialism in Eternity," p. 52; Britten, *Nineteenth Century Miracles*, pp. 198, 209–10; Nelson, *Spiritualism and Society*, p. 112; Podmore, *Modern Spiritualism*, 2:23–4.
56. Thomas P. Barkas, *Outlines of Ten Years' Investigations into the Phenomena of Modern Spiritualism* (London: Frederick Pitman; Newcastle: T. P. Barkas, 1862), p. 15.
57. [Robert Bell], "Stranger than Fiction," *Cornhill Magazine* 2 (August 1860):211–24; [Hayward], "Spiritualism, as related to Religion and Science," pp. 22–42; *Household Words* 11 (30 June 1855), 513–15, cited in Katherine H. Porter, *Through a Glass Darkly: Spiritualism in the Browning Circle* (1958; reprint ed., New York: Octagon Books, 1972), p. 18.
58. *The Mysteries of Mediumship. A Spirit Interviewed: Being an Account of the Life and Mediumship of J. J. Morse with a Full Report of an Interview with His Chief Control Tien Sien Tie* (London: Progressive Literature Agency, 1894), p. 14 (no. 6 in volume of collected pamphlets, *Spiritualism and Its Truth*, Houdini Collection). Also see Nelson, *Spiritualism and Society*, pp. 273–84.
59. Gauld, *Founders of Psychical Research*, p. 77. Barrow, "Socialism in Eternity," p. 39, thinks that spiritualism in all likelihood never numbered more than 10,000 activists.
60. *Mysteries of Mediumship*, p. 3. On Cogman, see Wallis, *Story of My Life*, p. 7.
61. Harrison, *Owen and the Owenites*, p. 251.
62. Nelson, *Spiritualism and Society*, p. 280; Podmore, *Modern Spiritualism*, 2:41–2.
63. Nelson, *Spiritualism and Society*, pp. 112, 114, 116; "North of England Spiritualists' Conference Committee," *Spiritualist Newspaper* 9 (11 August 1876):22.

64. Nelson, *Spiritualism and Society*, pp. 103, 111–16, 119.
65. British Association of Progressive Spiritualists, *Proceedings of the First, Second, and Third Conventions, 1865–1867*, in volume entitled *Tracts: Spiritual*, Harry Price Library; Nelson, *Spiritualism and Society*, pp. 100–2; Podmore, *Modern Spiritualism*, 2:164; Britten, *Nineteenth Century Miracles*, pp. 181–2.
66. Nelson, *Spiritualism and Society*, pp. 124–9.
67. Ibid., pp. 98–9; Podmore, *Modern Spiritualism*, 2:44.
68. Leigh Hunt, *The Story of the Marylebone Spiritualist Association, Its Work and Workers* (London: Marylebone Spiritualist Association, 1928); Roy Stemman, *One Hundred Years of Spiritualism: The Story of the Spiritualist Association of Great Britain, 1872–1972* (London: S.A.G.B., 1972), pp. 8–29.
69. Nelson, *Spiritualism and Society*, pp. 114, 116.
70. Britten, *Nineteenth Century Miracles*, p. 183; Harrison, *Spirit People*, p. 41; Rogers, *Life and Experiences*, pp. 43–4.
71. Declaration, as quoted in Davies, *Mystic London*, p. 401; also see Nelson, *Spiritualism and Society*, pp. 104–6; Podmore, *Modern Spiritualism*, 2:169.
72. Britten, *Nineteenth Century Miracles*, pp. 183–4; Podmore, *Modern Spiritualism*, 2:169, 177; Rogers, *Life and Experiences*, p. 44.
73. *DNB*, s.v. "Podmore, Frank"; Gauld, *Founders of Psychical Research*, pp. 142–3; for references to Podmore's activities with the BNAS, see Minute Books of the BNAS, Committees, no. 2 (24 November and 8 December 1879).
74. Earl of Dunraven, *Past Times and Pastimes*, 2 vols. (London: Hodder & Stoughton, 1922), 1:178; Fawn M. Brodie, *The Devil Drives: A Life of Sir Richard Burton* (New York: W. W. Norton, 1967), pp. 314–15.
75. Minute Books of the BNAS, Committees, no. 2, p. 126 (5 March 1879).
76. Britten, *Nineteenth Century Miracles*, p. 185; Nelson, *Spiritualism and Society*, p. 107.
77. See, for example, the meeting of the General Purposes Committee in Minute Books of the BNAS, Committees, no. 2, pp. 50–1 (3 October 1878).
78. Ibid., pp. 13–15 (19 June 1878); Massey's comment was made during Slade's trial. See "The Prosecution of Dr. Slade," *Spiritualist Newspaper* 9 (27 October 1876):153.
79. Nelson, *Spiritualism and Society*, pp. 107–10; Rogers, *Life and Experiences*, pp. 44–5.
80. Minute Books of the BNAS, Committees, no. 2 (June and July 1881, 17 April 1883); Central Association of Spiritualists, Council Minutes, no. 3, pp. 12–14 (11 April 1882), 83–4 (25 September 1883), College of Psychic Studies. The minutes of the CAS committee meetings are found at the end of the volume containing minutes of the BNAS committees.
81. Rogers, *Life and Experiences*, p. 45.
82. Ibid.; London Spiritualist Alliance, Minute Book of Council Meetings 1896–1909, p. 1 (7 September 1896), College of Psychic Studies.
83. Rogers, *Life and Experiences*, p. 46.

Part II. A surrogate faith

1. John S. Farmer, *Spiritualism as a New Basis of Belief* (London: E. W. Allen, 1880), pp. v–vi.

2. John Tulloch, *Movements of Religious Thought in Britain During the Nineteenth Century, Being the Fifth Series of St. Giles' Lectures* (1885; reprint ed., New York: Humanities Press, 1971), p. 128. Tulloch was Principal of St. Andrews.
3. Howard R. Murphy, "The Ethical Revolt Against Christian Orthodoxy in Early Victorian England," *American Historical Review* 60 (July 1955):800–17; Josef L. Altholz, "The Warfare of Conscience with Theology," in *The Mind and Art of Victorian England*, ed. Altholz (Minneapolis: University of Minnesota Press, 1976), pp. 58–77.
4. Tulloch, *Movements of Religious Thought*, p. 329.
5. In *Primitive Culture*, 1871. See Frank Miller Turner, *Between Science and Religion: The Reaction to Scientific Naturalism in Late Victorian England* (New Haven and London: Yale University Press, 1974), p. 32.
6. "Illusions and Hallucinations," *British Quarterly Review* 36 (October 1862):387.
7. "The Prosecution of Dr. Slade," *Spiritualist Newspaper* 9 (13 October 1876):129; C. Maurice Davies, *Unorthodox London: or, Phases of Religious Life in the Metropolis*, 3d ed. (1875; reprint ed., New York: Augustus M. Kelley, 1969), p. 302.
8. "Correspondence," *Spiritualist Newspaper* 9 (15 September 1876):81.

3. Spiritualism and Christianity

1. J. P. Hopps, *A Scientific Basis of Belief in a Future Life: or, The Witness Borne by Modern Science to the Reality and Pre-Eminence of the Unseen Universe* (London: Williams & Norgate, [1880], p. 5). (No. 8 in *Houdini Pamphlets: Future Life*, vol. 1, Houdini Collection.)
2. Archbishop Tait, *Some Thoughts on the Duties of the Established Church of England as a National Church* (1876), p. 28, quoted in P. T. Marsh, *The Victorian Church in Decline: Archbishop Tait and the Church of England 1868–1882* (London: Routledge & Kegan Paul, 1969), p. 59; Gladstone, quoted in Philip Magnus, *Gladstone, A Biography* (London: John Murray, 1954), p. 219.
3. S. C. Hall, *The Use of Spiritualism?* (London: E. W. Allen, 1884), p. 6 (no. 13 in *Houdini Pamphlets: Spiritualism*, vol. 13, Houdini Collection); Newton Crosland, *Apparitions; A New Theory. And Hartsore Hall, A Ghostly Adventure*, 2d ed. rev. (London: Effingham Wilson, & Bosworth & Harrison, 1856), pp. 9–10 (no. 5 in *Houdini Pamphlets: Spiritualism*, vol. 13, Houdini Collection); Thomas Shorter [Brevior], *The Two Worlds, The Natural and the Spiritual: Their Intimate Connexion and Relation Illustrated by Examples and Testimonies, Ancient and Modern* (London: F. Pitman, [1864]), p. 351.
4. "To Our Readers," *Spiritual Magazine*, 2 n.s. (January 1867):2.
5. Charles Maurice Davies, *Mystic London: or, Phases of Occult Life in the Metropolis* (London: Tinsley Brothers, 1875), pp. 395–6.
6. Frank Podmore, *Modern Spiritualism: A History and a Criticism*, 2 vols. (London: Methuen, 1902), 2:12–16, 168; Alan Gauld, *The Founders of Psychical Research* (London: Routledge & Kegan Paul, 1968), p. 68n; Rev. John Jones, *Spiritualism the Work of Demons* (Liverpool: Edward Howell, 1871), and Rev. Thomas Worrall, *Modern Spiritualism One Form of the Predicted Apostacy* (Liverpool: Gabriel Thompson, 1874). Jones's

and Worrall's pamphlets are numbers 14 and 15 in a bound volume, *Pamphlets*, in the Harry Price Library.
7. Geoffrey Rowell, *Hell and the Victorians. A Study of the Nineteenth-Century Theological Controversies Concerning Eternal Punishment and the Future Life* (Oxford: Clarendon Press, 1974), p. 195.
8. See J. F. C. Harrison, *The Second Coming: Popular Millenarianism 1780–1850* (New Brunswick, N. J.: Rutgers University Press, 1979), and Keith Thomas, *Religion and the Decline of Magic* (New York: Charles Scribner's Sons, 1971), pp. 253–79.
9. "Is It All Satanic?" *British Spiritual Telegraph* 1 (4 July 1857):19; W. S. Moses, "Some Things that I *Do* Know of Spiritualism, and Some that I do *Not* Know: A Chapter of Autobiography," p. 13, address to the London Spiritualist Alliance, 29 November 1887, which appears as no. 11 in volume entitled *Addresses Delivered Before the London Spiritualist Alliance during the Years 1884 to 1888* (London: LSA, 1889), Houdini Collection.
10. Alice E. Hacker, "Is Spiritualism Satanic?" *Christian Spiritualist* 1 (February 1871):19–20; Crosland, *Apparitions*, pp. 11–12; Howitt's letter to the Rev. G. H. Forbes, 9 May 1861, quoted in Thomas P. Barkas, *Outlines of Ten Years' Investigations into the Phenomena of Modern Spiritualism* (London: Frederick Pitman: Newcastle: T. P. Barkas, 1862), p. 154.
11. "Spiritual Manifestations," *Blackwood's Edinburgh Magazine* 73 (May 1853):633. *The Wellesley Index to Victorian Periodicals*, vol. 1, attributes the article to William Edmonstoune Aytoun, poet and rhetorician. The article in the *Sussex Daily News* is reprinted under "Objections to Spiritualism" in the *Spiritualist Newspaper* 9 (11 August 1876):23.
12. Hall, quoted in Hudson Tuttle and J. M. Peebles, *The Year-Book of Spiritualism for 1871* (London: James Burns, 1871), p. 25.
13. Arthur Christopher Benson, *The Life of Edward White Benson, sometime Archbishop of Canterbury*, 2 vols. (London: Macmillan, 1900), 1:98.
14. The views of Benson, Westcott, and Temple on spiritualism are presented in "The Response to the Appeal. From Prelates, Pundits and Persons of Distinction," *Borderland* 1 (July 1893):10–11. Even Archbishop Whately, who gave his name to a list of people who believed in the validity of D. D. Home's manifestations, concluded: "On the whole I think it is the safe course to have nothing to do with any necromantic practices." Whately, quoted in "Spiritualism, as related to Religion and Science," *Fraser's Magazine* 71 (January 1865):27, 37–9. *The Wellesley Index to Victorian Periodicals*, vol. 2, attributes this article to Abraham Hayward, barrister and essayist.
15. *DNB.*, s.v. "Newbould, William Williamson," and "The Spiritualist Scientific Research Committee," *Spiritualist Newspaper* 9 (11 August 1876):23.
16. "Dr. Monck," *Spiritualist Newspaper* 9 (3 November 1876):163.
17. For an account of some of Colley's séances with Monck, see William Stainton Moses ["M.A., (Oxon.)"], *Psychography: A Treatise on One of the Objective Forms of Psychic or Spiritual Phenomena* (London: W. H. Harrison, 1878), pp. 30–2, 106, 134–5.
18. Thomas Colley, *Phenomena, Bewildering, Psychological, Being a Lecture by the Ven. Archdeacon Colley (Dio. Natal), Rector of Stockton, Warwickshire, on Spiritualism. Delivered at Weymouth during the Week*

of the Church Congress, October 1905 (London: Published at the Office of *Light*, n.d.), pp. 4–6, 12, 19–22 (no. 5 in *Houdini Pamphlets: Spiritualism*, vol. 11, Houdiní Collection); Podmore, *Modern Spiritualism*, 2:252.

19. The *Times* ran a detailed report of the case, in the "Law Report" column, on 27 April 1907 (p. 14), 30 April 1907 (p. 3), and 1 May 1907 (p. 4).
20. The flyer is included, with a letter from Colley to Sir Oliver Lodge, 7 May 1908, in the Lodge Collection (box 2, no. 295).
21. Lodge to Crookes, 23 December 1916, Lodge Collection (box 2, no. 364).
22. Statement of 1878, quoted in Colley, *Phenomena*, p. 26; appointment to H.M.S. *Malabar* announced in the *Spiritualist Newspaper* 9 (6 October 1876):114. Colley's claim to the title of archdeacon was a subject of some dispute, and the lawsuit of 1907 was in fact sparked by Maskelyne's charge that Colley had no right to such designation at all. The trouble was that Colley had been initially nominated to the post by Colenso, whose own status as Bishop of Natal was the subject of one of the hotter controversies in Victorian ecclesiastical history. The complexities of the issue were resolved by the jury in the *Colley v. Maskelyne* case when it ruled that Maskelyne's charge was libelous. See the three accounts in *Times* (n. 19 above).
23. "Law Report," *Times*, 30 April 1907, p. 3; Colley's letter to the bishop of Salisbury is quoted in Colley, *Phenomena*, p. 27.
24. Charles Maurice Davies, *Orthodox London: or, Phases of Religious Life in the Church of England* (London: Tinsley Brothers, 1873), pp. 6, 20.
25. *DNB*, s.v. "Haweis, Hugh Reginald." Also see the biography of Haweis's wife, Mary Eliza Joy, by Bea Howe, *Arbiter of Elegance* (London: Harvill Press, 1967), p. 36.
26. H. R. Haweis, "Spiritualism and Christianity. An Address Delivered Before the London Spiritualist Alliance, Ltd., at St. James's Hall, April 20th, 1900," pp. 3, 13–15 (no. 7 in *Houdini Pamphlets: Spiritualism*, vol. 9, Houdini Collection); H. R. Haweis, *Thoughts for the Times. Sermons* (New York: Holt & Williams, 1872), p. 189; Owen Chadwick, *The Secularization of the European Mind in the Nineteenth Century* (Cambridge: Cambridge University Press, 1975), pp. 104–5, points out that "in 1864 a legal judgment in the English Judicial Committee allowed a clergyman of the Church of England to teach or at least to hope for the ultimate salvation even of the wicked."
27. Haweis, *Thoughts for the Times*, p. 77.
28. Charles Maurice Davies, *The Great Secret, and its Unfoldment in Occultism. A Record of Forty Years' Experience in the Modern Mystery*, 3d ed. (Blackpool: The Ellis Family, n.d.), p. 239; R. G. Medhurst, "Stainton Moses and Contemporary Physical Mediums. 3. Mr. and Mrs. Nelson Holmes," *Light* 83 (Autumn 1963):133. "Members, Associates, Honorary and Corresponding Members," *PSPR* 1 (1882-3):327;"Response to the Appeal," *Borderland* 1 (July 1893):12.
29. Haweis, "Spiritualism and Christianity," pp. 2, 5–7, 9–10.
30. Howe, *Arbiter of Elegance*, pp. 255–6.
31. Ibid.; *DNB*, s.v. "Haweis, Hugh Reginald."
32. "Obituary. The Rev. H. R. Haweis," *Times*, 30 January 1901, p. 7.
33. Davies, *Mystic London*, pp. 402–3.

34. Davies, *Great Secret*, pp. 261–2, 196.
35. London: Tinsley Brothers, 1876. His name appears in pencil on the title page of the copy in the Houdini Collection. Also see Eric J. Dingwall, *The Critics' Dilemma: Further Comments on Some Nineteenth Century Investigations* (Crowhurst, Sussex: E. J. Dingwall, 1966), p. 14; Trevor H. Hall, *The Spiritualists: The Story of Florence Cook and William Crookes* (New York: Helix Press, Garrett Publications, 1963), pp. 26n, 84n; and Davies's entry in the *DNB*.
36. Davies, *Great Secret*, pp. 16–18, 25–36, 57–62, 66–79, 95–97, 114–18, 160, 229–30; *Mystic London*, pp. 308–19, 399. Also see Trevor Hall, *The Spiritualists*, pp. 84–6.
37. Davies, *Great Secret*, pp. 253–64; Trevor Hall, *The Spiritualists*, pp. 137–8; See Minute Books of the BNAS, Committes, no. 2 (15 February 1881), for the General Purposes Committee's grant of BNAS facilities for Davies's use.
38. Davies, *Great Secret*, pp. 259–65.
39. Davies quoting from one of his addresses, "What Mean Ye By this Service?" in Ibid., pp. 270, 276–8.
40. Davies, *Great Secret*, pp. 285–91; *DNB*, s.v. "Davies, Charles Maurice"; "Obituary. The Rev. Dr. C. M. Davies," *Times*, 9 September 1910, p. 11.
41. Charles Maurice Davies, *Heterodox London: or, Phases of Free Thought in the Metropolis*, 2 vols. (London: Tinsley Brothers, 1874), 2:1.
42. Ghost Club Minutes, vol. 1 (1 December 1882), p. 6–8, B. M. Add. MSS. 52258, British Library. Stainton Moses served as the Club's secretary through 1890. Also see Podmore, *Modern Spiritualism*, 2:275–6.
43. See Moses's series of articles in *Human Nature*, 1874–1875, vols. 8–9, "Researches in Spiritualism During the Years 1872–73"; and "Our Gallery of Borderlanders. The Rev. W. Stainton Moses," *Borderland* 1 (April 1894):310–17, written by Ada Goodrich Freer under the pseudonym "Miss X."
44. Davies, *Great Secret*, p. 232; "Gallery of Borderlanders. Moses," p. 307.
45. William Stainton Moses ["M. A., (Oxon.)"], *Spirit-Identity*, new ed. (London: London Spiritualist Alliance, 1908), pp. 38, 43. *Spirit-Identity* was first published in 1879.
46. Gauld, *Founders of Psychical Research*, p. 78.
47. A. W. Trethewy, *The "Controls" of Stainton Moses ("M. A. Oxon.")* (London: Hurst & Blackett, [1923]), pp. 21–3.
48. W. H. Salter, *Trance Mediumship: An Introductory Study of Mrs. Piper and Mrs. Leonard* (London: SPR, 1950), pp. 9–10; Podmore, *Modern Spiritualism*, 2:288.
49. Frederic W. H. Myers, "The Experiences of W. Stainton Moses. – I," *PSPR* 9 (1893–4):246, 249, 250, 253; Podmore, *Modern Spiritualism*, 2:276, 280.
50. This is the hypothesis proposed in "Modern Mystics and Modern Magic," *Saturday Review* 77 (20 January 1894):75–6.
51. Podmore, *Modern Spiritualism*, 2:275; Myers, "Experiences of W. S. Moses. – I," pp. 249–50; Charlton Templeman Speer, introduction to *Spirit Teachings Through the Mediumship of William Stainton Moses (M.A., Oxon.)*, 7th ed. (London: London Spiritualist Alliance, 1912), p. viii.

412   *Notes to pp. 79–83*

52. Trethewy, *"Controls" of Stainton Moses*, p. 193; also see Frederic W. H. Myers, "The Experiences of W. Stainton Moses. – II," *PSPR* 11 (1895):113.
53. W. S. Moses, *A Spirit's Creed and Other Teachings* (London: *Light* Publishing Office, n.d.), p. 4 (no. 10 in *Houdini Pamphlets: Spiritualism*, vol. 4, Houdini Collection).
54. W. S. Moses ["M. A., (Oxon.)"], "Researches in Spiritualism During the Years 1872–73," *Human Nature* 8 (March 1874):110; Podmore, *Modern Spiritualism*, 2:284.
55. "Correspondence," *Spiritualist Newspaper* 9 (15 September 1876):81; Moses, "Some Things that I *Do* Know of Spiritualism," p. 19.
56. Moses, *Spirit-Identity*, pp. 44, 52–3.
57. W. S. Moses, "Spiritualism At Home and Abroad: Its Present Position and Future Work," p. 29, address to the LSA, 13 November 1885 (no. 4 in *Addresses Delivered Before the L.S.A.*, Houdini Collection).
58. R. G. Medhurst, "Stainton Moses and Contemporary Physical Mediums. 2. Florence Cook, Kate Cook," *Light* 83 (Summer 1963):72.
59. Podmore, *Modern Spiritualism*, 2:284.
60. The *Saturday Review*, for example, described him as a clergyman in 1894, twenty-four years after he left the ministry. See "Modern Mystics and Modern Magic," pp. 75–6 (n. 50 above).
61. "Ven. Albert Basil Orme Wilberforce, D.D.," *Who Was Who, 1916–1928*; addresses by Thornton and Wilberforce published in W. S. Moses ["M. A., (Oxon.)"], *Spiritualism at the Church Congress* (Chicago: Religio-Philosophical Journal, 1882), pp. 11, 13, 17 (no. 4 in *Houdini Pamphlets: Spiritualism*, vol. 10, Houdini Collection). E. J. Dingwall discusses Thornton and Wilberforce in *"Light* and the Farmer Mystery," *JSPR* 51 (February 1981):23.
62. In *Presidential Addresses to the Society for Psychical Research 1882–1911* (Glasglow: Robert Maclehose for the SPR, 1912), p. 4.
63. A. C. Benson, *Edward White Benson*, 1:98. In *PSPR* 1 (1882–3):326, "Benson, Mrs., The Palace, Lambeth, London, S. W.," is listed under "Associates, Honorary, and Corresponding Members." Some clerical members even contributed articles to the Society's *Proceedings*.
64. W. Boyd Carpenter, "Presidential Address. Delivered on May 23rd, 1912," *PSPR* 26 (September 1912):8.
65. "Obituary. The Rev. H. R. Haweis," *Times*, 30 January 1901, p. 7. Also see John Wilson, "British Israelism: The Ideological Restraints on Sect Organisation," in *Patterns of Sectarianism: Organisation and Ideology in Social and Religious Movements*, ed. Bryan R. Wilson (London: Heinemann, 1967), p. 360. John Wilson points out that spiritualism flourished in "a religious context in which there was tolerance and encouragement for free inquiry." Both Anglican and Nonconformist clergy could, if they wished, add spiritualism to their own creeds, whereas "in a more disciplined and more dogmatic religious system than the Protestant denominations of Victorian England, [such] accretions to accepted teaching would never have been able to find association with orthodoxy."
66. Davies, *Mystic London*, p. 291.
67. Statement in the *Nonconformist*, quoted in "The 'Nonconformist' on Spiritualism," *Christian Spiritualist* 1 (January 1871):6.
68. "The Religious Heresies of the Working Classes," *Westminster Review*, n.s., 21 (January 1862):90. *The Wellesley Index to Victorian Periodicals*,

vol. 3, attributes the article to William Binns, Unitarian minister. "First Words," *Christian Spiritualist* 1 (January 1871):1–2, where Young not only explains Christian spiritualism, but vigorously repudiates "all the forms of the doctrine of the Trinity known to him, . . ." F. R. Young is described as a minister of the Free Christian Church, Swindon, in "Healing Mediumship," *Daybreak* 1 (November 1868):10–12.

69. Jean Burton, *Heyday of a Wizard: Daniel Home the Medium* (London: George G. Harrap, 1948), pp. 155–7, and the flippant debate on the subject of Home's expulsion from Rome in *Parliamentary Debates* (Commons), 3d ser., 175 (30 May 1864):839–44; conciliar statement of 1866, quoted in Herbert Thurston, *The Church and Spiritualism* (Milwaukee: Bruce Publishing Co., 1933), p. 146.
70. "Response to the Appeal," *Borderland* 1 (July 1893):12
71. Wilberforce, quoted in Moses, *Spiritualism at the Church Congress*, pp. 14–15. Wilberforce subsequently joined the SPR.
72. Crosland, *Apparitions*, p. 11; J. H. Powell, *Spiritualism; Its Facts and Phases* (London: F. Pitman, 1864), p. vii.
73. Geoffrey K. Nelson, *Spiritualism and Society* (London: Routledge & Kegan Paul, 1969), pp. 136–7. Also see R. Laurence Moore, *In Search of White Crows: Spiritualism, Parapsychology, and American Culture* (New York: Oxford University Press, 1977), p. 54, for comments on the highly detailed descriptions of living conditions beyond the veil, which appeared in spiritualist literature.
74. Florence Marryat, *The Spirit World* (New York: Charles B. Reed, 1894), p. 140; John Jones, *The Natural and Supernatural: or, Man Physical, Apparitional, and Spiritual* (London: H. Bailliere, 1861), pp. 222, 225.
75. J. Burns, "Spiritualism and the Gospel of Jesus," p. 4 (no. 1 in *Houdini Pamphlets: Spiritualism*, vol. 2, Houdini Collection). This undated, four-page pamphlet was probably published by Burns in London.
76. Hugh Doherty, "Bible Spiritualism," *Spiritual Magazine*, n.s., 7 (April 1872):151.
77. On Johnson, St. George Stock, "A New Religion," *Human Nature* 8 (October 1874):432–4; on Gutteridge, Valerie E. Chancellor, ed., *Master and Artisan in Victorian England. The Diary of William Andrews and the Autobiography of Joseph Gutteridge* (New York: Augustus M. Kelley, 1969), pp. 163–71; on Uncle Arthur, V. S. Pritchett, *A Cab at the Door* (New York: Random House, 1968), pp. 42, 45, 233–4. Logie Barrow, "Socialism in Eternity: The Ideology of Plebeian Spiritualists, 1853–1913," *History Workshop: A Journal of Socialist Historians* 9 (Spring 1980), cites a number of "plebeian" spiritualists with a background in free thought. For example, see p. 52, for Seth Ackroyd, the machine sawyer, whose intellectual development led him from Wesleyan Methodism through Liberal Christianity, to close affiliation with secularism and, by 1890, to anti-Christian spiritualism. In the case of Ackroyd, as with uncounted others from a similar social milieu, Barrow (p. 43) found that "nonconformity, secularism and spiritualism formed a triangle of tensions." Also see pp. 39, 44–5, 50.
78. Stock, "New Religion," p. 436.
79. Publication information in the volumes of the *Spiritual Magazine*, Harry Price Library; "George Sexton," *Human Nature* 8 (January 1874):24–9; Barrow, "Socialism in Eternity," pp. 45, 50. Gutteridge, too, appears to

414 *Notes to pp. 87–91*

have returned to the Christian fold before his death. Chancellor, ed.,
Master and Artisan, p. 7.

80. [Binns], "Religious Heresies of the Working Classes," p. 90.
81. Barrow, "Socialism in Eternity," pp. 38–9, 46–8, deftly sketches the
    relationships among Owenism, rationalism, millennialism, spiritualism,
    and secularism, and discusses the transfer of millenarian hopes from this
    world to the "Summerland."
82. Gerald Massey, for example, gave Swedenborg credit for doing "more
    than any other to make the world of spirit solid ground for men to tread."
    Concerning Spiritualism (London: James Burns, [1871]), p. 31. E. D.
    Rogers, likewise, found that Swedenborg's life and writings "had a good
    effect in preparing [his] mind for Spiritualism, . . ." He was a Swed-
    enborgian after abandoning Wesleyan Methodism and before embracing
    spiritualism. E. D. Rogers, *Life and Experiences of Edmund Dawson
    Rogers, Spiritualist and Journalist* (London: Office of *Light*, [1911]), pp.
    11–12; "Transition of Mr. E. Dawson Rogers," *Light* 30 (8 October
    1910):483.
83. British Association of Progressive Spiritualists, *Proceedings of the First
    Convention*, 26 and 27 July 1865, p. iii, in *Tracts: Spiritual*, Harry Price
    Library. Barrow, "Socialism in Eternity," pp. 60–2, points out, however,
    that progressive spiritualists were actually ambivalent about emancipated
    women and tended to accept uncritically the "'angel in the home' ster-
    eotype" of the wife and mother.
84. R. W. Postgate, *Out of the Past. Some Revolutionary Sketches* (New
    York: Vanguard Press, 1926), pp. 88–93.
85. Barrow, "Socialism in Eternity," pp. 48–53.
86. If not long before then. See Thomas, *Religion and the Decline of Magic*,
    pp. 159–73, for the argument that organized religion was losing its popular
    grip well before the onset of industrialization and urbanization.
87. [Binns], "Religious Heresies of the Working Classes," p. 85. Also see
    Owen Chadwick, "The Established Church Under Attack," in Anthony
    Symondson, ed., *The Victorian Crisis of Faith: Six Lectures* (London:
    Society for Promoting Christian Knowledge, 1970), p. 93.
88. J. F. C. Harrison, *Robert Owen and the Owenites in Britain and America:
    The Quest for the New Moral World* (London: Routledge & Kegan Paul,
    1969), p. 251, observes: "Part of the appeal of Owenism had always been
    that it was a heresy, and spiritualism satisfied the same need."
89. On Cogman, "Passed to Spirit Life," *Spiritualist Newspaper* 9 (27 Oc-
    tober 1876):151; on Johnson, Stock, "New Religion," pp. 434–5; on Gut-
    teridge, Chancellor, ed., *Master and Artisan*, pp. 7, 163–71.
90. [Binns], "Religious Heresies of the Working Classes," p. 90–1.
91. Gutteridge, in Chancellor, ed., *Master and Artisan*, p. 171. For once,
    Conan Doyle was accurate when he wrote about "what may be termed
    anti-Christian, though not anti-religious, Spiritualism." Arthur Conan
    Doyle, *The History of Spiritualism*, 2 vols. (New York: George H. Doran,
    1926), 1:180. Several general histories of the period have recognized the
    role of spiritualism as a working-class religion. See, for example, Harold
    Perkin, *The Origins of Modern English Society 1780–1880* (London: Rout-
    ledge & Kegan Paul, 1969), p. 206. E. J. Hobsbawm, *The Age of Capital
    1848–1875* (London: Weidenfeld & Nicolson, 1975), pp. 273–4, observing
    that "even among free-thinkers, a nostalgia for religion remained," points

to the parallels between the ritual and paraphernalia of spiritualism on the one hand, and of trade unions and mutual aid societies on the other. This is not the place to join the argument about the role of socialism as a substitute religion for segments of the British working classes in the late nineteenth century, but, despite obvious differences, the affinities between socialism and progressive spiritualism are suggestive in this respect.

92. See Tables 3A and 4 in Nelson, *Spiritualism and Society*, pp. 281–4.
93. [Binns], "Religious Heresies of the Working Classes," p. 89; Nelson, *Spiritualism and Society*, pp. 143, 281–3; Edward Royle, *Victorian Infidels. The Origins of the British Secularist Movement 1791–1866* (Manchester: Manchester University Press, 1974), pp. 295–7. For the similar spread of provincial free thought societies subsequently, see Edward Royle, *Radicals, Secularists and Republicans: Popular Freethought in Britain, 1866–1915* (Manchester: Manchester University Press, 1980), pp. 337–42.
94. As Gauld, *Founders of Psychical Research*, p. 77, points out.
95. V. S. Pritchett's explanation is simpler. Describing his Yorkshire uncle's conversion from atheism to spiritualism, he commented: "There was a ruinous drift to religion in these Northerners." *Cab at the Door*, p. 45.
96. Moore, *In Search of White Crows*, p. 61, points to this significant problem.
97. E. W. Wallis, *The Story of My Life; and Development and Experiences as a Medium* (London: James Burns, n.d.), p. 4 (no. 3 in volume of collected pamphlets, *Spiritualism and its Truth*, Houdini Collection). E. W. Wallis, "An Open Letter to Christian Opponents of Spiritualism, to Rev. Fleming and Mr. Waldron, and the Public Generally," p. 3 (no. 6 in *Houdini Pamphlets: Spiritualism*, vol. 4, Houdini Collection); J. J. Morse, *Leaves from My Life: A Narrative of Personal Experiences in the Career of a Servant of the Spirits*; . . . (London: James Burns, 1877), p. 7; Burns, "Spiritualism and the Gospel of Jesus," p. 4. Wallis's "Open Letter," an undated, four-page pamphlet, was reprinted from *Two Worlds*.
98. Rowell, *Hell and the Victorians*, pp. vii, 2; Marsh, *Victorian Church in Decline*, p. 38. The Victorians were not, needless to say, the first to question and denounce the doctrine of eternal punishment. For much earlier versions of the same attitudes, see D. P. Walker, *The Decline of Hell: Seventeenth-Century Discussions of Eternal Torment* (Chicago: University of Chicago Press, 1964).
99. Princess Louise quoted in Elizabeth Longford, *Victoria R. I.* (London: Weidenfeld & Nicolson, 1964), p. 343; Rowell, *Hell and the Victorians*, pp. 146–8, 150; Chadwick, *Secularization of the European Mind*, pp. 104–5; Susan Budd, *Varieties of Unbelief. Atheists and Agnostics in English Society 1850–1960* (London: Heinemann Educational Books, 1977), p. 117. Barrow, "Socialism in Eternity," p. 62, comments on "the gradual softening of many denominations" over the question of eternal hellfire. In the United States, the Universalist refusal to accept the doctrine of eternal punishment figured as a prominent ingredient in American spiritualism.
100. British Association of Progressive Spiritualists, *Proceedings of the Second Convention*, 25–27 July 1866, pp. 4–5, 34, in *Tracts: Spiritual,* Harry

Price Library; Morse, *Leaves from My Life*, p. 7; Rogers, *Life and Experiences*, pp. 7, 10–11.

101. St. George Stock, *Attempts at Truth* (London: Trübner, 1882), pp. 134, 159, 137.
102. Declaration, quoted in Nelson, *Spiritualism and Society*, p. 127.
103. Stock, *Attempts at Truth*, p. 150.
104. Wallis, "Open Letter," p. 3.
105. Some opted for the question-begging theory of a "spirit–body" or, as St. George Stock explained, "a quasi-material envelope underlying the physical organism, and serving as the vehicle or garb of the spirit on decay of its old covering." Stock, *Attempts at Truth*, p. 166.
106. Massey quoted in B. O. Flower, *Gerald Massey: Poet, Prophet and Mystic* (Boston: Arena Publishing Co., 1895), p. 111. Unlike Shorter, Gerald Massey ultimately moved from Christian Socialism to anti-Christian spiritualism because "Truths stared him in the face which were in advance of an unadvancing Church." James Robertson, *Spiritualism: The Open Door to the Unseen Universe. Being Thirty Years of Personal Observation and Experience Concerning Intercourse between the Material and Spiritual Worlds* (London: L. N. Fowler, 1908), p. 210. Burns, "Spiritualism and the Gospel of Jesus," p. 4.
107. Moore, *In Search of White Crows*, pp. 52–3.
108. In its opening issue, for example, the *British Spiritual Telegraph* insisted that man "has a conscious individualized existence after the death of the physical body." "Introduction," *British Spiritual Telegraph* 1 (27 June 1857):4.
109. Stock, *Attempts at Truth*, p. 134; Burns, "Spiritualism and the Gospel of Jesus," p. 3. For an argument similar to Burns's, see Wallis, "Open Letter," p. 3.
110. Stock, *Attempts at Truth*, p. 247; Burns, "Spiritualism and the Gospel of Jesus," pp. 2, 4.
111. Burns, "Spiritualism and the Gospel of Jesus," pp. 1, 2, 4; John Nevil Maskelyne, *Modern Spiritualism. A Short Account of its Rise and Progress, with Some Exposures of So-Called Spirit Media* (London: Frederick Warne, for the author, 1875), p. 111; Moore, *In Search of White Crows*, p. 55; Louisa Lowe, "The Ends, Aims, and Uses of Modern Spiritualism," in *"Rifts in the Veil": A Collection of Inspirational Poems and Essays Given Through Various Forms of Mediumship; also of Poems and Essays by Spiritualists* (London: W. H. Harrison, 1878), p. 110.
112. [Binns], "Religious Heresies of the Working Classes," p. 84; Burns, "Spiritualism and the Gospel of Jesus," p. 1.
113. Lowe, "Ends, Aims, and Uses of Modern Spiritualism," pp. 111–12.
114. Wallis, "Open Letter," pp. 3–4.
115. British Association of Progressive Spiritualists, *Proceedings of the First Convention*, 26–27 July 1865, p. iii, in *Tracts: Spiritual*, Harry Price Library. This convention was held in a Darlington Mechanics' Institute.
116. Lowe, "Ends, Aims, and Uses of Modern Spiritualism," p. 113.
117. Burns, "Spiritualism and the Gospel of Jesus," pp. 3–4; Robert Cooper, 'Facts are Stubborn Things" (London: James Burns, n.d.), p. 1 (no. 3 in *Houdini Pamphlets: Spiritualism*, vol. 2, Houdini Collection); Stock, *Attempts at Truth*, p. 135. It is highly probable, nonetheless, that less rigorous anti-Christian spiritualists did tend to perceive mediums in the priestly role of intercessor between the worlds of matter and spirit.

118. Lowe, "Ends, Aims, and Uses of Modern Spiritualism," pp. 107–8. Barrow, "Socialism in Eternity," p. 50, places working-class spiritualism among the "Victorian and Edwardian currents of plebeian independence." Also see pp. 49, 53, 57, 63.
119. Emma Hardinge Britten, *Nineteenth Century Miracles; or, Spirits and Their Work in Every Country of the Earth* (New York: William Britten, 1884), pp. 222–5; "Services for Sunday, November 20, 1887," *Two Worlds* 1 (18 November 1887):13; Stock, "New Religion," p. 433; Nelson, *Spiritualism and Society*, pp. 143–4. The similarities between progressive spiritualist tenets and the doctrines held by British Unitarians in the nineteenth century deserve further exploration.
120. Britten, *Nineteenth Century Miracles*, p. 222. Nelson, *Spiritualism and Society*, p. 119, reports that, in 1900, the Sowerby Bridge choir was selected as the Spiritualist National Prize Choir in the Annual Choir Contest that had, by then, been organized. Also see Nelson's description, p. 119, of the neo-Gothic St. Paul's Spiritual Church, Bradford, opened in 1899.
121. Owen Chadwick, *The Victorian Church*, 2 vols. (New York: Oxford University Press, 1966–1970), 1:334. Also see Nelson, *Spiritualism and Society*, p. 147.
122. Royle discusses the secularists' adaptation of religious practices in *Radicals, Secularists and Republicans*; for a brief discussion of the Labour Churches established under the initiative of John Trevor, see Henry Pelling, *The Origins of the Labour Party 1880–1900*, 2d ed. (London: Oxford University Press Paperbacks, 1966), pp. 132–44.
123. See Alan D. Gilbert, *Religion and Society in Industrial England: Church, Chapel and Social Change, 1740–1914* (London: Longman, 1976), pp. 200–2.
124. Alfred Kitson, *Spiritualism for the Young. Designed for the Use of Lyceums, and the Children of Spiritualists in General Who Have no Lyceums at which They Can Attend* (Keighley, Yorkshire: S. Billows, 1889), pp. 47–8. That the pedagogical principles prevailing in the Summerland were publicized in great detail suggests the degree of thoroughness with which all aspects of life after death were made known to the progressive spiritualists.
125. "Children's Lyceums," *Daybreak* 1 (September 1868):13. Issues of this short-lived paper, edited by J. P. Hopps, can be found in the volume entitled *Tracts: Spiritual* in the Harry Price Library. Also see Nelson, *Spiritualism and Society*, pp. 174–87, for a useful discussion of the Lyceum movement in Britain from the 1860s to the 1960s.
126. Weatherhead combined his spiritualism with a zeal for temperance and dietetic reform. Anna Mary Howitt Watts, *The Pioneers of the Spiritual Reformation* (London: Psychological Press Association, 1883), p. 294n, and Britten, *Nineteenth Century Miracles*, pp. 198–9.
127. Emma Hardinge Britten, Alfred Kitson, and H. A. Kersey, comps., *The Lyceum Manual. A Compendium of Physical, Moral, and Spiritual Exercises for Use in Progressive Lyceums* (Manchester: British Spiritualists' Lyceum Union, 1935), first published in 1887. Also see Alfred Kitson, *Autobiography of Alfred Kitson* (Hanging Heaton: A. Kitson, [1922]). From 1890, the lyceum movement had its own periodical, the monthly *Lyceum Banner*, published in close conjunction with *Two Worlds*.
128. Kitson, *Spiritualism for the Young*, pp. 6, 9–12, 85–92, 127, 137.

129. Ibid., pp. 106–21, 137–8, 140–4; "Golden Chain Recitations," No. 108, in Britten, Kitson, and Kersey, comps., *Lyceum Manual* (not paginated).
130. Kitson, *Spiritualism for the Young*, pp. 138–40, 144, 104.
131. Nelson, *Spiritualism and Society*, pp. 179–81. Also see Barrow, "Socialism in Eternity," pp. 53–8, for a discussion of the lyceums, which the author rightly calls "the most organised aspect of plebeian spiritualism."
132. Thomas Shorter [Brevior], *Concerning Miracles* (London: James Burns, 1872) and *Immortality in Harmony with Man's Nature and Experience* (London: James Burns, [1875]). In his address to the Church Congress, 1881, Canon Wilberforce cited "a remarkable pamphlet by Rev. T. Colley, . . . published by Burns, 15 Southampton Row" (Moses, *Spiritualism at the Church Congress*, p. 14); Moses, "Researches in Spiritualism," *Human Nature* 8 (March 1874):107–8; *Proceedings of the Third Convention* of Progressive Spiritualists, 1867, bears the information: J. H. Powell, Printer, Grove Rd., Victoria Park, London (in *Tracts: Spiritual*, Harry Price Library).
133. Minute Books of the BNAS, Committees, no. 2, pp. 1, 12, 32, 106 (29 May, 19 June, 19 July 1878, and 29 January 1879).
134. LSA, Minute Book of Council Meetings 1896–1909, pp. 27–8 (8 March 1897), pp. 37–8 (14 June 1897), p. 54 (11 January 1898), and p. 172 (14 October 1903). Today, the Alliance's successor organization, the College of Psychic Studies, avoids religious affiliation and "welcomes enquiries from those of any religion or none." Paul Beard, *A Field of Enquiry. The College of Psychic Studies* (London: College of Psychic Studies, 1971), p. 5.
135. Cooper, "Facts Are Stubborn Things," p. 1.
136. "Spiritualism in Liverpool," *Spiritualist Newspaper* 9 (27 October 1876):156.
137. Chadwick, "Established Church Under Attack," in Symondson, ed., *Victorian Crisis of Faith*, p. 94.
138. A. O. J. Cockshut, *Truth to Life: The Art of Biography in the Nineteenth Century* (London: Collins, 1974), p. 153.
139. Nelson, *Spiritualism and Society*, pp. 118–19, discusses the similarity between Congregationalist and spiritualist ideas of organization.
140. After World War I, the similarities between the two grew ever more apparent, as Christian spiritualist groups launched their own churches and lyceums, and as the SNU, at first with great reluctance, gradually came to include Christian spiritualists. Ibid., pp. 148–9, 182, 188.
141. Countess of Caithness (Marie Sinclair), *Old Truths in a New Light or, An Earnest Endeavour to Reconcile Material Science with Spiritual Science, and with Scripture* (London: Chapman & Hall, 1876), pp. 13, 99; Shorter, *Two Worlds*, p. 346; John S. Farmer, *Spiritualism as a New Basis of Belief* (London: E. W. Allen, 1880), pp. 67–8; W. S. Moses ["M.A., (Oxon.)"], *Higher Aspects of Spiritualism* (London: E. W. Allen, 1880), p. 14. Also see the statement, "This Association is, therefore, not the enemy of true religion," in British Association of Progressive Spiritualists, *Proceedings of the First Convention*, p. iv, in *Tracts: Spiritual*, Harry Price Library.
142. For references to Friedrich Max Müller, see Farmer, *Spiritualism as a New Basis of Belief*, p. 67, and Stock, "New Religion," p. 435.

143. Katherine H. Porter, *Through a Glass Darkly: Spiritualism in the Browning Circle* (1958; reprint ed., New York: Octagon Books, 1972), pp. 29, 38; Lowe, "Ends, Aims, and Uses of Modern Spiritualism," p. 113.
144. James Smith, *The Coming Man*, 2 vols. (London: Strahan, 1873), 1:386; Burns, "Spiritualism and the Gospel of Jesus," p. 3; Wallis, *Story of My Life*, p. 28; Stock, *Attempts at Truth*, p. 146.
145. Stock, *Attempts at Truth*, p. 133.

4. Psychical research and agnosticism

1. Keynes quoted in R. F. Harrod, *The Life of John Maynard Keynes* (1951; reprint ed., New York: Augustus M. Kelley, 1969), p. 116. Keynes's comment appears in a letter to B. W. Swithinbank, 27 March 1906.
2. C. D. Broad, *Religion, Philosophy and Psychical Research: Selected Essays* (New York: Harcourt, Brace, 1953), pp. 114–15, sketches the relationship between telepathy and faith for Sidgwick and his colleagues in the SPR. A recent volume that deals less successfully with similar issues is John James Cerullo, *The Secularization of the Soul: Psychical Research in Modern Britain* (Philadelphia: Institute for the Study of Human Issues, 1982).
3. Arthur and Eleanor Mildred Sidgwick, *Henry Sidgwick, A Memoir* (London: Macmillan, 1906), Chapter 1.
4. Arthur Christopher Benson, *The Leaves of the Tree: Studies in Biography* (London: Smith, Elder, 1911), pp. 65–6.
5. Sidgwick and Sidgwick, *Henry Sidgwick*, pp. 47, 68, 74–5, 198.
6. Caroline Jebb, *Life and Letters of Sir Richard Claverhouse Jebb* (Cambridge: The University Press, 1907), p. 356.
7. See the discussion of Sidgwick in Alan Gauld, *The Founders of Psychical Research* (London: Routledge & Kegan Paul, 1968), pp. 47–57, and Sidgwick's own analysis of Biblical evidence in his lengthy review of J. R. Seeley's *Ecce Homo: a Survey of the Life and Works of Jesus Christ*. The review, which appeared in the *Westminster Review*, July 1866, is included in Sidgwick's posthumous collection, *Miscellaneous Essays and Addresses* (London: Macmillan, 1904), pp. 1–39. Frank Miller Turner, *Between Science and Religion: The Reaction to Scientific Naturalism in Late Victorian England* (New Haven and London: Yale University Press, 1974), p. 39, suggests that Sidgwick was finally convinced of his alienation from Christianity by "his absolute inability to accept or in some manner to rationalize the doctrine of the Virgin Birth."
8. Sheldon Rothblatt, *The Revolution of the Dons: Cambridge and Society in Victorian England* (London: Faber & Faber, 1968), pp. 133–54; Sidgwick and Sidgwick, *Henry Sidgwick*, p. 79, for Sidgwick's brief flirtation with Broad Church theology.
9. Of the secondary studies that analyze Sidgwick's ethical theory, see C. D. Broad, *Five Types of Ethical Theory* (London: Kegan Paul, Trench, Trübner, 1930), pp. 143–256; William C. Havard, *Henry Sidgwick and Later Utilitarian Political Philosophy* (Gainesville, Fla.: University of Florida Press, 1959), pp. 90–108; J. L. Mackie, "Sidgwick's Pessimism," *Philosophical Quarterly* 26 (October 1976): 317–27; Rothblatt, *Revolution of the Dons*, pp. 137–8; Turner, *Between Science and Religion*, pp. 38–67; J. B. Schneewind, *Sidgwick's Ethics and Victorian Moral Philosophy* (Oxford: The Clarendon Press, 1977).

10. James Bryce, *Studies in Contemporary Biography* (New York and London: Macmillan, 1903), pp. 334–5.
11. Sidgwick and Sidgwick, *Henry Sidgwick*, pp. 125, 205; Rothblatt, *Revolution of the Dons*, pp. 134, 193. Reba N. Soffer, *Ethics and Society in England: The Revolution in the Social Sciences 1870–1914* (Berkeley and Los Angeles: University of California Press, 1978), pp. 6–7, discusses "the intense sense of duty," which men like Sidgwick experienced.
12. Sidgwick and Sidgwick, *Henry Sidgwick*, pp. 346–8.
13. Henry Sidgwick, *The Methods of Ethics* (London: Macmillan, 1874), p. 473. As Turner points out, in *Between Science and Religion*, p. 47n, although Sidgwick altered this concluding paragraph in the second and subsequent editions of *The Methods*, adding more dependent clauses and blunting the impact of that final sentence, there is nothing in his writing to indicate that he ever changed his mind.
14. Sidgwick and Sidgwick, *Henry Sidgwick*, pp. 90, 357, 395, 538; for his admiration of Jesus, see ibid., pp. 145–7, and his review of *Ecce Homo*, in *Miscellaneous Essays and Addresses*, p. 39.
15. C. D. Broad, *Ethics and the History of Philosophy: Selected Essays* (London: Routledge & Kegan Paul, 1952), pp. 65–6.
16. Sidgwick and Sidgwick, *Henry Sidgwick*, pp. 466–7.
17. Broad, *Five Types of Ethical Theory*, pp. 253–6, discusses Sidgwick's "postulate of a benevolent and powerful God." Also see Turner, *Between Science and Religion*, pp. 46, 48, 50, 60.
18. Turner, *Between Science and Religion*, pp. 60–7; Broad, *Religion, Philosophy and Psychical Research*, pp. 109–10; Alan Willard Brown, *The Metaphysical Society: Victorian Minds in Crisis, 1869–1880* (1947; reprint ed., New York: Octagon Books, 1973), p. 283.
19. Turner, for example, deals with Sidgwick first among those Victorians, caught "Between Science and Religion," whom he describes. Also see David Gwilym James, *Henry Sidgwick: Science and Faith in Victorian England* (London: Oxford University Press, 1970).
20. Walter E. Houghton, *The Victorian Frame of Mind 1830–1870* (New Haven: Yale University Press, 1971), pp. 50–1. See Stephen's classic statement of the reasonableness of agnosticism, "An Agnostic's Apology," in *An Agnostic's Apology and Other Essays* (London: Smith, Elder, 1893), pp. 1–41; also Josef L. Altholz, "The Warfare of Conscience with Theology," in *The Mind and Art of Victorian England*, ed. Altholz (Minneapolis: University of Minnesota Press, 1976), p. 75.
21. "A Modern 'Symposium.' The Influence upon Morality of a Decline in Religious Belief," *Nineteenth Century* 1 (April 1877):355. Clifford was only one of several distinguished participants in this symposium.
22. Geoffrey Rowell, *Hell and the Victorians. A Study of the Nineteenth-Century Theological Controversies Concerning Eternal Punishment and the Future Life* (Oxford: Clarendon Press, 1974), p. 4; chapter 2 of Gauld's *Founders of Psychical Research* is entitled "The Genesis of Reluctant Doubt."
23. The first quotation is from Barbara W. Tuchman, *The Proud Tower: A Portrait of the World Before the War 1890–1914* (New York: Macmillan, 1966), p. 48; the second is from Marghanita Laski, "Domestic Life," in *Edwardian England 1901–1914*, ed. Simon Nowell-Smith (London: Oxford University Press, 1964), p. 211.

24. John Tulloch, *Movements of Religious Thought in Britain During the Nineteenth Century, Being the Fifth Series of St. Giles' Lectures* (1885; reprint ed., New York: Humanities Press, 1971), p. 171.
25. E. R. Dodds, *Missing Persons. An Autobiography* (Oxford: Clarendon Press, 1977), p. 103.
26. The phrase is Basil Willey's in *More Nineteenth Century Studies. A Group of Honest Doubters* (New York: Columbia University Press, 1956), p. 148.
27. Walter Leaf, *Walter Leaf 1852–1927. Some Chapters of Autobiography, with a Memoir by Charlotte M. Leaf* (London: John Murray, 1932), pp. 15, 20, 22–3, 27, 117.
28. Frederic W. H. Myers, *Fragments of Inner Life, an Autobiographical Sketch* (London: SPR, 1961), p. 15. This brief summary of Myers's emotional and spiritual development was first printed, privately, in 1893.
29. Leaf, *Walter Leaf*, p. 182.
30. Noel Annan, "Science, Religion, and the Critical Mind: Introduction," in *1859: Entering an Age of Crisis*, ed. Philip Appleman, William A. Madden, and Michael Wolff (Bloomington: Indiana University Press, 1959), p. 34.
31. Gauld, *Founders of Psychical Research*, p. 246, describes the distinction that some SPR leaders made between "mental" and "physical mediums," to the detriment of the latter.
32. Annan, "Science, Religion, and the Critical Mind: Introduction," pp. 33–4, discusses the search for a branch of "demonstrable" "knowledge about human beings." He argues, however, that history, not psychical research, was that branch.
33. Ethel Sidgwick, *Mrs. Henry Sidgwick, a Memoir by her Niece* (London: Sidgwick & Jackson, 1938), pp. 71–3. Eleanor Balfour and Sidgwick were married in 1876.
34. Helen de G. Salter, "Our Pioneers, I. Mrs. Henry Sidgwick," *JSPR* 39 (June 1958):236.
35. Ethel Sidgwick, *Mrs. Henry Sidgwick*, p. 66n. Alice Johnson was Mrs. Sidgwick's secretary and colleague, both at Newnham and the SPR.
36. See her massive article, "Phantasms of the Living. An Examination and Analysis of Cases of Telepathy between Living Persons printed in the 'Journal' of the Society since the publication of the book 'Phantasms of the Living,' by Gurney, Myers, and Podmore, in 1886," *PSPR 33* (October 1922):23–429 passim. The quotation appears on p. 23.
37. Henry Longueville Mansel, *The Limits of Religious Thought Examined in Eight Lectures, Preached before the University of Oxford, in the Year M.DCCC.LVIII on the Foundation of the Late Rev. John Bampton, M.A. Canon of Salisbury* (London: John Murray, 1858), p. xix, end of chapter summary for Lecture VIII in the Table of Contents.
38. W. H. Salter, "Our Pioneers, V: Edmund Gurney," *JSPR* 40 (June 1959): 49; Sidgwick and Sidgwick, *Henry Sidgwick*, p. 75.
39. Leaf, *Walter Leaf*, pp. 106–8, 92–4. While remaining a lifetime student and translator of Homer, Leaf also became a highly successful banker and City financier, and was elected president of the International Chamber of Commerce in 1925.
40. Sidgwick's first presidential address to the SPR, 1882, in *Presidential Addresses to the Society for Psychical Research 1882–1911* (Glasgow: Robert Maclehose for the SPR, 1912), p. 2.

41. Sir Charles Oman, "The Old Oxford Phasmatological Society," *JSPR* 33 (March–April 1946):208–9; Sidgwick and Sidgwick, *Henry Sidgwick*, pp. 43, 52; J. Fraser Nicol, "Philosophers as Psychic Investigators," *Parapsychology Review* 8 (March–April 1977):3. Nicol points out, p. 11, that Trinity's ties with psychical research go back to its origins during the reign of Henry VIII, when one of the first Fellows was Dr. John Dee. Also see Alan Gauld, "Psychical Research in Cambridge from the Seventeenth Century to the Present," *JSPR* 49 (December 1978):925–37.
42. Sidgwick and Sidgwick, *Henry Sidgwick*, pp. 53, 55, 104–6, 171.
43. Lodge Collection (box 2, no. 366). The observation occurs in some typed notes by Crookes, not in a letter to Lodge.
44. Sidgwick and Sidgwick, *Henry Sidgwick*, pp. 289–90.
45. F. W. H. Myers, "In Memory of Henry Sidgwick," *PSPR* 15 (December 1900):452.
46. Myers, *Fragments of Inner Life*, pp. 13–15, and "In Memory of Henry Sidgwick," p. 454.
47. Mary Reed Bobbitt, *With Dearest Love to All: The Life and Letters of Lady Jebb* (Chicago: Henry Regnery, 1960), p. 111. At the turn of the century, Mrs. became Lady Jebb, when Professor Jebb was knighted. In 1883, in a letter to C. C. Massey, Myers singled out Stainton Moses as the person to whom he owed his own belief in human immortality. Myers Papers 19[6].
48. Sidgwick and Sidgwick, *Henry Sidgwick*, p. 288. Gauld, *Founders of Psychical Research*, p. 89, persuasively questions "whether Sidgwick's investigations would have been so extensive or so prolonged as they in fact were had it not been for the influence upon him of a person whose temperament was far more eager than his own, namely Frederic Myers." Myers also helped Cox launch the Psychological Society of Great Britain in 1875, according to Fraser Nicol, "The Founders of the S.P.R.," *PSPR* 55 (March 1972):350.
49. Leaf, *Walter Leaf*, pp. 93–4, 121.
50. See Sidgwick's contrasting opinions about Fairlamb and Wood, in Sidgwick and Sidgwick, *Henry Sidgwick*, pp. 296, 298–9. His first, guardedly optimistic, view was expressed on 23 March 1875; his second, frankly suspicious of imposture, on 18 July 1875. Mrs. Henry Sidgwick, "Results of a Personal Investigation into the 'Physical Phenomena' of Spiritualism. With Some Critical Remarks on the Evidence for the Genuineness of Such Phenomena," *PSPR* 4 (1886–7):53–4. Also see pp. 46–7 of the same article for Mrs. Sidgwick's discussion of Mrs. Jencken, and p. 60 for Mrs. Sidgwick's disappointing séances with Eglinton. Leaf, *Walter Leaf*, p. 94; Gauld, *Founders of Psychical Research*, pp. 113–14, 127, 129; Eric J. Dingwall, *The Critics' Dilemma: Further Comments on Some Nineteenth Century Investigations* (Crowhurst, Sussex: E. J. Dingwall, 1966), chapters 2 and 3.
51. Myers, "In Memory of Henry Sidgwick," p. 455.
52. Mrs. Sidgwick, "Results of a Personal Investigation," pp. 70–2.
53. Leaf, *Walter Leaf*, pp. 85–6; Sidgwick and Sidgwick, *Henry Sidgwick*, pp. 34–5.
54. Rothblatt, *Revolution of the Dons*, pp. 138, 143.
55. R. H. Hutton, "'The Metaphysical Society.' A Reminiscence," *Nineteenth Century* 18 (August 1885):177–8; Brown, *Metaphysical Society*, pp. 26, 127–8.

56. Susan Budd, *Varieties of Unbelief. Atheists and Agnostics in English Society 1850–1960* (London: Heinemann Educational Books, 1977), p. 129.

57. Brown, *Metaphysical Society*, pp. 244, 254, 257; Sidgwick and Sidgwick, *Henry Sidgwick*, p. 556.

58. Brown, *Metaphysical Society*, pp. 32, 111; Sidgwick and Sidgwick, *Henry Sidgwick*, p. 556, report that Hutton was also a member of the Synthetic.

59. Robert H. Tener, "R. H. Hutton's Editorial Career. I. 'The Inquirer'," and "R. H. Hutton's Editorial Career. II. 'The Prospective' and 'National Reviews'," *Victorian Periodicals Newsletter* 7 (June 1974):3–10 and 7 (December 1974):6–13; Julia Wedgwood, "Richard Holt Hutton," *Contemporary Review* 72 (October 1897): 466–7.

60. John Hogben, *Richard Holt Hutton of 'The Spectator'. A Monograph*, 2d ed. (Edinburgh: Oliver and Boyd, 1900), pp. 18, 102–3, 108.

61. Brown, *Metaphysical Society*, pp. 206, 253. Also see Hogben, *Hutton*, pp. 97–102, and Wedgwood, "Hutton," p. 465.

62. Brown, *Metaphysical Society*, p. 111. Also see Hutton's contribution to "A Modern 'Symposium'," *Nineteenth Century* 1 (May 1877):545, and his essay, "The Resurrection of the Body," written in 1895, in Richard Holt Hutton, *Aspects of Religious and Scientific Thought*, ed. Elizabeth M. Roscoe (London: Macmillan, 1899), pp. 153–8.

63. John David Root, "The Philosophical and Religious Thought of Arthur James Balfour (1848–1930)," *Journal of British Studies* 19 (Spring 1980): 131–7, discusses Balfour's role in founding the Synthetic Society and his dedicated participation in its activities during the following years.

64. In late July 1911, at the most tense moments in the constitutional crisis over the Parliament Bill, Balfour invited Annie Besant "to dinner at Carlton Gardens to talk occultism." Kenneth Young, *Arthur James Balfour. The Happy Life of the Politician, Prime Minister, Statesman and Philosopher 1848–1930* (London: G. Bell, 1963), p. 319. On Balfour's absentee presidency of the SPR, see Nicol, "Founders of the S.P.R.," p. 345, and Balfour's correspondence with Myers on the subject in 1892 and 1893, Myers Papers 1[49–50]

65. Blanche E. C. Dugdale, *Arthur James Balfour*, 2 vols. (London: Hutchinson, 1936), 1:29.

66. Like his sister Eleanor, Balfour became a communicant in both the Church of England and of Scotland. Arthur James Balfour, *Retrospect, an Unfinished Autobiography 1848–1886*, ed. Blanche E. C. Dugdale (Boston and New York: Houghton Mifflin, 1930), pp. 19–20; Dugdale, *Balfour*, 1:51; Ethel Sidgwick, *Mrs. Henry Sidgwick*, p. 15; Young, *Balfour*, p. 178.

67. Bobbitt, *With Dearest Love to All*, pp. 160–1.

68. From a speech which Balfour delivered in London to the Pan-Anglican Congress, 1908, quoted in part in Wilfrid M. Short, ed., *Arthur James Balfour as Philosopher and Thinker* (London: Longmans, Green, 1912), p. 504.

69. Arthur James Balfour, "The Religion of Humanity," in *Essays and Addresses* (1893; reprint ed., Freeport, N.Y.: Books for Libraries Press, 1972), pp. 283–314.

70. Arthur James Balfour, *Theism and Humanism. Being the Gifford Lectures Delivered at the University of Glasgow, 1914* (London: Hodder & Stoughton, [1915]), p. 36; Hutton, "A Modern 'Symposium'," p. 541.

71. Balfour, "Religion of Humanity," pp. 295–6.
72. Letter to Lady Desborough, 5 August 1915, quoted in Dugdale, *Balfour*, 2:296–7; Young, *Balfour*, pp. xix, 18.
73. The quotations in this paragraph are taken from: Arthur James Balfour, "The Bible" (1903), in *Balfour as Philosopher and Thinker*, pp. 58–60; Balfour, "Religion of Humanity," p. 314, and speech to Pan-Anglican Congress, 1908, in *Balfour as Philosopher and Thinker*, p. 504.
74. Balfour's presidential address to the SPR, in *Presidential Addresses*, pp. 75–6. For his undergraduate views, see Balfour, *Retrospect*, p. 60.
75. Dr. Ivor Tuckett, however, insisted on purely naturalistic explanations for the cross correspondences. See his "Psychical Researchers and 'The Will to Believe'," *Bedrock. A Quarterly Review of Scientific Thought* 1 (July 1912):201, and "The Illogical Position of Some Psychical Researchers," the first essay in a three-part article on psychical research, *Bedrock* 1 (January 1913):486.

   All the interconnected aspects of the case were not fully elucidated until 1960. The fullest discussion is found in Jean Balfour, "The 'Palm Sunday' Case: New Light on an Old Love Story," *PSPR* 52 (February 1960):79–267. Also see Rosalind Heywood, "The Palm Sunday Case: A Tangle for Unravelling," *JSPR* 40 (June 1960):285–91, and in the same issue of the *Journal*, W. H. Salter, "The Palm Sunday Case: A Note on Interpreting Automatic Writings," pp. 275–85. For earlier analyses, see H. F. Saltmarsh, *Evidence of Personal Survival from Cross Correspondences* (1938; reprint ed., New York: Arno Press, 1975) and W. H. Salter, *The Society for Psychical Research: An Outline of its History* (London: SPR, 1948), pp. 33–7.
76. Gerald Balfour, presidential address to the SPR, *Presidential Addresses*, p. 250; *DNB* (1941–1950), s.v. "Balfour, Gerald"; Nicol, "Founders of the S.P.R.," p. 345; Leaf, *Walter Leaf*, p. 89; Bobbitt, *With Dearest Love to All*, p. 151.
77. Gerald Balfour, presidential address to the SPR, *Presidential Addresses*, p. 273, and "The Ear of Dionysius: Further Scripts affording Evidence of Personal Survival," *PSPR* 29 (March 1917):237.
78. Gauld, *Founders of Psychical Research*, p. 104; Michael Stenton, ed., *Who's Who of British Members of Parliament* (Hassocks, Sussex: Harvester Press, 1976–), 1:196; Brown, *Metaphysical Society*, pp. 27, 154; Sidgwick and Sidgwick, *Henry Sidgwick*, pp. 47–8.
79. After a visit to Hawarden in September 1885, Sidgwick recorded that Gladstone was "full of interesting talk on various topics. The geology of Norway and Psychical Research appeared to be the subjects that interested him most." Sidgwick and Sidgwick, *Henry Sidgwick*, p. 425. Gladstone's interest in psychical research certainly complemented his defense of belief in the unseen world and was, no doubt, part of his lifelong combat against the materialism and religious skepticism that he saw permeating British society in his day.
80. Except where otherwise stated, all information about membership in the SPR is derived from the membership lists in the successive volumes of the SPR *Proceedings*.
81. Henry Sidgwick to A. J. Balfour, 10 February 1882, Balfour Papers, B. M. Add. MSS. 49832, f. 150, British Library.
82. "Objects of the Society," *PSPR* 1 (1882–3):3; Sidgwick's presidential address, 1883, in *Presidential Addresses*, p. 12; Myers, "In Memory of

Henry Sidgwick," p. 458; W. Boyd Carpenter, "Presidential Address. Delivered on May 23rd, 1912," *PSPR* 26 (September 1912):3.
83. E. D. Rogers, *Life and Experiences of Edmund Dawson Rogers, Spiritualist and Journalist* (London: Office of *Light*, [1911]), p. 46; "Personal Testimonies," *Light* 30 (15 October 1910):496; Nicol, "Founders of the S.P.R.," pp. 341–2.
84. There is a bundle of fascinating replies to the invitation in the Barrett Papers (box 2, A2).
85. Gauld, *Founders of Psychical Research* p. 138; Myers, "In Memory of Henry Sidgwick," p. 458; Sidgwick and Sidgwick, *Henry Sidgwick*, p. 358.
86. W. S. Moses to W. F. Barrett, 12 January 1882, Barrett Papers (box 2, A4, no. 54).
87. Wedgwood is mentioned, although infrequently, in the Minute Books of the BNAS, Committees, no. 2. Also see *DNB*, s.v. "Wedgwood, Hensleigh," and "In Memoriam. Hensleigh Wedgwood," *Academy* 39 (27 June 1891):610.
88. London: T. Fisher Unwin, 1887. Theobald and his wife were converted to spiritualism in 1869 after losing three young children (*Spirit Workers*, pp. 18–19). See Frank Podmore, *Modern Spiritualism: A History and a Criticism*, 2 vols. (London: Methuen, 1902), 2:91–4, for a devastating description of Theobald's deficiencies as a psychical researcher.
89. Nicol, "Founders of the S.P.R.," pp. 342–3, makes a strong case for the actual preponderance of spiritualists in the SPR for the first couple of years.
90. Ethel Sidgwick, *Mrs. Henry Sidgwick*, p. 91.
91. "Society for Psychical Research," *Light* 2 (25 February 1882):92. Readers were told to request further information from Edward T. Bennett, first secretary of the SPR and still active at the time in the affairs of the BNAS.
92. "The S.P.R. and the C.A.S.," *Light* 3 (3 February 1883):54.
93. Ibid.
94. Eleanor Mildred Sidgwick, "Mr. Eglinton," *JSPR* 2 (June 1886):282. The incidents are described on pp. 282–7. Many examples of Eglinton's alleged feats can be found in John S. Farmer, *'Twixt Two Worlds: A Narrative of the Life and Work of William Eglinton* (London: The Psychological Press, 1886).
95. Gauld, *Founders of Psychical Research*, pp. 202–3; A. T. Baird, *Richard Hodgson: The Story of a Psychical Researcher and his Times* (London: Psychic Press, 1949) is an unsatisfactory, and often inaccurate, biography. There is also a brief *Memoir of Richard Hodgson 1855–1905*, by Mark Anthony DeWolfe Howe, read at the annual meeting of the Boston Tavern Club, of which Hodgson was a member, on 6 May 1906, and presumably printed for that club in the same year.
96. Podmore, *Modern Spiritualism*, 2:217–8, describes how Davey reproduced Eglinton's phenomena using such techniques. Also see E. M. Sidgwick, "Mr. Eglinton," pp. 282–334, "The Charges Against Mr. Eglinton," *JSPR* 2 (November 1886):467–9, and "Rejoinder by Mrs. H. Sidgwick to Mr. Wedgwood's Reply," *JSPR* 2 (December 1886):474–85; Richard Hodgson and S. J. Davey, "The Possibilities of Mal-Observation and Lapse of Memory from a Practical Point of View," *PSPR* 4 (1886–

7):381–495. Other articles that supported the Sidgwick–Hodgson–Davey position were Angelo J. Lewis ("Professor Hoffman," author of *Modern Magic*), "How and What to Observe in Relation to Slate-Writing Phenomena," *JSPR* 2 (August 1886):362–75; H. Carvill Lewis and others, "Accounts of Some So-Called 'Spiritualistic' Séances," *PSPR* 4 (1886–7):338–80.

97. Farmer, *'Twixt Two Worlds* passim; C. C. Massey, "The Possibilities of Mal-Observation in Relation to Evidence for the Phenomena of Spiritualism," *PSPR* 4 (1886–7):75–99; Nicol, "Founders of the S.P.R.,"p. 357; H. Wedgwood, "Reply to Mrs. Sidgwick," *JSPR* 2 (November 1886):455–60.

98. Mrs. Sidgwick, "Results of a Personal Investigation," p. 74; her letter to Sidgwick is quoted in Ethel Sidgwick, *Mrs. Henry Sidgwick*, p. 99.

99. W. S. Moses, "Spiritualism at Home and Abroad: Its Present Position and Future Work," pp. 20–1, no. 4 in *Addresses Delivered Before the London Spiritualist Alliance during the Years 1884 to 1888* (London: LSA, 1889), Houdini Collection. Gauld, *Founders of Psychical Research*, p. 204, states that "a considerable number of the more Spiritualistically inclined S.P.R. members" left the Society over this issue, but Nicol, "Founders of the S.P.R.," p. 357, argues that the number, in fact, was small.

100. Rogers, *Life and Experiences*, p. 47.

101. See, for example, her letter to Sir Oliver Lodge, 5 November 1909, where she busily arranged the selection of the SPR's next president, to succeed herself. Lodge Collection (box 8, no. 2125).

102. Gauld, *Founders of Psychical Research*, p. 153; the initial SPR committees and their honorary secretaries are listed in "Objects of the Society," pp. 3–5. The division of investigations into these six categories, with inquiry by entire committees, gradually ceased to be the Society's *modus operandi* after a few years.

103. Edmund Gurney, Frederic W. H. Myers, and Frank Podmore, *Phantasms of the Living*, 2 vols. (1886; facsimile reprint ed., Gainesville, Fla.: Scholars' Facsimiles & Reprints, 1970), 1:xxxv, lxv, lxvi; *Times*, 30 October 1886, p. 9; Gauld, *Founders of Psychical Research*, p. 137n.

104. Gauld, *Founders of Psychical Research*, p. 164. See pp. 160–74 for Gauld's summary of *Phantasms* and his discussion of the critical response to it.

105. W. H. Salter, "Our Pioneers, V: Edmund Gurney," p. 48.

106. These essays are included in Edmund Gurney, *Tertium Quid: Chapters on Various Disputed Questions*, 2 vols. (London: Kegan Paul, Trench, 1887), 1:151–203, 204–26, and 2:251–302.

107. Frederic W. H. Myers, "Edmund Gurney," in *Fragments of Prose & Poetry*, ed. Eveleen Myers (London: Longmans, Green, 1904), p. 58. Myers's obituary sketch of Gurney first appeared in *PSPR* 5 (1888–9).

108. Edmund Gurney, "The Human Ideal," in *Tertium Quid*, 1:46–7; Myers, "Edmund Gurney," p. 64.

109. This is Myers's interpretation of Gurney's life and work, and it is certainly supported by Gurney's own writings. Also see Gauld, *Founders of Psychical Research*, pp. 154–60.

110. Myers, "Edmund Gurney," pp. 56, 69–70.

111. Edmund Gurney, "Stages of Hypnotic Memory," *PSPR* 4 (1886–7):531.

112. *DNB*, s.v. "Gurney, Edmund."
113. Ethel Sidgwick, *Mrs. Henry Sidgwick*, p. 110.
114. Trevor H. Hall, *The Strange Case of Edmund Gurney* (London: Gerald Duckworth, 1964). A second edition of Hall's study was published by Duckworth in 1980.
115. Ibid., pp. 120–4.
116. Critics have been quick to pinpoint problems with that reconstruction. See, in particular, Fraser Nicol, "The Silences of Mr. Trevor Hall," *International Journal of Parapsychology* 8 (Winter 1966):5–59; also Alan Gauld, "Mr. Hall and the S.P.R.," *JSPR* 43 (June 1965):53–62. The SPR was especially offended by Hall's book because he used the incident of Gurney's death, and what he alleged to have been a cover-up of his suicide, to issue a sweeping condemnation of most of the Society's early leaders.
117. *DNB* (*1901–1911*), s.v. "Podmore, Frank." Also see Nicol, "Founders of the S.P.R.," pp. 344–5.
118. Norman and Jeanne MacKenzie, *The Fabians* (New York: Simon & Schuster, 1977), pp. 16–17, 21–7. Also see Edward R. Pease, *The History of the Fabian Society*, 2d ed. (London: Fabian Society and George Allen & Unwin, 1925), pp. 26–35.
119. Podmore to Davidson, 31 December 1883 and 16 December 1884, Thomas Davidson Papers, Yale University Library.
120. But his two-volume biography of *Robert Owen* (London: Hutchinson, 1906) attests to his continuing interest in the complementary nature of spiritualism and socialism. For information in this paragraph, see Pease, *History of the Fabian Society*, p. 57, and MacKenzie and MacKenzie, *Fabians*, p. 93.
121. Frank Podmore, *Apparitions and Thought-Transference. An Examination of the Evidence for Telepathy* (London: Walter Scott, 1894), p. xi.
122. Podmore's contribution to A. H. Pierce and F. Podmore, "Subliminal Self or Unconscious Cerebration?" *PSPR* 11 (1895):332.
123. New York and London: G. P. Putnam's Sons, 1908. See pp. 10–14 for Podmore's discussion of the problems connected with existing theories of the physical basis of telepathy.
124. Frank Podmore, *The Newer Spiritualism* (London: T. Fisher Unwin, 1910), pp. 145–6.
125. Ibid., pp. 300, 316.
126. Podmore's letter to Sidgwick, 27 August 1900, is quoted in Gauld, *Founders of Psychical Research*, p. 316. His letter to the dying Myers, written on 15 November of the same year, was, however, far more guarded in its hopefulness. "This world will be much the poorer to me for your loss," Podmore told Myers. "I wish that I could share to the full your confidence that some other will be the gainer." Myers Papers 3[137].
127. Podmore, *Newer Spiritualism*, p. 316.
128. Hall, *Strange Case of Edmund Gurney*, pp. 200–6. It is certainly curious that Podmore resigned from the SPR council in 1909 and that, Hall reports, no one from the SPR attended his funeral. Mrs. Sidgwick, however, wrote a warm tribute to him in "Frank Podmore and Psychical Research," *PSPR* 25 (March 1911):5–10. She insisted, p. 7, that Podmore had no doubts about "the reality of telepathy between living persons."
129. The study was undertaken by W. T. Stead's assistant at *Borderland*, Ada Goodrich Freer, whose research methods were, apparently, none too

scrupulous. See John L. Campbell and Trevor H. Hall, *Strange Things: The Story of Fr. Allan McDonald, Ada Goodrich Freer, and the Society for Psychical Research's Enquiry into Highland Second Sight* (London: Routledge & Kegan Paul, 1968), where Goodrich Freer is accused of basing much of her reports to the SPR – of which she was then a member – on the folklore notes collected by Fr. Allan McDonald.

130. Dingwall, *Critics' Dilemma*, p. 67, among others, discusses the preference of the Trinity group for mediums of good social standing. For some reason, they thought that social respectability guaranteed a sterling character.

131. See Sidgwick's letter to Lodge concerning Palladino, 5 August 1894, in the Sidgwick Papers, Add. MS. c. 105. 41, Wren Library, Trinity College, Cambridge. At the time, Lodge had written an enthusiastic report of his séances with the medium, and the Sidgwicks were about to embark on their own, with high hopes.

132. See Podmore's chapters on Palladino, pp. 87–144, in *The Newer Spiritualism*, where he treats her as very much part of the old suspect spiritualism. Sidgwick and Myers also contributed to Hodgson's salary as secretary of the ASPR, which became a branch of the London SPR in 1890. Nicol, "Philosophers as Psychic Investigators," p. 4. Also see Hodgson to Myers, 27 September 1886, Myers Papers 2[109], where Hodgson expresses gratitude for Sidgwick's and Myers's ongoing financial support.

133. Bennett to Lodge, 16 October 1895, Lodge Collection (box 1, no. 179). Bennett was another former BNAS member who stayed with the SPR long after the dispute over Eglinton.

134. Myers to Lodge, 2 and 3 December 1898, postcards in the Lodge Collection (box 6, no. 1524 and 1525); Sidgwick to Myers, 15 December 1898, with note added from Myers to Lodge, Lodge Collection (box 6, no. 1527).

135. E. J. Dingwall, *Very Peculiar People: Portrait Studies in the Queer, the Abnormal and the Uncanny* (London: Rider, [1950?]), p. 194; Gauld, *Founders of Psychical Research*, p. 242.

136. Hodgson died of a heart attack in Boston, in December 1905.

137. Feilding to Johnson, 6 December 1908; quoted in Gauld, *Founders of Psychical Research*, p. 244.

138. On Eusapia Palladino, in general and with regard to the SPR, see Hereward Carrington, *Eusapia Palladino and Her Phenomena* (New York: B. W. Dodge, 1909); Everard Feilding, *Sittings with Eusapia Palladino and Other Studies*, with an Introduction by E. J. Dingwall (New Hyde Park, N.Y.: University Books, 1963); the chapter on "Eusapia Palladino: Queen of the Cabinet," in Dingwall, *Very Peculiar People*, pp. 178–217; and Gauld, *Founders of Psychical Research*, pp. 223–45. Two reports on the Feilding sittings were published in the SPR *Proceedings*: Everard Feilding, W. W. Baggally, and Hereward Carrington, "Report on a Series of Sittings with Eusapia Palladino," *PSPR* 23 (November 1909):306–569; and "Report on a Further Series of Sittings with Eusapia Palladino at Naples," *PSPR* 25 (March 1911):57–69.

139. A. O. J. Cockshut, *Truth to Life: The Art of Biography in the Nineteenth Centruy* (London: Collins, 1974), p. 181.

140. See Turner, *Between Science and Religion*, p. 157, for important observations along these lines.

141. Myers, *Fragments of Inner Life*, p. 12, and title of last chapter.
142. "Provisional Sketch of a Religious Synthesis," reprinted in Frederic W. H. Myers, *Human Personality and Its Survival of Bodily Death*, 2 vols. (New York and London: Longmans, Green, 1903), 2:284–92. Also see James Webb's Introduction to F. W. H. Myers, *The Subliminal Consciousness* (New York: Arno Press, 1976).
143. Myers's relationship with Mrs. Marshall was sketched by W. H. Salter, in "F. W. H. Myers's Posthumous Message," *PSPR* 52 (October 1958):1–32. Gauld, *Founders of Psychical Research*, pp. 116–124, provides a concise summary of the affair, or friendship, and of the controversy that arose after the publication of Salter's article. Whether Myers's relationship with his cousin's wife was sexual or platonic evidently matters a great deal to those people who want to know if he was merely a gross sensualist or a platonist through and through. He was both, it seems, in alternating moods.
144. Myers to Barrett, 23 September 1877, Barrett Papers (box 2, A4, no. 61); F. W. H. Myers, "Resolute Credulity," *PSPR* 11 (July 1895):213–34.
145. F. W. H. Myers, "The Drift of Psychical Research," *National Review* 24 (October 1894):191, 203, 208–9.
146. Myers, "Provisional Sketch of a Religious Synthesis," p. 284, and Frederic W. H. Myers, "Science and a Future Life," in *Science and a Future Life, with Other Essays* (London: Macmillan, 1893), p. 40.
147. Myers's ideal natural religion was, of course, very different from the enlightened natural theology which the Anglican church had long upheld. Nonetheless, as Frank M. Turner suggested to the author, the Anglican tradition may have influenced both Myers and Sidgwick. The Church of England had, in one sense, already naturalized the supernatural.
148. Lodge's presidential address to the SPR, in *Presidential Addresses*, p. 128.
149. Myers, "Provisional Sketch of a Religious Synthesis," p. 286.
150. Myers, "In Memory of Henry Sidgwick," p. 460.
151. Rogers, *Life and Experiences*, p. 47.
152. Sidgwick to Barrett, 8 December 1894?, Barrett Papers (box 2, A4, no. 130); Sidgwick and Sidgwick, *Henry Sidgwick*, pp. 357, 365, 466–8.
153. Sidgwick and Sidgwick, *Henry Sidgwick*, p. 494. Also see pp. 392, 435–6, 473–4, for subtle shifts in his attitudes toward telepathy as a proven phenomenon.
154. Sidgwick to Mozley, Cardinal Newman's nephew and a colleague of Sidgwick's at Cambridge in the 1860s, 11 January 1891, ibid., p. 508.
155. Broad, *Religion, Philosophy and Psychical Research*, p. 115.
156. Gore's address to the Synthetic Society, November 1900, as quoted in Sidgwick and Sidgwick, *Henry Sidgwick*, pp. 557–8; Sidgwick to Lord Tennyson, son of the Poet Laureate, undated, but probably mid-1890s, ibid., pp. 538–40; Sidgwick to Wilfrid Ward, 4 March 1898, ibid., p. 560.
157. Mrs. Sidgwick's presidential address to the SPR, in *Presidential Addresses*, pp. 290–1.
158. Mrs. Henry Sidgwick, "The Society for Psychical Research. A Short Account of its History and Work on the Occasion of the Society's Jubilee, 1932," and Lord Balfour's remarks, *PSPR* 41 (July 1932):16, 26. Ethel Sidgwick, *Mrs. Henry Sidgwick*, pp. 171–7, traces Mrs. Sidgwick's growing belief in the survival of physical death. Also see Alice Johnson, "Mrs.

Henry Sidgwick's Work in Psychical Research," *PSPR* 44 (June 1936):53–93.
159. Tuckett, "Illogical Position of Some Psychical Researchers," p. 470.
160. Gerald Balfour's presidential address to the SPR, in *Presidential Addresses*, pp. 259, 265.
161. Baird, *Hodgson*, p. 3; letter of 1901, as quoted in Howe, *Memoir of Hodgson*, p. 7.
162. Myers's presidential address to the SPR was reprinted in *Human Personality*, 2:292–307. The quotation appears on pp. 297–8.

5. Theosophy and the occult

1. James Robertson, *Spiritualism: The Open Door to the Unseen Universe. Being Thirty Years of Personal Observation and Experience Concerning Intercourse between the Material and Spiritual Worlds* (London: L. N. Fowler, 1908), p. 169.
2. E. R. Dodds, *Missing Persons. An Autobiography* (Oxford: Clarendon Press, 1977), pp. 97–8.
3. R. Laurence Moore, *In Search of White Crows: Spiritualism, Parapsychology, and American Culture* (New York: Oxford University Press, 1977), p. 236, discusses the refusal of American spiritualists in the nineteenth century "to accept a marginal status for their ideas." Much of Moore's treatment of this subject is, of course, highly pertinent here. See especially pp. 6–7, 224–7.
4. The quarterly *Borderland* is a perfect example from the 1890s of an indiscriminate blending of spiritualism and the occult.
5. See George Mills Harper, *Yeats's Golden Dawn* (New York: Barnes & Noble, 1974), pp. 9–13; Joseph Hone, *W. B. Yeats 1865–1939* (London: Macmillan, 1942), p. 71; John Symonds, *The Great Beast: The Life and Magick of Aleister Crowley* (London: Macdonald 1971), pp. 17–35; and R. G. Torrens, *The Inner Teachings of the Golden Dawn* (London: Neville Spearman, 1969), pp. 193–4.
6. Hargrave Jennings, *The Rosicrucians, Their Rites and Mysteries* (London: John Camden Hotten, 1870), p. 1.
7. A. P. Sinnett, *Nature's Mysteries and How Theosophy Illuminates Them* (London: Theosophical Publishing Society, 1913), pp. 3–4, 21, 23–4, 30. The 1913 edition of *Nature's Mysteries* is a condensed and somewhat revised version of a work first published in 1901. Jennings, *Rosicrucians*, pp. 337–8, thought alchemy a plausible art that ought not to be rejected out of hand.
8. A. P. Sinnett, ed., *Incidents in the Life of Madame Blavatsky* (1886; reprint ed., New York: Arno Press, 1976), p. vi. Also see A. P. Sinnett, *The Early Days of Theosophy in Europe* (London: Theosophical Publishing House, 1922), p. 8.
9. Mrs. A. P. Sinnett, *The Purpose of Theosophy* (London: Chapman & Hall, 1885), p. 3; Emily Kislingbury, "Spiritualism in its Relation to Theosophy," *Theosophical Siftings* 5, no. 3 (1892):15. Each number of *Theosophical Siftings* is paginated separately.
10. William Ashton Ellis, "Richard Wagner as Poet, Musician, and Mystic." Paper read to the Society for the Encouragement of the Fine Arts, 1887, as quoted in Anne Dzamba Sessa, *Richard Wagner and the English* (Lon-

don: Associated University Presses, 1979), p. 133. For information about Ellis, see ibid., pp. 39, 132, and Constance Wachtmeister, et al., *Reminiscences of H. P. Blavatsky and The Secret Doctrine* (Wheaton, Ill.: Theosophical Publishing House, 1976), pp. 59–64.

11. See the discussion of radical-eccentric organizations "opposed to conventional Victorian ideals," in Samuel Hynes, *The Edwardian Turn of Mind* (Princeton, N.J.: Princeton University Press, 1968), pp. 135–6.

12. Many of the points raised by Mircea Eliade, in *Occultism, Witchcraft, and Cultural Fashions: Essays in Comparative Religions* (Chicago: University of Chicago Press, 1976), pp. 63–5, to explain the mushrooming of occult sects in the past few decades are actually less dated than the author implies and help to elucidate similar developments in the late nineteenth century. Eliade stresses, in particular, the failure of the Christian churches to satisfy a broad, vague, but profound, need "for sacramental experiences" and soul-renewing insights into the meaning of modern life.

13. J. N. Maskelyne, *The Fraud of Modern 'Theosophy' Exposed. A Brief History of the Greatest Imposture ever Perpetrated under the Cloak of Religion*, 2d ed. (London: George Routledge, [1913]), p. 11.

14. F. Max Müller, "Esoteric Buddhism," *Nineteenth Century* 33 (May 1893):767–88; W. E. Gladstone, "True and False Conceptions of the Atonement," *Nineteenth Century* 36 (September 1894):317–31.

15. C. C. Massey, *Thoughts of a Modern Mystic. A Selection from the Writings of the Late C. C. Massey*, ed. W. F. Barrett (London: Kegan Paul, Trench, Trübner, 1909), p. 17n; Hargrave Jennings, *The Indian Religions; or, Results of the Mysterious Buddhism, concerning that also which is to be Understood in the Divinity of Fire*, 2d ed. (London: George Redway, 1890), pp. 136–7. Looking for even earlier roots, Alvin Boyd Kuhn, in *Theosophy, A Modern Revival of Ancient Wisdom* (New York: Henry Holt, 1930), p. 7, finds theosophical ideas in the work of Thales of Miletus. On Besant's alleged kinship with Bruno, see Arthur H. Nethercot, *The Last Four Lives of Annie Besant* (Chicago: University of Chicago Press, 1963), p. 180.

16. F. Edmund Garrett, *Isis Very Much Unveiled, being the Story of the Great Mahatma Hoax*, 3d ed. (London: *Westminster Gazette* Office, [1895]), p. 14.

17. Max Müller, "Esoteric Buddhism," p. 775.

18. Maskelyne, *Fraud of Modern 'Theosophy,'* p. 71, labels Blavatsky "the greatest impostor in history." For the other descriptive phrases, see Arthur Lillie, *Madame Blavatsky and Her "Theosophy". A Study* (London: Swan Sonnenschein, 1895), pp. x, 118.

19. Max Müller, "Esoteric Buddhism," pp. 770–2.

20. Warren Sylvester Smith writes, in *The London Heretics 1870–1914* (New York: Dodd, Mead, 1968), p. 142: "Theosophy could not have made a serious bid for attention among Londoners – or elsewhere – if a revival of Spiritualism had not preceded it." Frank Podmore, *Studies in Psychical Research* (London: G. P. Putnam's Sons, 1897), p. 40, refers to the Theosophical Society as "that vigorous offshoot of the spiritualist movement."

21. Kislingbury, "Spiritualism in its Relation to Theosophy," p. 6. Blavatsky's links with spiritualism are as murky as everything else in her past, but it seems that she first became acquainted with séance phenomena

when she met D. D. Home in Paris in 1858, that she tried her hand at a short-lived *Société spirite* in Cairo in the early 1870s, and that the middle of the decade found her in Philadelphia, associated with the fraudulent mediums Mr. and Mrs. Nelson Holmes. For Blavatsky and the Holmeses, see R. G. Medhurst, "Stainton Moses and Contemporary Physical Mediums. 3. Mr. and Mrs. Nelson Holmes," *Light* 83 (Autumn 1963):130–1. There are numerous biographies of Blavatsky, all necessarily vague about much of her life prior to her arrival in America in 1873. There appears, however, to be general agreement about her involvement with spiritualism before the founding of the Theosophical Society. See, for example, Gertrude Marvin Williams, *Priestess of the Occult: Madame Blavatsky* (New York: Alfred A. Knopf, 1946), pp. 31, 52–3, 75–9. More recent biographies include the "authorized" study by Howard Murphet, *When Daylight Comes. A Biography of Helena Petrovna Blavatsky* (Wheaton, Ill.: Theosophical Publishing House, 1975) and Marion Meade, *Madame Blavatsky: The Woman Behind the Myth* (New York: G. P. Putnam's Sons, 1980). Bruce F. Campbell, *Ancient Wisdom Revived: A History of the Theosophical Movement* (Berkeley, Calif.: University of California Press, 1980), is less concerned with Blavatsky's flamboyant personality than with her place in the history of occult thought.

22. H. P. Blavatsky, *The Key to Theosophy, being a Clear Exposition, in the form of Question and Answer, of the Ethics, Science, and Philosophy for the Study of which the Theosophical Society has been Founded* (1889; reprint ed., London and Bombay: The Theosophy Company, 1948), p. 31. Sinnett, *Early Days of Theosophy in Europe*, p. 28, writes of Blavatsky's "bitter detestation of spiritualism."
23. This is the interpretation proposed by Kuhn, *Theosophy*, pp. 90ff.
24. Nor is the debate by any means over today. While Geoffrey K. Nelson, *Spiritualism and Society* (London: Routledge & Kegan Paul, 1969), p. 93, describes Theosophy as "outside the body of Spiritualism," Moore, *In Search of White Crows*, pp. 229–34, underscores the beliefs which the two groups shared in common.
25. Kislingbury, "Spiritualism in its Relation to Theosophy," pp. 12, 13, 15, 16; Kuhn, *Theosophy*, pp. 94–5.
26. Kislingbury, "Spiritualism in its Relation to Theosophy," pp. 6, 7, 9. A belief in astral spirits attached to the material body, which remain earthbound after the latter's decay, was part of the occult teaching of the neo-Platonists and Paracelsians. See Keith Thomas, *Religion and the Decline of Magic* (New York: Charles Scribner's Sons, 1971), pp. 591–2. For more about the elusive elementals, see Sinnett, *Nature's Mysteries*, pp. 45–6, and, about the astral plane, Campbell, *Ancient Wisdom Revived*, p. 67.
27. "More About the Theosophists: An Interview with Mdme. Blavatsky," *Pall Mall Gazette* 39, pt. 2 (26 April 1884):4; Kuhn, *Theosophy*, pp. 96–7.
28. H. P. Blavatsky, *Isis Unveiled: A Master-Key to the Mysteries of Ancient and Modern Science and Theology*, 2 vols. (1877; reprint ed., Pasadena, Calif.: Theosophical University Press, 1972), 1:573, for example, speaks of "hopeless materialism" as "the mortal epidemic of our century."
29. Blavatsky, *Key to Theosophy*, p. 5.
30. Mrs. Sinnett, *Purpose of Theosophy*, pp. 3–4. Also see *The Theosophical Society: Constitution, etc.*, drawn up in 1901 by Thomas Green, "Sec-

retary in England of the Theosophical Society," pp. 1–3 (no. 11 in *Houdini Pamphlets: Theosophy*, vol. 1, Houdini Collection).

31. See, for example, *Theosophical Society: Constitution*, p. 2, where the first of the Society's objects is stated as: "To form a nucleus of Universal Brotherhood without distinction of race, creed, sex, caste or colour." Kislingbury, "Spiritualism in its Relation to Theosophy," p. 15, reminded the spiritualists: "Above all, we stand on the common platform of Universal Brotherhood."

This is not to say that spiritualists were unanimously impressed by Theosophy. An acerbic article in *Two Worlds* outlined the reasons for animosity between Theosophists and spiritualists, claiming that Theosophy was founded upon "baseless antique myths," while spiritualism rested squarely on "FACTS." "Mrs. Besant on Theosophy and Spiritualism," *Two Worlds* 3 (2 May 1890):287–8.

32. "More About the Theosophists," *Pall Mall Gazette*, 26 April 1884, p. 4; Henry Steel Olcott, *Old Diary Leaves. The True Story of the Theosophical Society* (New York: G. P. Putnam's Sons, 1895), p. 462.

33. "Mrs. Besant on Theosophy and Spiritualism," *Two Worlds*, 2 May 1890, p. 287; Gertrude Marvin Williams, *The Passionate Pilgrim: A Life of Annie Besant* (New York: Coward-McCann, 1931), p. 308.

34. For the "'after-glow' of church-worship," see Smith, *London Heretics*, p. 142. Williams, *Passionate Pilgrim*, pp. 308–9, describes the changes wrought in Theosophical meetings by C. W. Leadbeater, who introduced a full religious service into the proceedings. Leadbeater was Besant's cohort in the leadership of Theosophy for most of the period from the mid-1890s until their deaths in the early 1930s. He became, in time, a Theosophical bishop. Mrs. Besant further steered Theosophy away from the course set by Blavatsky when she and Leadbeater began searching for a Messiah early in the twentieth century and claimed to have found him in the young Indian boy, Jiddu Krishnamurti. Ibid., pp. 303–7; James Webb, *The Occult Underground* (La Salle, Ill.: Open Court Publishing Co., 1974), pp. 94–104 (first published in 1971 as *The Flight from Reason*). The second volume of Nethercot's biography of Besant – *The Last Four Lives of Annie Besant* – deals extensively with the impact of the Besant–Leadbeater team on Theosophy.

35. Blavatsky, *Key to Theosophy*, pp. 64, 72. She heatedly denied that the Theosophical definition of God was equivalent to pantheism, as anyone would know, she snapped, who bothered to "etymologise the word Pantheism esoterically" (ibid., p. 63). Also in the same volume, see the section on "Prayer Kills Self-Reliance," pp. 71–4.

36. Charles W. Leadbeater, "What Theosophy Does for Us: An Address Delivered at Aeolian Hall, Buffalo, N.Y., U.S.A., Monday Evening, October 29, 1900," p. 2 (no. 5 in *Houdini Pamphlets: Theosophy*, vol. 1, Houdini Collection).

37. Robertson, *Spiritualism: The Open Door*, p. 170.

38. "What is Theosophy?" Reprinted from the *New York Herald*, 18 August 1889, in *Theosophical Siftings* 2, no. 11 (1889):9.

39. See the section "On Eternal Reward and Punishment; and on Nirvana," in Blavatsky, *Key to Theosophy*, pp. 108–15; also Campbell, *Ancient Wisdom Revived*, pp. 69–71.

40. Kuhn, *Theosophy*, p. 233. The distinction between spirit and soul is not the least confusing of Theosophy's tenets. In Theosophical literature, the

term "soul" is used apparently to designate the personal identity of an individual through his or her various incarnations, and, as the thinking portion of the human being, can also be dubbed the Ego. Spirit, by unclear contrast, is "the divine portion of the soul" and has existed throughout time. The soul undergoes recurring reincarnations, but the spirit does not. After purifying itself "through cyclic transmigrations," the soul may look forward to "a reunion with its spirit which alone confers upon it immortality." See Blavatsky, *Key to Theosophy*, pp. 74ff, 100–15.

41. A. P. Sinnett, *The Growth of the Soul*, 2d enl. ed. (London and Benares: Theosophical Publishing Society, 1905), p. 59. The volume was first published in 1896. For an earlier discussion of Karma and the mechanism of reincarnation, see A. P. Sinnett, *Esoteric Buddhism*, 2d ed. (London: Trübner, 1883), pp. 66ff.

42. See Kuhn, *Theosophy*, pp. 232–3, 237, 242–52.

43. Moore, *In Search of White Crows*, p. 233, points out, furthermore, that the spiritualist vision of "life in the heavenly spheres" was so concrete, specific, and indeed "earthlike," that it was all the closer to the Theosophical doctrine of successive reincarnations.

44. Frank Podmore, *Modern Spiritualism: A History and a Criticism*, 2 vols. (London: Methuen, 1902), 2:161, 166, 168. Kardec's outstanding English disciple, and translator, was Anna Blackwell. She found ready access to Burns's *Human Nature*, where she published several articles, such as "The Philosophy of Re-Incarnation," *Human Nature* 4 (January 1870).

45. Allan Kardec, *L'Evangile selon le spiritisme, contenant l'explication des maximes morales du Christ, leur concordance avec le spiritisme et leur application aux diverses positions de la vie*, 3d ed. rev. (Paris: Dentu, 1866). W. S. Moses, "Spiritualism at Home and Abroad: Its Present Position and Future Work," p. 17, comments on the ties between Kardec's works and Christ's teachings. Moses's pamphlet is no. 4 in *Addresses Delivered Before the London Spiritualist Alliance during the Years 1884 to 1888* (London:LSA, 1889), Houdini Collection.

46. Known in the spiritualist literature of the period both as Lady Caithness and as the Duchesse de Pomar, she signed herself, in hybrid fashion, Marie Caithness de Pomar. For biographical information, see the obituary of the Earl of Caithness in the *Times*, 30 March 1881, p. 12; *Burke's Peerage, Baronetage, and Knightage* (under the Sinclair family, Earls of Caithness); *DNB*, s.v. "Sinclair, James, 14th Earl of Caithness"; and, for a brief paragraph on the Countess herself, *Enciclopedia Universal Ilustrada Europeo-Americana* (Barcelona: Hijos de J. Espasa, 1922), 46:184.

47. The priceless lace was observed by Miss Bates when she traveled abroad in 1894. She was an associate of the SPR. E. Katharine Bates, *Seen and Unseen* (New York: Dodge Publishing Co., 1908), pp. 146, 150. The impressive diamonds appear in Williams, *Priestess of the Occult*, p. 215.

48. Nethercot, *Last Four Lives of Annie Besant*, p. 30, argues that the Duchess "was quite sure she was the reincarnated spirit of Mary Queen of Scots." C. E. Bechhofer-Roberts, *The Mysterious Madame. Helena Petrovna Blavatsky: The Life and Work of the Founder of the Theosophical Society* (New York: Brewer & Warren, 1931), p. 290, and Webb, *Occult Underground*, p. 278, take the same point of view. Williams, *Priestess of the Occult*, p. 215, advocates the guardian angel theory.

49. Bates, *Seen and Unseen*, pp. 145, 150–1.
50. Countess of Caithness (Marie Sinclair), *Old Truths in a New Light or, An Earnest Endeavour to Reconcile Material Science with Spiritual Science, and with Scripture* (London: Chapman & Hall, 1876), pp. 7–11, 14–17, 443, and chapters 22–8 for the full discussion of reincarnation.
51. Williams, *Priestess of the Occult*, pp. 214–17, 284; Vsevolod Sergyeevich Solovyoff, *A Modern Priestess of Isis*, trans. and abridged by Walter Leaf for the SPR (1895; reprint ed., New York: Arno Press, 1976), pp. 2–3; Arthur H. Nethercot, *The First Five Lives of Annie Besant* (Chicago: University of Chicago Press, 1960), p. 369; Bechhofer-Roberts, *Mysterious Madame*, p. 290.
52. See Myers's remarks about Lady Caithness's new society in "General Meeting," *JSPR* 6 (May 1894):242–4 and "The Anglo-French Psychological Society," *JSPR* 6 (June 1894):263–4.
53. "The Anglo-French Psychological Society," pp. 263–4.
54. "More About the Theosophists," *Pall Mall Gazette*, 26 April 1884, p. 4; A. P. Sinnett described various occult wonders, performed by Blavatsky in India, throughout *The Occult World*, 4th ed. (London: Trübner, 1884) – first published in 1881.
55. "Report of the Committee Appointed to Investigate Phenomena Connected with the Theosophical Society," *PSPR* 3 (1885):218–19; Williams, *Priestess of the Occult*, pp. 201–5.
56. Solovyoff, *Modern Priestess of Isis*, p. 7.
57. Wachtmeister, et al., *Reminiscences of H. P. Blavatsky*, pp. 29–30.
58. Kuhn, *Theosophy*, p. 246. Theosophists believed that the etheric body might be photographed and that it was composed of ectoplasm. The etheric body, according to Theosophy, is the layer identical in appearance to the physical body; it remains for a while after death to enclose the less substantial astral and mental bodies. In this respect, it was not unlike the "spirit-body" in which some spiritualists believed, as a kind of transitional protective covering for the pure spirit after the death of the physical body.
59. LSA, Minute Book of Council Meetings 1909–1921, inserted between pp. 16 and 17, in the printed report of the LSA Council, presented at the Annual General Meeting, 17 March 1910.
60. SPR membership list, December 1883, at the end of *PSPR*, vol. 1; Sinnett, *Early Days of Theosophy in Europe*, pp. 11, 42, 45, 48, 50; George Wyld, *Notes of My Life* (London: Kegan Paul, Trench, Trübner, 1903), pp. 71–2; Wyld to Barrett, undated, Barrett Papers (box 2, A2, no. 24). Myers's certificate of membership in the Theosophical Society (Western Division) can be found in Myers Papers 12[223].
61. Arthur and Eleanor Mildred Sidgwick, *Henry Sidgwick, A Memoir* (London: Macmillan, 1906), p. 385; C. D. Broad, "Our Pioneers, VI. Henry Sidgwick," *JSPR* 40 (September 1959):106. For another brief account of Blavatsky's visit to Cambridge in August 1884, see J. J. Thomson, *Recollections and Reflections* (London: G. Bell, 1936) pp. 153–4.
62. *Theosophical Society: Constitution*, p. 2.
63. Sinnett, *Early Days of Theosophy in Europe*, pp. 47–8.
64. C. D. Broad, *Religion, Philosophy and Psychical Research: Selected Essays* (New York: Harcourt, Brace, 1953), p. 95; W. H. Salter, "Our Pioneers, VIII. Richard Hodgson (1855–1905)," *JSPR* 40 (September 1960):329–31.

65. Podmore, *Studies in Psychical Research*, pp. 171, 186; Sidgwick's journal entry, 22 March 1885, in Sidgwick and Sidgwick, *Henry Sidgwick*, p. 405.
66. That Blavatsky wrote the Mahatma letters was, likewise, the conclusion of Harold Edward Hare and William Loftus Hare, in their exhaustive study, *Who Wrote the Mahatma Letters? The First Thorough Examination of the Communications Alleged to Have Been Received by the Late A. P. Sinnett from Tibetan Mahatmas* (London: Williams & Norgate, 1936). The Hares' inquiry was prompted by the publication, in 1923, of *The Mahatma Letters to A. P. Sinnett from the Mahatmas M. & K. H.*, compiled by A. T. Barker. Barker, a staunch Theosophist, did not question the authenticity of the letters. On the Mahatmas in general, and Koot Hoomi and Morya in particular, see Campbell, *Ancient Wisdom Revived*, pp. 53–6.
67. For points covered in this summary of the contents of Hodgson's report, see "Report of the Committee Appointed to Investigate Phenomena Connected with the Theosophical Society," pp. 203 -26, 241, 250, 276–313, 381–2; Campbell, *Ancient Wisdom Revived*, pp. 88–90; and Williams, *Priestess of the Occult*, pp. 112, 294–6.
68. Hare and Hare, *Who Wrote the Mahatma Letters?* p. 19.
69. "Report," p. 207.
70. Sinnett, *Early Days of Theosophy in Europe*, p. 69; A. P. Sinnett, *The "Occult World Phenomena," and The Society for Psychical Research, with a Protest by Madame Blavatsky* (London: George Redway, 1886), pp. 43–4, 49.
71. From Sidgwick's journal, 30 April 1885, in Sidgwick and Sidgwick, *Henry Sidgwick*, p. 410.
72. Back in 1878, Moses had called the two volumes of *Isis Unveiled* "masterpieces of industry and erudition" (*Psychography: A Treatise on One of the Objective Forms of Psychic or Spiritual Phenomena* [London: W. H. Harrison, 1878], p. 141), and Blavatsky returned the compliment, in *Key to Theosophy* (p. 31), by referring to him as "one of the very few philosophical Spiritualists." For Moses's change of view, see "Modern Mystics and Modern Magic," *Saturday Review* 77 (20 January 1894):76. C. M. Davies offered his opinion of Blavatsky in *The Great Secret, and its Unfoldment in Occultism. A Record of Forty Years' Experience in the Modern Mystery*, 3d ed. (Blackpool: The Ellis Family, n.d.), p. 300.
73. Sinnett, *Early Days of Theosophy in Europe*, pp. 9–11, 19, 29, 69–71; members listed at the front of Minute Books of the BNAS, Committees, no. 2. On Dr. Blake, see Wachtmeister, et al., *Reminiscences of H.P. Blavatsky*, pp. 103–5. Also see "Report of the Committee Appointed to Investigate Phenomena Connected with the Theosophical Society," pp. 207, 397–400.
74. Wyld, in his memoirs, recollected joining the Society in 1879 (*Notes of My Life*, pp. 71–2), but Sinnett, in *Early Days of Theosophy in Europe*, p. 11, citing the Society minute book, listed Wyld among the people present at the first meeting, on 27 June 1878. Wyld himself pointed out that he became the Society's second president "after some two years," and since he was holding that office by January 1880, the earlier date seems plausible.
75. Wyld, *Notes of My Life*, pp. 72–4. (*The Theosophist* was published in India, under Blavatsky's editorship.) For the fullest statement of Wyld's

evolved religious views, see his *Christo-Theosophy* (London: Kegan Paul, 1895), which includes an expanded version of his pamphlet, *Miracles as Not Contrary to Nature* (London: Kegan Paul, 1892).

76. Sinnett, *Early Days of Theosophy in Europe*, p. 71.
77. Ibid., pp. 21, 23–7, 38–41; Nethercot, *First Five Lives of Annie Besant*, p. 195; Sinnett, ed., *Incidents in the Life of Madame Blavatsky*, pp. 222–5, 227; Williams, *Priestess of the Occult*, pp. 161–4, 167–73, 179–80. Sinnett joined the Ghost Club in April 1883 – having been proposed by Stainton Moses – and at the Club's meeting of 2 November 1883, he "gave an interesting account of his introduction to occultism thro' the instrumentality of Madam Blavatsky." Ghost Club Minutes, vol. 1 (2 November 1883), pp. 14, 23–4, B.M. Add. MSS. 52258.
78. Sinnett, *Early Days of Theosophy in Europe*, pp. 51–5, 82, 86–93, 97. By the early 1890s, Sinnett was, according to Nethercot, at the center of "the anti-Blavatsky faction – though of course it would have been suicidal to accept such a name openly." *Last Four Lives of Annie Besant*, p. 26. See the attack on Sinnett's personal ambition in Alice Leighton Cleather, *H. P. Blavatsky as I Knew Her* (Calcutta: Thacker, Spink, 1923), pp. 8–9, 41–74.
79. Sinnett, *Early Days of Theosophy in Europe*, pp. 27–8, 111–17.
80. James Robertson, *A Noble Pioneer. The Life Story of Mrs. Emma Hardinge Britten* (Manchester: "Two Worlds" Publishing Co., n.d.), p. 10 (no. 5 in *Houdini Pamphlets: Spiritualism*, vol. 2, Houdini Collection); Kislingbury, "Spiritualism in its Relation to Theosophy," p. 6; Kuhn, *Theosophy*, pp. 105–6.
81. See *The Theosophical Movement 1875–1925. A History and a Survey* (New York: E. P. Dutton, 1925), pp. 207–8, for a full account of the Blavatsky–Collins relationship, told from a pro-Blavatsky perspective. The documents from which the author draws the narrative in this anonymous and lengthy volume suggest that he or she had a close familiarity with many of the events and personalities that figured in the early history of the Theosophical Society. Nethercot, *First Five Lives of Annie Besant*, p. 395n, claims that the book was written by a committee of the United Lodge of Theosophists, which had severed all relations with the Theosophical Society by 1925. Williams, *Priestess of the Occult*, pp. 288–90, 300, gives a brief summary of Collins's departure from Blavatsky's inner circle, as does Kuhn, *Theosophy*, pp. 301–3.

On Yeats's experiences, see W. B. Yeats, *Memoirs*, ed. Denis Donoghue (New York: Macmillan, 1973), pp. 23–6, 281–2. Yeats was curious to test whether, "if you burned a flower to ashes, and then put the ashes under a bellglass in the moonlight, the phantom of the flower would rise before you."
82. On Kingsford, see Lillie, *Madame Blavatsky and Her "Theosophy,"* p. 121, and Sinnett, *Early Days of Theosophy in Europe*, p. 41. On Besant: Estelle W. Stead, *My Father: Personal and Spiritual Reminiscences* (New York: George H. Doran, 1913), p. 155; Theodore Besterman, *Mrs. Annie Besant: A Modern Prophet* (London: Kegan Paul, Trench, Trübner, 1934), pp. 135–6. A colleague of Annie Besant in her various secularist and social crusades who also migrated to Theosophy was Herbert Burrows. See Nethercot, *First Five Lives of Annie Besant*, pp. 285, 287, 292–3, and passim.

83. John Carswell, *Lives and Letters: A. R. Orage, Beatrice Hastings, Katherine Mansfield, John Middleton Murry, S. S. Koteliansky, 1906–1957* (London: Faber & Faber, 1978), p. 20. My thanks to Dr. Ann McLaughlin for calling this book to my attention. Also see Philip Mairet, *A. R. Orage* (1936; reprint ed. with new introduction by author, New Hyde Park, N.Y.: University Books, 1966), pp. 11–19; and Samuel Hynes, *Edwardian Occasions: Essays on English Writing in the Early Twentieth Century* (New York: Oxford University Press, 1972), pp. 39–47.

84. Edward Carpenter, *My Days and Dreams* (1916), quoted in Hynes, *Edwardian Turn of Mind*, pp. 134–5. Also see Hynes's comments on pp. 9, 150.

85. Wachtmeister, et al., *Reminiscences of H. P. Blavatsky*, pp. 12–68. By contrast, Williams, *Priestess of the Occult*, p. 288, describes the house where Madame resided during the summer of 1887 as "the Keightleys' suburban villa." Also see Cleather, *Blavatsky as I Knew Her*, p. 22, and Nethercot, *First Five Lives of Annie Besant*, pp. 303, 331–3, 357.

86. Blavatsky, *Key to Theosophy*, p. 287; Sinnett, *Early Days of Theosophy in Europe*, p. 19; "Report of the Committee Appointed to Investigate Phenomena Connected with the Theosophical Society," p. 202; "What a Mahatma Is," *Borderland* 1 (July 1894):467–8. Also see Kuhn, *Theosophy*, pp. 101, 110, 112, and Smith, *London Heretics*, pp. 149–50.

87. Lillie, *Madame Blavatsky and Her "Theosophy"*, pp. 118–19. See A. P. Sinnett, "Apollonius of Tyana," *Transactions of the London Lodge of the Theosophical Society*, no. 32, January 1898. This address, delivered to the London Lodge on 6 November 1897, appears as the third pamphlet in a bound volume of Sinnett's *Theosophical Tracts*, in the Houdini Collection.

88. Hodgson's suggestion that she was a Russian spy fomenting native Indian "disaffection towards British rule" is farfetched ("Report of the Committee Appointed to Investigate Phenomena Connected with the Theosophical Society," p. 314). Myers suggested that Blavatsky came "within an ace of founding a world-religion, merely to amuse herself & to be admired" (draft of letter from Myers to Lord Acton, 28 April 1892), quoted in Alan Gauld, *The Founders of Psychical Research* (London: Routledge & Kegan Paul, 1968), p. 367. Hare and Hare, *Who Wrote the Mahatma Letters?* pp. 303–7, suggest several different possible motives behind Blavatsky's establishment of Theosophy: "the power-seeking motive," "a special animus against Christianity," and an irresponsible fondness for mischief, malice, and "controversial muckraking."

89. Thomas, *Religion and the Decline of Magic*, p. 138.

90. The suggestion that a few women may have found in Theosophy an authoritative voice that they could not otherwise command in late nineteenth-century Britain also generally applies to the attraction of mediumship for certain women. It may, furthermore, provide an underlying, and even subconscious, motive for the visionary writings of women like Mrs. Newton Crosland and Mrs. Augustus De Morgan. Podmore, *Modern Spiritualism*, 2:39, observes that Mrs. Crosland gave women a "lofty part" in her New Moral Order. Harper, *Yeats's Golden Dawn*, p. 12, refers to the comparatively large number of women who joined the Golden Dawn in its early years.

91. Edward Maitland, *Anna Kingsford: Her Life, Letters, Diary and Work*, 2d ed., 2 vols. (London: George Redway, 1896), 1:15, 57–9; *DNB*, s.v.

"Kingsford, Mrs. Anna." Kingsford and Maitland posed as niece and uncle in Paris.

92. Maitland, *Kingsford*, 1:18–19, quotes from a pamphlet on female suffrage which she published in 1868, at the age of twenty-two, in which Kingsford tellingly described women as "shackled by the chains of ignorance, a helpless prey to that terrible monster whose name is 'Ennui'." On women, boredom, Theosophy, and women's suffrage, see Webb, *Occult Underground*, p. 105.

93. Maitland, *Kingsford*, 1:14, 33, 49. She suffered dreadfully from asthma (an attack of which after her wedding night was "so violent . . . as to endanger her life"), "acute accesses of neuralgia, nervous panics, and sudden losses of consciousness."

94. Lillie, *Madame Blavatsky and Her "Theosophy,"* pp. 120–1. Kingsford's thesis was a manifesto on behalf of vegetarianism: *De l'alimentation végétale chez l'homme*. It was published in England as *The Perfect Way in Diet; a Treatise Advocating a Return to the Natural and Ancient Food of our Race* (London: Paul, Trench, 1881). The University of Paris medical degree was opened to women in the late 1860s.

95. Or at least her own and Maitland's.

96. Anna (Bonus) Kingsford and Edward Maitland, *The Perfect Way or, The Finding of Christ*, 5th ed. (New York: Metaphysical Publishing Co., 1901), pp. 49, 53–7, 60. *The Perfect Way* was first published in 1882. On possible links with the doctrine of the Woman-Messiah, see Webb, *Occult Underground*, p. 342. Alternatively, Kingsford's theories about the cosmic feminine principle may have had more venerable roots in neo-Platonic visions of a pervasive, universal sexual dualism.

97. Kingsford to Lady Caithness, 12 May 1884, in Maitland, *Kingsford*, 2:168.

98. Maitland, *Kingsford*, 2:168, 246. Webb, *Occult Underground*, p. 356, suggests that this expert was MacGregor Mathers, soon to be involved in founding the Golden Dawn.

99. Maitland, *Kingsford*, 2:246, 267–8. Kingsford did not have the satisfaction of seeing Pasteur succumb before her death in 1888. He survived until 1895. She had confessed in her diary, in November 1885, that throughout all her previous incarnations she had been driven by "the desire for greatness, the desire to achieve." Ibid., p. 216.

100. Ibid., pp. 216–19, for an example of Kingsford's reliance on the doctrine of Karma; Kingsford and Maitland, *Perfect Way*, pp. 50, 60.

101. Sinnett, *Esoteric Buddhism*, p. xiii. Kingsford, needless to say, never acknowledged any debt to Blavatsky.

102. Maitland, *Kingsford*, 2:167–8, 174.

103. Ibid., 1:454; 2:158–65, 176, 179, 185.

104. After Kingsford's death, Maitland founded an Esoteric Christian Union that, apparently, remained inactive. Webb, *Occult Underground*, p. 279. Maitland, who has received scant attention in this account of Kingsford's life and work, was a religious seeker in his own right who, as a young man, had abandoned the evangelicalism of his family. Twenty-two years Kingsford's senior, he seems to have devoted his life to her from 1874 until 1888, and to her memory from 1888 until his death in 1897. A professional writer and journalist, he was the author of several romances. He, too, joined the London Lodge of the Theosophical Society in 1883–4 and was for many years an associate of the SPR. See Chapter 3, "Some Ac-

count of Myself," in Maitland, *Kingsford*, 1:36–45; *DNB*, s.v. "Maitland, Edward," and membership lists at back of SPR *Proceedings*.
105. Besterman, *Mrs. Annie Besant*, p. 130.
106. See Nethercot, *First Five Lives of Annie Besant*, p. 19, for her early attraction to the Oxford Movement.
107. Annie Besant, *Why I Became a Theosophist* (London: Theosophical Publishing Society, 1891), p. 30. The pamphlet was written in July 1889. Besterman, *Mrs. Annie Besant*, p. 130, points to the incompatibility between her temperament and "the dull, level atmosphere of nineteenth-century Freethought." Actually the atmosphere of Victorian free thought could be quite heady at times, but not in the aesthetically rich, even exotic, way which Besant seemed to crave. Norman and Jeanne MacKenzie, *The Fabians* (New York: Simon & Schuster, 1977), p. 47, describe as "an inverted form of revivalist religion" the secularist services that Besant led at Bradlaugh's Hall of Science in London and for which she composed a *Secular Song and Hymn Book*.
108. Besant, *Why I Became a Theosophist*, pp. 17, 30.
109. Annie Besant, *Esoteric Christianity or The Lesser Mysteries* (New York: John Lane, 1902), pp. 21, 36–9, 120–1. The book was first published in 1901, but was based on an earlier version that appeared in 1898.
110. Ibid., pp. 383–4. Smith, *London Heretics*, p. 163, gives another example of Besant's Christian imagery and what he calls "Pauline overtones."
111. Nethercot, *Last Four Lives of Annie Besant*, pp. 126–9 and passim, for the training of young Krishnamurti to be World Teacher, a role for which he ultimately proved unsuited. Besant did not formally become president of the Theosophical Society until after Olcott's death, but for years she had been rehearsing for the job.
112. Kuhn, *Theosophy*, p. 263, suggests one line of comparison in this respect.
113. Blavatsky, *Isis Unveiled*, 1:ix, xi, xii, xv, xxii, and passim.
114. To a Theosophist, of course, Blavatsky's prophecies were not clever guesswork, but the result of her insights into nature's secrets. See, for example, Kuhn, *Theosophy*, pp. 263–4, and Murphet, *When Daylight Comes*, p. 248.
115. That is to say, human development did not have to await a certain stage of animal development before it could commence. Apes, however, apparently emerged when one of the earlier root races of man cohabited with animals. Kuhn, *Theosophy*, pp. 253–8, summarizes the evolutionary argument of *The Secret Doctrine*. See A. P. Sinnett, "The Beginnings of the Fifth Race," *Transactions of the London Lodge of the Theosophical Society*, no. 31, February 1897 (no. 2 in Sinnett, *Theosophical Tracts*, Houdini Collection). On root races, also see Campbell, *Ancient Wisdom Revived*, pp. 44–5, 64–5.
116. "More About the Theosophists," *Pall Mall Gazette*, 26 April 1884, p. 4.
117. Wyld, *Notes of My Life*, p. 75 (Wyld's italics); Sinnett, *Nature's Mysteries*, pp. 6ff, 45, 58–9, for examples of both the airs and the scattering; Annie Besant, *Theosophy* (London: T. C. & E. C. Jack, [1912]), p. 21.
118. Hugh Kearney, *Science and Change 1500–1700* (New York: McGraw-Hill, 1971), p. 41.
119. Besant first published her chemical researches in *Lucifer*, November 1895, and subsequently printed them as a pamphlet: *Occult Chemistry* (London and Benares: Theosophical Publishing Society, 1905). A much

enlarged version was produced with C. W. Leadbeater as coauthor: *Occult Chemistry. A Series of Clairvoyant Observations on the Chemical Elements* (Adyar: Theosophist Office, [1909]). Nethercot, *Last Four Lives of Annie Besant*, pp. 49–52, 124, discusses how Leadbeater and Besant were able "to depict the molecule and to split the atom" in their fashion, long before orthodox science, and he remarks on William Crookes's sympathetic response to their "occult chemistry."
120. Besant, *Theosophy*, pp. 21–42.
121. William Kingsland, "The Higher Science," *Theosophical Siftings* 1, no. 11 (1888–9):1; Massey, *Thoughts of a Modern Mystic*, pp. 40–1.
122. Blavatsky, *Isis Unveiled*, 1:306.
123. C. C. Massey, Preface to *The Philosophy of Mysticism*, by Carl Du Prel, trans. C. C. Massey, 2 vols. (London: George Redway, 1889), 1:xii.

Part III. A pseudoscience

1. Publisher's advertisement, quoted in S. C. Hall, *The Use of Spiritualism?* (London: E. W. Allen, 1884), p. 49n. The publishers of the *Spiritual Record* were Hay Nisbet of Glasgow and E. W. Allen of London.
2. Basil Willey, *More Nineteenth Century Studies. A Group of Honest Doubters* (New York: Columbia University Press, 1956), p. 145.
3. Charles Maurice Davies, *Heterodox London: or, Phases of Free Thought in the Metropolis*, 2 vols. (London: Tinsley Brothers, 1874), 2:41.
4. W. S. Moses, "Some Things that I *Do* Know of Spiritualism, and Some that I do *Not* Know: A Chapter of Autobiography," p. 8, which appears as no. 11 in *Addresses Delivered Before the London Spiritualist Alliance during the Years 1884 to 1888* (London: LSA, 1889), Houdini Collection.
5. [G. H. Lewes], "Seeing is Believing," *Blackwood's Edinburgh Magazine* 88 (October 1860):386. The *Wellesley Index to Victorian Periodicals*, vol. 1, attributes the article to Lewes.
6. Logie Barrow, "Socialism in Eternity: The Ideology of Plebeian Spiritualists, 1853–1913," *History Workshop: A Journal of Socialist Historians* 9 (Spring 1980):55.
7. "Spiritual Phenomena and Men of Science," *Spiritualist Newspaper* 9 (4 August 1876):1.
8. A. W. Drayson, "Science and the Phenomena Termed Spiritual," an address to the London Spiritualist Alliance, 23 October 1884, pp. 5–6 (no. 2 in *Addresses Delivered Before the LSA*).

6. Concepts of mind

1. Letter of 25 January 1850, quoted in Clement John Wilkinson, *James John Garth Wilkinson; a Memoir of his Life, with a Selection from his Letters* (London: Kegan Paul, Trench, Trübner, 1911), p. 81.
2. The changed orientation of psychology is the main theme traced in Robert M. Young's indispensable study, *Mind, Brain and Adaptation in the Nineteenth Century: Cerebral Localization and its Biological Context from Gall to Ferrier* (Oxford: Clarendon Press, 1970).
3. R. J. Cooter, "Phrenology and British Alienists, c. 1825–1845. Part 1: Converts to a Doctrine," *Medical History* 20 (January 1976):13; Lorraine

J. Daston, "British Responses to Psycho-Physiology, 1860–1900," *Isis* 69 (June 1978):192, 194–5.

4. Alexander Bain, *Mind and Body. The Theories of their Relation*, 2d ed. (London: Henry S. King, 1873), p. 4. The volume was first published in 1872.

5. Daston, "British Responses to Psycho-Physiology," p. 196. Also see pp. 198–9, 202.

6. For a brief summary of earlier theories of the seat of the soul, from the ancient Greeks through the Renaissance, see H. W. Magoun, "Development of Ideas Relating the Mind with the Brain," in Chandler McC. Brooks and Paul F. Cranefield, eds. *The Historical Development of Physiological Thought* (New York: Hafner Publishing, 1959), pp. 81–9. Also see the first chapter, "Gall and Phrenology," in Young, *Mind, Brain and Adaptation*, to which the remarks here are heavily indebted.

7. R. J. Cooter, "Phrenology and British Alienists, c. 1825–1845. Part 2: Doctrine and Practice," *Medical History* 20 (April 1976):144.

8. "Materialism and Scepticism," *Phrenological Journal and Miscellany* 1 (1823–1824):120.

9. Letter from Combe to David George Goyder, 14 December 1847, quoted in A. Cameron Grant, "Combe on Phrenology and Free Will: A Note on XIXth-Century Secularism," *Journal of the History of Ideas* 26 (January–March 1965):146; John Tulloch, *Movements of Religious Thought in Britain During the Nineteenth Century, Being the Fifth Series of St. Giles' Lectures* (1885; reprint ed., New York: Humanities Press, 1971), p. 128.

10. Robert MacNish, *An Introduction to Phrenology*, 2d ed. (Glasgow: John Symington, 1837), p. 22.

11. (London: Houlston & Wright, 1868). Chapter 6, "The Ungodly Error," in David de Giustino, *Conquest of Mind: Phrenology and Victorian Social Thought* (London: Croom Helm, 1975) fully traces the conflict within phrenological ranks between those who leaned toward materialism on the one hand and the "Christian Phrenologists" on the other.

12. Alfred Kitson, *Spiritualism for the Young. Designed for the Use of Lyceums, and the Children of Spiritualists in General Who Have no Lyceums at which They Can Attend* (Keighley, Yorkshire: S. Billows, 1889), p. 138. On the "dual appeal" of phrenology as science and faith, see Terry M. Parssinen, "Popular Science and Society: The Phrenology Movement in Early Victorian Britain," *Journal of Social History* 8 (Fall 1974):8, and John D. Davies, *Phrenology, Fad and Science. A 19th-Century American Crusade* (New Haven: Yale University Press, 1955), pp. 8–9.

13. Robert Darnton persuasively sketches the place of mesmerism among the pseudosciences that merged with occultism in the late eighteenth century, in *Mesmerism and the End of the Enlightenment in France* (Cambridge, Mass.: Harvard University Press, 1968), pp. 33–9. For a contemporary Victorian discussion of the subject, see J. C. Colquhoun, *An History of Magic, Witchcraft, and Animal Magnetism*, 2 vols. (London: Longman, Brown, Green, & Longmans, 1851).

14. Mesmer was not, of course, the first to assert the universal influence of a magnetic fluid. For the views of Paracelsus and Van Helmont along these lines in the sixteenth and seventeenth centuries, see Edwin G. Boring, *A History of Experimental Psychology*, 2d ed. (New York: Appleton-Century Crofts, 1950), p. 116.

15. Darnton, *Mesmerism and the End of the Enlightenment in France*, p. 4.
16. According to Mesmer, it was possible to collect the invisible magnetic fluid, store it away, and redistribute it as necessary. The redistribution could take place from one person directly to another, as from mesmerizer to mesmerized, or through the help of such devices as the *baquet*. Some of Mesmer's patients actually bathed in water that Mesmer had touched, thereby conveying the magnetic fluid from his own body to the water, which in turn conveyed the force to the afflicted sufferer. Among the numerous books that discuss mesmerism, usually as the first step toward the development of modern psychiatric theory, see Henri F. Ellenberger, *The Discovery of the Unconscious: The History and Evolution of Dynamic Psychiatry* (New York: Basic Books, 1970), pp. 57–69.
17. Darnton, *Mesmerism and the End of the Enlightenment in France*, pp. 10–11, 13, 18, 26, 30, 38–9, 44, 93, 107. Mesmer's medical theory that bodily health depended on the balanced distribution of fluid continued, however, to hark back to ancient Greek beliefs in the four bodily humors.
18. For mesmerism in Britain before 1837, see Fred Kaplan, " 'The Mesmeric Mania': The Early Victorians and Animal Magnetism," *Journal of the History of Ideas* 35 (October–December 1974):694–7; and Frank Podmore, *From Mesmer to Christian Science: A Short History of Mental Healing* (New Hyde Park, N.Y.: University Books, 1963), pp. 122–6. Podmore's volume was first published in 1909 as *Mesmerism and Christian Science*.
19. On Elliotson, in addition to the *DNB*, see the biography in J. H. Harley Williams, *Doctors Differ: Five Studies in Contrast* (Springfield, Illinois: Charles C. Thomas, 1952), pp. 13–80; Kaplan, " 'Mesmeric Mania,' " pp. 697–701; Boring, *History of Experimental Psychology*, pp. 119–23; George Rosen, "John Elliotson, Physician and Hypnotist," *Bulletin of the Institute of the History of Medicine* 4 (July 1936):600–3; Jacques M. Quen, "Case Studies in Nineteenth Century Scientific Rejection: Mesmerism, Perkinism, and Acupuncture," *Journal of the History of the Behavioral Sciences* 11 (April 1975):149–56, and "Mesmerism, Medicine, and Professional Prejudice," *New York State Journal of Medicine* 76 (December 1976):2219–20.
20. See, for example, Elliotson's pamphlet, *Numerous Cases of Surgical Operations without Pain in the Mesmeric State; with Remarks upon the Opposition of Many Members of the Royal Medical and Chirurgical Society and Others to the Reception of the Inestimable Blessings of Mesmerism* (London: H. Baillière, 1843). The *Zoist* was filled with accounts of medical cases – of palsy, St. Vitus dance, deafness, dumbness, epilepsy, and delirium, for example – all cured by mesmerism. It ceased publication in 1856. Dupotet's name is sometimes spelled du Potet or Du Potet.
21. Terry M. Parssinen, "Professional Deviants and the History of Medicine: Medical Mesmerists in Victorian Britain," in Roy Wallis, ed., *On the Margins of Science: The Social Construction of Rejected Knowledge*, Sociological Review Monograph, no. 27 (Keele, Staffordshire: University of Keele, 1979), pp. 109, 115; Quen, "Mesmerism, Medicine, and Professional Prejudice," p. 2221.
22. R. K. Webb, *Harriet Martineau, A Victorian Radical* (London: Heinemann, 1960), pp. 19–22, draws together the slight available information

about Atkinson in a sketch that contemptuously dismisses all claims to intellectual distinction on Atkinson's part.

23. See the review of *Mesmerism the Gift of God* in the *Zoist* 1 (July 1843):217.

24. Darnton, *Mesmerism and the End of the Enlightenment in France*, pp. 52, 64.

25. James Braid, *Neurypnology, or the Rationale of Nervous Sleep, Considered in Relation with Animal Magnetism* (London: John Churchill, 1843), reprinted in Arthur Edward Waite, ed., *Braid on Hypnotism: The Beginnings of Modern Hypnosis* (New York: The Julian Press, 1960), pp. 101–2. Also see Boring, *History of Experimental Psychology*, pp. 124–8; Quen, "Mesmerism, Medicine, and Professional Prejudice," p. 2220, and Franklin Fearing, *Reflex Action: A Study in the History of Physiological Psychology*, new ed. (Cambridge, Mass,: M.I.T. Press, 1970), p. 215.

26. James Braid, *Magic, Witchcraft, Animal Magnetism, Hypnotism, and Electro-Biology; Being a Digest of the Latest Views of the Author on these Subjects*, 3d ed. enl. (London: John Churchill, 1852), p. 10. Braid had already published a pamphlet entitled *The Power of the Mind over the Body: an Experimental Inquiry into the Nature and Cause of the Phenomena Attributed by Baron Reichenbach and Others to a "New Imponderable"* (London: Churchill, 1846). Also see J. H. Conn's Foreword to *Braid on Hypnotism*, pp. v–vii. The fullest discussion of Braid's work appears in J. Milne Bramwell, "James Braid: His Work and Writings," *PSPR* 12 (June 1896):127–66. Bramwell points out that, when Braid in time realized that the essential phenomena of hypnotism could be induced while the subject remained alert, he eventually rejected the term "hypnotism" as misleading and used "monoideism" instead.

27. He had been preceded in France by the work of the Abbé Faria and Alexandre Bertrand. Ellenberger, *Discovery of the Unconscious*, pp. 75–6.

28. Even its efficacy as a form of anaesthesia was undermined by the introduction of ether and chloroform in the late 1840s.

29. On Atkinson, see *Zoist* 1:134, 142–3, 224, 247–8, 263, 277; William Gregory, *Letters to a Candid Inquirer on Animal Magnetism* (London: Taylor, Walton, & Maberly, 1851), ch. 11; John Forbes, *Mesmerism True – Mesmerism False: A Critical Examination of the Facts, Claims, and Pretensions of Animal Magnetism* (London: John Churchill, 1845), pp. 59–61. Forbes's work is no. 7 in *Houdini Pamphlets: Hypnotism*, vol. 4, Houdini Collection. His phrenological interests are mentioned in Charles Gibbon, *The Life of George Combe*, 2 vols. (London: Macmillan, 1878), 2:135.

30. "What is Mesmerism?" *Blackwood's* 70 (July 1851):74. *The Wellesley Index to Victorian Periodicals*, vol. 1, attributes the article to John Eagles, artist and writer.

31. Terry M. Parssinen, "Mesmeric Performers," *Victorian Studies* 21 (Autumn 1977):87–104.

32. *DNB*, s.v. "Hall, Spencer Timothy." The *Zoist*, understandably, registered some uneasiness over the activities of sympathizers such as S. T. Hall. "From all we have heard of Mr. Hall, and if we may judge from the style of his lectures, we believe him to be a gentleman influenced by

good motives; but not having enjoyed the advantage of a scientific education, he is evidently inclined to follow the promptings of an imaginative brain, rather than the calm, persevering, philosophical course essential to the cultivation of inductive science." "The Lecture Mania," *Zoist* 1 (April 1843):99.

33. "Mr. Spencer T. Hall's Phreno-Magnetic Lectures," *Phreno-Magnet* 1 (February 1843):22–3.

34. *Phreno-Magnet* 1 (July 1843):161. For descriptions of some of Hall's other demonstrations see (March 1843):46–51, (April 1843):68–72, (December 1843):334–42.

35. *Phreno-Magnet* 1 (November 1843):294, 298, and (June 1843):129–30. See, also, Parssinen, "Popular Science and Society," p. 13.

36. "Mesmerism," *British Spiritual Telegraph* 4 (June 1859):12. Scholars today likewise underscore the external similarities between mesmerism and spiritualism. See, for example, Anthony A. Walsh, "A Note on the Origin of 'Modern' Spiritualism," *Journal of the History of Medicine and Allied Sciences* 28 (April 1973):168.

37. Harley Williams, *Doctors Differ*, pp. 40, 57–8. Okey also claimed to have a familiar, whom she called her Negro, who frequently appeared when she was mesmerized and who bore the same relationship to her as spiritualist familiars subsequently bore to their mediums. Ibid., pp. 38, 56.

38. Charles Maurice Davies, *The Great Secret, and its Unfoldment in Occultism. A Record of Forty Years' Experience in the Modern Mystery*, 3d ed. (Blackpool: The Ellis Family, n.d.), pp. 75–6.

39. Mrs. De Morgan, quoted in A. M. W. Stirling, *William De Morgan and His Wife* (London: Thornton Butterworth, 1922), p. 108. William was the son of Augustus and Sophia De Morgan.

40. James Victor Wilson, *How to Magnetize, or Magnetism and Clairvoyance* (London: L. N. Fowler, 1878). The manual appears as no. 11 in *Houdini Pamphlets: Hypnotism*, vol. 4, Houdini Collection. Also see Frank Podmore, *Modern Spiritualism: A History and a Criticism*, 2 vols. (London: Methuen, 1902), 1:112, 123–4, 141–153, and Slater Brown, *The Heyday of Spiritualism* (New York: Hawthorn Books, 1970), pp. 25–42.

41. William Stainton Moses ["M. A., (Oxon.)"], *Psychography: A Treatise on One of the Objective Forms of Psychic or Spiritual Phenomena* (London: W. H. Harrison, 1878), p. 135.

42. "Objects of the Society," *PSPR* 1 (1882–3):3. Newton Crosland, *Apparitions; A New Theory. And Hartsore Hall, A Ghostly Adventure*, 2d ed. rev. (London: Effingham Wilson, & Bosworth & Harrison, 1856), p. 13, observed: "The Spirits appear to work their marvels by using the vital magnetic fluid of the medium" (no. 5 in *Houdini Pamphlets: Spiritualism*, vol. 13, Houdini Collection).

Baron von Reichenbach was a German metallurgist and chemist whose brainchild was the concept of odylic, or odic, force. According to him, this force was a kind of effluence from virtually every material substance in the universe. On sensitive people, it produced sensations – feelings of heat or cold, well-being, or unpleasantness – and could equally dazzle the percipient with visions of light and color radiating from objects. William Gregory was largely responsible for introducing Reichenbach's theories into Britain in the 1840s and 1850s. Gregory was convinced that the mesmeric force and Reichenbach's odyle were the same imponderable fluid, and, indeed, while Reichenbach may have added a few flourishes to Mesmer's ideas, the Baron's fundamental debt to the original propounder remained evident.

The notion of some sort of nervous fluid at work transmitting impulses and sensations was, in the eighteenth century, a reputable hypothesis concerning the nervous system. It lingered on in the nineteenth, until neurological studies obviated the need for such a fluid.

43. Podmore, *Modern Spiritualism*, 1:131; Elizabeth Longford, *Victoria R. I.* (London: Weidenfeld & Nicolson, 1964), p. 339; "Conversazione at Manchester," *Times*, 13 June 1853, p. 8. Parssinen, "Mesmeric Performers," p. 100, discusses public confusion over the distinction between galvanism and mesmerism.

44. Darnton, *Mesmerism and the End of the Enlightenment in France*, pp. 8, 58; Brown, *Heyday of Spiritualism*, pp. 9–11.

45. Emma Hardinge Britten, *Nineteenth Century Miracles; or, Spirits and Their Work in Every Country of the Earth* (New York: William Britten, 1884), p. 125. For her recollections of childhood somnambulism, see Emma Hardinge Britten, *Autobiography of Emma Hardinge Britten*, ed. Margaret Wilkinson (London: John Heywood, 1900), p. 6.

46. Podmore, *Modern Spiritualism*, 2:5, 20–1, 68. Hardinge's sickly subject may have been Emma herself. Also see p. 24, where Podmore describes a mesmerist named Carpenter whose weekly Sunday gatherings at Greenwich featured discourses by a trance medium.

47. Charles Maurice Davies, *Mystic London: or, Phases of Occult Life in the Metropolis* (London: Tinsley Brothers, 1875), pp. 260–8; "Services for Sunday," *Two Worlds* 1 (18 November 1887):13.

48. Advertisement quoted in Parssinen, "Mesmeric Performers," p. 103.

49. Davies, *Mystic London*, p. 378, and *Great Secret*, p. 79; "Spiritual Manifestations," *Blackwood's Edinburgh Magazine* 73 (May 1853):632. The *Wellesley Index to Victorian Periodicals*, vol. 1, attributes the article to William Edmonstoune Aytoun, poet and rhetorician.

50. Valerie E. Chancellor, ed., *Master and Artisan in Victorian England. The Diary of William Andrews and the Autobiography of Joseph Gutteridge* (New York: Augustus M. Kelley, 1969), p. 168; E. D. Rogers, *Life and Experiences of Edmund Dawson Rogers, Spiritualist and Journalist* (London: Office of *Light*, [1911]), pp. 12–13, 19; William Henry Harrison, *Spirit People* (London: W. H. Harrison, 1875), p. 10; Anna Mary Howitt Watts, *The Pioneers of the Spiritual Reformation* (London: Psychological Press Association, 1883), pp. 217–21. Also see Carl Ray Woodring, *Victorian Samplers: William and Mary Howitt* (Lawrence, Kansas: University of Kansas Press, 1952), pp. 95–7.

51. Wallace is discussed more fully in the next chapter. On Mrs. De Morgan, see Stirling, *William De Morgan and His Wife*, pp. 31, 108–9; on Wyld, George Wyld, *Notes of My Life* (London: Kegan Paul, Trench, Trübner, 1903), pp. 30–1, 58–63. Wyld gives 1854 as the year of his first meeting with Home in London. He must have meant 1855. Also see "Dr. George Wyld. Some More Opinions on the Study of Borderland," *Borderland* 1 (October 1893):109.

52. Katherine H. Porter, *Through a Glass Darkly: Spiritualism in the Browning Circle* (1958; reprint ed., New York: Octagon Books, 1972), discusses Bulwer-Lytton's interest in somnambulism, clairvoyance, and mesmer-

ism in the 1840s, pp. 9–10; Elizabeth Barrett Browning's attraction to mesmerism, pp. 34–40; the shared enthusiasm for mesmerism of Mrs. Frances Trollope and her son, Thomas Adolphus, pp. 105–10; and Alfred Tennyson's experiments with mesmerism in the early 1850s, pp. 121–2. Also see Fred Kaplan, *Dickens and Mesmerism: The Hidden Springs of Fiction* (Princeton, N.J.: Princeton University Press, 1975), p. 26. There were, of course, mesmerists who altogether repudiated the spiritualist connection, such as Dickens and Harriet Martineau.

53. John Ashburner, *On the Connection Between Mesmerism and Spiritualism, with Considerations on their Relations to Natural and Revealed Religion and to the Welfare of Mankind* (London: W. Horsell, 1859), pp. 61–2, 65, 70, 84. The six lectures that constitute this work were published as a supplement to the *British Spiritual Telegraph*, vol. 3. William Howitt, *The History of the Supernatural in all Ages and Nations, and in all Churches, Christian and Pagan: Demonstrating a Universal Faith*, 2 vols. (London: Longman, Green, Longman, Roberts, & Green, 1863), 1:15.

54. John Ashburner, *Notes and Studies in the Philosophy of Animal Magnetism and Spiritualism* (London: H. Baillière, 1867), p. 151; D. D. Home, *Incidents in My Life*, vol. 2 (New York: Carleton, 1872), pp. 61–6, describes his meeting with Elliotson and the doctor's conversion to spiritualism. Porter, *Through a Glass Darkly*, p. 26, discusses Dickens's concern for Elliotson's mental health in 1862.

55. Mesmer knew much about Father Gassner, whose faith healing provoked controversies during the 1770s in Austria, Germany, and Switzerland. Darnton, *Mesmerism and the End of the Enlightenment in France*, pp. 48, 71; Ellenberger, *Discovery of the Unconscious*, pp. 53–60; Keith Thomas, *Religion and the Decline of Magic* (New York: Charles Scribner's Sons, 1971), pp. 202–4, on Greatrakes.

56. Davies, *Great Secret*, pp. 75–6; "Spiritualism, as related to Religion and Science," *Fraser's Magazine* 71 (January 1865):30–1 – attributed to Abraham Hayward, barrister and essayist, in *The Wellesley Index to Victorian Periodicals*, vol. 2. On Young, see "Facts," *Daybreak* 1 (October 1868):6–7; on Newton, C. Maurice Davies, *Unorthodox London: or, Phases of Religious Life in the Metropolis*, 3d ed. (1875; reprint ed., New York: Augustus M. Kelley, 1969), pp. 310–13.

57. See Jerome D. Frank, *Persuasion and Healing. A Comparative Study of Psychotherapy* (Baltimore: Johns Hopkins Press, 1961), p. 62; Thomas, *Religion and the Decline of Magic*, pp. 208–11.

58. See Parssinen, "Professional Deviants," pp. 109–10.

59. After Mesmer lost control of the movement that bore his name, in France, at least, it fell into the hands of wilder visionaries who joined it with such beliefs as metempsychosis and spirit hierarchies in what was definitely an antimaterialist program. See Darnton, *Mesmerism and the End of the Enlightenment in France*, pp. 138, 140–1, 155, 157.

60. Parssinen, "Mesmeric Performers," pp. 97–8; Kaplan, *Dickens and Mesmerism*, p. 26; Rogers, *Life and Experiences*, p. 12; "Phreno-Magnetic Facts and Deductions," *Phreno-Magnet* 1 (May 1843):107; "Letter from Mr. J. Inwards," *Phreno-Magnet* 1 (November 1843):303; "Report of a Case Before the Northampton Phreno-Mesmeric Society," *Phreno-Magnet* 1 (December 1843):329.

61. Mechanics' Institutes did not serve an exclusively working-class audience and indeed were often virtually taken over by the middle-class members of the community. Nonetheless, lectures at Mechanics' Institutes remained the major route by which phrenology reached a working-class audience. See Parssinen, "Popular Science and Society," pp. 1–2, 9. Christopher Hibbert, *The Royal Victorians: King Edward VII, His Family and Friends* (Philadelphia: J. B. Lippincott, 1976), p. 10, describes Combe's examination of the future Edward VII.
62. de Giustino, *Conquest of Mind*, p. 114.
63. While spiritualists argued that such improvement would occur beyond the veil of death, phrenologists looked forward to reaping the rewards on earth; over several centuries, as successive generations exercised their moral faculties ever more vigorously, the results would eventually be evident in the "noble heads" of subsequent offspring (ibid., pp. 137–9). Also see Angus McLaren, "Phrenology: Medium and Message," *Journal of Modern History* 46 (March 1974):89, 95.
64. George Combe, *The Constitution of Man Considered in Relation to External Objects* (Boston: Carter & Hendee, 1829), pp. 17, 28–9. The book was first published in Edinburgh, in 1828. The article on George Combe in the *DNB* gives some indication of its sales record. It is true that one element in Combe's philosophy seemed to contradict the rest: The determinist theme running through his work apparently argued for man's irresponsibility in the face of his moral inadequacies. Man, Combe seemed to be saying at times, was simply what his mental organs made him. Although he could try to strengthen his morally valuable mental organs, and seek to weaken the morally harmful ones, nature placed inherent limits on what each individual could achieve in terms of self-improvement. Although this latent determinism in Combe's phrenological theory attracted the attention of philosophers and theologians, it did not receive much emphasis in the popular treatments of phrenology, where the optimistic note prevailed and where responsibility for personal conduct fell squarely on the individual, who possessed feelings to balance and faculties to cultivate. See Grant, "Combe on Phrenology and Free Will," and Parssinen, "Popular Science and Society," p. 5.
65. On phrenology, see, for example, Cooter, "Phrenology and British Alienists," pt. 1, pp. 10–11; on mesmerism, Webb, *Harriet Martineau*, pp. 239–40.
66. See Kaplan, "'Mesmeric Mania,'" pp. 692–3, 700. Darnton, throughout *Mesmerism and the End of the Enlightenment in France*, describes numerous radical social and political programs that grew out of mesmerism, from Mesmer's own day down to the mid-nineteenth century.
67. Susan Budd, *Varieties of Unbelief. Atheists and Agnostics in English Society 1850–1960* (London: Heinemann Educational Books, 1977), p. 24.
68. A. Cameron Grant, "New Light on an Old View," *Journal of the History of Ideas* 29 (April–June 1968):293–301, explores Combe's ambivalent views about Owen. Also see the article "Phrenological Analysis of Mr. Owen's New Views of Society," in the *Phrenological Journal* 1 (1823–4):218–37, which Grant attributes to Combe. There were further thoughts on the subject from Combe in "Phrenology and Mr. Owen," in the same volume, pp. 463–6.

69. Phrenologists participated in a wide range of social crusades, including temperance reform and the abolition of slavery. See de Giustino, *Conquest of Mind*, pp. 136, 140–5, and McLaren, "Phrenology: Medium and Message," p. 88. J. F. C. Harrison, *Robert Owen and the Owenites in Britain and America: The Quest for the New Moral World* (London: Routledge & Kegan Paul, 1969), pp. 87, 239–40, discusses the attraction of Owenites to phrenology, despite some fundamental disagreements: Owenites believed human character to be largely the result of social environment, while phrenologists looked to the size of mental organs.

70. For S. T. Hall's treatment of Harriet Martineau, see his *Mesmeric Experiences* (London: H. Baillière and J. Ollivier, 1845), pp. 63–75. He mentions his mesmeric experiments at the Howitts' home on pp. 30, 59–60. Also see the *DNB* article on Hall.

71. Quoted in Davies, *Mystic London*, p. 262. Phrenology, too, remained an integral part of Burns's enthusiasms. In 1874, for example, on a trip to Halifax, he offered "a phrenological entertainment on Saturday evening," and "two discourses on Spiritualism, on Sunday." "Mr. Burns's Visits," *Medium and Daybreak* 5 (13 November 1874):725. Also see Logie Barrow, "Socialism in Eternity: The Ideology of Plebeian Spiritualists, 1853–1913," *History Workshop: A Journal of Socialist Historians* 9 (Spring 1980):55, 58–9.

72. See Parssinen, "Professional Deviants," p. 113, and Cooter, "Phrenology and British Alienists," pt. 1, p. 10.

73. Edward Theodore Withington, *Medical History from the Earliest Times. A Popular History of the Healing Art* (London: Scientific Press, 1894), p. 342. See John A. Lee's interesting essay, "Social Change and Marginal Therapeutic Systems," in Roy Wallis and Peter Morley, eds., *Marginal Medicine* (New York: Free Press, 1976), pp. 23–41.

74. "Medical Books," *Athenaeum*, 29 March 1856, p. 394. The book under review was William Neilson, *Mesmerism in its Relation to Health and Disease and the Present State of Medicine* (Edinburgh: Shepherd & Elliot, 1855). On the od-force, see note 42 above.

75. Parssinen, "Professional Deviants," p. 109.

76. Lee, "Social Change and Marginal Therapeutic Systems," pp. 25–6. Also see Brian Inglis, *Fringe Medicine* (London: Faber and Faber, 1964), pp. 74–81. Well into the nineteenth century, allopathic doctors regularly prescribed large doses of drugs for patients, usually without full realization of their impact on the human body. It was partly Elliotson's realization, by the late 1830s, that he had reached the limits of what chemical drugs could perform that sparked his interest in animal magnetism. Harley Williams, *Doctors Differ*, pp. 26–8.

77. *DNB*, s.v. "Quin, Frederic Hervey Foster."

78. Needless to say, apothecaries and chemists likewise anathematized homoeopathy.

79. Bruce Haley, *The Healthy Body and Victorian Culture* (Cambridge, Mass.: Harvard University Press, 1978), pp. 16–17. The excellent discussion of hydropathy continues on pp. 33–5.

80. Wyld, *Notes of My Life*, pp. 33–5. The title of his pamphlet was *Homoeopathy: An Attempt to State the Question with Fairness* (London: J. Walker, 1853). In 1860, he published *Diseases of the Heart and Lungs, Their Physical Diagnosis, and Homoeopathic and Hygienic Treatment*.

81. Wyld, *Notes of My Life*, pp. 34–6.
82. In the printed list of names in the beginning of the Minute Books of the BNAS, Committees, no. 2, George Wyld is listed as a member of the General Purposes Committee for 1878. Also see Britten, *Nineteenth Century Miracles*, p. 184.
83. "Evidence in Defence of Dr. Slade," and "Spiritualists' Defence Fund," *Spiritualist Newspaper* 9 (3 November 1876):163–4; Wyld, *Notes of My Life*, pp. 66–9.
84. Wyld, *Notes of My Life*, p. 91, and "Dr. George Wyld. Some More Opinions on the Study of Borderland," p. 110.
85. Wyld, *Notes of My Life*, pp. 100–15, 47–55. He did, however, think that the House of Lords had to be made more efficient.
86. Wilkinson's letter of 12 February 1869 to the Committee of the London Dialectical Society, in the *Report on Spiritualism, of the Committee of the London Dialectical Society* (London: Longmans, Green, Reader, & Dyer, 1871), p. 234.
87. C. J. Wilkinson, *J. J. G. Wilkinson*, pp. 1, 15, 17. The opening sentence about J. J. G. Wilkinson (1812–99) in the *DNB* describes him simply as "Swedenborgian."
88. C. J. Wilkinson, *J. J. G. Wilkinson*, pp. 16, 32–3, 40.
89. In fact, Wilkinson embraced Braid's reinterpretation of mesmerism over the older fluid theory that teemed with occult possibilities. Ibid., p. 74.
90. Ibid., pp. 17, 263–4.
91. Ibid., pp. 112–13, 279. He contributed two pamphlets to the effort to repeal the Contagious Diseases Acts: "The Forcible Introspection of Women for the Army and Navy by the Oligarchy considered physically," and "A Free State and Free Medicine." For recent discussions of the Contagious Diseases Acts, see Paul McHugh, *Prostitution and Victorian Social Reform* (London: Croom Helm, 1980) and Judith R. Walkowitz, *Prostitution and Victorian Society: Women, Class, and the State* (Cambridge: Cambridge University Press, 1980).
92. C. J. Wilkinson, *J. J. G. Wilkinson*, pp. 265, 267–8. He was also coeditor of the series of sixteen pamphlets, *Vaccination Tracts*, published in London from 1877 to 1880. The titles of some of the tracts suggest the nature of their contents: "The Vaccination Laws a Scandal to Public Honesty and Religion," "Vaccination a Sign of the Decay of the Political and Medical Conscience in the Country," and "Compulsory Vaccination a Desecration of Law, a Breaker of Homes and Persecutor of the Poor." His resentment of the tyranny exercised by cruel experts at the expense of helpless sufferers also prompted him to oppose vivisection. Ibid., pp. 121–2, 275.

    On resistance to compulsory vaccination, see R. J. Lambert, "A Victorian National Health Service: State Vaccination 1855–1871," *Historical Journal* 5 (1962):1–18, and Roy M. MacLeod, "The Frustration of State Medicine, 1880–1899," *Medical History* 11 (1967):15–40.
93. J. J. G. Wilkinson, *War, Cholera, and the Ministry of Health. An Appeal to Sir Benjamin Hall and the British People* (London: R. Theobald, 1854). See C. J. Wilkinson, *J. J. G. Wilkinson*, pp. 41, 79, 93, 246–9. Unlike Wyld, Wilkinson did not perceive vaccination in the guise of homeopathic therapy. After all, vaccination involves the forceful introduction of actual disease into the body – something that can be viewed as opposed

to the homeopathic emphasis on harmony and balance within a healthy organism.

94. In *English Traits*, 1856. Quoted in W. G. H. Armytage, *Heavens Below: Utopian Experiments in England 1560–1960* (Toronto: University of Toronto Press, 1961), p. 195. C. J. Wilkinson, *J. J. G. Wilkinson*, pp. 251–2, discusses Wilkinson's homeopathic unorthodoxy. His deviation from Hahnemann was, in part, influenced by the Swedenborgian doctrine of correspondences, for with herbal medicines he sought "to see in the growth and character of a plant the very double of the effects which it produces upon the animal body."

95. C. J. Wilkinson, *J. J. G. Wilkinson*, pp. 78, 80–4, 275; Wilkinson's letter of 12 February 1869 to the Committee of the London Dialectical Society, in its *Report*, p. 235.

96. The *Spiritual Herald*, which only appeared for several months in 1856, was published in London by a group of Swedenborgian spiritualists. See Podmore, *Modern Spiritualism*, 2:23. Thomas Shorter mentions Garth Wilkinson's contributions to spiritualist literature in *The Two Worlds, The Natural and the Spiritual: Their Intimate Connexion and Relation Illustrated by Examples and Testimonies, Ancient and Modern* (London: F. Pitman, [1864]), pp. 211, 339, 342–3.

97. C. J. Wilkinson, *J. J. G. Wilkinson*, pp. 89–90, 95–100, 102.

98. See Barrow's discussion of working-class spiritualist views on medicine and the medical profession in "Socialism in Eternity," pp. 57–60. Early in the twentieth century, Barrow reports, non-Christian spiritualists participated in a "People's League of Medical Freedom."

99. C. J. Wilkinson, *J. J. G. Wilkinson*, pp. 54–64; Armytage, *Heavens Below*, p. 194; J. F. C. Harrison, *Owen and the Owenites*, p. 115. Wilkinson, Armytage, and Harrison all mention Hugh Doherty, Wilkinson's friend and companion on the 1848 trip to Paris. Doherty was another doctor who endorsed Fourier's blueprint for social reform in the late 1840s, was drawn to Swedenborgianism, and moved on to spiritualism. His article, "Bible Spiritualism," in the *Spiritual Magazine*, n.s. 7 (April 1872):151–4, was a plea for the "Bible Spiritualists," among whom he counted himself, to extend tolerance to the "anti-Bible Spiritualists."

100. Ashburner, *On the Connection Between Mesmerism and Spiritualism*, pp. 15, 39. Also see Ashburner, *Notes and Studies*, pp. 76–7.

101. (London: W. Horsell, [1859]). See Shorter, *Two Worlds*, p. 338; the London Dialectical Society's *Report*, pp. 175, 233, 243–5; Ashburner, *On the Connection Between Mesmerism and Spiritualism*, p. 61; Podmore, *Modern Spiritualism*, 2:24, 29–30, 44–5.

102. Earl of Dunraven, *Past Times and Pastimes*, 2 vols. (London: Hodder & Stoughton, 1922), 1:10. As Viscount Adare, Dunraven first met Home in 1867 when both were taking the hydropathic cure at Malvern under Gully's supervision. Gully also apparently subscribed to phrenology, as suggested in his article, "Hygiene," in *Water Cure Journal and Hygienic Magazine* 1 (November 1847):121–7.

103. Francis Darwin, ed., *The Life and Letters of Charles Darwin, including an Autobiographical Chapter*, 2 vols. (New York: D. Appleton, 1888), 1:341; Haley, *Healthy Body and Victorian Culture*, p. 34; also see *DNB*, s.v. "Gully, James Manby." Gully's reputation collapsed in 1876, four years after his professional retirement, when his name became involved

in a celebrated murder case, and he was rumored to be the lover of Florence Bravo, whose husband died of poisoning. The case is fully analyzed in Mary S. Hartman, *Victorian Murderesses. A True History of Thirteen Respectable French and English Women Accused of Unspeakable Crimes* (New York: Schocken Books, 1977), pp. 136–73.

104. William L. Courtney, "The New Psychology," *Fortnightly* n.s. 26 (September 1879):318–19. Also see Frank Miller Turner, *Between Science and Religion: The Reaction to Scientific Naturalism in Late Victorian England* (New Haven and London: Yale University Press, 1974), pp. 212–13, and Roger Smith's very useful survey, "The Background of Physiological Psychology in Natural Philosophy," *History of Science* 11 (June 1973):75–123.

105. In Cyril Burt's, at the University of Liverpool, between 1908 and 1913. L. S. Hearnshaw, *Cyril Burt, Psychologist* (Ithaca, N.Y.: Cornell University Press, 1979), p. 26. In *A Short History of British Psychology 1840–1940* (New York: Barnes & Noble, 1964), p. 180, Hearnshaw suggests that Burt's "adventurous courses" at Liverpool "included probably the first lectures on psychoanalysis in any university precinct in Great Britain."

106. This is the main contention, persuasively argued, in Daston, "British Responses to Psycho-Physiology." See, especially, pp. 197, 201, 206–7.

107. Boring, *History of Experimental Psychology*, pp. 460, 489–94.

108. Hearnshaw, *Short History of British Psychology*, pp. 182–4. As late as 1918, the membership of the British Psychological Society still did not exceed one hundred.

109. Samuel Hynes, *The Edwardian Turn of Mind* (Princeton, N.J.: Princeton University Press, 1968), p. 138. Also see Turner, *Between Science and Religion*, pp. 210–11. The ninth edition of the *Encyclopaedia Britannica* appeared in twenty-five volumes between 1875 and 1889. James Ward's essay, "Psychology," was included in vol. 20.

110. Daston, "British Responses to Psycho-Physiology," p. 200, and Ward quoted in Turner, *Between Science and Religion*, p. 215. See Ward's discussion concerning the problem of defining psychology's subject matter in "Psychology," *Encyclopaedia Britannica* (Edinburgh: Adam & Charles Black, 1875–89), vol. 20 (1886), p. 37.

111. J. Arthur Hill, comp., *Letters from Sir Oliver Lodge, Psychical, Religious, Scientific and Personal* (London: Cassell, 1932), pp. 38–40. Letter from Lodge to Hill, 7 November 1914.

112. J. C. Colquhoun, *Isis Revelata: An Inquiry into the Origin, Progress, and Present State of Animal Magnetism*, 2d ed., 2 vols. (London: Baldwin & Cradock, 1836), 1:79; Ashburner, *Notes and Studies*, p. 78.

113. T. P. Barkas, "Recent Investigations in Psychology," *Psychological Review* 1 (October 1878):215–42.

114. Anna Mary Howitt Watts, *Pioneers of the Spiritual Reformation*, pp. v, 238; John Beattie, "Psychological Phenomena. Manifestations Through Mr. Home at Clifton," *Human Nature* 4 (1870).

115. BNAS prospectus quoted in Charles Maurice Davies, *Heterodox London: or, Phases of Free Thought in the Metropolis*, 2 vols. (London: Tinsley Brothers, 1874), 2:19.

116. On the Ghost Club, see chapter 3, p. 77, above; "Dr. George Wyld. Some More Opinions on the Study of Borderland," p. 110.

117. The physician and psychologist James Crichton-Browne supported these assertions, after a fashion, in an essay he wrote as a young man. Commenting on mesmerism, physiognomy, and table turning, he explained that they had inadvertently given "an impetus to the study of psychology, by inducing many to think of the subject, and by impelling research into its extremest obscurities." "The History and Progress of Psychological Medicine," *Journal of Mental Science* 7 (April 1861):26.
118. W. K. Clifford, "Body and Mind," *Fortnightly*, n.s. 16 (December 1874):734. Daston, "British Responses to Psycho-Physiology," p. 201, points out, however, the ways in which Clifford's materialism was not so absolute as this quotation would suggest.
119. Courtney, "New Psychology," p. 327.
120. *Report of the Seventy-Sixth Meeting of the British Association for the Advancement of Science, York, August 1906* (London: John Murray, 1907), p. 25.
121. *DNB*, s.v. "Carpenter, William Benjamin," and Cooter, "Phrenology and British Alienists," pt. 2, pp. 136, 146. For Carpenter's characterization of mesmerism and spiritualism as "Epidemic Delusions," see his *Mesmerism, Spiritualism, &c., Historically and Scientifically Considered. Being Two Lectures Delivered at the London Institution* (New York: D. Appleton, 1877).
122. Carpenter to Barrett, 2 November 1876, Barrett Papers (box 2, A4, no. 12). Carpenter completed his medical education at Edinburgh, where he submitted a thesis entitled "The Physiological Inferences to be Deduced from the Structure of the Nervous System of Invertebrated Animals."
123. William B. Carpenter, *Principles of Mental Physiology, with their Applications to the Training and Discipline of the Mind, and the Study of its Morbid Conditions* (London: H. S. King, 1874), p. xii. The seventh edition came out in 1896. Among Carpenter's other writings that touch on free will, see "On the Doctrine of Human Automatism," first published in the *Contemporary Review* in 1875, and included in William B. Carpenter, *Nature and Man: Essays Scientific and Philosophical* (London: Kegan Paul, Trench, 1888). Also see the discussion of Carpenter's psychology in Daston, "British Responses to Psycho-Physiology," pp. 198, 202–6; Haley, *Healthy Body and Victorian Culture*, pp. 36–7, 40–1; and Alan Gauld, *The Founders of Psychical Research* (London: Routledge & Kegan Paul, 1968), pp. 58–9.
124. Carpenter, *Mesmerism, Spiritualism, &c.*, argued the point from start to finish. See pp. 4, 115, for examples. In his letter to Barrett, cited above, Carpenter claimed particular knowledge of "the extraordinary *self*-deception of those who go into the enquiry prepossessed with an idea."
125. "Spiritualism and its Recent Converts," *Quarterly Review* 131 (October 1871):310. The *Wellesley Index to Victorian Periodicals*, vol. 1, attributes the article to Carpenter. The glowing references to Carpenter's previous writings must have given readers a strong suspicion concerning authorship.
126. Address to the Royal Institution, in Carpenter, *Nature and Man*, p. 172. "Electro-Biology and Mesmerism," *Quarterly Review* 93 (September 1853):501–57. The *Times* discussed "what Dr. Carpenter called the ideomotor power" in "Table-Moving," 13 June 1853, p. 8. Carpenter reminded Barrett of "Ideo-motor action" in his letter of 2 November 1876.

127. Carpenter, "Spiritualism and its Recent Converts," p. 317. Also see Carpenter's letter to the London Dialectical Society, 24 December 1869, in its *Report*, p. 266.
128. See, for example, Carpenter's contribution to a discussion on thought reading that took place at the BAAS meeting at Glasgow in 1876, reported in the *Spiritualist Newspaper* 9 (22 September 1876):90–1.
129. Ibid., and letter to Barrett cited above.
130. *The Psycho-Physiological Sciences, and their Assailants.* Being a Response by Alfred R. Wallace, of England; Professor J. R. Buchanan, of New York; Darius Lyman, of Washington; Epes Sargent, of Boston; to the Attacks of Prof. W. B. Carpenter, of England, and Others (Boston: Colby & Rich, 1878), p. 7.
131. See, for example, Thomas Laycock, *A Treatise on the Nervous Diseases of Women: comprising an Inquiry into the Nature, Causes, and Treatment of Spinal and Hysterical Disorders* (London: Longman, Orme, Brown, Green, & Longmans, 1840), pp. 318–25.
132. Henry Maudsley, *The Physiology and Pathology of the Mind* (London: Macmillan, 1867), p. 116. Maudsley's extended treatise on this subject was *Natural Causes and Supernatural Seemings*, first published in 1886. Henry Holland, *Recollections of Past Life* (London: Longmans, Green, 1872), p. 324. Also see Holland's *Chapters on Mental Physiology* (1852). I am indebted to J. P. Williams for allowing me to read a draft version of his article "Psychical Research and Psychiatry in Late Victorian Britain: Trance as Ecstasy or Trance as Insanity," in William F. Bynum, Roy Porter, and Michael Shepherd, eds., *Essays in the History of Psychiatry*, 2 vols. (London: Tavistock Publications), forthcoming.
133. Blackwell to Sidgwick, 24 October 1889, Sidgwick Papers, Add. MS. c. 93. 38.
134. For example, see F. C. S. Schiller, "Psychology and Psychical Research," *PSPR* 14 (July 1899):348–65, in reply to an article, "Psychology and Mysticism," by Professor Hugo Münsterberg of Harvard, which had appeared in the *Atlantic Monthly*, January 1899. Schiller, the Oxford philosopher whose thought was deeply influenced by William James, shared James's interest in psychical research and was president of the SPR in 1914.
135. Henry Sidgwick and F. W. H. Myers, "The Second International Congress of Experimental Psychology," *PSPR* 8 (1892):601–9. Ethel Sidgwick, *Mrs. Henry Sidgwick, a Memoir by her Niece* (London: Sidgwick & Jackson, 1938), pp. 120–1; also see the account of "The Fourth International Congress of Psychology," in *PSPR* 15 (October 1900):445–8.
136. In "The Mechanism of Hysteria," *PSPR* 9 (1893–4):12–15. The essay is the sixth in a series by Myers, published in volumes 7–11 (1891–95) of the *PSPR*, under the title "The Subliminal Consciousness." The nine essays, or chapters, have been published as *The Subliminal Consciousness* (New York: Arno Press, 1976), with an introduction by James Webb. Not all members of the SPR favorably assessed Freud's work, however. Lodge wrote to J. A. Hill in May 1914: "As to Freud, I am not impressed with his Dream book, . . . I do not believe that any one hypothesis will explain all dreams; and he seems to me to press his hypothesis to death" (*Letters from Sir Oliver Lodge*, p. 36). T. W. Mitchell discusses Freud's connection with the SPR in "The Contributions of Psychical Research

to Psychotherapeutics," *PSPR* 45 (February 1939):183. William James and G. Stanley Hall were also active in the American Society for Psychical Research. Corresponding membership lists, 1882–1914, can be found at the end of successive volumes of the SPR *Proceedings*.

137. Membership and officer lists at back of *PSPR* volumes; Frederic Boase, *Modern English Biography*, 6 vols. (Truro: Netherton & Worth, for the author, 1892–1921), 3:204.

138. Courtney, "New Psychology," p. 320.

139. Moses, *Psychography*, p. 15.

140. Selborne's contribution to "A Modern 'Symposium,'" *Nineteenth Century* 1 (April 1877):337.

141. Carpenter, "On the Doctrine of Human Automatism," in *Nature and Man*, pp. 282–3.

142. Turner, *Between Science and Religion*, pp. 201–45, provides an excellent interpretation of Ward's intellectual development, including Sidgwick's influence on him.

143. C. Lloyd Tuckey, "Faith Healing as a Medical Treatment," *Nineteenth Century* 24 (December 1888):839. The following year, Tuckey published *Psycho-Therapeutics; or, Treatment by Sleep and Suggestion* (London: Baillière, Tindall, & Cox, 1889).

144. Frank Podmore, "Reviews," *PSPR* 21 (October 1907):151–4; J. Milne Bramwell, *Hypnotism: Its History, Practice and Theory* (London: Grant Richards, 1903), p. 437.

145. Eleanor Sidgwick, describing Gerald Balfour's views to W. F. Barrett, 18 July 1911, Barrett Papers (box 2, A4, no. 120).

146. Bramwell, *Hypnotism*, pp. 413–14, 438. Also see his article, "What is Hypnotism?" *PSPR* 12 (December 1896):252–8.

147. Walter Leaf's review of Bramwell's *Hypnotism*, in *PSPR* 18 (October 1904):486, 488.

148. Henry Sidgwick, "On Historical Psychology," *Nineteenth Century* 7 (February 1880):357; Leaf review of Bramwell's *Hypnotism*, p. 490; W. F. Barrett, Introduction to Beckles Willson, *Occultism and Common-Sense* (New York: R. F. Fenno, [1908?]), p. xiii.

149. Edmund Gurney and Frederic W. H. Myers, "Some Higher Aspects of Mesmerism," *National Review* 5 (July 1885):703. They published an almost identical article, under the same title, in *PSPR* 3 (1885):401–23. In "The Drift of Psychical Research," *National Review* 24 (October 1894):191, Myers described psychical research as "the left wing of Experimental Psychology."

150. Edmund Gurney, "Peculiarities of Certain Post-Hypnotic States," *PSPR* 4 (1886–7):323.

151. Edmund Gurney, *The Power of Sound* (New York: Basic Books, 1966), p. xx. The volume was first published in London, by Smith, Elder, in 1880.

152. Edmund Gurney, "The Stages of Hypnotism," *PSPR* 2 (1884):72. Also see p. 62, for Gurney's characterization of the alert and deep stages of hypnotic trance, and p. 67, for another rebuttal of the opinion that the actions of a hypnotized person "are purely reflex and unconscious."

153. Edmund Gurney, "The Problems of Hypnotism," *Mind* 9 (October 1884):481. Also see pp. 494, 496, 498, 500–3. The article also appeared in *PSPR* 2 (1884):265–92.

154. Edmund Gurney, "Recent Experiments in Hypnotism," *PSPR* 5 (1888–9): 3–5. See pp. 5–6 for Gurney's experimental data. He had used the concept of "intelligent automatism" earlier, in "Peculiarities of Certain Post-Hypnotic States," pp. 268–323.
155. Gurney, "Stages of Hypnotism," pp. 69–70.
156. Gurney, "Peculiarities of Certain Post-Hypnotic States," pp. 292–3, 323.
157. Edmund Gurney, *Tertium Quid: Chapters on Various Disputed Questions*, 2 vols. (London: Kegan Paul, Trench, 1887), 1:319n, 358–9.
158. Ibid., pp. 362–4, and Edmund Gurney, "Hypnotism and Telepathy," *PSPR* 5 (1888–9):258–9.
159. Gurney, "Recent Experiments in Hypnotism," pp. 14–17, and "An Account of Some Experiments in Mesmerism," *PSPR* 2 (1884): 203–5. For a full listing of these finger experiment reports in the *PSPR*, see Gauld, *Founders of Psychical Research*, p. 359. Gurney acknowledged that purely physical or purely mental agencies might sometimes induce hypnosis, but at other times he felt sure that something else was involved.
160. Edmund Gurney, "Stages of Hypnotic Memory," *PSPR* 4 (1886–7): 530–1.
161. Frederick W. H. Myers, "Edmund Gurney," in *Fragments of Prose & Poetry*, ed. Eveleen Myers (London: Longmans, Green, 1904), pp. 66–7, 72–3, 75, and Mitchell, "Contributions of Psychical Research to Psychotherapeutics," p. 179.
162. For some of the criticism leveled against Gurney's methods and assumptions, see Bramwell, *Hypnotism*, pp. 409–13, and R. Laurence Moore, *In Search of White Crows: Spiritualism, Parapsychology, and American Culture* (New York: Oxford University Press, 1977), pp. 143–4. Two American critics, C. S. Peirce and G. Stanley Hall, both involved with the American Society for Psychical Research, were particularly acerbic in their comments.
163. Daniel N. Robinson, *An Intellectual History of Psychology*, rev. ed. (New York: Macmillan, 1981), p. 389.
164. Frederic W. H. Myers, "Human Personality in the Light of Hypnotic Suggestion," *PSPR* 4 (1886–7):6. For a full discussion of the ways in which Myers and Gurney deviated from the dominant medical interpretation of automatism, hypnotism, and hallucinations, see Williams, "Psychical Research and Psychiatry."
165. William James, *Memories and Studies* (London and New York: Longmans, Green, 1911), pp. 156, 158. James's obituary essay on Myers first appeared as "Frederic Myers's Service to Psychology," *PSPR* 17 (May 1901):13–23. Also see James's review of Myers's *Human Personality and its Survival of Bodily Death*, in *PSPR* 18 (June 1903):22; Flournoy's review of *Human Personality*, in *PSPR* 18 (June 1903):46, 51. Oliver Lodge, too, reviewing the same book in the same volume of SPR *Proceedings*, p. 36, commented on the "consistent comprehensive scheme" that Myers imposed on all the "disintegrations, abnormalities, and supernormalities of personality."
166. James, *Memories and Studies*, p. 170, and review of *Human Personality*, p. 30.
167. Ellenberger, *Discovery of the Unconscious*, p. 314.
168. J. Fraser Nicol, "Philosophers as Psychic Investigators," *Parapsychology Review* 8 (March–April 1977):10. Nicol points out that Myers was

also the first to bring Henri Bergson to the attention of psychical researchers. Also see Gardner Murphy, "The Life Work of Frederic W. H. Myers," *Tomorrow* 2 (Winter 1954):36.
169. Myers, "Human Personality in the Light of Hypnotic Suggestion," p. 6, and "French Experiments on Strata of Personality," *PSPR* 5 (1888–9):374–97.
170. James, *Memories and Studies*, pp. 170, 152.
171. Frank Podmore, *Studies in Psychical Research* (London: G. P. Putnam's Sons, 1897), p. 410. For the place of Piper and Thompson in Myers's thought, see Myers's article, "On the Trance-Phenomena of Mrs. Thompson," *PSPR* 17 (June 1902):67–74; Walter Leaf's review of *Human Personality*, in *PSPR* 18 (June 1903):57, where he points to the crucial importance of Mrs. Piper for Myers's whole theory; and Flournoy, in the same review article, p. 51, where both Piper and Thompson are mentioned in that significant capacity.
172. Myers, "Human Personality in the Light of Hypnotic Suggestion," p. 5, and "Multiplex Personality," *Nineteenth Century* 20 (November 1886):648. The latter article also appeared in *PSPR* 4 (1886–7):496–514. See, too, Leaf's review of *Human Personality*, p. 59.
173. F. W. H. Myers, "General Characteristics of Subliminal Messages," Chapter 1 of "The Subliminal Consciousness," *PSPR* 7 (1891–2):301, 305. Murphy, "Life Work of Myers," pp. 36–7, uses the iceberg simile.
174. F. W. H. Myers, "Automatic Writing. –IV– The Daemon of Socrates," *PSPR* 5 (1888–9):537, 539; and Myers's review of William James, *The Principles of Psychology* in *PSPR* 7 (1891–2):125. Gauld, *Founders of Psychical Research*, p. 287, finds that, according to Myers, the same, "or pretty much the same," stratum of the subliminal self appears in hysteria, dreams, hypnosis, and somnambulism. Myers himself is not explicit on this point.
175. Myers, "General Characteristics of Subliminal Messages," p. 306.
176. Myers, "Daemon of Socrates," p. 524. Also see Myers "General Characteristics of Subliminal Messages," pp. 312–13, 319. It would be significant to ascertain what, if any, influence the work of John Hughlings Jackson had on Myers at the time when he was formulating the theory of the subliminal self. Williams, "Psychical Research and Psychiatry," mentions a link between the two men in the person of Myers's younger brother, Arthur, a medical doctor and an epileptic. See E. Stengel, "Hughlings Jackson's Influence in Psychiatry," *British Journal of Psychiatry* 109 (1963):350, for Jackson's views on conscious and subconscious mental activities, which might have stimulated Myers's far less restrained hypotheses.
177. Myers to Lodge, 13 September and 16 October 1893, Lodge Collection (box 6, nos. 1372, 1374); Myers, "Trance-Phenomena of Mrs. Thompson," pp. 68, 73.
178. Myers, "Trance-Phenomena of Mrs. Thompson," p. 68. Also see Frederic W. H. Myers, "On Some Fresh Facts Indicating Man's Survival of Death," *National Review*, 32 (October 1898):233–4, and "Science and a Future Life," in *Science and a Future Life, with Other Essays* (London: Macmillan, 1893), pp. 35–6. James discussed Myers's theory of psychical invasion in his review of *Human Personality*, pp. 28–9, as did Leaf, in his review of the same book, pp. 56–8.

179. On Myers's confusing and loose use of "spirit," see Gauld, *Founders of Psychical Research*, p. 305.
180. Leaf's review of *Human Personality*, pp. 54–6, 59.
181. Gauld, *Founders of Psychical Research*, pp. 281, 295, discusses Myers's muddled handling of the concepts of "personality," "self," and "soul." On this problem, also see Andrew Lang, "'The Nineteenth Century' and Mr. Frederic Myers," *PSPR* 18 (June 1903):70–1.
182. Myers, "Human Personality in the Light of Hypnotic Suggestion," p. 20.
183. Frederic W. H. Myers, *Human Personality and Its Survival of Bodily Death*, 2 vols. (New York and London: Longmans, Green, 1903), 1:xxvi.
184. Webb, in his unpaginated Introduction to Myers, *Subliminal Consciousness*, suggests that Myers's attempts to convey his reconfirmed certainty of individual immortality in the pages of *Human Personality* were thwarted by Alice Johnson and Hodgson, who edited the manuscript for publication after Myers's death. Both had learned to treat Myers's enthusiasms cautiously, and there was some reason to suspect Mrs. Thompson of fraud.
185. On cosmic memory, see Myers's review of James's *Principles of Psychology*, p. 119. He spoke of "a transcendental Self," for example in "Automatic Writing.–III," *PSPR* 4 (1886–7):260. Gauld attempts a detailed explanation of "Myers' Cosmology and Theory of the Soul," in *Founders of Psychical Research*, pp. 300–12, and Turner discusses Myers's "panpsychism" in *Between Science and Religion*, p. 249.
186. Myers, "Provisional Sketch of a Religious Synthesis," in *Human Personality*, 2:291. See p. 153, above, for a discussion of the 1899 Synthetic Society paper.
187. Myers to Lodge, 14 February 1893, Lodge Collection (box 6, no. 1353).
188. Lang, "'Nineteenth Century' and Myers," p. 77; G. Balfour, in *Presidential Addresses to the Society for Psychical Research 1882–1911* (Glasgow: Robert Maclehose for the SPR, 1912), p. 267.
189. G. F. Stout, "Mr. F. W. H. Myers on 'Human Personality and Its Survival of Bodily Death,'" *Hibbert Journal* 2 (October 1903):45–56, 64. Gauld, *Founders of Psychical Research*, pp. 294–5, tries to defend Myers from Stout's critique, but repeatedly acknowledges the obscurity of Myers's ideas. Gauld also points out, p. 299, that Myers never successfully countered Janet's contention that secondary streams of consciousness are not normal, but pathological. The burden of proof lay on Myers, and, while empirical demonstration in large numbers of cases would, of course, have been exceedingly difficult, it nonetheless behooved Myers at least to address himself to the problem. He failed to do so.
190. W. McDougall, "Critical Notices: *Human Personality and Its Survival of Bodily Death*," *Mind*, n.s. 12 (October 1903):520, 524–6. McDougall doubted, however, that future generations would "remember the hypothesis of the 'subliminal self' as a part of [Myers's] achievement."
191. For McDougall's biography, see *DNB 1931–40*, s.v. "McDougall, William," and Major Greenwood and May Smith, "William McDougall 1871–1938," *Obituary Notices of Fellows of the Royal Society* 3 (January 1940):39–53. There is also a useful, brief summary of McDougall's life and career in Hearnshaw, *Short History of British Psychology*, pp. 185–95.

192. William McDougall, *Psychology, the Study of Behaviour* (London: Williams & Norgate, [1912]), pp. 19–20, 23, 104, 112.
193. William McDougall, *Body and Mind: A History and a Defense of Animism* (Boston: Beacon Press, 1961), pp. 349–53.
194. William McDougall, "In Memory of William James," *PSPR* 25 (March 1911): 12, 16, 19, 23.
195. Gauld, *Founders of Psychical Research*, p. 279, discusses James's influence on Myers's theory of the subliminal self.
196. McDougall, "In Memory of William James," pp. 21, 26.
197. Reba N. Soffer, *Ethics and Society in England: The Revolution in the Social Sciences 1870–1914* (Berkeley and Los Angeles: University of California Press, 1978), pp. 156, 161, discusses the old and new forms of determinism at the turn of the century.

7. The problem of evolution

1. Peter J. Bowler, "The Changing Meaning of 'Evolution,'" *Journal of the History of Ideas* 36 (January–March 1975):95, points out that the word "evolution" only assumed its current application to the theory of species transmutation some time after 1859. In earlier usage, the word applied to "the embryological development of a single individual, rather than the overall development of life on the earth." In this chapter, nonetheless, "evolution" designates theories of the general emergence of terrestrial life, including the arguments for transmutation of species.
2. As in *The Expression of the Emotions in Man and Animals* (London: John Murray, 1872).
3. Bowler, "Changing Meaning of 'Evolution,'" pp. 106–9, argues that the growing popularity of Spencer's work, from 1870 on, had much to do with the progressive connotation that the theory of evolution carried in the latter decades of the century.
4. *Report of the Sixty-Eighth Meeting of the British Association for the Advancement of Science, Held at Bristol in September 1898* (London: John Murray, 1899), p. 32.
5. In his presidential address to the SPR, January 1894, Balfour hazarded the "unverifiable guess . . . that in these cases of the individuals thus abnormally endowed, we really have come across faculties which, had it been worth Nature's while, had they been of any value or purpose in the struggle for existence, might have been normally developed, and thus become the common possession of the whole human race." In *Presidential Addresses to the Society for Psychical Research 1882–1911* (Glasgow: Robert Maclehose for the SPR, 1912), p. 70. Frank Podmore, "Subliminal Self or Unconscious Cerebration?" *PSPR* 11 (1895):332.
6. J. Milne Bramwell, *Hypnotism: Its History, Practice and Theory* (London: Grant Richards, 1903), pp. 359, 362–3, 415–16, explains and criticizes the way in which Myers applied evolution to his theory of the subliminal self.
7. Frederic W. H. Myers, *Fragments of Inner Life, an Autobiographical Sketch* (London: SPR, 1961), p. 37. This pamphlet was first printed, privately, in 1893.
8. Frederic W. H. Myers, "The Drift of Psychical Research," *National Review* 24 (October 1894):206–8. Also see his "Provisional Sketch of a

Religious Synthesis," in *Human Personality and Its Survival of Bodily Death*, 2 vols. (New York and London: Longmans, Green, 1903), 2:290.

9. Gerald Massey, *Concerning Spiritualism* (London: James Burns, [1871]), p. 55; prospectus of the BNAS, quoted in Charles Maurice Davies, *Heterodox London: or, Phases of Free Thought in the Metropolis*, 2 vols. (London: Tinsley Brothers, 1874), 2:19; J. P. Hopps, "Faithful Unto Death," *Light* 30 (15 October 1910):495.

10. J. P. Hopps, "The Ideal Holy Ghost," p. 7, address to the LSA, 2 February 1888, which appears as no. 12 in *Addresses Delivered Before the London Spiritualist Alliance during the Years 1884 to 1888* (London: LSA, 1889), Houdini Collection; John Page Hopps, *The Future Life* (London: Simpkin, Marshall, [1884]), pp. 12, 21–2.

11. Henry Maudsley, "Materialism and its Lessons," *Fortnightly*, n.s. 26 (August 1879):253; *Report of the Seventy-Sixth Meeting of the British Association for the Advancement of Science, York, August 1906* (London: John Murray, 1907), p. 26. Also see C. Lloyd Morgan's review of *Mind in Evolution*, by L. T. Hobhouse, in *Mind*, n.s. 12 (January 1903):103–9. For a concise discussion of the Great Chain of Being, see Stephen Toulmin and June Goodfield, *The Discovery of Time* (Harmondsworth: Penguin Books, 1967), pp. 116–22.

12. William McDougall, *Psychology, the Study of Behaviour* (London: Williams & Norgate, [1912]), p. 145; on Myers, Bramwell, *Hypnotism*, p. 363.

13. Henry Sidgwick, "On Historical Psychology," *Nineteenth Century* 7 (February 1880):357. The question of emergence in nature particularly intrigued C. Lloyd Morgan after 1910. For a brief summary of his arguments, see L. S. Hearnshaw, *A Short History of British Psychology 1840–1940* (New York: Barnes & Noble, 1964), pp. 99–100.

14. *DNB*, s.v. "Chambers, Robert"; William Chambers, *Memoir of Robert Chambers with Autobiographic Reminiscences of William Chambers* (New York: Scribner, Armstrong, 1872); Gavin de Beer, "Introduction," to Robert Chambers, *Vestiges of the Natural History of Creation* (1844; reprint ed., Leicester: Leicester University Press, 1969), pp. 31–2. The latter volume is a reprint of the first edition of 1844. There were frequent, and substantial, changes in the text of subsequent editions.

15. See, for example, Chambers, *Vestiges*, pp. 331–4. Milton Millhauser, *Just Before Darwin: Robert Chambers and Vestiges* (Middletown, Conn.: Wesleyan University Press, 1959), pp. 107–11, discusses the reluctance of the Victorian public to accept man's inextricable place in the animal kingdom, and Chambers's challenge to that reluctance. Chambers was by no means the first to deny fixity of species. That debate had been flourishing for some decades before he took up cudgels against the Linnean view of rigidly distinct and unchanging species.

16. "The Spirits Come to Town," *Chambers's Edinburgh Journal*, n.s. 19 (21 May 1853):321–4. Although the article was signed "A. R.," Frank Podmore, in *Modern Spiritualism: A History and a Criticism*, 2 vols. (London: Methuen, 1902), 2:5, asserts that the author was "understood to be Robert Chambers himself," and Millhauser, *Just Before Darwin*, pp. 209–10n, follows Podmore in thus interpreting both this article and one published subsequently, in the same journal, on 11 June 1853. George Wyld, *Notes of My Life* (London: Kegan Paul, Trench, Trübner, 1903),

p. 59, reports that Robert Chambers was interested in mesmerism during the 1830s, while the language and ideas of many passages in *Vestiges* reveal Chambers's debt to phrenology.
17. "Spirits Come to Town," p. 324; Millhauser, *Just Before Darwin*, pp. 182–3. The date of composition of the *Superstition* manuscript is uncertain, but Millhauser makes a strong case for the years 1850–3.
18. Mme. Dunglas Home, *D. D. Home: His Life and Mission* (1888; reprint ed., New York: Arno Press, 1976), pp. 9, 146–50.
19. Podmore, *Modern Spiritualism*, 2:6, describes Chambers's change of heart, manifested in the *Chambers's Journal* article of 11 June 1853.
20. This is Millhauser's scenario, *Just Before Darwin*, pp. 176–83, 209–10n.
21. Thomas P. Barkas, *Outlines of Ten Years' Investigations into the Phenomena of Modern Spiritualism* (London: Frederick Pitman; Newcastle: T. P. Barkas, 1862), pp. 16–17. Mme. Home offers the same story in *D. D. Home*, p. 146.
22. Mme. Home, *D. D. Home*, pp. 149–50. Chambers had by this time received an honorary LL.D. degree from the University of St. Andrews. Millhauser, *Just Before Darwin*, pp. 174, 185, mentions the depth of Chambers's bereavement after the two deaths in 1863.
23. The Newcastle spiritualist, Thomas P. Barkas, cited Chambers among "several gentlemen of eminence" who endorsed spiritualism as early as 1862. Barkas *Outlines of Ten Years' Investigations*, pp. 16–17. Millhauser, *Just Before Darwin*, p. 181, mentions the manuscript papers about spiritualism.
24. Mme. Home, *D. D. Home*, pp. 147, 268; Podmore, *Modern Spiritualism*, 2:145.
25. *Report on Spiritualism, of the Committee of the London Dialectical Society* (London: Longmans, Green, Reader, & Dyer, 1871), p. 246; Hudson Tuttle and J. M. Peebles, *The Year-Book of Spiritualism for 1871* (London: James Burns, 1871), pp. 156, 178.
26. Alfred Russel Wallace, *My Life. A Record of Events and Opinions*, 2 vols. (London: Chapman & Hall, 1905), 2:285–6.
27. Millhauser, *Just Before Darwin*, pp. 179–80, points to Chambers's rapidly decreasing interest in geology, for example, after 1855.
28. See Leonard Huxley, *Life and Letters of Thomas Henry Huxley*, 2 vols. (New York: D. Appleton, 1900), 1:202.
29. To make Chambers's conversion by her husband all the more extraordinary, no doubt. Mme. Home, *D. D. Home*, p. 147.
30. William Chambers, *Memoir of Robert Chambers*, p. 288.
31. Millhauser, *Just Before Darwin*, pp. 174–85, to which much of the following argument is indebted.
32. Chambers, *Vestiges*, pp. 353, 354, 361.
33. Hearnshaw, *Short History of British Psychology*, p. 92.
34. George John Romanes, *Mental Evolution in Animals. With a Posthumous Essay on Instinct by Charles Darwin* (London: Kegan Paul, Trench, 1883), p. 62.
35. Letter from Romanes to Darwin, 17 December 1880, and letter to the *Times*, 19 September 1888, quoted in Ethel Romanes, *The Life and Letters of George John Romanes, M.A., LL.D., F.R.S.* (London and New York: Longmans, Green, 1896), pp. 105, 241–7.
36. Hearnshaw, *Short History of British Psychology*, pp. 94–5.

37. For criticisms of Romanes's tendency to "anthropomorphize" the animal, see ibid.; Edwin G. Boring, *A History of Experimental Psychology*, 2d ed. (New York: Appleton-Century Crofts, 1950), pp. 472–3; and Daniel N. Robinson, *An Intellectual History of Psychology*, rev. ed. (New York: Macmillan, 1981), pp. 363–5.

38. "My dear Romanes," Darwin wrote on 2 September 1878, "Many thanks for your letter. I am delighted to hear that you mean to work the comparative psychology well." E. Romanes, *Life and Letters*, p. 78. When Romanes published *Mental Evolution in Animals* in 1883, it included a posthumous essay on instinct which Darwin had left in Romanes's care.

39. E. Romanes, *Life and Letters*, pp. 162–6, 168. Also see pp. 169–77, 200–17. Romanes tried to get around a major problem with his theory by explaining: "I believe most of all in what I have called 'collective variation' of the reproductive system in the way of physiological selection, whereby, owing to some common influence acting on a large number of individuals similarly and simultaneously, they all become sexually coadapted *inter se* while physiologically isolated from the rest." (Ibid., p. 217.)

40. George John Romanes [Physicus], *A Candid Examination of Theism* (London: Trübner, 1878). Long quotations from this work, which Romanes actually wrote in 1876, can be found in Frank Miller Turner, *Between Science and Religion: The Reaction to Scientific Naturalism in Late Victorian England* (New Haven and London: Yale University Press, 1974), pp. 142–4.

41. E. Romanes, *Life and Letters*, p. 7.

42. Letter of 31 March 1878, in C. C. Massey, *Thoughts of a Modern Mystic. A Selection from the Writings of the Late C. C. Massey*, ed. W. F. Barrett (London: Kegan Paul, Trench, Trübner, 1909), pp. 20–1. See E. Romanes, *Life and Letters*, p. 69, for the death of Romanes's sister, Georgina, in April 1878.

43. E. Romanes, *Life and Letters*, p. 46.

44. The correspondence is contained in Wallace, *My Life*, 2:317–26. Wallace had been shown drafts of Romanes's letters to Darwin when he visited Kingston, Canada, in March 1887. Romanes was born in Kingston, and a woman whom Wallace met there was friendly with Romanes's brother. She and the brother shared an interest in spiritualism, and he gave her the drafts of the letters which, he claimed, he had jointly written with Romanes (ibid., 2:126).

45. E. Romanes, *Life and Letters*, pp. 48–9.

46. Romanes's two letters are quoted in Wallace, *My Life*, 2:310–13. The first letter was dated 17 February 1880; Wallace did not provide the date of the second. On Williams's exposure, see Podmore, *Modern Spiritualism*, 2:111.

47. Wallace, *My Life*, 2:314–15.

48. E. Romanes, *Life and Letters*, p. 97. In his letter to Romanes of 18 July 1890, Wallace denied ever professing to believe in astrology (Wallace, *My Life*, 2:317). In E. Romanes, *Life and Letters*, p. 97, the name of the spiritualist whom Romanes visited in April 1880 was left blank. Wallace filled in the blank when he published his autobiography. Malcolm Jay Kottler, "Alfred Russel Wallace, the Origin of Man, and Spiritualism," *Isis* 65 (June 1974):180–2, summarizes the several exchanges between Romanes and Wallace on the subject of spiritualism.

49. C. C. Massey to Barrett, 24 December 1881, Barrett Papers (box 2, A2, no. 11).
50. Edward T. Bennett, *The Society for Psychical Research: Its Rise & Progress & A Sketch of Its Work* (London: R. Brimley Johnson, 1903), p. 6. On 21 December 1881, Romanes wrote to Barrett: "I shall be happy to attend the meeting in January." Barrett Papers (box 2, A4, no. 108).
51. Romanes to Barrett, 28 October 1882, Barrett Papers (box 2, A4, no. 109). Letters no. 104–7, from Romanes to Barrett, show that Romanes had been following Barrett's work with the Creery sisters since June 1881.
52. Romanes to Barrett, 8 December 1882, Barrett Papers (box 2, A4, no. 110).
53. Crichton-Browne's account, in the *Westminster Gazette* of 29 January 1908, is quoted in Trevor H. Hall, *The Strange Case of Edmund Gurney* (London: Gerald Duckworth, 1964), pp. 111–15. It is also reproduced, in abbreviated form, in Ivor Tuckett, "Psychical Researchers and 'The Will to Believe,'" *Bedrock. A Quarterly Review of Scientific Thought* 1 (July 1912):197–8. Crichton-Browne himself recollected the experiments in his book of memoirs, *The Doctor's Second Thoughts* (London: Ernest Benn, 1931), pp. 58–64. He was over ninety years old by 1931 and assigned the wrong date to the event.
54. E. Romanes, *Life and Letters*, p. 345.
55. Constance Wachtmeister, et al., *Reminiscences of H. P. Blavatsky and The Secret Doctrine* (Wheaton, Ill.: Theosophical Publishing House, 1976), p. 71. The precise date of Romanes's visit is unclear.
56. E. Romanes, *Life and Letters*, p. 156.
57. Turner, *Between Science and Religion*, p. 147.
58. Ibid., pp. 134–8, 161–2, makes it clear that Romanes did not return to the Anglican fold before his death in a spirit of meek acceptance, but as an "'experiment of faith.'" Turner warns that Mrs. Romanes's comments, in *Life and Letters*, concerning her husband's rediscovery of Christian faith must be taken with the knowledge that she herself was "a devout Christian and the author of several devotional works" (p. 136). In brief, she overstated the case.
59. George John Romanes, "The Fallacy of Materialism (No. I): Mind and Body," *Nineteenth Century* 12 (December 1882):871–88.
60. Letters to Asa Gray, dated 16 May and 30 December 1883, in E. Romanes, *Life and Letters*, pp. 154–5.
61. The twistings are charted with characteristic clarity in Turner, *Between Science and Religion*, pp. 157–62. Also see pp. 151–3, for Romanes's impatience with the dogmatism of contemporary science.
62. Romanes's later religious and philosophical speculations are primarily contained in two posthumous works: George John Romanes, *Mind and Motion and Monism* (New York and London: Longmans, Green, 1895), and *Thoughts on Religion*, ed. Charles Gore (Chicago: Open Court Publishing Co., 1895).
63. The lecture was republished in *Mind and Motion and Monism*.
64. See, for example, E. Romanes, *Life and Letters*, pp. 249–52; also see Turner, *Between Science and Religion*, pp. 150, 156, 158–9, 162.
65. Concluding Note by the Editor, in Romanes, *Thoughts on Religion*, p. 184.
66. Romanes, *Thoughts on Religion*, p. 31n.

67. Karl Pearson, *The Life, Letters and Labours of Francis Galton*, 3 vols. in 4 (Cambridge: The University Press, 1914–30), 2:62.

68. In this context, it is not surprising to find the name of William Bateson, founder of the study of genetics in England, among the earliest associate members of the SPR. "Associate, Honorary and Corresponding Members," *PSPR* 1 (1882–3):326. Bateson, the champion of Mendelian genetics, and Pearson, the biometrician, were bitter opponents in English biology around the turn of the century.

69. L. Huxley, *Life and Letters*, 1:451. Leonard Huxley reported that his father did not date the paper on which he recorded his experiments with Mrs. Hayden – which Leonard spelled Haydon – but believed that "it must have been before 1863." In fact, it must have been in 1852 or 1853 when Mrs. Hayden was in England.

70. Thanks, among others, to G. H. Lewes, who wrote to the *Leader* in March to explain Mrs. Hayden's trick. Podmore, *Modern Spiritualism*, 2:5.

71. Huxley to Wallace, quoted in James Marchant, *Alfred Russel Wallace: Letters and Reminiscences* (New York and London: Harper & Bros., 1916), p. 418. The undated letter was in reply to one from Wallace, 22 November 1866.

72. Letter from Huxley, 29 January 1869, quoted in *Report on Spiritualism, of the Committee of the London Dialectical Society*, pp. 229–30. He wrote the committee a more civil letter on 2 January 1870 (ibid., pp. 278–9), but still declined to have anything to do with its inquiry.

73. L. Huxley, *Life and Letters*, 1:452–6 (report to Darwin, 27 January 1874). Kottler, "Wallace, the Origin of Man, and Spiritualism," p. 171, identifies the medium as Williams. On that occasion, the medium was unaware of Huxley's identity. Huxley may have also attended a séance as late as 1877, if Wallace's recollection was correct. In December of that year, Wallace wrote to Barrett that Huxley was present at a séance "a few months ago," with Tyndall and Carpenter, and the Fox sisters (Mrs. Kane and Mrs. Jencken) as mediums. Marchant, *Wallace*, p. 427.

74. Pearson, *Life, Letters and Labours of Galton*, 2:63–5. Also see Kottler, "Wallace, the Origin of Man, and Spiritualism," pp. 170–1, and Francis Darwin, ed., *The Life and Letters of Charles Darwin, including an Autobiographical Chapter*, 2 vols. (New York: D. Appleton, 1888), 1:341–2.

75. Darwin to unnamed correspondent, in F. Darwin, ed., *Life and Letters*, 2:364–5; Pearson, *Life, Letters and Labours of Galton*, 2:66–7.

76. Darwin to Huxley, 29 January 1874, in F. Darwin, ed., *Life and Letters*, 2:365.

77. Francis Galton, *Memories of My Life* (New York: E. P. Dutton, 1909), first published in London in 1908. There is a useful summary of Galton's work in Robert I. Watson, *The Great Psychologists*, 3d ed. (Philadelphia: J. B. Lippincott, 1971), pp. 311–20. Also see Boring, *History of Experimental Psychology*, pp. 460–1, 476–88.

78. Francis Galton, *English Men of Science: Their Nature and Nurture* (London: Macmillan, 1874).

79. Galton, *Memories*, pp. 244–5, 294–5, and Appendix, p. 326.

80. Boring, *History of Experimental Psychology*, p. 477, and Galton, *Memories*, pp. 302–3.

81. Galton, *Memories*, pp. 310–23.
82. Boring, *History of Experimental Psychology*, pp. 483–4. In a letter to Sidgwick, dated 25 June 1890, Galton pondered a scheme to provide financial encouragement for capable young women (those who gained honors at university) to marry early and produce several children. Sidgwick Papers, Add. MS. c. 94. 1.
83. Galton, *Memories*, p. 322. Galton gave money to found both a research fellowship and a scholarship in eugenic research at University College, London, and in his will left his residual estate to establish a chair of eugenics at the University. See Samuel Hynes, *The Edwardian Turn of Mind* (Princeton, N.J.: Princeton University Press, 1968), pp. 15–53, for a discussion of the fears of racial degeneration, common in the Edwardian decade, which helped to give urgency to the eugenicist message during that period.
84. Boring, *History of Experimental Psychology*, p. 484.
85. Galton to Darwin, 28 March 1872, 19 April 1872, and May–November 1872, in Pearson, *Life, Letters and Labours of Galton*, 2:63–6; see p. 67 for Galton's agnostic attitude.
86. Stainton Moses's séance notes, 18 January 1873, in R. G. Medhurst, "Stainton Moses and Contemporary Physical Mediums. 3. Mr. and Mrs. Nelson Holmes," *Light* 83 (Autumn 1963):133–7.
87. Pearson, *Life, Letters and Labours of Galton*, 2:67. Crookes's tests with Fay are discussed in the next chapter.
88. Romanes to Barrett, 18 July 1881, Barrett Papers (box 2, A4, no. 106). The committee was never convened.
89. George J. Romanes, "Thought-Reading," *Nature* 24 (23 June 1881):171–2. Galton read and approved Romanes's report before publication.
90. Crichton-Browne's account, as quoted in T. Hall, *Strange Case of Edmund Gurney*, p. 115.
91. Francis Galton, "Free-Will – Observations and Inferences," *Mind* 9 (July 1884):406–13.
92. Galton, *Memories*, p. 80.
93. See Galton's chapters on mental imagery, number-forms, color associations, and visionaries, in *Inquiries into Human Faculty and its Development* (London: Macmillan, 1883).
94. Galton's discussion, for example, of the self in his 1884 essay on "Free-Will," p. 409, was not the work of a complete materialist.
95. "Law Report," *Times*, 27 April 1907, p. 14; Alfred Russel Wallace, "Slate-Writing Extraordinary," *Spectator* 50 (6 October 1877):1239–40; "Evidence in Defence of Dr. Slade," *Spiritualist Newspaper* 9 (3 November 1876):161, 164.
96. Alfred Russel Wallace, *Miracles and Modern Spiritualism* (1896; reprint ed., New York: Arno Press, 1975), pp. 188–90, and Marchant, *Wallace*, pp. 423–4. The 1896 edition of *Miracles and Modern Spiritualism* was the third, the book having been initially published in 1874.
97. Wallace, *Miracles and Modern Spiritualism*, pp. 211–12. The text here reprints "A Defence of Modern Spiritualism," which Wallace published in the *Fortnightly*, in May and June 1874.
98. A. R. Wallace, *The Psycho-Physiological Sciences, and their Assailants*. Being a Response by Alfred R. Wallace, of England; Professor J. R. Buchanan, of New York; Darius Lyman, of Washington; Epes Sargent,

of Boston; to the Attacks of Prof. W. B. Carpenter, of England, and Others (Boston: Colby & Rich, 1878), p. 17. Wallace had a good point when he challenged the canon of repeatability as a major criterion for determining the scientific validity of an hypothesis in all cases. Many of the phenomena of biology, for example, simply do not lend themselves to the same kinds of tests as the phenomena of physics and chemistry.

99. F. W. H. Myers, "Resolute Credulity," *PSPR* 11 (July 1895):218n. Upon her death, Faustina, wife of Marcus Aurelius, was deeply mourned by her husband. Subsequently, however, her good character was cast into serious doubt.

100. Myers to Lodge, 27 March 1894, Lodge Collection (box 6, no. 1392). Such condescension was scarcely fair, coming from Myers. In April 1890, he had written to Wallace, asking the naturalist to put in a good word for him in "the next world," as Myers was certain that Wallace would "have much influence there." Quoted in Marchant, *Wallace*, p. 431.

101. Lodge to J. Arthur Hill, 17 February 1914, in J. Arthur Hill, comp., *Letters from Sir Oliver Lodge, Psychical, Religious, Scientific and Personal* (London: Cassell, 1932), p. 34. Also see Kottler, "Wallace, the Origin of Man, and Spiritualism," pp. 185–6, and Turner, *Between Science and Religion*, p. 90.

102. William B. Carpenter, "Psychological Curiosities of Spiritualism," *Fraser's*, n.s. 16 (November 1877):545–51; Alfred R. Wallace, "Psychological Curiosities of Scepticism. A Reply to Dr. Carpenter," *Fraser's*, n.s. 16 (December 1877):694, 699, 703.

103. London Dialectical Society, *Report on Spiritualism*, pp. vi, 83–6.

104. Wallace to Barrett, 28 January 1882, marked *Private*, Barrett Papers (box 2, A4, no. 135).

105. See, for example, Marchant, *Wallace*, p. 432, where, in a letter to a correspondent, dated 14 September 1896, Wallace regretted that the majority of the SPR's active members "are so absurdly and illogically sceptical . . ."

106. Wallace to Barrett, 17 February 1901, quoted in Marchant, *Wallace*, pp. 435–6. Barrett tried again in 1905, when he urged Wallace to accept the SPR presidency in order to keep Podmore from that office. Barrett to Wallace, 3 November 1905, ibid., pp. 437–8. Wallace's reply on that occasion is not known, but one may surmise that he turned down the proposal for the same reasons that he gave in 1901. In fact, Gerald Balfour became the president of the SPR for 1906–7.

Wallace's refusal to accept offices that required his presence in London extended even to spiritualist organizations. In the autumn of 1910, he turned down an invitation to assume the honorary presidency of the London Spiritualist Alliance. LSA, Minute Book of Council Meetings 1909–1921, pp. 22–5 (6 and 27 October 1910). Many years earlier, Wallace had been a member of the BNAS, but the existing minute books bear no evidence of his attendance at meetings. His name only appears under the list of Experimental Research Committee members in 1878, which is found at the front of the BNAS Minute Book for Committees, no. 2.

107. Alfred Russel Wallace, "What Are Phantasms, and Why Do They Appear?" reprinted in *Miracles and Modern Spiritualism*, pp. 262–3. Also see pp. xiv–xvii, 273, 278.

108. Wallace to Barrett, 15 and 20 February 1911, quoted in Marchant, *Wallace*, pp. 439–40.

109. Wallace to Lodge, 18 January 1908, Lodge Collection (box 10, no. 2736).
110. Kottler, "Wallace, the Origin of Man, and Spiritualism," p. 183, takes Wallace's word on this important point, and I see no reason to doubt it either.
111. Turner, *Between Science and Religion*, in the chapter on Wallace, stresses Wallace's dissatisfaction with the ethical values and standards of his day, and his appreciation of spiritualism as a moral guide. Similarly, Roger Smith traces the links between Wallace's humanitarian values and spiritualist beliefs, in "Alfred Russel Wallace: Philosophy of Nature and Man," *British Journal for the History of Science* 6 (December 1972):177–99.
112. Wallace, *My Life*, 1:77–8, 226.
113. Ibid., pp. 79, 87, 104.
114. Ibid., pp. 87–9.
115. Ibid., pp. 227–8.
116. Ibid., p. 228.
117. Wallace, *Miracles and Modern Spiritualism*, pp. vi–vii.
118. Ibid., p. x, and Wallace, *My Life*, 1:232–4. Also see ch. 17, "The Opposition to Hypnotism and Psychical Research," in Alfred Russel Wallace, *The Wonderful Century. Its Successes and Its Failures* (London: Swan Sonnenschein, 1898), pp. 194–212.
119. Wallace quoted in "The British Association at Glasgow," *Spiritualist Newspaper* 9 (22 September 1876):93. For a quite different explanation of mesmerism, as spirit acting on spirit, written after Wallace became deeply interested in spiritualism, see Wallace, *Miracles and Modern Spiritualism*, p. 109.
120. Combe's *Constitution of Man* (1828) figured among the advanced literature which Wallace obtained from his brother.
121. Wallace, *My Life*, 1:234–6.
122. Ibid., pp. 257–62. Also see Wallace, *Wonderful Century*, pp. 159–93 (ch. 16, "The Neglect of Phrenology"). Robert M. Young, *Mind, Brain and Adaptation in the Nineteenth Century: Cerebral Localization and its Biological Context from Gall to Ferrier* (Oxford: Clarendon Press, 1970), p. 44, points out an important analogy between the theory of evolution in its early stages and phrenology. Both rested on "great numbers of naturalistic observations and pieces of anecdotal evidence." Just as Gall piled up observations to "prove" his craniology, so did Wallace and Darwin amass observations from nature to support the theory of natural selection. In neither case, however, did the accumulation of such observations amount to demonstration of a causal relationship.
123. Wallace, *My Life*, 1:254–5. Bates was an amateur entomologist with whom Wallace enjoyed collecting beetles.
124. H. Lewis McKinney, *Wallace and Natural Selection* (New Haven: Yale University Press, 1972), p. 147, and Turner, *Between Science and Religion*, pp. 82–3. The actual year during which Wallace discussed *Vestiges* in two letters to Bates is uncertain. Marchant, *Wallace*, p. 73, gives 1847, but other Wallace scholars have placed the letters as early as 1845. Barbara G. Beddall, "Wallace, Darwin, and the Theory of Natural Selection: A Study in the Development of Ideas and Attitudes," *Journal of the History of Biology* 1 (Fall 1968):265–6n., follows Marchant. In *My Life*, 1:254, Wallace quoted from the letters after a reference to the summer of 1847, but his memory for dates was not infallible.

125. The 1855 essay was reprinted, unaltered, in Alfred Russel Wallace, *Contributions to the Theory of Natural Selection. A Series of Essays*, 2d ed. (New York: Macmillan, 1871), pp. 1–25. The volume of essays was first published in 1870. Also see Beddall, "Wallace, Darwin, and the Theory of Natural Selection," pp. 269–70.
126. "On the Tendency of Varieties to Depart Indefinitely from the Original Type," 1858, in Wallace, *Contributions to the Theory of Natural Selection*, pp. 36–7, 41–2. The essay was first published in 1858 in the *Journal of the Proceedings of the Linnean Society*. Wallace's argument, of course, applied only to animals in the wild, not domesticated, state.
127. Ronald Rainger, "Race, Politics, and Science: The Anthropological Society of London in the 1860s," *Victorian Studies* 22 (Autumn 1978):51–60.
128. Alfred R. Wallace, "The Origin of Human Races and the Antiquity of Man Deduced from the Theory of 'Natural Selection,'" *Journal of the Anthropological Society of London* 2 (1864):clviii–clxii, clxxxiv–clxxxv. While in the Malay Archipelago, Wallace traveled to Borneo primarily to study the orangutan, from which – or a closely allied species – he thought man had evolved. Marchant, *Wallace*, pp. 43–4; McKinney, *Wallace and Natural Selection*, p. 150; Turner, *Between Science and Religion*, p. 83. In the previous century, the close relationship of man and orangutan had figured prominently in the theory of the Great Chain of Being. Lord Monboddo, a Scottish author, had in fact "contemplated educating orang-outangs, and hoped that they might eventually take over the more menial forms of housework." Toulmin and Goodfield, *Discovery of Time*, p. 118.
129. Wallace, "Origin of Human Races," p. clix.
130. Ibid., p. clxii.
131. Ibid., pp. clxv, clxvii.
132. Ibid., pp. clxii, clxiv, clxvii, clxix. This interpretation of Wallace's 1864 paper basically agrees with Kottler's summary in "Wallace, the Origin of Man, and Spiritualism," pp. 147–9. For a somewhat different emphasis, see Turner, *Between Science and Religion*, pp. 74–8, 83–4. Turner quite rightly points to the phrenological emphasis in the paper, but sees it, prematurely I think, as evidence that Wallace was already repudiating scientific naturalism in 1864.
133. Wallace, "Origin of Human Races," pp. clxii, clxv, clxix–clxx.
134. Wallace, *My Life*, 2:276; Kottler, "Wallace, the Origin of Man, and Spiritualism," pp. 167–8.
135. Wallace recorded his early experiences among the "Notes of Personal Evidence," in *Miracles and Modern Spiritualism*, pp. 125–44.
136. Carpenter to Barrett, 2 November 1876, Barrett Papers (box 2, A4, no. 12). Also see Wallace, *My Life*, 2:277.
137. Wallace, *Miracles and Modern Spiritualism*, pp. 168–71. Also see Podmore, *Modern Spiritualism*, 2:65–7, who reports that Miss Nichol could likewise produce live eels and lobsters, and Kottler, "Wallace, the Origin of Man, and Spiritualism," pp. 168–9. In "Psychological Curiosities of Spiritualism," pp. 547–50, Carpenter made mincemeat of Miss Nichol's mediumship, and in the letter to Barrett cited in the preceding note, he complained of wasting two hours at a seance with her. In his presence, she "was found to be utterly powerless."
138. Wallace, *My Life*, 2:349–50. The quotation within the quotation echoes sentiments expressed by Sir David Brewster, the prominent Scottish scientist.

139. Ibid., 2:276.
140. Ibid., 2:286–7, 330; Kottler, "Wallace, the Origin of Man, and Spiritualism," pp. 174–5.
141. Wallace to Dr. Edwin Smith, 19 October 1899, quoted in Marchant, *Wallace*, p. 437.
142. Wallace, *My Life*, 1:236.
143. Wallace particularly articulated this vision of the natural world catering to man's development in the volume published very late in his life, *The World of Life. A Manifestation of Creative Power, Directive Mind and Ultimate Purpose* (London: Chapman & Hall, 1910).
144. Alfred Russel Wallace, "Sir Charles Lyell on Geological Climates and The Origin of Species," *Quarterly Review* 126 (April 1869):359–94.
145. Wallace to Darwin, 18 April 1869, quoted in Marchant, *Wallace*, p. 200. Among the men who could corroborate his findings, Wallace cited Robert Chambers in this letter. Also see Darwin to Wallace, 14 April 1869, 26 January [1870], 31 March 1870, ibid., pp. 199, 206.
146. Wallace, *Contributions to the Theory of Natural Selection*, pp. 330–1.
147. Both Kottler, "Wallace, the Origin of Man, and Spiritualism," pp. 153–4, and Turner, *Between Science and Religion*, p. 93, discuss the significance of the changed ending which Wallace gave his 1864 essay in 1870. As Kottler observes, the new conclusion wreaked havoc with the argument of the rest of the essay.
148. Wallace, *Contributions to the Theory of Natural Selection*, pp. 333–51. This discussion of significant physical traits amplified comments which Wallace had made in his *Quarterly Review* article of 1869.
149. Wallace, *Contributions to the Theory of Natural Selection*, pp. 351–5, 359. Also see Turner, *Between Science and Religion*, pp. 96–7.
150. Indeed, as Turner points out, Joseph Priestley had proposed similar explanations a century earlier. Karl Pearson was one of a very few late nineteenth-century scientists who took Wallace's metaphysics seriously enough to refute his arguments. Turner also observes that Wallace distorted Huxley's assertion by using it out of context (*Between Science and Religion*, p. 98).
151. Wallace, *Contributions to the Theory of Natural Selection*, pp. 361–8, 370, 372A–C, and *Miracles and Modern Spiritualism*, pp. vii–viii. Also see Smith, "Wallace: Philosophy of Nature and Man," p. 178.
152. Many of Wallace's critics are summarized in Kottler, "Wallace, the Origin of Man, and Spiritualism," pp. 157–60.
153. Wallace to Lodge, 8 March 1898, Lodge Collection (box 10, no. 2732).
154. See, for example, Wallace, "Psychological Curiosities of Scepticism," p. 704.
155. Marchant, *Wallace*, p. 417.
156. A fuller list can be found in Turner, *Between Science and Religion*, p. 71.
157. Wallace, *Contributions to the Theory of Natural Selection*, pp. 372, 372A, and *Miracles and Modern Spiritualism*, pp. 33, 38–9, 48–9, 205.
158. Wallace, *Miracles and Modern Spiritualism*, pp. vii, 13, 39, 46–9, 108, 216–18, 220. Wallace to Lodge, 23 December 1906, Lodge Collection (box 10, no. 2734).
159. Wallace to Huxley, 22 November 1866, quoted in Marchant, *Wallace*, p. 418. Kottler, "Wallace, the Origin of Man, and Spiritualism," pp. 169–74, describes Wallace's "missionary" efforts with other scientists.

160. Wallace to Dr. Edwin Smith, 19 October 1899, quoted in Marchant, *Wallace*, p. 437.
161. Wallace to Mrs. Fisher, 9 April 1897, quoted in Marchant, *Wallace*, p. 432. Yet W. G. H. Armytage, *Heavens Below: Utopian Experiments in England 1560–1960* (Toronto: University of Toronto Press, 1961), pp. 308–9, claims that Blavatsky "attracted Alfred Russel Wallace," and Bruce F. Campbell, *Ancient Wisdom Revived: A History of the Theosophical Movement* (Berkeley, Calif.: University of California Press, 1980), p. 87, states that Wallace was an early member of the Theosophical Society in England. Certainly Wallace's views on progress after death were very close to Blavatsky's, but little else about Theosophy would have been likely to hold Wallace's interest for long.
162. Wallace, *Miracles and Modern Spiritualism*, pp. 228–9.
163. See Wallace's letter to the Rev. J. B. Henderson, 10 August 1893, quoted in Marchant, *Wallace*, p. 436.
164. Wallace, *Miracles and Modern Spiritualism*, pp. 91, 115, 221–5.
165. Wallace, *My Life*, 1:87–8, 2:236–8; and Turner, *Between Science and Religion*, p. 101.
166. Together with Mill's. Wallace, *My Life*, 1:104, 2:266. Spencer's writing exerted considerable influence on the development of Wallace's own concepts of evolution and society. For example, the idea for his paper to the Anthropological Society in 1864 arose from "the perusal of Mr. Herbert Spencer's works, especially *Social Statics*." Wallace, "Origin of Human Races," p. clxx.
167. Armytage, *Heavens Below*, p. 344; Wallace, *My Life*, 2:235–40.
168. Wallace, *My Life*, 2:266–9. Subsequently, Wallace was even more impressed with Bellamy's *Equality*, published in 1897.
169. Ibid., p. 267. In his opposition to the medical establishment and his conviction that "orthodox medical men" knew little about the intricacies of the human body, Wallace, too, underscored his membership in the progressive wing of British spiritualism. See his letter to Miss Buckley, 24 April 1874, quoted in Marchant, *Wallace*, p. 422.
170. Kottler mentions Wallace's refusal to apply the same arguments to animals as part of a general conclusion in favor of spiritualism as the root cause of Wallace's deviation from natural selection in the case of man. "Wallace, the Origin of Man, and Spiritualism," pp. 188–92.
171. C. Lloyd Morgan, *An Introduction to Comparative Psychology* (London: W. Scott, 1894), quoted in Robinson, *Intellectual History of Psychology*, p. 366.

8. Physics and psychic phenomena

1. E. E. Fournier D'Albe, *New Light on Immortality* (London: Longmans, Green, 1908), pp. vii–viii.
2. E. J. Hobsbawm, *The Age of Capital 1848–1875* (London: Weidenfeld & Nicolson, 1975), p. 269. Also see pp. 254–6.
3. Reba N. Soffer, *Ethics and Society in England: The Revolution in the Social Sciences 1870–1914* (Berkeley and Los Angeles: University of California Press, 1978), pp. 112–13. Of the many other studies that deal with these general themes of contemporary physics, see Noel Annan, "Science, Religion, and the Critical Mind: Introduction," in *1859: Entering*

an Age of Crisis, ed. Philip Appleman, William A. Madden, and Michael Wolff (Bloomington: Indiana University Press, 1959), p. 45; Samuel Hynes, *The Edwardian Turn of Mind* (Princeton, N.J.: Princeton University Press, 1968), pp. 134, 136; Frank Miller Turner, *Between Science and Religion: The Reaction to Scientific Naturalism in Late Victorian England* (New Haven and London: Yale University Press, 1974), p. 228.

4. John Tyndall, *Fragments of Science for Unscientific People: A Series of Detached Essays, Lectures, and Reviews*, 2d ed. (London: Longmans, Green, 1871), p. 435.

5. Michael Faraday, "Professor Faraday on Table-Moving," *Athenaeum*, 2 July 1853, p. 801. The article in the *Athenaeum* was an expanded version of a letter from Faraday which appeared in the *Times*, 30 June 1853. In it (p. 803), Faraday referred readers to Carpenter's lecture on unconscious muscular movement, given the previous year at the Royal Institution.

6. William Thomson (later Lord Kelvin), *Popular Lectures and Addresses*, 2d ed. (London: Macmillan, 1891), vol. 1, *Constitution of Matter*, p. 265. Thomson's remark about spiritualism is found in "The Six Gateways of Knowledge," his presidential address to the Birmingham and Midland Institute, Birmingham, 3 October 1883.

7. British Association for the Advancement of Science, *Report of the Forty-First Meeting of the British Association for the Advancement of Science; Held at Edinburgh in August 1871* (London: John Murray, 1872), pp. 4, 121–2 of "Notices and Abstracts of Miscellaneous Communications to the Sections."

8. John Tyndall, "The Belfast Address," in *Fragments of Science. A Series of Detached Essays, Addresses, and Reviews*, 2 vols. (New York: D. Appleton, 1900), 2:191, 197.

9. Balfour Stewart to Barrett, 26 December 1881, Barrett Papers (box 2, A2, no. 19).

10. See SPR membership lists at back of the volumes of its *Proceedings*.

11. J. J. Thomson, *Recollections and Reflections* (London: G. Bell, 1936), pp. 112, 397–401; Robert John Strutt, fourth Baron Rayleigh, *Life of John William Strutt, Third Baron Rayleigh, O.M., F.R.S.* (Madison: University of Wisconsin Press, 1968), pp. 187–219. The fourth Lord Rayleigh's biography of his father was first published in 1924.

12. Edward VII quoted in Roger Fulford, "The King," in *Edwardian England 1901–1914*, ed. Simon Nowell-Smith (London: Oxford University Press, 1964), p. 21.

13. Rayleigh, *Life of J. W. Strutt*, pp. 312–15, 320–1, 340.

14. "Presidential Address to the SPR, Delivered on April 11th, 1919," ibid., Appendix II, pp. 380, 391; Eric J. Dingwall, *The Critics' Dilemma: Further Comments on Some Nineteenth Century Investigations* (Crowhurst, Sussex: E. J. Dingwall, 1966), pp. 55, 61.

15. Rayleigh, *Life of J. W. Strutt*, pp. 66–7.

16. "Presidential Address to the S.P.R.," ibid., pp. 383–4, 388.

17. Robert Bruce Lindsay, *Lord Rayleigh: The Man and His Work* (Oxford and London: Pergamon Press, 1970), p. 15.

18. Rayleigh, *Life of J. W. Strutt*, pp. 218, 359.

19. Reply in 1910 to a correspondent collecting information about the religious beliefs of scientists, quoted ibid., p. 360.

20. Rayleigh, *Life of J. W. Strutt*, pp. 29, 67. David B. Wilson, "The Thought of Late Victorian Physicists: Oliver Lodge's Ethereal Body," *Victorian*

Studies 15 (September 1971):29–48, contains a useful summary of Rayleigh's religious opinions.

21. See note 19 above. Rayleigh, *Life of J. W. Strutt*, p. 361.
22. Rayleigh, *Life of J. W. Strutt*, pp. 358–9.
23. See his "Presidential Address to the S.P.R.," ibid., pp. 389–90.
24. Quoted in Rayleigh, *Life of J. W. Strutt*, p. 361.
25. Thomson, *Recollections and Reflections*, p. 158.
26. Ibid., pp. 147–52. Thomson was elected FRS in 1884, just months before his election to the Cavendish Professorship at the age of twenty-eight. See his obituary in *PSPR* 46 (November 1940):209.
27. Thomson, *Recollections and Reflections*, pp. 18, 158–63; Robert John Strutt, fourth Baron Rayleigh, *The Life of Sir J. J. Thomson, O.M., Sometime Master of Trinity College, Cambridge* (Cambridge: The University Press, 1942), pp. 79–95.
28. Thomson, *Recollections and Reflections*, pp. 153–8, 298.
29. Rayleigh, *Life of Thomson*, pp. 283–5.
30. [Sophia E. De Morgan], *From Matter to Spirit. The Result of Ten Years' Experience in Spirit Manifestations*. By C. D. with a Preface by A. B. [Augustus De Morgan] (London: Longman, Green, Longman, Roberts, & Green, 1863), pp. v–vi.
31. Sophia E. De Morgan, *Memoir of Augustus De Morgan* (London: Longmans, Green, 1882), p. 192.
32. Arthur Schuster, *Biographical Fragments* (London: Macmillan, 1932), p. 206; also see Thomson, *Recollections and Reflections*, p. 21.
33. B. Stewart, "Mr. Crookes on the 'Psychic' Force," *Nature* 4 (27 July 1871):237. See Robert H. Kargon, *Science in Victorian Manchester: Enterprise and Expertise* (Baltimore: Johns Hopkins University Press, 1977), p. 215, for details of Stewart's career.
34. See, for example, Stewart's 1887 presidential address to the SPR, in *Presidential Addresses to the Society for Psychical Research 1882–1911* (Glasgow: Robert Maclehose for the SPR, 1912), p. 28.
35. Schuster, *Biographical Fragments*, pp. 214–15. Schuster, Stewart's successor as Langworthy Professor of Physics at Owens College after Stewart's death in 1887, points out that Tait played a very secondary role in the writing of *The Unseen Universe*. The work was largely Stewart's. Also see P. M. Heimann, "The *Unseen Universe*: Physics and the Philosophy of Nature in Victorian Britain," *British Journal for the History of Science* 6 (June 1972):73–9.
36. See his 1887 address to the SPR in *Presidential Addresses*, p. 31, and Stewart, "Mr. Crookes on the 'Psychic' Force," p. 237.
37. [G. H. Lewes], "Seeing is Believing," *Blackwood's Edinburgh Magazine* 88 (October 1860):394, commented sardonically on those who, rejecting "with scorn the suggestion of spirits, accept, on no better evidence, the suggestion of 'electricity.'" The *Wellesley Index to Victorian Periodicals*, vol. 1, attributes the article to Lewes.
38. Schuster, *Biographical Fragments*, p. 215, and P. G. T[ait], "Obituary Notices of Fellows Deceased: Dr. Balfour Stewart," *Proceedings of the Royal Society of London* 46 (1889):xi.
39. Fournier D'Albe, *New Light on Immortality*, pp. 145–52.
40. W. H. Brock, "William Crookes," *Dictionary of Scientific Biography*, 15 vols. (New York: Charles Scribner's Sons, for the American Council of Learned Societies, 1970–1978), 3:475.

41. Frank Greenaway, "A Victorian Scientist: The Experimental Researches of Sir William Crookes (1832–1919)," *Proceedings of the Royal Institution of Great Britain*, vol. 39, pt. 2 (1962):198.
42. Examples of this writing paper can still be admired in Crookes's correspondence with Oliver Lodge, Lodge Collection (box 2, nos. 325–65, passim).
43. In addition to the sources cited above, for biographical information about Crookes, see E. E. Fournier D'Albe, *The Life of Sir William Crookes, O.M., F.R.S.* (London: T. Fisher Unwin, 1923); W. A. T. [William Augustus Tilden], "Obituary Notices of Fellows Deceased: Sir William Crookes," *Proceedings of the Royal Society of London* 96A (November 1919):i–ix; and P. Zeeman, "Scientific Worthies. Sir William Crookes, F.R.S.," *Nature* 77 (7 November 1907):1–3.
44. Thomson, *Recollections and Reflections*, pp. 91, 378. For Crookes's reliance on Lodge, see their correspondence, particularly during the period when Crookes was preparing his presidential address to the BAAS, November 1897–August 1898. Lodge Collection (box 2, nos. 326–42).
45. Thomas A. Dardis, *Some Time in the Sun* (New York: Charles Scribner's, 1976), p. 214. The director was George Cukor. I am indebted to my father, Charles J. Oppenheim III, for calling this information to my attention. Unfortunately, the film was never made.
46. Crookes's address on "Radiant Matter," delivered at the 1879 meeting of the BAAS, quoted in Fournier D'Albe, *Life of Crookes*, p. 290.
47. But he had, apparently, been attending occasional séances with her since 1872, or late 1871. Fournier D'Albe, *Life of Crookes*, pp. 180–1, 231. Crookes met the fifteen-year-old Florence in October 1871, although the register of his correspondence shows no arrangements for séances with her until the following year. R. G. Medhurst and K. M. Goldney, in their invaluable article "William Crookes and the Physical Phenomena of Mediumship," *PSPR* 54 (March 1964):62, dispel the fog surrounding Cook's age by locating her birth certificate at Somerset House and reporting her date of birth as 3 June 1856.
48. Crookes made these claims in three lengthy letters to the *Spiritualist Newspaper* on 6 February, 3 April, and 5 June 1874. The letters are reprinted in M. R. Barrington, K. M. Goldney, and R. G. Medhurst, eds., *Crookes and the Spirit World. A Collection of Writings by or Concerning the Work of Sir William Crookes, O.M., F.R.S., in the Field of Psychical Research* (London: Souvenir Press, 1972), pp. 130–41.
49. Trevor H. Hall, *The Spiritualists: The Story of Florence Cook and William Crookes* (New York: Helix Press, Garrett Publications, 1963), pp. 172–3, gleefully identifies the source of Crookes's quotation. Also see Trevor H. Hall, "Florence Cook and William Crookes," in *New Light on Old Ghosts* (London: Gerald Duckworth, 1965), pp. 67–85. Ruth Brandon, *The Spiritualists: The Passion for the Occult in the Nineteenth and Twentieth Centuries* (London: Weidenfeld & Nicolson, 1983), pp. 123–4, follows Hall in assuming that Crookes and Cook were lovers.
50. T. Hall, *The Spiritualists*, p. 171. Medhurst and Goldney persuasively demolish a number of Hall's theories and pieces of "evidence," in "Crookes and the Physical Phenomena of Mediumship."
51. Letter to the *Spiritualist Newspaper*, 5 June 1874, reprinted in Barrington, Goldney, and Medhurst, eds., *Crookes and the Spirit World*, p. 140;

Frank Podmore, *Modern Spiritualism: A History and a Criticism*, 2 vols. (London: Methuen, 1902), 2:154.

52. William Crookes, "Some Further Experiments on Psychic Force," first published in the *Quarterly Journal of Science*, October 1871, and reprinted in Barrington, Goldney, and Medhurst, eds., *Crookes and the Spirit World*, p. 35. Medhurst and Goldney discuss Crookes's "astonishingly casual reporting" in "Crookes and the Physical Phemonena of Mediumship," p. 137.

53. Crookes wrote to the *Spiritualist Newspaper* on 3 April 1874: "Enough has taken place to thoroughly convince me of the perfect truth and honesty of Miss Cook." Letter reprinted in Barrington, Goldney, and Medhurst, eds., *Crookes and the Spirit World*, p. 132.

54. Reprinted in Ibid., pp. 138–9.

55. Fournier D'Albe, *Life of Crookes*, pp. 181, 190.

56. Diary extracts quoted in Ibid., pp. 141–2, 170–1.

57. William Crookes, "Spiritualism Viewed by the Light of Modern Science," *Quarterly Journal of Science*, July 1870, reprinted in Barrington, Goldney, and Medhurst, eds., *Crookes and the Spirit World*, pp. 15–16, 21.

58. Medhurst and Goldney, "Crookes and the Physical Phenomena of Mediumship," p. 35.

59. William Crookes, "Experimental Investigation of a New Force," *Quarterly Journal of Science*, July 1871, reprinted in Barrington, Goldney, and Medhurst, eds., *Crookes and the Spirit World*, p. 22.

60. Michael Petrovo-Solovovo, "Correspondence," *JSPR* 9 (November 1900):322. Also see Frank Podmore, *Studies in Psychical Research* (London: G. P. Putnam's Sons, 1897), p. 53.

61. Crookes's articles in the *Quarterly Journal of Science*, 1870–71, were republished in 1874 under the title *Researches in the Phenomena of Spiritualism*, published by J. Burns. Also included in the volume were a further contribution, by Crookes, to the *Quarterly Journal of Science*, January 1874, entitled "Notes of an Enquiry into the Phenomena Called Spiritual, During the Years 1870–73," and, finally, Crookes's letters to the spiritualist press about Florence Cook's mediumship.

62. Crookes, "Experimental Investigation of a New Force," in Barrington, Goldney, and Medhurst, eds., *Crookes and the Spirit World*, p. 31. Also see Fournier D'Albe, *Life of Crookes*, pp. 135, 202.

63. See, for example, the description of séance vi, 23 June 1871, in William Crookes, "Notes of Séances with D. D. Home," *PSPR* 6 (1889):114. Count Perovsky-Petrovo-Solovovo, "Some Thoughts on D. D. Home," *PSPR* 39 (March 1930):258, underscores sharply the difference between Crookes's description of the Home séances that appeared in the *Quarterly Journal of Science* and the picture that emerged from Crookes's séance notes in 1889.

64. "A Scientific Examination of Mrs. Fay's Mediumship," *Spiritualist Newspaper*, 12 March 1875, reprinted in Medhurst and Goldney, "Crookes and the Physical Phenomena of Mediumship," pp. 95–100. For a discussion of the electrical tests with Cook, see C. D. Broad, "Cromwell Varley's Electrical Tests with Florence Cook," *PSPR* 54 (March 1964):158–72. The inventor of the galvanometer was Cromwell Varley, FRS, the prominent electrical engineer whose first wife was a private

medium. Having become a convinced spiritualist himself, he had influenced Crookes's decision to investigate spiritualist phenomena after the death of Philip Crookes. Fournier D'Albe, *Life of Crookes*, p. 133.
65. Podmore, *Modern Spiritualism*, 2:158–9; Harry Houdini, *A Magician Among the Spirits* (1924; reprint ed., New York: Arno Press, 1972), p. 204n; Colin Brookes-Smith, "Cromwell Varley's Electrical Tests," *JSPR* 43 (March 1965):26–31. By contrast, Broad (preceding note) emphasizes the difficulties involved in trying to fool the galvanometer.
66. Crookes, "Spiritualism Viewed by the Light of Modern Science," in Barrington, Goldney, and Medhurst, eds., *Crookes and the Spirit World*, p. 17.
67. The letters are quoted in Medhurst and Goldney, "Crookes and the Physical Phenomena of Mediumship," pp. 113–15. The authors discuss Crookes's tests with Showers on pp. 105–23. T. Hall, *The Spiritualists*, pp. 75–84, and *New Light on Old Ghosts*, pp. 72–6, puts a very different interpretation on the Crookes-Showers relationship and suggests that Crookes behaved improperly toward the young woman. Also see Dingwall, *Critics' Dilemma*, pp. 54–69.
   A letter from Rayleigh to Myers, 15 January 1875, implies rather cryptically that Crookes also acknowledged unsatisfactory séance conduct on Mrs. Fay's part. Rayleigh, however, seemed to have felt that Crookes's conduct likewise left much to be desired. It is not clear from the letter whether Rayleigh was commenting on Crookes's scientific carelessness or dishonorable behavior towards women. Myers Papers 2⁴².
68. Crookes to Lodge, 5 July 1909, Lodge Collection (box 2, no. 357).
69. E. T. Bennett, reporting on the meeting, to Lodge, 2 November 1895, Lodge Collection (box 1, no. 180).
70. Fournier D'Albe, *Life of Crookes*, p. 215.
71. Letter of 13 April 1871, reprinted ibid., p. 198.
72. He joined the Theosophical Society in 1883 and remained a member until his death in 1919 (Barrington, Goldney, and Medhurst, eds., *Crookes and the Spirit World*, p. 230). Adding the Ghost Club to his list of memberships in 1901, Crookes was elected its president for the following year. Ghost Club Minutes, vol. 4, pp. 81, 132, B.M. Add. MSS. 52261.
73. Crookes to Barrett, 15 May 1871, Barrett Papers (box 2, A4, no. 15) – also reprinted in Fournier D'Albe, *Life of Crookes*, p. 199; letter to Huggins, 6 November 1871, reprinted in Fournier D'Albe, *Life of Crookes*, p. 227.
74. Reprinted in Barrington, Goldney, and Medhurst, eds., *Crookes and the Spirit World*, pp. 237–8.
75. Crookes, "Notes of Séances with D. D. Home," p. 100; Crookes to Lodge, 16 November 1897, Lodge Collection (box 2, no. 326).
76. Crookes to Lodge, 6 February 1915, Lodge Collection (box 2, no. 362). There is an intriguing letter from Galton to Darwin, 28 March 1872, in which Galton described experiments with Kate Fox that involved the movement of needles "*in vacuo* in little bulbs of glass." Although we do not know whether psychical, rather than physical, investigations first led Crookes to invent the radiometer, it seems, from Galton's account, that Crookes was following similar lines of inquiry in both fields at this time, as part of an ongoing attempt to connect the natural forces of physics with the seemingly unnatural psychic phenomena. For Galton's letter,

see Karl Pearson, *The Life, Letters and Labours of Francis Galton*, 3 vols. in 4 (Cambridge: The University Press, 1914–30), 2:63.

77. *Report of the Sixty-Eighth Meeting of the British Association for the Advancement of Science, Held at Bristol in September 1898* (London: John Murray, 1899), p. 30.

78. Crookes's presidential address to the SPR, in *Presidential Addresses*, pp. 99–100; *Report of the 68th Meeting of the BAAS*, p. 31.

79. Ghost Club Minutes, vol. 3 (7 April 1899), pp. 268–9, B.M. Add. MSS. 52260.

80. Crookes's 1897 presidential address to the SPR, in *Presidential Addresses*, pp. 90–1. Crookes then proceeded to assert, p. 91, in occult vocabulary rarely heard from the Society's presidents, that "these intelligent centres of the various spiritual forces . . . in their aggregate go to make up Man's character or Karma."

81. See footnote 74.

82. Crookes's letter to a female correspondent, dated 10 May 1871, in which he claimed strong evidence for believing in "invisible intelligent beings distinct from the human race," is reprinted in Barrington, Goldney, and Medhurst, eds., *Crookes and the Spirit World*, p. 235; Massey's letter to Olcott is quoted in Medhurst and Goldney, "Crookes and the Physical Phenomena of Mediumship," pp. 130–1; they also quote, p. 68, Crookes's letter to C. H. Gimingham, his laboratory assistant, concerning the alleged exorcism (29 August 1874). On "4-dimensional space," see Crookes to Lodge, 22 May 1909, Lodge Collection (box 2, no. 356).

83. Fournier D'Albe, *Life of Crookes*, p. 404.

84. Crookes to Lodge, 22 December 1916; Lodge to Crookes, 23 December 1916; Crookes to Lodge, 27 December 1916; Lodge Collection (box 2, nos. 363–5). These letters are also reprinted in Barrington, Goldney, and Medhurst, eds., *Crookes and the Spirit World*, pp. 243–5.

85. Fournier D'Albe, *Life of Crookes*, pp. 405–6. The letter to Alice Bird was dated 24 January 1917. Also see Barrington, Goldney, and Medhurst, eds., *Crookes and the Spirit World*, pp. 231, 245–8.

86. See his letter to Lodge, 6 February 1915, Lodge Collection (box 2, no. 362), for some indication that Crookes was already modifying his attitude of suspended judgment.

87. W. B. Carpenter, "The Radiometer and its Lessons," *Nineteenth Century* 1 (April 1877):256. Among Carpenter's other attacks on Crookes, see the really vicious "Spiritualism and its Recent Converts," *Quarterly Review* 131 (October 1871):301–53, and "Psychological Curiosities of Spiritualism," *Fraser's Magazine* n.s. 16 (November 1877):541–64. For Crookes's reply to Carpenter, see William Crookes, "Letters to the Editor," *Nature* 17 (1 November 1877):7–8, and the lengthier rebuttal in "Another Lesson from the Radiometer," *Nineteenth Century* 1 (July 1877):879–87. Surprisingly, Fournier D'Albe, *Life of Crookes*, pp. 195–6, seems to subscribe to the theory of the "two Crookeses."

88. Greenaway, "A Victorian Scientist," p. 189.

89. Ibid., p. 182; Fournier D'Albe, *Life of Crookes*, p. 245; Robert K. DeKosky, "William Crookes and the Fourth State of Matter," *Isis* 67 (March 1976):36–43. DeKosky, p. 36, comments on Crookes's "burning desire for fundamental discovery and consequent eminence."

90. Crookes's address on "Radiant Matter," August 1879, quoted in Fournier D'Albe, *Life of Crookes*, pp. 284–5. Faraday had first suggested a fourth

state of matter in 1819, but Crookes gave specific contours to the idea on the basis of his own experiments. See DeKosky, "Crookes and the Fourth State of Matter," p. 51n.

91. Again borrowing a term from Faraday (DeKosky, "Crookes and the Fourth State of Matter," p. 51). DeKosky's article is extremely helpful in guiding the baffled through the confusing terminology of physics in the 1870s. He points out, p. 43, that Crookes used the term "fourth state of matter" to designate both the actual "very rarefied state in which the mean free path of the molecules has enlarged considerably," and also the potential "condition in which an energized surface can convert the fortuitous motions of molecules into a directed, mechanically efficacious stream of corpuscles."

92. DeKosky argues, ibid., p. 36, that Crookes's "tenacious adherence to his conception of a fourth state of matter" was in part responsible for "his failure to become a latter-day Faraday." For the connection between Crookes's work and Röntgen's discovery of X rays, see Greenaway, "A Victorian Scientist," p. 193, and William Barrett, "In Memory of Sir William Crookes," *PSPR* 31 (May 1920):15–16.

93. William Crookes, "Molecular Physics in High Vacua," Lecture delivered 4 April 1879, *Proceedings of the Royal Institution of Great Britain* 9 (December 1879):158.

94. DeKosky, "Crookes and the Fourth State of Matter," p. 38.

95. Crookes's address on "Radiant Matter," in Fournier D'Albe, *Life of Crookes*, p. 290.

96. See the opening and closing paragraphs of his 1897 SPR presidential address, in *Presidential Addresses*, pp. 86, 103.

97. William Barrett, "Some Reminiscences of Fifty Years' Psychical Research," *PSPR* 34 (December 1924):281–2. Barrett received his earliest scientific training, like Crookes, at the Royal College of Chemistry, also studying under Professor Hofmann. See Barrett's obituary of Crookes, "In Memory of Sir William Crookes," p. 13. Although Barrett does not give the dates of his own attendance at the Royal College of Chemistry, they would appear to be c. 1860–2. According to Tyndall's biographers, the great man had to discharge Barrett, "who was conscious of grievances which he was inclined to exaggerate. He was also claiming credit for some of Tyndall's results." A. S. Eve and C. H. Creasey, *Life and Work of John Tyndall* (London: Macmillan, 1945), p. 118. J. H. Gladstone, by contrast, describes the young Barrett's harmonious relationship with Faraday at the Royal Institution. *Michael Faraday* (London: Macmillan, 1872), pp. 72–3, 139.

98. Barrett, "Some Reminiscences," pp. 282–4, 287, and "On Some Phenomena Associated with Abnormal Conditions of Mind," *Spiritualist Newspaper* 9 (22 September 1876):87–8. Also see R. J. Campbell, Foreword to *Personality Survives Death. Messages from Sir William Barrett*, ed. Florence Barrett (London: Longmans, Green, 1937), p. viii.

99. Lane Fox, who subsequently assumed the surname Pitt-Rivers, was not himself impressed with allegedly spiritualist phenomena, but believed that science had no right to taboo a subject about which it knew virtually nothing. See his letter to the *Times*, 22 September 1876, p. 10, and "The British Association at Glasgow," *Spiritualist Newspaper* 9 (15 September 1876):78–9. On Wallace's role, see Alan Gauld, *The Founders of Psych-*

ical Research (London: Routledge & Kegan Paul, 1968), p. 137; James Marchant, *Alfred Russel Wallace: Letters and Reminiscences* (New York and London: Harper & Bros., 1916), p. 425; and Malcolm Jay Kottler, "Alfred Russel Wallace, the Origin of Man, and Spiritualism," *Isis* 65 (June 1974):178.

100. All quotations from Barrett's paper, and the discussion following it, are taken from the report printed in the *Spiritualist Newspaper* 9 (22 September 1876):85–94. A revised version of the paper was also printed in *PSPR* 1 (1882–3):238–44.

101. See his letter to the *Times*, 22 September 1876, p. 10.

102. "The British Association on Professor Barrett's Paper," *Spectator* 49 (30 September 1876):1209.

103. Myers to Barrett, 9 October 1877, Barrett Papers (box 2, A4, no. 62).

104. Barrett, "Some Reminiscences," p. 294, and *Seeing Without Eyes* (Halifax: Spiritualists' National Union, 1911), p. 37. This pamphlet is no. 22 in *Houdini Pamphlets: Spiritualism*, vol. 1, in the Houdini Collection. It appeared as no. 4 of the SNU Propaganda Publications, and was the printed version of a lecture which Barrett gave at a SNU conference in Leicester, July 1910.

105. Eleanor M. Sidgwick, "In Memory of Sir William Barrett, F.R.S.," *PSPR* 35 (July 1925):413. Gurney, Myers, and Podmore acknowledged Barrett's importance for their work in the preface to *Phantasms of the Living*, 2 vols. (1886; facsimile reprint ed., Gainesville, Fla.: Scholars' Facsimiles & Reprints, 1970), 1:vi.

106. W. F. Barrett, "Mind-Reading versus Muscle-Reading," *Nature* 24 (7 July 1881):212.

107. Balfour Stewart, "Note on Thought-Reading," *PSPR* 1 (1882–3):36. Hopkinson was another early member of the SPR.

108. Edmund Gurney, F. W. H. Myers, and W. F. Barrett, "Second Report on Thought-Transference," *PSPR* 1 (1882–3):70. Gurney, Myers, and Barrett were, likewise, responsible for the "First Report on Thought-Reading," in *PSPR* 1, pp. 13–34. The trio also coauthored an article on "Thought-Reading," in *Nineteenth Century* 11 (June 1882):890–900. For a sardonic description of the experiments with the Creery sisters, see Trevor H. Hall, *The Strange Case of Edmund Gurney* (London: Gerald Duckworth, 1964), pp. 55–63. Hall suggests the basic unreliability of the Reverend Creery's character by observing that, over a period of years, he changed his religious affiliation from Anglican, to Unitarian, to Presbyterian, and back again to Anglican.

109. "Second Report on Thought-Transference," p. 70.

110. Sidgwick to Barrett, 5 October 1887, Barrett Papers (box 2, A3, letter c). By October 1887, Mary and Alice Creery were in their twenties, and Reverend Creery had moved the family from Buxton to Manchester.

111. Barrett to Sidgwick, 31 October 1887, Barrett Papers (box 2, A3, letter f).

112. W. F. Barrett, *Psychical Research* (London: Williams & Norgate, [1911]), pp. 53, 55–6, 63; Barrett's letter is on p. 8 of the *Times* for that date.

113. See Myers's letters to Barrett, Barrett Papers (box 2, A4, nos. 64–83). Also see Gauld, *Founders of Psychical Research*, p. 148.

114. Barrett to Lodge, Lodge Collection (box 1, nos. 72, 73, 75–8).

115. E. Sidgwick, "In Memory of Barrett," pp. 413–14; Campbell, Foreword to *Personality Survives Death*, p. ix; Gurney to Barrett, 18 November [1886], and Myers to Barrett, 2 February 1899 and 18 July 1900, Barrett Papers (box 2, A4, nos. 32, 86, 92).
116. Sidgwick to Barrett, 14 July 1890, Barrett Papers (box 2, A4, no. 129); W. F. Barrett, "Water-Finding," *Times*, 21 January 1905, p. 14.
117. Although he only served a one-year term. Most of his predecessors in office had served for at least two years. See Eveleen Myers to Barrett, 10 November 1904, Barrett Papers (box 2, A4, no. 58).
118. William Barrett and Theodore Besterman, *The Divining-Rod. An Experimental and Psychological Investigation* (London: Methuen, 1926).
119. Barrett, "Water-Finding," *Times*, 21 January 1905, p. 14.
120. Barrett, *Seeing Without Eyes*, p. 37, and "On the So-called Divining Rod. Book II," *PSPR* 15 (October 1900):303, 306, 308–9, 311.
121. Barrett to Wallace, 3 November 1905, quoted in Marchant, *Wallace*, p. 438.
122. Barrett, "Some Reminiscences," p. 291.
123. "Obituary. Sir W. F. Barrett," *Times* 28 May 1925, p. 16; *Who Was Who, 1916–1928*, s.v., "Barrett, Sir Wm. Fletcher"; E. Sidgwick, "In Memory of Barrett," pp. 416–17.
124. Barrett to Lodge, 14 October 1893, Lodge Collection (box 1, no. 62).
125. "Transference of Thought. Sir W. Barrett's Conclusions," *Times*, 20 December 1924, p. 8.
126. Barrett to Lodge, 18 October 1890, Lodge Collection (box 1, no. 60).
127. Eleanor Sidgwick to Barrett, 7 April 1908, Barrett Papers (box 2, A4, no. 118).
128. Barrett, "Some Reminiscences," p. 286. A few paragraphs later, pp. 288–9, Barrett explained that his old skepticism about spirit photographs had, thanks to new "*indubitable* evidence," recently yielded to belief in the validity of "supernormal psychic photography." He also (p. 290) endorsed the existence "under suitable conditions" of the so-called Reichenbach phenomena.
129. W. F. Barrett, Introduction to Beckles Willson, *Occultism and Common-Sense* (New York: R. F. Fenno, [1908?]), p. xi. See a similar acknowledgment in W. F. Barrett, *On the Threshold of a New World of Thought: An Examination of the Phenomena of Spiritualism* (London: Kegan Paul, Trench, Trübner, 1908), p. xiv.
130. Barrett, "In Memory of Crookes," p. 27; W. F. Barrett, "Memorial Notices," in C. C. Massey, *Thoughts of a Modern Mystic. A Selection from the Writings of the late C. C. Massey*, ed. W. F. Barrett (London: Kegan Paul, Trench, Trübner, 1909), p. 7.
131. W. F. Barrett, "Swedenborg: The Savant and the Seer," *Contemporary Review* 102 (July 1912):37, and *Swedenborg: The Savant & the Seer* (London: John M. Watkins, 1912), p. 70. Both article and book were based on a lecture given to the Swedenborg Society in March 1912.
132. Campbell, Foreword to *Personality Survives Death*, pp. vii, xii–xiv. Barrett's obituary in the *Times* (28 May 1925, p. 16) pointed out that he was "a regular attendant" at St. Martin-in-the-Fields, after he moved from Dublin to London, following his marriage in 1916 to the eminent gynecologist Dr. Florence Willey.
133. Barrett, *Seeing Without Eyes*, p. 37.

134. Barrett, "Some Reminiscences," p. 296.
135. Barrett, Preface to Massey, *Thoughts of a Modern Mystic*, pp. iii–iv. Barrett's religious predilections also apparently included an interest in gnosticism, as his correspondence with the architect Frederick Bond in 1918 suggests. Barrett Papers (box 2, A4, nos. 4–10).
136. Barrett to Samuel Clemens, 19 October 1884, photostat copy in Barrett Papers, Box 2. Barrett entered into correspondence with Clemens when he invited the famous author to join the SPR. Also see Barrett to Lodge, 18 October 1890 and 23 December 1904, Lodge Collection (box 1, nos. 60, 71).
137. Barrett, "Some Reminiscences," pp. 286, 294, 296. In his letter on thought transference to the *Times*, 20 December 1924, p. 8, Barrett suggested that "our conscious lives emerge like peaks from a dense mist, a mist that covers and hides the vast plain which unites all sensient beings with one another and with the Universal."
138. W. F. Barrett, "The Deeper Issues of Psychical Research," *Contemporary Reivew* 113 (February 1918):174. The article contained the substance of a talk which Barrett had given to the Clerical Society of the Diocese of Birmingham.
139. Ibid., p. 173.
140. *Presidential Addresses*, p. 183.
141. Ibid., p. 179, and William F. Barrett, *On the Threshold of the Unseen. An Examination of the Phenomena of Spiritualism and of the Evidence for Survival After Death*, 2d ed. rev. (London: Kegan Paul, Trench, Trübner; New York: E. P. Dutton, 1917), p. xix.
142. Barrett, "Some Reminiscences," pp. 276, 275n.
143. Ibid., pp. 276–8.
144. Joseph Glanvill, *The Vanity of Dogmatizing: or Confidence in Opinions Manifested in a Discourse of the Shortness and Uncertainty of our Knowledge, and its Causes* . . . (1661; reprint ed., New York: Columbia University Press for The Facsimile Text Society, 1931), pp. 196–201. Also see Basil Willey, *The Seventeenth Century Background: Studies in the Thought of the Age in Relation to Poetry and Religion* (Garden City, N.Y.: Doubleday, Anchor Books, 1953), pp. 191–2; Leslie Stephen's article on Glanvill in the *DNB*; and the discussion of Glanvill throughout Richard S. Westfall, *Science and Religion in Seventeenth-Century England* (Ann Arbor: University of Michigan Press, Ann Arbor Paperbacks, 1973).
145. See, for example, Barrett, "Some Reminiscences," p. 278, and "Deeper Issues of Psychical Research," p. 178.
146. Ivor Tuckett, "Psychical Researchers and 'The Will to Believe,'" *Bedrock. A Quarterly Review of Scientific Thought* 1 (July 1912):204.
147. Oliver Lodge, "On Telepathy as a Fact of Experience," corrected typescript of an article, c. 1912–13, in the Lodge Collection (box 5, no. 1095, p. 9); Lodge to Rev. E. O. Davies, 19 October 1912, in Lodge Collection (box 2, no. 377); and Oliver Lodge, *Continuity. The Presidential Address to the British Association for 1913* (New York and London: G. P. Putnam's Sons, 1914), p. 105.
148. Oliver Lodge, *Past Years, an Autobiography* (London: Hodder & Stoughton, 1931), pp. 65–6, 70, 77–8, 82, 85–6, 109, 270–2, 345. Gurney attended Lodge's classes to enhance his understanding of the physics of sound.

149. Ibid., p. 273.
150. Malcolm Guthrie and James Birchall, "Record of Experiments in Thought-Transference, at Liverpool," *PSPR* 1 (1882–3):263–83. In the second volume of SPR *Proceedings* (1884), Guthrie and Birchall, honorary secretary of the Liverpool Literary and Philosophical Society, continued their report, pp. 24–42, while Gurney, Myers, and Barrett focused on the Liverpool experiments in the "Fourth Report of the Committee on Thought-Transference," pp. 1–11. Gurney and Myers had visited Guthrie's establishment in September 1883 and observed the thought transference tests there.
151. Oliver J. Lodge, "An Account of Some Experiments in Thought-Transference," *PSPR* 2 (1884):190–1. Lodge began his report with a statement of Guthrie's social standing and civic prominence in Liverpool, as though Lodge had already embraced the cardinal SPR fallacy that education, breeding, wealth, or public responsibility ipso facto made their possessor an astute observer and a man (or woman) of unimpeachable integrity. Also see Lodge, *Past Years*, pp. 274–5.
152. For a general survey of Mrs. Piper's mediumship, see Alta L. Piper, *The Life and Work of Mrs. Piper*, with an Introduction by Oliver Lodge (London: Kegan Paul, Trench, Trübner, 1929).
153. Lodge, *Past Years*, p. 279. Also see p. 345.
154. Lodge believed by this time that conscious, purposeful thought transference between persons not in contact, but in close proximity, had been indubitably established, and thus he could turn to telepathy between the living as a known cause, rather than merely a working hypothesis. He acknowledged, however, that the kind of thought transference that would have to operate in order to explain Mrs. Piper's trance mediumship – where the agents were not even aware of the ideas which they had transmitted to her – had not come anywhere near to experimental proof and was purely hypothetical. In his sincere conviction of Mrs. Piper's integrity, Lodge carefully distinguished "fishery on the part of Dr. Phinuit" from "trickery on the part of Mrs. Piper." See F. W. H. Myers, O. J. Lodge, Walter Leaf, and William James, "A Record of Observations of Certain Phenomena of Trance," *PSPR* 6 (1889–90):446–53.
155. Ibid., pp. 648–9.
156. Lodge to Myers, 21 October 1890, Lodge Collection (box 5, no. 1309).
157. Richard Hodgson, "A Further Record of Observations of Certain Phenomena of Trance," *PSPR* 13 (February 1898):284–582, especially pp. 370–406.
158. C. E. M. Hansel, *ESP: A Scientific Evaluation* (New York: Charles Scribner's, 1966), pp. 224–7; Sidgwick to Roden Noel, in Sidgwick and Sidgwick, *Henry Sidgwick*, pp. 502, 507.
159. Quoted in R. Laurence Moore, *In Search of White Crows: Spiritualism, Parapsychology, and American Culture* (New York: Oxford University Press, 1977), p. 163. Walter Leaf, by contrast, never abandoned the belief, stated in his section of the 1890 report (*PSPR*, 6:558–68), that Mrs. Piper's controls were secondary personalities and that their surprising information about strangers could be explained by the power of mind reading, heightened in the abnormal trance state. See Walter Leaf, *Walter Leaf 1852–1927. Some Chapters of Autobiography, with a Memoir by Charlotte M. Leaf* (London: John Murray, 1932), p. 155. Tuckett, "Psychical

Researchers and 'The Will to Believe,'" pp. 200–1, denigrates Piper's mediumship and suggests that the SPR investigators did not control her séances stringently enough.

Hodgson, it seems, was emotionally prepared to accept the reality of spirit communication several years before he could acknowledge it intellectually. After Gurney's death, he wrote to Myers in July 1888: "I think that his urgent soul is not yet done with *us* and *here* . . . Give him time to find the broadest ways from his domain to ours . . . More forceful and radiant and abiding than ever yet he will be with you." Myers Papers 2[110].

160. *Presidential Addresses*, p. 144.
161. Lodge, *Past Years*, p. 285. Shortly after Myers's death in January 1901, his close friend Lodge received communications through an English trance medium, Rosina (Mrs. Edmond) Thompson, purporting to come from Myers, and Lodge was very struck by "the dramatic verisimilitude" of the utterances. Oliver Lodge, "Report on Some Trance Communications Received Chiefly Through Mrs. Piper," *PSPR* 23 (June 1909):283. Frank Podmore, *The Newer Spiritualism* (London: T. Fisher Unwin, 1910), contains chapters on the mediumship of both Piper and Thompson. On these mediums, also see Gauld, *Founders of Psychical Research*, pp. 251–74.
162. J. Arthur Hill, comp., *Letters from Sir Oliver Lodge, Psychical, Religious, Scientific and Personal* (London: Cassell, 1932), p. 15.
163. "Sir Oliver Lodge. A Great Scientist," *Times*, 23 August 1940, p. 7, underscores the degree to which Lodge became associated with the war-inspired resurgence of popular interest in spiritualism. One of the mediums through whom Raymond allegedly contacted his family was Mrs. Osborne Leonard, who subsequently became a prominent subject of SPR investigation. For an analysis of Mrs. Leonard, as well as Mrs. Piper, see W. H. Salter, *Trance Mediumship: An Introductory Study of Mrs. Piper and Mrs. Leonard* (London: SPR, 1950).
164. For example, see Ray Lankester, "Sir Oliver Lodge on the Investigation of Consciousness Apart from Brain," and Bryan Donkin, "Science and Spiritualism," *Bedrock* 1 (January 1913):488–503; Charles A. Mercier, *Spiritualism and Sir Oliver Lodge* (London: Mental Culture Enterprise, 1917).
165. Lodge, *Continuity*, p. 103; also see Lodge, *Past Years*, p. 348.
166. See, for example, the opening paragraphs of Oliver Lodge, "The Christian Idea of God. A Plea for Simplicity," *Hibbert Journal* 9 (July 1911):697–9.
167. Lodge, *Past Years*, pp. 349–51, and *Continuity*, pp. 83, 75.
168. Oliver Lodge, "In Memory of F. W. H. Myers," in *Presidential Addresses*, p. 122; and *My Philosophy. Representing My Views on the Many Functions of the Ether of Space* (London: Ernest Benn, 1933), p. 25.
169. R. A. Gregory and Allan Ferguson, "Oliver Joseph Lodge, 1851–1940," *Obituary Notices of Fellows of the Royal Society* 3 (December 1941):557–9, 562–3; "Sir Oliver Lodge," *Times*, 23 August 1940, p. 7; Lodge, *Past Years*, pp. 174–93; and Hill, comp., *Letters from Sir Oliver Lodge*, pp. 46–8.
170. Lodge, *Past Years*, p. 199.
171. Ibid., p. 350.

172. Gregory and Ferguson, "Oliver Joseph Lodge," p. 557. Wilson, "Thought of Late Victorian Physicists," underscores the degree to which late Victorian physics relied on mechanical models, while Harold I. Sharlin, *The Convergent Century: The Unification of Science in the Nineteenth Century* (London: Abelard-Schuman, 1966), p. 2, states that such reliance was far more characteristic of British than of French scientists.
173. Lodge, *Continuity*, pp. 64–5, and Lodge to J. A. Hill, 21 February 1915, in Hill, comp., *Letters from Sir Oliver Lodge*, p. 60.
174. J. J. Thomson, *On the Light Thrown by Recent Investigations on Electricity on the Relation between Matter and Ether. The Adamson Lecture Delivered at the University on November 4, 1907* (Manchester: The University Press, 1908), p. 21. On Faraday's dual theories of the ether, see P. M. Harman, *Energy, Force and Matter: The Conceptual Development of Nineteenth-Century Physics* (Cambridge: Cambridge University Press, 1982), p. 72. Brian Wynne, "Physics and Psychics: Science, Symbolic Action, and Social Control in Late Victorian England," in *Natural Order: Historical Studies of Scientific Culture*, ed. Barry Barnes and Steven Shapin (Beverly Hills, Calif.: Sage Publications, 1979), pp. 167–86, provides a useful summary of late Victorian theories of the ether, although his attempt to apply sociology of knowledge theories to the late Victorian psychical researchers is not persuasive.
175. Lodge to J. A. Hill, 21 February 1915, in Hill, comp., *Letters from Sir Oliver Lodge*, pp. 60–1.
176. Lodge, *Past Years*, p. 349.
177. Lodge, *Continuity*, pp. 33–5, 44–6.
178. Lodge, *My Philosophy*, pp. 115, 136, 229.
179. Ibid., pp. 49, 58. In a footnote, Lodge added: "This was written before the verification of relativity theory, but I leave it as a confession of prejudice."
180. Crookes's presidential address to the BAAS, in *Report of the 68th Meeting of the B.A.A.S.*, p. 31. The same phrase appears in his 1897 presidential address to the SPR, in *Presidential Addresses*, p. 97.
181. W. P. Jolly, *Sir Oliver Lodge* (London: Constable, 1974), p. 142. "Brooding" was Lodge's word for his activity.
182. For one among many statements of this belief in continuity through time, see Oliver Lodge, *The Immortality of the Soul* (Boston: Ball Publishing, 1908), pp. 26–9.
183. Oliver Lodge, *Raymond, or Life and Death* (London: Methuen, 1916), p. 83.
184. Lodge, *Continuity*, pp. 21, 29.
185. Hill, comp., *Letters from Sir Oliver Lodge*, p. 16.
186. John D. Root, "Science, Religion, and Psychical Research: The Monistic Thought of Sir Oliver Lodge," *Harvard Theological Review* 71 (July–October 1978):247, argues that it was Lodge's involvement with the Synthetic Society after 1896 that prompted his "voluminous outpouring of religious and philosophical publications." Lodge's increasing certainty of human survival, thanks to Mrs. Piper and the cross correspondences, must have proved at least an equal source of inspiration.
187. Lodge, *Immortality of the Soul*, pp. 18–22, 12–16.
188. Lodge, *My Philosophy*, pp. 209, 213, 222–3. Lodge admitted, p. 222, that his theory of the etheric body did not answer the initial problem of how mind acted upon the ether.

189. Ibid., pp. 220–2, 235–6, 258. Also see Wilson, "Thought of Late Victorian Physicists," pp. 32–4.
190. Oliver Lodge, "On the Difficulty of Proving Individual Survival," *PSPR* 40 (January 1932):123, and *Why I Believe in Personal Immortality* (London: Cassell, 1928), p. vi.
191. See, for example, Oliver Lodge, *The Survival of Man. A Study in Unrecognised Human Faculty* (London: Methuen, 1909), pp. 78, 96, and *Man and the Universe. A Study of the Influence of the Advance in Scientific Knowledge upon Our Understanding of Christianity*, 5th ed. (London: Methuen, 1909), pp. 187–9. *Man and the Universe* was first published in 1908.
192. Oliver Lodge, *Reason and Belief* (London: Methuen [1910]), p. 47, and p. x, for the "lifetime of scientific study."
193. Lodge, "Christian Idea of God," p. 698.
194. Oliver Lodge, *The Substance of Faith Allied with Science. A Catechism for Parents and Teachers* (New York and London: Harper & Brothers, 1907), pp. iii–vi. Root, "Science, Religion, and Psychical Research," p. 257, reports that Lodge's catechism "went through seven editions in the first year." Clearly Lodge had an eager audience well before *Raymond*.
195. Lodge, *Substance of Faith*, p. v.
196. Lodge, *Man and the Universe*, pp. 213–21. Lodge, p. 219, was careful to point out that the pre-Christian venerability of certain doctrines did not alone mandate their revision. Immortality of the soul, after all, was a doctrine that predated Christian teaching, and in that case Lodge was willing "to appeal to the antiquity of human tradition as tending in favour of some sort of truth underlying this perennial and protean faith." With "vicarious punishment and bloody atonement," however, Lodge could find no "real and helpful" underlying truth, and no ethical significance whatsoever. *Man and the Universe* went through five editions in four months. Root, "Science, Religion, and Psychical Research," pp. 253–4, 263, discusses what Lodge found essential and nonessential in Christianity.
197. Lodge to Hill, 3 June 1915, in Hill, comp., *Letters from Sir Oliver Lodge*, p. 82; Lodge, *Past Years*, p. 351. Lodge also rejected any notion of a resurrection of the *physical* body, but for biological, rather than ethical, reasons. See Lodge, *Immortality of the Soul*, pp. 7–16.
198. Lodge, *Man and the Universe*, pp. 93, 138, 300.
199. Ibid., p. 51; also see Gregory and Ferguson, "Oliver Joseph Lodge," p. 561. Lodge went so far as to describe the ether as "the living garment of God," in *Ether and Reality* (1925) – quoted in Root, "Science, Religion, and Psychical Research," p. 262.
200. The drive to unify knowledge into a single overarching system of thought was another legacy of Victorian science which Lodge never abandoned. This is one of the themes of Wilson's article, "Thought of Late Victorian Physicists."
201. Lodge, *Continuity*, p. 97, and *Past Years*, pp. 313, 351–2. For an earlier statement of his belief in universal design and purpose, see Oliver Lodge, *Life and Matter: A Criticism of Professor Haeckel's 'Riddle of the Universe'* (New York and London: G. P. Putnam, 1905), pp. 102–3.
202. Lodge, *Reason and Belief*, p. 43.
203. Lodge, *Continuity*, pp. 90–2, 99.

204. Lodge, *Immortality of the Soul*, pp. 22–3, and *Continuity*, p. 106.
205. Lodge, *Past Years*, p. 291.
206. "Sir Oliver Lodge," *Times*, 23 August 1940, p. 7.
207. Hynes, *Edwardian Turn of Mind*, p. 136.
208. Fournier D'Albe, *Life of Crookes*, p. 238.

Conclusion

1. Susan Faye Cannon, *Science in Culture: The Early Victorian Period* (New York: Dawson and Science History Publications, 1978), p. 2, discusses the emergence of the limited modern sense of "physical science" in England during the 1830s. Yet throughout the century the idea of the "moral sciences" continued to be current – as in the new Cambridge tripos – while by the end of the century the concept of the social sciences was taking hold.
2. For an illuminating discussion of many of the points touched upon in this and the preceding paragraph, see Robert H. Kargon, *Science in Victorian Manchester: Enterprise and Expertise* (Baltimore: Johns Hopkins University Press, 1977). Jon Palfreman, "Between Scepticism and Credulity: A Study of Victorian Scientific Attitudes to Modern Spiritualism," in Roy Wallis, ed., *On the Margins of Science: The Social Construction of Rejected Knowledge*, Sociological Review Monograph, no. 27 (Keele, Staffordshire: University of Keele, 1979), p. 215, speaks of "the official criticism of the scientific establishment." His only example, however, is Carpenter's attack on Crookes's psychical research. Dangerous though Carpenter was when aroused, he did not alone constitute a British scientific establishment.
3. Crookes to Lodge, 20 October 1894, Lodge Collection (box 2, no. 319). In the summer of 1871, the Royal Society declined to publish a paper on psychical research which Crookes had submitted; nor would the BAAS pay any attention to his psychic investigations at this time. "News of the Week," *Spectator* 44 (22 July 1871):879; Crookes's reply, "Mr. Crookes and the Royal Society," *Spectator* 44 (29 July 1871):917, and Editor's note following.
4. For Barrett's use of those terms, see his Introduction to Beckles Willson, *Occultism and Common-Sense* (New York: R. F. Fenno, [1908?]), p. xii; *Seeing Without Eyes* (Halifax: Spiritualists' National Union, 1911), p. 17 – no. 22 in *Houdini Pamphlets: Spiritualism*, vol. 1, Houdini Collection; and his 1904 presidential address to the SPR, in *Presidential Addresses to the Society for Psychical Research 1882–1911* (Glasgow: Robert Maclehose for the SPR, 1912), p. 179.
5. Communication from the BAAS to Barrett, September 1883, Barrett Papers (box 2, A4, no. 11); Barrett to Lodge, 7 December 1894, Lodge Collection (box 1, no. 63); William Barrett, "Some Reminiscences of Fifty Years' Psychical Research," *PSPR* 34 (December 1924):281, 284n; Oliver Lodge, *Past Years, an Autobiography* (London: Hodder & Stoughton, 1931), p. 272.
6. Oliver Lodge, *Man and the Universe. A Study of the Influence of the Advance in Scientific Knowledge upon Our Understanding of Christianity*, 5th ed. (London: Methuen 1909), p. 25 – first published in 1908; *Past Years*, p. 348; Lodge to Hill, 23 December 1914, in J. Arthur Hill,

comp., *Letters from Sir Oliver Lodge, Psychical, Religious, Scientific and Personal* (London: Cassell, 1932). p. 49.

7. J. J. Thomson, *Recollections and Reflections* (London: G. Bell, 1936), p. 298. Organized science in the United States likewise greeted spiritualism with a notable lack of enthusiasm. For a few comments on the American scientific response, see R. Laurence Moore, *In Search of White Crows: Spiritualism, Parapsychology, and American Culture* (New York: Oxford University Press, 1977), p. 26.

8. R. Laurence Moore discusses the problems involved in trying "'to mainstream' spiritualist belief," in "Insiders and Outsiders in American Historical Narrative and American History," *American Historical Review* 87 (April 1982):390–1.

9. Lankester's presidential address, 1906, in *Report of the Seventy-Sixth Meeting of the British Association for the Advancement of Science, York, August 1906* (London: John Murray, 1907), p. 13.

10. On this point, see Leonard G. Wilson, "Science by Candlelight," in *The Mind and Art of Victorian England*, ed. Josef L. Altholz (Minneapolis: University of Minnesota Press, 1976), pp. 102–3.

11. Morton Hunt raises this point in "How the Mind Works," *New York Times Magazine*, 24 January 1982, p. 68.

12. Nor did the even more rigorous attempts to make parapsychology scientific after World War I result in its acceptance by the community of scientists at large. The work of J. B. Rhine, paramount in these endeavors, forms the central focus of Seymour H. Mauskopf and Michael R. McVaugh, *The Elusive Science: Origins of Experimental Psychical Research* (Baltimore: Johns Hopkins University Press, 1980).

13. "The British Association at Glasgow," *Spiritualist Newspaper* 9 (22 September 1876):93.

14. See Lodge's presidential address to the SPR, January 1902, in *Presidential Addresses*, p. 138.

15. Lodge, "On Telepathy as a Fact of Experience," pp. 11–12. This typescript in the Lodge Collection (box 5, no. 1095) is undated, but internal evidence indicates that Lodge wrote the essay in 1912.

16. Robert Jastrow, *The Enchanted Loom: Mind in the Universe* (New York: Simon & Schuster, 1981), p. 165.

17. J. J. Thomson, *On the Light Thrown by Recent Investigations on Electricity on the Relation between Matter and Ether. The Adamson Lecture Delivered at the University on November 4, 1907* (Manchester: The University Press, 1908), pp. 5–6.

INDEX

Adams, John Couch, 330
Adare, Lord, *See* Dunraven, Earl of
afterlife, spiritualist views on, 35–6,
    88, 101
  progressive evolution and, 72, 85,
    94–5, 108–9, 270–1, 322
agnosticism, 118, 152; *see also* Leaf,
    Walter; Myers, F. W. H.;
    Romanes, George John;
    Sidgwick, Henry
Albert, Prince, 29, 224
alphabet rapping, 8, 273, 290
American Society for Psychical
    Research, 150, 175
Ames, Julia, 34
Anderson, J. H., 25
Anglican church, 62, 65, 89, 114,
    168, 186
  Anglican clergy in SPR, 81–2
  Anglican position on spiritualism,
    68–9, 82
  Church of England Congresses,
    (1881) 80, 84; (1888) 130; (1905)
    71, 240
  psychical researchers and, 111–
    12, 119, 128, 145, 289
  spiritualist attitudes toward, 27,
    303, 304, 387
  see also Christian spiritualism
animal magnetism, *see* mesmerism
Anthropological Society of London,
    302, 308–9, 311, 314, 315
antivivisection, 37, 44, 183, 187,
    188–9
apparitions, 186, 223, 266, 297, 370

of the dying, 142
as evidence of psychopathological
    condition, 37, 244
Gurney's work on, 120, 144, 249,
    254
Myers's work on, 245, 255, 258,
    259
SPR committee on, 141
Ashburner, John, 221, 236, 239
atheism, 61, 86, 87, 90, 91, 96, 104,
    169
Atkinson, Henry George, 213, 216,
    224
automatic writing, 8, 120, 123, 148,
    239, 242, 266
  individual spiritualists and, 34, 35,
    37, 75, 77, 79
  subliminal self and, 258
  in Theosophy, 173
  see also cross correspondences;
    slate writing

BAAS, *see* British Association for
    the Advancement of Science
Baggally, W. W., 151
Bagshawe, Edward Gilpin, 84
Bain, Alexander, 237
Balfour, Arthur James, 125, 128,
    129–34, 135, 136, 269, 332, 386
Balfour, Francis Maitland, 133
Balfour, Gerald, 133, 134–5, 157,
    248, 262
Barkas, T. P., 239
Barrett, Lady (Dr. Florence Willey),
    479n132